DESPERATE REMEDIES

Desperate Remedies

PSYCHIATRY'S TURBULENT QUEST TO CURE MENTAL ILLNESS

ANDREW SCULL

THE BELKNAP PRESS OF HARVARD UNIVERSITY PRESS

Cambridge, Massachusetts

Published in the United Kingdom as *Desperate Remedies: Psychiatry and the Mysteries of Mental Illness* by Allen Lane, an imprint of Penguin Books Ltd., a Penguin Random House Company.

Printed in the United States of America

First Harvard University Press paperback edition, 2024
First printing

Library of Congress Cataloging-in-Publication Data

Names: Scull, Andrew, 1947- author.
Title: Desperate remedies : psychiatry's turbulent quest to cure mental illness / Andrew Scull.
Description: Cambridge, Massachusetts : The Belknap Press of Harvard University Press, 2022. | Includes bibliographical references and index.
Identifiers: LCCN 2021047502 | ISBN 9780674265103 (hardcover) | ISBN 9780674295513 (pbk.)
Subjects: LCSH: Mental illness—Treatment—United States—History. | Psychiatry—United States—History. | Psychiatric ethics—United States—History.
Classification: LCC RC443 .S394 2022 | DDC 362.20973—dc23/eng/20211008
LC record available at https://lccn.loc.gov/2021047502

This book is dedicated to the memory of our son,

Alexander Anthony Scull (1990–2016),

greatly missed every day.

~

CONTENTS

PART TWO | DISTURBED MINDS

PART THREE | A PSYCHIATRIC REVOLUTION

PREFACE

FEW OF US ESCAPE THE RAVAGES OF MENTAL ILLNESS. We may not suffer from it ourselves, but even then we feel the pain it inflicts on friends or family. And no one escapes its social burdens. Psychiatry seeks to lessen these afflictions, but too often it has increased them. In this book, I have attempted to provide a skeptical assessment of the psychiatric enterprise—its impact on those it treats and on society at large. I have focused most of my attention on the United States, because it is here that these interventions stand out in the starkest relief and because, by the closing decades of the twentieth century, American psychiatry had achieved a worldwide hegemony, its categorizations of and approaches to mental illness sweeping all before it. But many of the interventions I examine had their origins in Europe, including the drugs that are now so central to psychiatric identity and operations. So European developments loom large in this story and are woven through the narrative that follows.

For two centuries and more, most informed opinion has embraced the notion that disturbances of reason, cognition, and emotion—the sorts of things we used to gather under the umbrella of "madness"—properly belong in the domain of the medical profession. More precisely, such maladies are seen to be the peculiar province of those whom we now call psychiatrists. Mental illness, we are informed, is an illness like any other—one that is treated by a specialist group of doctors whose primary goals are to relieve suffering and, more ambitiously, to restore the alienated to the ranks of the sane. These are worthy goals, to be sure. How have psychiatrists sought to realize them? How have psychiatrists attacked the problem of mental illness? What weapons have they chosen and why? And have those treatments succeeded in relieving suffering and curing those consigned to psychiatrists' tender mercies? These are the central questions I propose to explore in the pages that follow.

My focus is on the therapeutics of mental illness and on the professionals who advanced them. But I seek constantly to keep in mind that these interventions are not abstractions but rather organized forms of action carried out upon our fellow human beings. Patients are the constant subtext of my story. They are the people whose bodies and minds are subjected to each of these therapies—sometimes several of them successively—and not always with success.

Psychiatry emerged in the nineteenth century as a specialized branch of medicine claiming expertise in the management and cure of what was then called insanity or lunacy. Psychiatry's rise was intimately linked to the emergence of the asylum, and for a long time psychiatry and the asylum were locked in a symbiotic embrace. Prior to the Civil War, these institutions housed an almost exclusively white population. When institutional provision for African Americans began to be provided after the war, it took the form of either segregated wards or entirely separate institutions for the "colored insane."

Psychiatry's marginalization of the Black population was, of course, of a piece with the exclusion and discrimination they faced in the larger society, and in many ways such marginalization persists into the present. Racism continues to have an impact on life chances, whichever sector of society one attends to: the economy, housing, education, or the criminal justice system, to mention only some of the more obvious arenas. And, of course, with respect to health—both mental and physical—where the cumulative disadvantages of poverty and racial prejudices are vividly demonstrated by, for example, data concerning maternal mortality and life expectancy.[1] Unsurprisingly, when the Office of the US Surgeon General examined mental health services in 2001, it found that racial and ethnic minorities had less access to mental health services than whites and that the care that minorities did receive was more likely to be of poor quality.[2] That makes the lack of contemporary research on racial disparities in psychiatric treatment even more dismaying. Organized psychiatry has belatedly awoken to the problem; in January 2021 the American Psychiatric Association issued an official "Apology to Black, Indigenous and People of Color for Its Support of Structural Racism in Psychiatry," following on the formation of a task force on the problem in 2020. There is an obvious need for more research on these issues.[3]

Mental illness haunts us, frightens us, and fascinates us. Its depredations are a source of immense suffering and often embody threats, both symbolic and practical, to the very fabric of the social order. Ironically, the stigma that surrounds those who exhibit a loss of reason has often extended to those who have claimed expertise in its identification and treatment. Of all the major branches of medicine, psychiatry, throughout its history, has been the least respected, not just by those to whom it ministers but also by physicians and the public at large.

Vast resources have been devoted over time to efforts to intervene in, ameliorate, and perhaps cure the mysterious conditions that constitute mental disorder. Yet, two centuries after the psychiatric profession first struggled to be born, the roots of most serious forms of mental disorder remain as enigmatic as ever. The wager that mental pathologies have their roots in biology was firmly ascendant in the late nineteenth century, but that consensus was increasingly challenged in the decades that followed. Then, a little less than a half century ago, the hegemony of psychodynamic psychiatry rapidly disintegrated, and biological reductionism once again became the ruling orthodoxy. But to date, neither neuroscience nor genetics have done much more than offer promissory notes for their claims, as I shall show in later chapters. The value of this currency owes more to faith and plausibility than to much by way of widely accepted science.

For what it is worth, I should be astonished if several of the major varieties of mental disturbance do *not* turn out to be at least partially explained by biological factors. But I should be equally astonished if biology turns out to be the whole story. Indeed, in some respects I think the whole debate about nature versus nurture rests on a serious category mistake, because our brains are extraordinarily plastic organs, jointly constituted by the social and the physical—by the biological endowment we are born with and by the psychosocial environment within which our brains grow and develop, a point I shall return to in my conclusion.

But the fundamental point remains: the limitations of the psychiatric enterprise to date rest in part on the depths of our ignorance about the etiology of mental disturbances. Psychiatry's deficits also reflect the profound limitations of the treatments psychiatrists can offer patients even in our own times. For the most severe forms of mental disturbance—schizophrenia, bipolar disorder, and grave depressions—which are the overwhelming focus of this book, it is important to be clear-eyed: not to deny that there has been some progress, but equally not to ignore the price that is sometimes paid for such relief as psychiatry can now provide. Periodically, as we shall see, enthusiasts have proclaimed that decisive breakthroughs are at hand or that miraculous cures have been discovered. To date, these supposed revolutions have proved evanescent and are often the harbinger of distinctly damaging interventions.

The continuing difficulties in understanding and treating the forms of mental illness have to some extent been masked by the great broadening of the problems claimed by the profession and embraced by the public. If our best efforts to treat schizophrenia, bipolar disorder, and melancholia have advanced slowly and fitfully, much of the energy and efforts of psychiatrists and clinical psychologists are now directed elsewhere. Anxiety disorders, the milder forms of depression, panic disorders, the impact of all forms of trauma, eating disorders, and substance abuse have all become the major preoccupations of psychiatric and psychological professionals. Many patients have welcomed this recognition of their troubles and embraced the combination of psychotherapy and medication that emerged to treat them.

For none of these forms of mental distress does psychiatry possess a magic wand, and the ability to treat this heterogeneous collection of disorders successfully varies considerably. Eating disorders and disorders associated with substance abuse and PTSD are particularly, but not completely, resistant to successful interventions.[4] But substantial numbers of people who are anxious or depressed do seem to be helped by psychotherapy or by psychotherapy combined with drug treatments.

Anxiety disorders, the apparent incidence of which has spiked in recent decades, provide a useful example of the value and the limits of existing therapeutics. As a recent comprehensive review notes, "Only 60–85% of patients with anxiety disorders respond (experience at least a 50% improvement) to current biological and psychological treatments. In addition, only about half of the responders achieve recovery. . . . [P]atients with anxiety disorders . . . have high rates of recurrence and / or experience persistent anxiety symptoms."[5] For the patients who improve, these results are obviously welcome, but the limitations of our current therapies are sobering.

The picture is similar elsewhere. Psychotherapy has been shown to make a real difference in many cases of depression. At the same time, as a recent review in *Current Psychiatry Reports* notes,

> We have to remember that the effects are still modest as are those of antidepressant medication. The majority of patients improve during treatment, but a considerable number of these would also have improved without treatment. Improvement rates without treatment have been estimated to be about a quarter after 3 months and 50% after 1 year. And on the other side of the spectrum, there is a considerable minority of about 30% of patients who do not respond to any treatment.[6]

There are, of course, a variety of competing sorts of psychotherapy, focused on very different approaches to the disorders they seek to treat. These approaches include cognitive-behavioral therapy, interpersonal therapy, and psychodynamic therapies—time-limited and less-intensive interventions based on psychoanalytic principles—to mention three of the most prominent. One relatively robust finding from evaluation research is that it is hard to differentiate among these techniques when it comes to assessing the degree of improvement they bring about in patients.[7] None is a panacea, but almost all seem to provide a measure of relief to a considerable number of patients with less serious forms of mental distress, even when treatment is provided in primary care settings rather than by mental health specialists.

For a brief period toward the close of the twentieth century, the profession persuaded itself that it had found a reliable way to identify those suffering from mental disorder and to divide their pathologies into separate and distinct subtypes. But that assurance is now vanishing. Lacking any biological markers or tests to identify the disorders they claim to treat, psychiatrists have been forced to rely on symptoms and patient self-reports to construct their categories and decide who belongs where. But the diagnostic manuals psychiatrists have created and relied on have become increasingly unwieldy, and the consensus about the categorization of mental illness has threatened to fall apart. Influential voices have dismissed the whole enterprise as fatally flawed, and the legitimacy of the undertaking, and of the profession that relies on it, has come under fire.[8]

My discussion in the pages that follow will strike many, quite correctly, as a deeply critical account of the psychiatric enterprise. But I hope it is not a wholly unsympathetic one. The puzzles that psychiatry wrestles with are profoundly difficult to resolve, and the desperation of many of those it seeks to treat is unmistakable. That is not to excuse some of the history that follows, but it does provide a crucial context for understanding how desperate remedies for mental illness have so often surfaced, and why they have been so broadly adopted.

ONE OF THE MOST DISTURBING features of the treatment of the mentally ill over the past century and more has been their extraordinary vulnerability to the therapeutic enthusiasms of the profession that purports to help them. Why is that? Psychosis has been presumed to rob patients of their capacity to make informed choices about their treatment. Legally and morally, patients have often been regarded as nonpersons. For much of the twentieth century, those suffering from major mental illnesses were locked up in institutions that deliberately isolated them from their families and from society, adding to their vulnerability. As wards of the state, they represented an enormous economic burden, while their pain and suffering remained immense. The pressures on psychiatrists to do something about conditions they understood poorly, if at all, were correspondingly great, and restraints on the zeal for therapeutic experimentation were largely absent or readily circumvented.

Hence the emergence of a host of dubious interventions aimed at these recalcitrant disorders. To mention only a handful: there were programs to induce fevers by deliberately infecting patients with malaria, by injecting horse serum into spinal canals to induce meningitis, or by placing patients in diathermy machines that broke down the body's homeostatic mechanism; there was the surgical removal of teeth and tonsils, followed by the evisceration of stomachs, spleens, cervixes, and colons; the use of the newly discovered insulin to create artificial comas that often brought patients to the brink of death; the induction of artificial epileptic seizures, first with drugs, then with electricity passed through the brain; and, most dramatically of all, the severing of brain tissue, either through surgical operations on the frontal lobes or by thrusting an ice pick through the eye socket into the brain—so-called transorbital lobotomies. Virtually all of these were disproportionately visited on women, though the best data we have indicate that mental illness afflicts men and women almost equally. The pattern of disparate treatment is indisputable, however, and I shall both document it and examine why it surfaces repeatedly.

In the middle decades of the twentieth century, Freudian psychoanalysis appeared to provide an alternative treatment for mental disorders, and some psychiatrists claimed that the reach of the talking cure extended even to the major psychoses. But the claimed efficacy of psychoanalysis in cases of serious mental illness proved chimerical, and in the final years of the last century, the Freudian enterprise lost the dominant position it had once held. These days, classical Freudian psychoanalysis survives in only the most vestigial of forms. Its influence can still be traced in some of the many contemporary versions of psychotherapy that seek to treat the kinds of mental distress that Freud sought to engage with: anxiety disorders, mild and moderate depression, somatoform disorders, and other nonpsychotic disorders. In the United States, though, Freudian techniques face severe competition from new approaches with a quite different intellectual lineage.

A national survey of psychiatrists in private practice undertaken in the early 1970s showed a heavy preference in their ranks for psychotherapy when treating their patients. A majority of patient visits were for fifty or fifty-five minutes—the classic analytic hour, and 86 percent of all outpatient encounters lasted forty minutes or more. The survey found that less than 30 percent of patients receiving nonpsychoanalytic care received "chemical treatment," a category that liberally included the so-called minor tranquilizers such as Valium and Librium. Among patients seeing psychoanalysts, that figure fell to only 14 percent.

For both groups, "chemical [i.e., drug] therapy was usually an adjunct to psychotherapy."[9] By 1996, only 44.4 percent of visits to a psychiatrist's office involved psychotherapy and by 2005 there had been a further decline to 28.9 percent. By the latter date, only 10.8 percent of psychiatrists routinely offered psychotherapy of any kind to their patients, overwhelmingly to those who paid for their own therapy.[10] Conversely, "a large and increasing proportion of mental health outpatients received psychotropic medication without psychotherapy."[11]

The incentives driving this process are clear. With the advent of managed care, a practice based on psychopharmacology was far more lucrative, and insurance companies paid less and less for psychotherapy. In substantial measure, this pattern reflected insurers' endorsement of new forms of therapy developed and popularized by clinical psychologists. These mental health professionals had emerged in competition with psychoanalysis after the Second World War. Their techniques of cognitive-behavioral therapy, for which there appeared to be statistically based evidence of efficacy, were directly aimed at the relief of patients' symptoms rather than at the more elaborate and open-ended reconstruction of personality that psychoanalysis proclaimed as its mission.

Cognitive-behavioral therapy and its variants have a definitive end point, making them far more attractive to insurance companies than the years-long process psychoanalysts proclaimed was necessary to achieve results. Even more attractive to the proponents of managed care, these therapies could be supplied by psychologists and psychiatric social workers at far lower reimbursement rates than would satisfy most medically trained psychiatrists. Many psychiatrists decided that they could no longer afford to practice psychotherapy on these financial terms, a development given further impetus by the spread of drug company–inspired claims that mental illness was simply brain disease, by the associated shift of academic psychiatry toward biology, and by the availability of psychotropic medications on which to base the treatment of a broad range of mental distress.[12]

More and more marginalized in a profession they had once dominated, psychoanalysts were left with a niche market, catering to a minority who could afford to pay for their services out of pocket and who remained convinced of the value of long-term talk therapy. Other analysts have tried to adapt to market pressures by providing a shorter and less intensive kind of psychotherapy that nevertheless draws on the psychoanalytic emphasis on affect and the expression of emotion, the relationship of past experience to the present, and the ways that patients knowingly and unknowingly seek to avoid distressing thoughts and feelings. Their comparative advantage, psychoanalysts claim, is that they not only treat symptoms but also have a broader positive influence on a person's psychological resources and capacities.

In important respects the demise of psychoanalysis as a dominant force in contemporary psychiatry was brought about, directly and indirectly, by the psychopharmacological revolution that began in the early 1950s. The Freudian enterprise survived and apparently thrived until the late 1970s, though, like the gnawing of termites proceeding unnoticed till a structure is fatally compromised, the new drugs and the transformations that flowed from their discovery undermined the supports of psychoanalysis and eventually led to its startlingly rapid collapse.

The psychopharmacological revolution that began in the years after the Second World War has swept all before it. It has transformed psychiatric practice, the image of the profession, and the patient experience. Biology has been reembraced as the source of mental illness and other possible dimensions of mental disorder largely cast aside. It is to faulty brain biochemistry, and perhaps to genetic factors, that we are supposed to look to explain the origins of mental illness, and with its primary reliance on pills, psychiatric practice now more closely resembles much of mainstream medicine. Progress has arrived, it would seem, but whether we really have a better grasp of the etiology of psychosis and whether the lot of the seriously mentally ill has improved markedly since the mid-1950s are matters to which we shall attend.

Families grappling with the suffering and trauma of mental illness mostly embraced a biologized psychiatry that, instead of blaming refrigerator mothers and ineffectual fathers, assured them mental illness was a real physical illness, the product of genetic flaws and biochemical abnormalities in the brain. Politicians anxious to rid themselves of the immense fiscal burden that updating and operating Victorian museums of madness had become rushed to embrace these magic potions that could return mental patients to the tender mercies of the community. But neither the antipsychotics nor the antidepressants were psychiatric penicillin. Though marketed as such by drug companies and many psychiatrists, these drugs were in fact no more than Band-Aids, sources of symptomatic relief that often carried with them a heavy price in side effects. Many patients found taking them to be intolerable: between two-thirds and four-fifths of those in a recent, carefully designed comparative study of antipsychotics' efficacy refused to continue taking their pills.[13]

Certainly, for some patients drugs offered a more tolerable existence than had been their previous lot, and that symptomatic relief is important. Nor is psychiatry unique in treating symptoms rather than causes. We should also remember that all general medicine can offer for such diseases as diabetes, Parkinson's, autoimmune disorders, and AIDS is palliative treatment, and yet no one would doubt the importance and value of those interventions. For many mental patients, though, modern psychiatry's drugs have proved to be a Faustian bargain, piling serious side effects on top of the mental turmoil from which they already suffer. For a not-insignificant fraction of the patient population, those complications can prove life-threatening.

A Hollywood producer who once contemplated making a movie based on one of my books informed me that it provided the basis for a great first and second act. But where, he asked me, was the third act? By this he meant, where was the happy ending? The story he was mulling *had* no happy ending, as is true of much of our human experience. Nor does the history I examine here have one. Mental illness remains a baffling collection of disorders, many of them resisting our most determined efforts to probe their origins or to relieve the suffering they bring in their train.

The profession to which I (half) belong, sociology, once embraced the foolish and romantic notion that mental illness was simply a matter of labels and exclusionary societal reactions.[14] Renegade psychiatrists like Thomas Szasz proclaimed that mental illness was a myth.[15] Both stances are profoundly mistaken. They constitute a failure to come to grips with the severe sufferings of so many people and with the intellectual challenges that face those trying to comprehend and ameliorate, let alone solve, the challenges posed by the protean world of unreason. One must hope that, in the future, serious progress will be made. For the present, we need to be honest about the dismal state of affairs that confronts us rather than deny reality or retreat into a world of illusions. Those, after all, are classically seen as signs of serious mental disorder.

PART ONE

THE ASYLUM ERA

CHAPTER ONE

Mausoleums of the Mad

AMERICAN PSYCHIATRY, like its counterparts in Europe, was a profession born of the asylum. From the 1820s onward, an almost utopian optimism had taken hold in reformist circles, inspiring a growing conviction that, under a carefully calibrated regimen, the lunatic could be treated without violence or threats and thereby restored to sanity. Enthusiasts for this doctrine of moral treatment proclaimed that insanity, when properly treated in appropriate physical and moral surroundings, was a readily curable condition, more easily cured, some asserted, than the common cold.

Particularly vocal proponents of this view were Samuel Woodward, the superintendent of the Worcester State Asylum in Massachusetts, and Dr. William Awl, superintendent of the Ohio State Asylum. (Awl's extravagant claims led to his being dubbed "Dr. Cure-Awl.") But the heads of the McLean Asylum in Boston, the Hartford Retreat in Connecticut, and the Bloomingdale Asylum in New York spoke equally forcibly about the curability of insanity, and the statistics that these men published were widely used to promote the construction of asylums at taxpayer expense.[1]

A veritable "cult of curability" swept through the ranks of the bien-pensants—men like the New England reformers Horace Mann and Samuel Gridley Howe. The upshot, fueled by the tireless efforts of one of the great moral entrepreneurs of the age, the Boston crusader Dorothea Dix, was the creation of a vast network of asylums all across the United States.[2] As the asylum system developed, those running the new institutions often referred to themselves as "medical superintendents of asylums for the insane," a clumsy title that captured the central role of the institution in the development of their profession. Subsequently, many Americans embraced the term "alienist," borrowed from the French and derived from one of the French terms for madness, *aliéné*.

By the closing decades of the nineteenth century, the small, therapeutically inclined institutions of the earlier period had been transformed into mausoleums of the mad, a captive population of several thousand and assorted support staff. Trapped almost as surely as the patients they presided over and experimented on in this series of Potemkin villages, America's alienists, now increasingly embracing the German label of "psychiatrists," were an isolated, insular lot.[3] The claims of this first generation of alienists, whose early promises to cure 60, 70, even 80 percent of cases proved wildly off the mark, were debunked by one of their own, Pliny Earle, the first superintendent at Northampton State Hospital in Massachusetts and a founder of the American Medical Association, who repented his earlier enthusiasms.[4] The extraordinary optimism that had marked the 1830s and 1840s had been replaced by an equally profound pessimism.

The numbers of patients confined in mental hospitals exceeded 150,000 by 1903, and would grow to a half million by 1950.[5] With the decline in almshouses, an ever-larger fraction of the whole was composed of the senile and demented, for whom few alternative sources of support and care existed.[6] To these patients, who would depart the asylum only in caskets, one could add the cases of tertiary syphilis—"pitiable . . . wrecks of humanity sitting in a row, their heads on their breasts, grinding the teeth, saliva running out of the angles of the mouth, oblivious to their surroundings with expressionless faces and solid, livid, immobile hands."[7] Such patients, diagnosed as suffering from General Paralysis of the Insane (GPI), made up as many as 25 percent of male admissions. The syphilitic origins of GPI, long suspected, would finally be confirmed in 1913, when Hideo Noguchi and J. W. Moore of the Rockefeller Institute for Medical Research in New York demonstrated the syphilitic spirochete in the brains of paretics.[8]

To the senile and the syphilitic (and alcoholic), one must add the legions of those who lapsed into chronic insanity. In the mammoth asylums scattered across the late nineteenth-century landscape, a large proportion of the population came to be composed of long-stay patients, and it was this specter of chronicity, this horde of the hopeless, that was to haunt the public imagination, to constitute the public identity of asylums, and to dominate late nineteenth- and early twentieth-century psychiatric theorizing and practice.

In reality, a substantial fraction (perhaps two-fifths or one-half) of those admitted each year were released as recovered or improved within twelve months, and another 10 percent (many of them elderly) died during that period. But those who were not discharged within a year lingered and in most cases became part of the ever-larger fraction of permanent residents. We have no reliable statistics for 1900, but those from 1950 paint a grim picture. Of those then confined in state mental hospitals, "one quarter . . . have been hospitalized for more than sixteen years, one-half for more than eight years, and three-quarters for more than two and a half years." This was all the more remarkable since "the turnover of senile cases is very rapid because of their high death rate." Among schizophrenics, who constituted almost half of the hospital's population, the median length of stay was 10.5 years.[9]

The stark contrast between the early promise of the asylum and the dismal reality that it served for many as little more than a living tomb did little for the reputation of the psychiatric profession. When asylum superintendents formed their professional association and held their first meeting in Philadelphia in 1844, it was at the height of optimism about the curability of insanity. Supremely confident that they had unlocked the keys to the therapeutics of madness, and ruling over a rapidly expanding asylum system, alienists saw themselves (and were seen by others) as more efficacious at treating mental illness than their counterparts in the rest of the medical profession were at treating physical illness. That self-confidence was bolstered by the financial security of the asylum superintendent, a situation that compared favorably with the highly uncertain prospects most physicians faced in a desperately overcrowded medical marketplace.[10] When regular physicians formed the American Medical Association in 1847, the alienists scornfully rejected an invitation to join forces with their less well-placed professional brethren. A quarter century later, the tables were turned.

Unable to deliver the cures they had blithely promised, alienists found themselves increasingly marginalized by the rest of the medical profession. In the last third of the nineteenth century, neurologists, a rival group of specialists who laid claim to expertise in diseases of the nervous system, were unsparing in their reproaches. It was thus distinctly odd when, on the occasion of the fiftieth anniversary of the American Medico-Psychological Association in 1894, psychiatry's leadership had the temerity to invite Silas Weir Mitchell, an eminent Philadelphia neurologist, to address their annual conference. Mitchell at first demurred, and then, when pressed, agreed to speak at their annual meeting in nearby Princeton. Twice bidden, not shy: he used the occasion to issue a scathing indictment of the state of American psychiatry.

AMERICAN NEUROLOGISTS EMERGED as subspecialists immediately after the American Civil War. Alongside the mass slaughter, the conflict produced large numbers of casualties who had suffered trauma to their brains, spines, and extremities. Taken together, the wrecked bodies of these soldiers provided a series of naturalistic experiments that elucidated important facets of the human nervous system. War, as always, visited unspeakable horrors on those who fought it, but proved invaluable to the medics who treated them.[11]

When army surgeons set themselves up as "nerve specialists" in the war's aftermath, still another set of military casualties crowded their waiting rooms. These were men with mysterious nervous complaints that failed to obey what the neurologists had learned about the structure of the brain and central nervous system, but who nonetheless insisted that their suffering was real and combat-related. Their ranks were joined by civilians, many of them female, equally difficult to diagnose and equally importunate, loudly insisting that their troubles were real and rooted in the nervous system.

These "hysterics," as they were popularly known, and "neurasthenics" (suffering from weakness or overtaxing of the nervous system) came to constitute a large fraction of the new nerve doctors' clientele. In exasperation, Weir Mitchell once referred to hysteria as "mysteria," but, like most of his colleagues, he could not afford to turn these patients aside.[12] Not disturbed enough to warrant confinement in an asylum, the ambulatory neurasthenics and hysterics had the wherewithal to pay for their treatment, and their social standing more closely matched that of the professionals whom they consulted. Their demographic profile contrasted markedly with the population that crowded the wards of the state hospitals, helping the neurologists to escape the stigma that asylum doctors increasingly shared with their patients.

Like their eminent French colleague Jean-Martin Charcot, American neurologists insisted that these patients' troubles had an organic origin and were the result of wear and tear caused by overtaxing of the brains. As specialists in disorders of the brain, neurologists initially laid claim (at least in principle) to expertise in the diagnosis and treatment of insanity—a jurisdictional claim that inevitably brought them into direct conflict with asylum doctors, who for half a century had proclaimed themselves uniquely qualified to treat such patients. Openly scornful of their rivals' capacities, neurologists dismissed asylum doctors as mere boardinghouse keepers and curators of dead souls, willfully ignorant of the latest scientific advances. Fierce claims and counterclaims of this sort marked the 1870s and 1880s.[13]

The brash New York neurologist Edward Spitzka was particularly scathing, condemning asylum superintendents as experts in roofing and drainpipes, farms and mechanical restraint, "experts at everything except the diagnosis, pathology and treatment of insanity." "Psychiatry," he declared, "is but a subsidiary branch of neurology."[14] His fellow New Yorker William Hammond, who had served as surgeon general of the Union Army in the Civil War, insisted that "there is nothing surprisingly difficult, obscure, or mysterious about diseases of the brain which can only be learned within the walls of the asylum." General practitioners with a modicum of training in "cerebral pathology and physiology," he contended, were "more capable of treating successfully a case of insanity than the average asylum physician."[15]

Asylum doctors responded vituperatively to such withering assaults. Eugene Grissom, the superintendent of the North Carolina State Asylum, attacked Hammond as "a moral monster whose baleful eyes gleamed with a delusive light," willing to skew his science on the witness stand to whatever position those paying his large fees desired, an adulterer who had been court-martialed and dismissed from the army for misconduct.[16] (Hammond had indeed been court-martialed on trumped-up charges from which he was later cleared, despite a stellar record as surgeon general.)[17]

Prominent neurologists along the Eastern Seaboard allied themselves with critics of the asylum, denouncing "the absolute and irresponsible power of the superintendents" and attacking "the grated windows, crib-beds, bleak walls, gruff attendants, narcotics and insane surroundings of an asylum."[18] With their lay allies, these critics formed the National Association for the Protection of the Insane and the Prevention of Insanity, and sought to investigate the scandals they alleged to be lurking behind asylum walls. But this breach of professional solidarity brought censure from the medical profession, leading to a rapid, if unstable, cessation of hostilities.

In practice, the institutionalized remained firmly beyond the reach of the new nerve specialists, isolated behind the walls of America's asylums. And perhaps this was just as well, since, for all their bluster, neurologists had no new therapies to offer this stigmatized group and could not even reach agreement on a scientific description of the terrain—limitations they reluctantly came to concede in their textbooks.[19] By the early 1890s, animosities had begun to subside as neurologists distanced themselves from the more extreme forms of disturbance that warranted asylum care.

WEIR MITCHELL'S SPEECH SEEMED likely to reopen old wounds. Striding to the podium, he conceded that what he was about to say violated the usual expectations for such celebratory occasions: "It is customary on birthdays to say only pleasant things," he began, but those who had asked him to speak had persisted in asking him to do so after he had warned them that he would offer criticism "without mercy." He had given in to their importunities. "That was a momentary insanity; I have been sorry ever since," for he now had to face up to "the uncongenial task of being disagreeable."[20]

Sorry he may or may not have been, but disagreeable he most certainly was. Medicine and surgery, he pointed out, had made extraordinary progress over the past half century. Psychiatry, he insisted, had been stagnant. "Your hospitals are not our hospitals; your ways are not our ways." Psychiatrists lived in isolation, cutting themselves off from healthy criticism and scientific progress, and had become "almost a sect apart." The consequences for profession and patient alike were "evil." The very title to which the more senior members of his audience clung, "medical superintendent," he suggested, was "absurd." Many listening to him had won their jobs through political patronage, still another "grave evil."[21]

The invective rained down for an hour and more. Weir Mitchell denounced the "senile characteristics" of asylum boards; the repetition of "old stupidities in brick and stone"; the "neat little comedy" of outside inspections known about in advance and thus incapable of getting at the truth. But this was the least of the problems. The medical superintendent was the "monarch" of all he surveyed, but he was an emperor with no clothes. "Where, we ask, are your annual reports of scientific study, of the psychology and pathology of your patients[?] . . . We commonly get as your contributions to science, odd little statements, reports of a case or two, a few useless pages of isolated post-mortem records, and these are sandwiched among incomprehensible statistics and farm balance sheet." Asylum case records put on display an appalling state of affairs, an "amazing lack of complete physical study of the insane, . . . the failure to see obvious lesions," and a complete ignorance of the diagnostic technologies indispensable to the practice of modern medicine. These problems were as visible in the most prominent and best-endowed asylums for the rich, as in the meanest, most overcrowded state hospital.[22]

Psychiatrists had spent a half century attempting to persuade the public of the "superstition . . . that an asylum is in itself curative. You hear the regret in every report that patients are not sent soon enough, as if you had ways of curing which we have not. Upon my word, I think asylum life is deadly to the insane." Far from being therapeutic environments, mental hospitals (as they were ironically beginning to be called) had the air of being a prison, with "grated windows and locked doors. . . . I presume that you have, by habit, lost the sense of jail and jailor which troubles me when I walk behind one of you and he unlocks door after door. Do you think it is not felt by some of your patients?"[23]

The upshot of such conditions, Weir Mitchell announced, was what he himself had observed in the wards of an asylum in his home city of Philadelphia, where "the insane, who have lost even the memory of hope, sit in rows, too dull to know despair, watched by attendants; silent, grewsome machines which eat and sleep, sleep and eat." Nor were they the only victims of prolonged confinement: their captors, the psychiatrists, had fallen into the same sort of paralysis. "The cloistral lives you lead give rise, we think, to certain mental peculiarities. . . . [Y]ou are cursed by that slow atrophy of the energizing faculties that is the very malaria of asylum life." Indeed, he concluded, "I cannot see how, with the lives you lead, it is possible for you to retain the wholesome balance of mental and moral faculties."[24]

It was a remarkable performance, a savage assault in which Weir Mitchell seemed to have sought, point by point, to demolish the entire psychiatric enterprise. Unsurprisingly, when he finally sat down, he was greeted with token applause. Oddly enough, the association then proceeded to elect its distinguished guest an honorary member.

Much of the inevitable grumbling and complaint that followed took place in private. The Swiss transplant Adolf Meyer, then occupying a lowly position as the staff pathologist at Kankakee State Hospital in Illinois but soon to emerge as America's most prominent psychiatrist, recalled his superintendent, Clarke Gapen, returning from the Philadelphia meeting full of "resentment."[25] Some months later, a semi-official public response to Mitchell's criticisms appeared in the pages of the *American Journal of*

Insanity. Written by the New England alienist Walter Channing, it was notable for its defensive tone and for statements that inadvertently suggested how close to the mark Weir Mitchell had been. Far from seeking to show that the superintendents were men of science concerned with curing their patients, Channing went out of his way to dismiss the idea that asylum superintendents could perform such roles as an irrelevant fantasy.

Like their predecessors, the current generation of psychiatrists was rescuing the insane "from the tortures of the damned"—the vile treatment they otherwise faced in the community, in prisons, or in almshouses. "The medical superintendent," Channing pointed out, "is an executive officer. . . . His real specialty is insane-hospital management." That meant that, unlike the erudite neurologists whom Weir Mitchell represented, he was not capable of writing an abstruse volume "on the cerebral anatomy of a spider." But "no man can do everything, and only in a very few cases are scientific and executive talent combined." Channing acknowledged that "the usual medical superintendent has little taste for science"—but then, he contended, "scientific men, put in charge of institutions, are apt to be failures." Asylum doctors are rather "able, efficient physicians of business instincts." Their work, he posited modestly, consists "of giving rest and succor to as many of a wretched and neglected class as a niggardly and ignorant public will allow, and . . . will never be done till every insane pauper is the ward of the state."[26]

For all Weir Mitchell's invocations of science, Channing quite correctly pointed out that, where mental illnesses were concerned, "their treatment and cure is both unsatisfactory and baffling." He could have pointed out that neurologists dealt only with milder forms of mental disorder, dispensing an array of tonics and animal extracts that purported to stimulate the nerves, and adding treatment with shiny electrical machines that dispensed static electricity to the same presumed effect. Some made use of the intervention Weir Mitchell himself was famous for inventing, the so-called rest cure, confining nervous patients to bed for weeks at a time, denying them all mental stimulation or outside company, and feeding them a high-calorie diet to build up "fat and blood" and thereby restore their shattered nerves. Affluent but nervous patients who feared their troubles might lead to consignment to an asylum sought out such treatments, and many professed themselves grateful for interventions later generations saw as profoundly misguided.[27]

In reality rather than in Weir Mitchell's world of make-believe (for the concluding pages of his indictment had conjured up the vision of a curative, scientifically oriented mental hospital), Channing insisted that "the inmates of an insane hospital are . . . helpless children" needing to be watched over and protected. And as for the eminent neurologist's complaints about bolts and bars and mechanical restraint, "hospital medical officers . . . refuse to deny plain facts, in order to make a good showing."[28]

The complacency on display in Channing's response was hardly universal, but it was widespread. Asylum superintendents could rightly assert that in sheltering those suffering from tertiary syphilis, alcoholism, and the troubles of old age they were relieving society of serious burdens. But mental hospitals purported to be therapeutic institutions. Tightening state controls and sharp cuts to their budgets signaled the impatience of their political masters, and those issues were of far greater moment than the verbal barbs of a jealous rival.

Massachusetts had pioneered the creation of a body charged with uniform oversight of its institutions for "the dependent and vicious classes" in the 1860s. Its State Board of Charities was widely emulated, and while these entities at the outset were somewhat toothless, they were increasingly asserting themselves. Declining cure rates in state asylums provided the rationale for centralizing and rationalizing oversight of the superintendents' activities and also reining in the costs of institutions, which by then were the largest single item in state budgets. (As late as 1950, state hospitals for the mentally ill absorbed as much as a third of many states' revenues.)[29]

As almshouses emptied and were closed down, the decrepit and senile who had previously found refuge there migrated, *faute de mieux*, to the local state asylum. Charles Wagner, the superintendent of the Binghamton State Hospital in New York, lamented:

> We are receiving every year a large number of old people, some of them very old, who are simply suffering from the mental decay incident to extreme old age. A little mental confusion, forgetfulness and garrulity are sometimes the only symptoms exhibited, but the patient is duly certified to us as insane and has no-one at home capable or possessed of means to care for him. We are unable to refuse these patients without creating ill-feeling in the community where they reside, nor are we able to assert that they are not insane within the meaning of the statute, for many of them, judged by the ordinary standards of sanity, cannot be regarded as entirely sane.[30]

Thus, the ability of the asylum to discharge some fraction of its patients as "cured" suffered a further blow, and the pressures on the asylum superintendents intensified. Calculating "cures" on the total number of patients resident, the measure that ensured the lowest announced cure rate and the one insisted upon by the state bureaucracy, led Utica State Hospital's recovery rate to fall from 20 percent in 1889 to between 7 and 9 percent over the course of the following decade.

Caught in a vice, New York's mental hospital superintendents found their budgets slashed—an 1893 statute cut the average cost per patient from $208 to $184 per year—and their powers curtailed. A new three-person lunacy commission insisted on statewide standards, particularly with respect to financial records, and began to intrude into a whole range of decisions, from which personnel to hire to how many paying (and thus desirable) patients the superintendents could admit.

Peter Wise, the superintendent of a new state hospital at Ogdensburg, New York, was soon regretting his accession to a job most of his junior doctors still longed for. As he wrote to his fellow superintendent at Utica, George Alder Blumer, "If only I could give up this unpleasant work, which has become so distasteful to me that I dread the day that dawns and retire with the feeling intensified."[31] A year later, he had managed the trick, moving to be the president of the lunacy commission, whence he could issue the orders, not receive them, and escape the daily grind of running a custodial institution. It was an escape his colleagues could only long for.[32]

Once small enough to be overseen by a single physician, a state asylum (or state hospital, as it was being relabeled in a desperate attempt to shed the stigma that clung to the term "asylum") had relentlessly increased in size. In time, a hospital had required the addition of one more physician, and then a phalanx of assistant medical men—and, by the closing years of the nineteenth century, even the odd woman who could be assigned to the special task of coping with the female side of the asylum. Provided with food and lodging and a slender stipend, though often forbidden to marry, these drudges spent years on the dreary routines of life on mental hospital wards. As the growth in the number of asylums slowed in the latter part of the century, the opportunities for advancement to the plum position of superintendent grew ever more remote, prompting the exodus of anyone with ambition and further cementing the low reputation of medical psychologists among the medical profession at large. Soon, superintendents and state bureaucrats were bemoaning "their growing inability to attract young medical graduates to a career in institutional psychiatry."[33]

In the words of British assistant asylum doctors, who labored under similar misfortunes, their situation was like that of poor relations waiting impatiently for the demise of a rich, childless relation. The contrast of "the fat salaries of the Superintendents with the lean ones of the assistants" had perhaps been bearable "when the Assistant Medical Officers were few and the Superintendencies ripened in four or five years, but [their elders' counsel to be patient] loses all its sweet reasonableness when we have to wait ten, twelve, or more years for the golden fruit, and even run the risk of its being plucked by some [politically connected] outsider from over the wall just as we thought it about to drop."[34] Those who could fled, and for the most part only dull time-servers remained.[35] A profession once convinced that it could cure had mostly subsided into somnolence, bound by the dull rounds of administrative routine.

CHAPTER TWO

Disposing of Degenerates

THE NOTION THAT INSANITY was somehow the product of an interaction between a vaguely defined inherited constitution and the environmental pressures of daily life had long been a cliché among those who sought to understand the origins of mental illness—and, indeed, of illness more generally. That acquired characteristics, even patterns of behavior, could be inherited—a view that came to be formally associated with the French biologist Jean-Baptiste Lamarck—was widely accepted, and these beliefs would continue to be held by the public at large into the early twentieth century. The biological constitution that parents handed on to their children was held to be the unique and peculiar combination of their own biological inheritances and the experiences, habits, and modes of life they had adopted over their lifetimes. Intemperance, slothfulness, immorality, poor diet, indulgence of the baser emotions, overstrain, and overwork—all these and more could have a deleterious effect on one's offspring, giving rise to what was referred to as a "diathetic weakness" that produced lifelong troubles. Mothers in particular were thought to bear a heavy responsibility for the quality of the offspring, and they were expected to nurture their fetuses (and infants) both morally and physically—injunctions that provided a "scientific" warrant for the enforcement of Victorian limitations on women's lives.

As the gloom attending the failures of asylum doctors spread, a new account of the roots of insanity emerged that simultaneously explained away their apparent therapeutic impotence and provided an alternative rationale for the isolation and containment of the insane in vast institutions. First articulated by the French alienist B. A. Morel in his *Traité des dégénérescences* in 1857, the new doctrine rapidly secured converts across Europe and then spread to America.

The first generation of asylum doctors had argued that insanity was a catastrophe that could potentially be visited on anyone. Indeed, in some respects, the prominent and successful members of society had particular reasons to fear this affliction, for they had the finest and most refined nervous systems, stretched tight as their ambitions placed greater strains on their brains. In the last third of the nineteenth century, psychiatrists increasingly adopted the directly contrary view, insisting that insanity was a symptom of degeneration and biological inferiority, a form of evolution run in reverse to be found for the most part among "the dregs of society." Where madness had once been a disease of civilization, now it was held to mark its absence, the return to a lower state of being of an almost animalistic sort.[1]

In the hands of those developing the idea of degeneration, the influence of hereditary factors assumed a far darker and more deterministic form. Americans of earlier generations had bought William Buchan's *Domestic Medicine* in vast quantities between 1775 and 1830, finding their beliefs about the importance of heredity repeatedly reinforced as they sought his advice on how to maintain their health. But Buchan's message could be read optimistically, even as he warned of the consequences of dissipation. "Family constitutions," he proclaimed, "are as capable of improvement as family estates"—provided one lived a life of prudence and discretion. To be sure, the contrary was equally true: "the libertine who impairs the one, does greater injury to his posterity, than the prodigal, who squanders away the other." Progress ought to be the order of the day. Hence, odd as it might seem, for Americans of the Jacksonian Age, a belief in heredity and an optimistic attitude toward the possibilities of improving the human race went hand in hand.[2]

Not so in the closing decades of the nineteenth century, especially with respect to the burgeoning troublesome classes, among whom the insane and feeble-minded (a group that substantially overlapped) loomed large. Charles Darwin's ideas about evolution, however controversial at first, had popularized the notion of the survival of the fittest as the indispensable engine of progress. But as Darwin's cousin Francis Galton argued, in civilized societies the operation of this natural law was often suspended, prompting overbreeding by the less fit and underbreeding by the more thoughtful and conscientious, an observation that led him to embrace eugenics.[3] It was a theme Darwin himself took up in *The Descent of Man and Selection in Relation to Sex*, published in 1871:

> With savages the weak in body or mind are soon eliminated; and those that survive commonly exhibit a vigorous state of health. We civilized men, on the other hand, do our utmost to check the process of elimination; we build

> asylums for the imbecile, the maimed, and the sick; we institute poor laws; and our medical men exert their utmost skill to save the life of everyone to the last moment. . . . Thus the weak members of civilised societies propagate their kind. No one who has attended to the breeding of domestic animals will doubt that this must be highly injurious to the race of man.[4]

If one accepted this line of reasoning, its implications for those charged with running asylums and other institutions designed to cope with the dangerous and defective seemed obvious. Left to its own devices, a ruthless Nature might be expected to eliminate the diseased and defective. But since a misplaced charity had intervened to slow or even reverse the inexorable pathway to extinction, they threatened to return society's kindness by reproducing their kind in ever-larger numbers.[5] Quite what the hereditarian mechanism might be that produced such an outcome remained undefined, but that defects were passed from one generation to the next, and that this defective germ plasm was the primary explanation for feeblemindedness, crime—and soon enough the whole panoply of social pathologies that afflicted late nineteenth-century societies—was accepted by experts and laymen alike. By the closing decades of the nineteenth century, what had once been the speculations of isolated asylum doctors acquired a broader cultural significance. Critics of those toiling in the asylums agreed with alienists on this point. As the New York neurologist Edward Spitzka put it, the "most important predisposing cause of insanity is undoubtedly . . . hereditary transmission of structural and physiological defects of the central nervous apparatus."[6]

Biological explanations of criminality, most famously advanced in the work of the Italian phrenologist and criminal psychologist Cesare Lombroso, were widespread. Lombroso argued that the propensity to violence and crime was visible in a carefully cataloged variety of malformations of the body: long arms, asymmetrical heads, sloping foreheads, projecting jaws. These defects were thought to match the impulsive behavior, vindictiveness, and absence of remorse and civilized morality among the lower orders, for they were two sides of the same coin.[7] As the notion of deviants as evolutionary throwbacks gained broad acceptance, it became linked in the United States to rising nativist feelings.

Morel and his French colleagues were the first to proclaim the degenerate "a morbid deviation from the primitive human type."[8] Hysteria, epilepsy, and nervous disorders of all types, ranging all the way to insanity, were increasingly held to constitute a single class of afflictions. The epileptic and insane were thought to recapitulate in their bodies the pathologies of past generations, and to risk contaminating those around them. As Morel put it, as early as 1857, "The degenerate human being, if he is abandoned to himself, falls into a progressive degeneration. He becomes . . . not only incapable of forming part of the chain of progress in human society, he is the greatest obstacle to this progress through his contact with the healthy part of the population."[9]

Morel's doctrines fell on receptive ears. There were soon mutterings about the need to penetrate "the mystery of the propagation of hereditary diatheses" so that medicine might move "towards purging the blood of races of living poisons."[10] In a France recently defeated by Prussia, suffering from a sharply declining birth rate and a pressing sense of national decline, the hereditary account of mental disorder had become entrenched. Defective biology was invoked to explain a whole host of putatively closely interrelated forms of pathology: crime, drunkenness, epilepsy, and hysteria, with madness and feeble-mindedness the final way station on the road to extinction.

In Germany, as early as the 1840s, Wilhelm Griesinger had introduced the notion that mental illness was brain disease. His successors were even more enamored of biological reductionism. German psychiatrists were wedded to laboratory research on the brains of deceased mental patients, ensconced in university clinics that kept them at a safe distance from the custodial realities of an asylum system that provided them with an endless supply of anatomical specimens. Almost to a man, they enthusiastically embraced degenerationist doctrines. Their convictions did not waver in the least when their minute histological studies failed to find the evidence of degeneration that their theories assured them must be there. As for living patients, they were completely uninterested in them. Convinced that insanity was an incurable disorder, they disdained those who suffered from it.

TO THE EXTENT THAT AMERICAN psychiatry drew on foreign sources in developing its own notions of the connections between mental illness and degeneration, it was to British counterparts that it turned—most notably to the work of the misanthropic but enormously influential Henry Maudsley; and to Charles Mercier, Daniel Hack Tuke, and the Edinburgh alienist Thomas Clouston. Their textbooks were those that American asylum doctors consulted, if they troubled with book learning at all. In the absence of a substantial American text, publishers promptly made US editions of their books available.

For Maudsley, anatomy was destiny. Mental illness "has its foundation in a definitive physical cause," he wrote in *The Physiology and Pathology of Mind,* and the madman "is the necessary organic consequence of certain organic antecedents: and it is impossible that he should escape the tyranny of his organization."[11] His (or her) defects were written upon the body, in "malformations of the external ear . . . tics, grimaces, or other spasmodic movements of the face, eyelids or lips . . . stammering and defects of pronunciation." The eyes were notable for "a vacantly-abstracted, or half-fearful, half-suspicious, and distrustful look. . . . These marks are, I believe, the outward and visible signs of an inward and invisible peculiarity of cerebral organization."[12]

Science, Maudsley claimed, had demonstrated that madness was the penalty to be paid for vice and immorality. "The so-called moral laws are laws of nature," he wrote, which men "cannot break, any more than they can break physical laws, without avenging consequences. . . . As surely as the raindrop is formed and falls in obedience to physical law, so surely do causality and law reign in the production and distribution of morality and immorality on earth." Excess of any sort threatened the mental integrity of future generations, and "the wicked man" must be brought "to realize distinctly that his children and his children's children will be the heirs of his iniquities."[13]

Moral causes wrought physical changes on people's bodies, and these degenerative modifications were successively "transmitted as evil heritages to future generations: the acquired ill of the parent becomes the inborn infirmity of the offspring. It is not that the child necessarily inherits the particular disease of the parent . . . but it does inherit a constitution in which there is a certain aptitude to some kind of morbid degeneration . . . an organic infirmity which shall be determined in its special morbid manifestations according to the external conditions of life."[14] Tuke, whose great-grandfather had founded the York Retreat in 1796, the institution that inspired the construction of reformed asylums in the English-speaking world, was blunter still: "Recklessness, drunkenness, poverty, misery characterise the class," he insisted. "No wonder that from such a source spring the hopelessly incurable lunatics who crowd pauper asylums, to the horror of the [tax]payers." They were most emphatically "an infirm type of humanity.... 'No good' is plainly inscribed on their foreheads."[15]

The implication was that it was positively misguided to attempt to cure the mad and restore them to society. Such apparently humane and well-intentioned efforts "prevent, so far as is possible, the operation of those laws which weed out and exterminate the diseased and otherwise unfit in every grade of natural life." Irrationally, the insane "are not only permitted, but are aided by every device known to science to propagate their kind." They are "turned loose to act as parents to the next generation . . . centres of infection deliberately laid down, and yet we marvel that nervous disease increases."[16]

The embrace of a rigid biological determinism, while it appeared to leave little scope for positive intervention, had the compensating virtue of explaining away the profession's dismal therapeutic performance. Far from lamenting their failure to cure the mad, the citizenry should see it as a blessing in disguise, "for no human power can eradicate from insanity its terrible hereditary nature, and every so-called 'cure' in one generation will be inclined to increase the tale of lunacy in the next. . . . [I]t is evident that the higher the percentage of recoveries in the present, the greater will be the proportion of recoveries in the future."[17] Asylums thus had a vital role to play in warding off the evils of insanity, even if it was by providing custody rather than a cure.

WELL-TO-DO AMERICANS HAD NEVER regarded the asylum with much favor. Hospitals were for the poor and friendless, not those of means, an attitude that only began to change once the combination of anesthesia and the aseptic revolution in surgery brought a change in the hospital's reputation, and a therapeutic advantage to being treated in an in-patient setting. Still, if general hospitals were becoming essential to the practice of advanced medicine, the same could not be said of the asylum. The stigma attached to certification as a lunatic made confinement of a near relation in such places a potent source of shame, and inflicted serious damage on one's social standing.

Those who possessed the resources preferred to cope with mental infirmity in domestic surroundings, consulting neurologists or other nerve doctors. In the aftermath of the Civil War, the newly emerging specialty of neurology had found that many of the patients crowding into their waiting rooms complained of a variety of nervous disorders. Some were veterans of combat who carried no signs of physical pathology. Others were society matrons (or their daughters) or captains of industry suffering from the ancient disease of hysteria or the newly fashionable disorder of neurasthenia. They turned to nerve doctors after general practitioners had failed them.

Victims of train accidents soon joined them, and "railway spine" became part of an expanding lexicon of functional disorders—complaints that manifested themselves via physical symptoms but seemed at odds with what anatomy and physiology taught about the makeup of the body. Nervous patients not disturbed enough to require seclusion in an asylum created a demand for a new kind of nerve doctor, and neurologists willing to accept the challenge (and the income) soon found their ranks augmented by refugees from the asylum sector. A heterogeneous and disorganized group, these practitioners existed alongside (though at a considerable intellectual remove) from traditional alienists.

To treat their patients' nervous prostration, these doctors invented an enormous variety of nerve tonics whose ingredients remained trade secrets. Many contained dangerous substances, some of which proved addictive: strychnine, morphine, cocaine, and lithium salts, for example. Others proffered animal extracts designed to fortify the nerves.[18] Hydrotherapy was another treatment they frequently employed; while in their consulting rooms, patients connected to elaborate machines received jolts of electricity to provide painful stimuli to the body. Electrotherapy had been popular in some circles from the eighteenth century onward, but for many it had the odor of the charlatan and the quack. As neurologists began to demonstrate the role of electricity in transmitting nervous impulses, however, it was not difficult to persuade themselves and their clientele that electrical charges had healing powers.[19] Manufacturers rushed to fill the void.[20]

The shocks were painful, of course, and were particularly touted as remedies for one of the commonest symptoms of hysteria, a loss or impediment of speech, or hysterical aphonia, where electrodes could directly be applied to the larynx. Few remained mute for long when subjected to these interventions.[21] More generally, neurasthenics and hysterics were informed that overuse and overstrain had disturbed the equilibrium of the nervous system. A commonly employed metaphor was that overstimulation and overuse had run down one's battery, an image that suggested why the application of electricity could have therapeutic effects. Within a generation, though, having originally insisted that electrical treatment worked directly on the nervous system, there was a growing consensus that its efficacy, such as it was, rested on the power of suggestion.

The most famous therapy the nerve doctors employed was Silas Weir Mitchell's "rest cure." Like most of his fellow neurologists, Weir Mitchell attributed the debility of his nervous patients to the wear on their systems caused by the pace of modern life. *Wear and Tear: Or, Hints for the Overworked,* was the title of one of his monographs intended for popular consumption. The title of a subsequent volume in 1877 hinted at the solution: *Fat and Blood.*[22] The book's (and the treatment's) popularity is suggested by the appearance of an eighth edition as late as 1911. Those subjected to the rest cure included the novelists Charlotte Perkins Gilman and Virginia Woolf, both of whom wrote devastating fictional accounts of the experience.[23] The rest cure lasted six or eight weeks, sometimes longer, all designed to build up the "fat and blood" of the patient. Prescribed a high-calorie diet and subjected to forced bed rest, patients were isolated and denied access to any and all forms of psychical and mental stimulation, save for the electrotherapy, massage, and hydrotherapy that were prescribed as part of the treatment. It was a treatment only available to the well-to-do, and its primary "beneficiaries" were overwhelmingly women. Presumably some pronounced themselves better, and certainly nervous invalids from near and far flocked to Philadelphia to be treated by Weir Mitchell. For Gilman and Woolf, the experience practically drove them mad. As for men, Weir Mitchell prescribed virtually the opposite regimen. Dubbed the "West cure," it sent neurasthenic men out West to rope cattle, hunt, and compete in a sturdy contest with Nature. Thomas Eakins, Theodore Roosevelt, and Walt Whitman were among those who tried it.

It is little wonder that modern feminists have denounced the rest cure as a prototype of Victorian patriarchal oppression. Certainly, these gender-based curative regimes uncannily echo the stereotypes of the broader culture of the time: women subjected to a hyperexaggerated version of domesticity, and men sent forth to strengthen nervous systems that their sedentary lives had enfeebled.[24] By the early twentieth century, these earlier interventions were gradually giving way to various forms of psychotherapeutics.

EVEN AMONG THE RICH, however, many found their mentally disturbed relations too disruptive, too wearying, or too dangerous to themselves or others to be allowed to remain at large. From the earliest years of the asylum era, a handful of institutions had emerged whose clientele came from more affluent social strata than those who crowded the halls of the state hospitals. To be rich and mad might require shelter not just from the gaze of the world, but also from any intimate association with what the Massachusetts authorities called the "odor of pauperism."[25]

At the McLean Asylum in Boston, the Butler Hospital in Rhode Island, and the Hartford Retreat in Connecticut, as well as at the Bloomingdale in New York and the Pennsylvania Hospital for the Insane in Philadelphia, continuous efforts were under way to compete for a limited number of wealthy patients whose families could choose where their relations would be confined. Though secure confinement of the mad might be the key requirement for such an establishment, it had the potential to alienate the asylum's true clientele, the patients' families, and so these features had to be disguised beneath a veneer of good taste and cheerfulness. Manicured grounds could play a vital role in massaging the impressions of families faced with the difficult task of confining their nearest and dearest.[26]

The Hartford Retreat, which opened its doors in 1824, initially admitted a mix of public and private patients, there being no other asylum in the state. But once Connecticut constructed its first state asylum in 1858, the Retreat rapidly upgraded its facilities to attract wealthy patients. In 1860, it hired Frederick Law Olmsted to landscape its thirty-five acres of grounds (Olmsted also submitted a design for the campus of its rival, the McLean—though it was never carried out—and sadly ended his days as a patient there). After the Civil War, the Hartford Retreat began a rapid program to redesign and upgrade its physical fabric, aided by the fact that the war, and the fortuitous presence of the Colt armaments factory, had turned Hartford into the richest city in the country.

If opulent surroundings did not suffice, families could be reminded that the asylum provided a range of other amenities—French lessons, drawing classes, singing classes, theater, and the like. Staffing levels were high—the McLean boasted that "its sane population is about half as numerous as the insane patients"—and patient numbers remained constrained. If all these amenities meant that these private establishments cost six or eight times as much as their public counterparts, their clientele—the patients' families—were not disposed to object. In these social circles, the linkage of mental illness and degeneracy was a sensitive subject. Just how carefully these doctors felt they had to proceed is suggested by the intellectual evolution of George Alder Blumer, who had previously been one of the most vocal psychiatrists endorsing eugenic ideas.

As superintendent of the Utica State Asylum in Upstate New York, Blumer was intimately familiar with the dire conditions of his state's hospitals. He grew increasingly frustrated at the attempted micromanagement of his institution by Carlos MacDonald, who had been appointed head of a reconstituted state commission on lunacy in 1889. MacDonald, who previously served as superintendent of three state hospitals, immediately set about enforcing uniform state standards, and in 1893 he endorsed state legislation that sharply cut hospital budgets, from $208 to $184 per patient per year. Instructions to hire a female assistant physician and limitations placed on his ability to admit and charge higher fees for private patients exacerbated Blumer's frustration. His initial instinct was to challenge MacDonald, whom other asylum superintendents also resented for his interference. In such circles, MacDonald was increasingly viewed as a "Prince of Darkness."[27] But it soon became apparent who had the governor's ear, and when neither a change of administration nor MacDonald's resignation brought any relief, Blumer's disillusionment reached the breaking point. In 1899, he accepted a new position as head of the private Butler Hospital in Rhode Island.

At first, Blumer was as vocal as ever about the links between mental illness, vice, and the biological defects that ran in families. In 1903, four years into his tenure at Butler, he was elected president of the American Medico-Psychological Association. He used the occasion of his presidential address to warn once again of the links between inherited biological defect and mental defect, speaking darkly of the "infinite disaster" that awaited society if the mad were allowed to give free rein to their instincts. He added that the insane were "notoriously addicted to matrimony and by no means satisfied with one brood of defectives."[28]

The relatives of his new clientele were not shy when it came to communicating their unhappiness with such notions. It was all very well to dismiss the pauperized masses who crowded the state hospitals as degenerate fiends; not so when the blame was attached to the captains of industry and their close relations. Very quickly, Blumer found himself forced to soften and then almost repudiate his earlier views. In an attempt to soothe the anger of patients' families, he informed them that "in these New England communities of ours, where brains are more apt to be highly organized than in less favored parts of the country, it may even be a mark of distinction to possess a mind of sufficient delicacy to invite damage under the stresses of life."[29]

When the nineteenth-century neurologist George Beard had popularized the term "neurasthenia," he had employed similar arguments to suggest that this weakness of the nerves was a condition peculiarly likely to afflict the rich and successful, whose nervous systems were more refined and delicate than those of the poor, and stretched by the stresses of their ultracivilized lifestyles. It was a well-judged piece of flattery that had salutary effects in expanding the market for these nerve doctors' wares, and Blumer was not slow to recognize that such notions could easily be adapted to placate rich relatives of asylum inmates. He returned to the theme on later occasions, reminding them that "there cannot be complexity of the nervous system without what the world calls nervousness."[30]

In private, and in communications solely intended for his fellow psychiatrists, Blumer made it clear that in reality his views had not changed at all. "Insanity is to a large extent a degeneracy," he wrote to one correspondent in 1916; four years later, he told another to pay no mind to the "comforting view that there is less in heredity that most of us believe."[31] As for those "refined" New England elites, "Those old families have been breeding and inter-breeding ever since [they arrived in colonial times], insomuch that there are few of the old families in Rhode Island today which do not reveal the unhappy consequence in neuroses of some sort, and the evil work is still going on."[32]

Most of Blumer's colleagues worked in the ever-expanding state asylum system, and had come under little or no pressure to disguise their view that their charges were a biologically defective lot. Paid large sums to minister to rich patients, those who ran asylums for the well-to-do were keener than their colleagues in the public sector to seize upon innovations that promised cures. Besides, they faced great pressure from a clientele accustomed to the power of the purse, who expected results.

The sorts of patients who formed the bulk of the population of asylums like the Hartford Retreat and the McLean were by no means the only members of the well-to-do classes who came to the attention of psychiatrists. Precisely because of the stigma that attached itself so firmly to the "lunatic" or "mentally ill," and the social erasure that accompanied commitment to an asylum or mental hospital, whenever it proved possible—unless management difficulties or the threat of scandal grew too powerful—the wealthy sought alternatives.

PRIVATE ESTABLISHMENTS FOR THE NERVOUS proliferated in the late nineteenth century: spas for those suffering from the strains imposed by the pace of modern life, and sanitoriums for successful men of business (and their wives) to recharge their run-down nervous batteries. Drawing patients by means of discreet advertisements or word of mouth, they occupied a shadowy existence in attractive but secluded locations. The most successful of these began life as a small institution for members of one of those new varieties of Christianity that have been a recurrent feature of the American scene. The Western Health Reform Institute was set up in 1866 to serve members of the Seventh Day Adventist Church who were experiencing mental or spiritual crises. Within a few decades, it had become the largest and most famous sanitarium in the world.

The Adventists traced their origins back to the religious revival of the Second Great Awakening and to the expectations of the End Time and the Second Coming of Christ that gripped the followers of William Miller in the early 1840s. Members of the church made wholeness and health (and an emphasis on "pure" food) a central part of their daily lives. Ellen White, the Church's spiritual adviser, had received a series of visions from angels, Christ, and God, enjoining certain dietary constraints to be adopted by the Adventists while awaiting the Second Coming. (As the historian Ronald Numbers has pointed out, she borrowed her prescriptions from other "health reformers" active at the time.)[33] On Christmas Day in 1865, a new vision told her to found an institution for members of the Church, lest they fall victim to "the sophistry of the devil" in hydropathic institutions run by nonbelievers. Though the divine injunction was rapidly put into practice, the institution struggled throughout its first decade.

Two brothers who were members of the faith, John Harvey and Will Keith Kellogg, took over the floundering institution and transformed it, with John attending to the medical side, and Will to the business end. The 106 patients at what was in 1876 called the Western Health Reform Institute were swamped by the 7,006 who patronized its successor in 1906. Masters of marketing, the Kellogg brothers modified the existing word for a health resort, "sanitorium," and renamed the establishment the Battle Creek

Sanitarium, nicknamed "the San." A disastrous fire in 1902 consumed most of the buildings. The catastrophe provided the means for the Kelloggs to separate their enterprise from the Church. They secured their own financing for its reconstruction, and expanded still further. Staff grew to 800, including thirty physicians, and Battle Creek's treatment rooms could accommodate as many as a thousand patients at a time.

All sorts of affluent patients—"nervous" invalids, those plagued by insomnia and headaches or by a multitude of aches and pains—came to Battle Creek to reinvigorate their nervous systems and restore their health. John Harvey Kellogg introduced them to a regimen based on a cleansing vegetarian diet (meat was held to rot in the gut and poison the system), frequent enemas to cleanse the colon, abstinence from alcohol and sex, cold showers, electrotherapy and phototherapy, and plentiful open-air exercise. Whole grains, fiber, and nuts were the order of the day, often transformed into flaked cereals (the secret of which was appropriated by one of their patients, C. W. Post, much to Will's dismay).[34] Beyond this, peanut butter and vegetarian substitutes were offered for meat and fish (skallops—a fake version of the real thing—were a particular favorite). To aid in the emptying of the colon, Kellogg even designed a special toilet, lower than most and sloping backward, so that the civilized were forced to squat in a position that he claimed facilitated elimination.

Clean bowels and defecation several times a day were the route to health and happiness, and an in-house laboratory for fecal analysis provided a scientific check on the patients' progress. It was a formula that attracted both the great and the not-so-good: Lincoln's widow, Mary Todd Lincoln; Amelia Earhart; Alfred Du Pont, Henry Ford, and John D. Rockefeller; presidents Warren G. Harding and William Howard Taft; Thomas Edison and the original Tarzan, Johnny Weissmuller. In the long run, though, it was the Kelloggs' breakfast-cereal business—originally a spin-off for those unable to afford a sanitarium visit—that made them multimillionaires.[35]

So great was the demand for space that, in the late 1920s, John Harvey Kellogg decided to expand again, building a fourteen-story tower block across the street from the main complex. The temptation was obvious: in the years between 1927 and 1929, Battle Creek was grossing $4,000,000 a year, and making a profit of $700,000. Kellogg took out a loan for $3,000,000 to finance what turned out to be a folly. The huge profits of earlier years had largely been frittered away, and the opening of the new facility turned out to be the harbinger of financial ruin. The stock market crash of 1929 dried up the flow of new patients. Over the next four years, the anticipated profits evaporated and, ultimately, Kellogg found himself unable to pay the interest on the loan. A last-ditch appeal to the Rockefeller Foundation in 1934 met with emphatic rejection: Alan Gregg, the director of medical sciences, noted that "J.H. Kellogg is 83 yrs old and there is no one to succeed him—it is impossible for the RF to interfere or advise in a situation like this . . . we should keep out."[36] Within weeks, the Sanitarium was forced to declare bankruptcy, and though it limped along in a much-reduced state for the rest of the decade, it was purchased for use as an army hospital in 1942.

Battle Creek was a special case, but a host of far smaller establishments more closely resembling the original Western Health Reform Institute were created in the closing decades of the nineteenth and early decades of the twentieth century. In many cases, physicians who had developed a nodding acquaintance with the treatment of insanity in state asylums left the depressing environment of these mausoleums of the mad and established private homes where they took in well-to-do patients not so disturbed as to require confinement in an asylum.

In the Boston suburb of Brookline, for example, Walter Channing opened his "private hospital for mental diseases" in 1879. Channing was a direct descendant of Walter Channing, an officer in the Revolutionary army, the grand-nephew of William Ellery Channing (one of the founders of Unitarianism), and eldest son of the poet William Ellery Channing—establishment credentials that must have stood him in good stead when it came to attracting patients. Unusually for someone of his background, after finishing his training at Harvard Medical School he served as an assistant physician at New York's Asylum for the Criminally Insane, and then as first assistant physician at the Massachusetts State Insane Hospital in Danvers.[37] This gave him the practical knowledge he needed to open his own establishment.

Taking fewer than twenty-five patients, mostly women, Channing provided discreet treatment in secluded surroundings. When Brookline began to be overrun by development, he relocated his enterprise in 1916 to a large wooded estate in Wellesley. Here he offered the standard remedies "nerve doctors" relied on—hydrotherapy, massage, electrotherapy using shiny static electricity machines—coupled with a soothing environment and gentle admonitions to improve one's self-control. If not particularly efficacious, neither were these interventions particularly challenging, and genteel invalids unfit for society could idle away their days.

FAR FROM THESE MANICURED ESTATES, those shunted off to state hospitals could expect no such tolerant treatment. As the theory of degeneracy acquired a patina of respectability, some began to argue that even lifelong confinement or involuntary sterilization were half measures, the product of a failure of nerve. Attentive readers of the Anglo-American psychiatric literature would have encountered hints that a more robust solution was worthy of consideration. "Every year," the English alienist Samuel Strahan reminded his audience, "thousands of children are born with pedigrees that would condemn puppies to the horsepond." Lunatics were waste products of the evolutionary process, "morbid varieties fit only for excretion."[38]

On occasion, the euphemism of "excretion" or "extrusion" was dispensed with entirely. Blumer had well-thumbed copies of both Maudsley's and Clouston's books, and proved to be an adept disciple.[39] "Our modern hospitals for the insane are in some measure responsible for the increase of insanity by promoting, not the survival of the fittest, but the survival of the unfit," he lamented, "as well as by permitting unstable persons to leave institutions and mate themselves with their kind, instead of allowing an affinity of contrasts to prevail in selecting their wives."[40] The ancient Scots may have been condemned by modern sentimentalists for their "rough and ready method" of burying alive babies and their epileptic or mentally disturbed mothers, but "from the point of view of science the cruel and remorseless Scot was more advanced than his descendants of our day."[41]

Blumer's wistful comment was not meant as a serious policy proposal. Others were not so circumspect. The prominent New York physician W. Duncan McKim, heir to a banking fortune and contemptuous of his social inferiors, warned darkly of "the ever-strengthening torrent of defective and criminal humanity." He urged that "a gentle and painless death" was "the most humane means" of resolving the societal problem that they presented. "This should be administered not as a punishment," he clarified, "but as an expression of enlightened pity for the victims—too defective by nature to find true happiness in life." Euthanasia was a remedy, he acknowledged, that "is in part a very old idea, but so modified and expanded as to differ profoundly from its prototype": "If we view broadly the evil which these individuals engender, we find not only that it thwarts the best purposes of men, but that it lies at the very root of human misery. . . . When we reflect upon the vast amount of wealth and affection these semi-human automata absorb . . . it would hardly seem that we are justified in preserving them merely because of an abstract sentiment for which reason can give no warrant." Time to be rid of "many an unworthy life," those whose existence has "proved a curse for the race and the individual." "The idiot and the low-grade imbecile," he hastened to reassure his readers, "are not true *men*, for certain essential human elements have never entered into them, and never can; nor is the moral idiot truly a man, nor, while the sad condition lasts, the lunatic. These beings live amongst us as men, but if we reckon them as human we shall fare much as if we bargained with the dead or with beasts of prey." They should be exterminated en masse with "carbonic gas."[42] The numbers marked out for extermination would inevitably be large at first, but they would diminish with each generation, as the defective germ plasm was eliminated.

Though McKim's discussion reads today like a twentieth-century equivalent of Jonathan Swift's satirical *Modest Proposal*, it was meant in most deadly earnest. *The Nation*, in a review published on November 1, 1900, recommended McKim's book to its readers and to "all good citizens interested in human progress." Who could doubt that the public had every right to protect itself from "the ravages, one may say the compulsory and automatic ravages, of the physically, mentally, and morally diseased [?] . . . There seems no logical objection to the absolute removal of those whose unsoundness is complete and irremediable, particularly when they are a public charge."[43]

For good measure, McKim's publisher, Scribner's, more than a decade and a half later brought out the popular *The Passing of the Great Race*, by Madison Grant, director of the Bronx Zoo and president of the Immigration Restriction League. Here the proposed "obliteration of the unfit" was extended to an assault on the "inferior races." Grant complained that the "mistaken regard for what are believed to be divine laws and a sentimental belief in the sanctity of human life tend to prevent the elimination of defective infants and the sterilization of such adults as are themselves of no value to the community."[44] Such well-known progressives as Clarence Darrow joined in the chorus, advocating efforts to "chloroform unfit children" so as to "show them the same mercy that is shown to beasts that are no longer fit to live."[45]

As a practical matter, the opposition of religious groups and constraints of a democratic polity meant that the chances of actually instituting such policies were essentially nil. Charles Davenport, the Harvard-trained biologist and member of the National Academy of Sciences, did what he could as head of the Eugenics Records Office in Cold Spring Harbor, New York, to persuade the American public to put aside its concerns, but in the end he could only lament that "it seems to be against the mores to burn any considerable part of our population."[46] It would prove otherwise, of course, in Nazi Germany where, with the enthusiastic participation of German psychiatrists, Hitler's T-4 program would see mental patients herded into gas chambers in tens, even hundreds of thousands, "useless eaters" who became the first victims of a policy of mass extermination that would soon extend to other groups defined as subhuman and a menace to the purity of the race.[47] But the harsh and condemnatory language of these authors and the insistence that mental illness was an irredeemable biological condition were not without serious consequences for American social policy.

THE THEORIES OF DEGENERACY that psychiatrists had done so much to develop and propagate were seen to justify a whole series of legislative changes. The easiest to pass were attempts to prevent marriage and reproduction among the unfit, first codified into law in Connecticut in 1895, launching the fashion for laws prohibiting the marriage of the defective. If one of the marriage partners was determined to be "unfit"—feeble-minded, epileptic, or an imbecile—both parties could be imprisoned for a term of up to three years. Statutes of this sort proved irresistible to politicians, and by the mid-1930s, as many as thirty-one states prohibited the mentally ill and "feeble-minded" from wedding.[48] But if such statutes were politically popular, their practical effects were slight, and even those broadly supportive of the eugenic cause came to see them as symbolic gestures, rather than effective interventions to stem the burgeoning number of defectives.

These were years of massive immigration from southern and eastern Europe. The influx of Jews and Catholics aroused powerful nativist sentiments. Critics charged that these migrants were biologically inferior to the Nordic types who were the hope of America, and threatened the nation's future. Worse still, these were "races" with high rates of mental illness and defect and biologically based criminal propensities. The need for restrictive immigration laws to shut out the mentally defective entirely, and to limit the influx of these inferior ethnicities, was widely discussed. Francis Walker, a prominent economist and president of the Massachusetts Institute of Technology, had written in the *Atlantic* in 1896 of the necessity not just of "straining out from the vast throngs of foreigners arriving at our ports . . . [the] deaf, dumb, blind, idiotic, insane, pauper, or criminal," but also of excluding "the tumultuous access of . . . hordes of ignorant and brutalized peasantry from the countries of eastern and southern Europe." These were "beaten men from beaten races; representing the worst failures in the struggle for existence"—"foul and loathsome" creatures, whose presence would serve only to degrade and debase America's culture.[49]

Two years earlier, the Boston bien-pensants Charles Warren, Robert DeCourcy Ward, and Prescott Hall had founded the Immigration Restriction League. The spread of eugenic ideas among "progressive" elites combined with hostility to competition from the new migrants among the working classes ensured that the agitation to exclude those considered biologically inferior increased in the decades that followed. Ultimately it met with legislative success, a process helped along by the expert testimony of sympathetic psychiatrists and social scientists. Asian immigrants were barred in 1917, and the Immigration Act of 1924 imposed strict quotas on immigration from southern and eastern Europe, inaugurating a system that would remain in place for more than four decades.

This was not the first time questions of race and ethnicity had aroused nativist sentiment. Between 1847 and 1854, the Irish potato famine prompted a mass immigration of Catholics to a still largely Protestant country. Two-fifths of the nearly two and three-quarter million immigrants in these years were Irish Catholics, and disproportionate numbers of them soon began to show up in public mental hospitals. This pattern was particularly marked in the major cities: in Boston, New York, Philadelphia, Cincinnati, and St. Louis, where unskilled Irish immigrants clustered in urban slums. Sharing the larger culture's distaste for the new arrivals, asylum superintendents found them "more noisy, destructive, and troublesome," "very ignorant, uncultivated people [who] from some cause or another, seldom recover."[50] There was talk of establishing separate institutions for them, and a frequent resort to segregating them into separate wards from those that housed the native-born.[51]

If the prospect of mingling Irish and other foreign-born patients with the native-born was problematic, mixing white and Black lunatics was unthinkable. In the pre–Civil War North, the African American population was small, those who applied were often turned away, and the few who were admitted to asylums were simply placed in segregated cells. In the South, slaves were largely excluded from the asylums, except where their owners were willing to pay for them, in which case they were separately provided for in grossly inferior accommodation. After 1865, either completely separate provision was made in existing asylums or separate and distinctly unequal "colored" asylums were established in which to confine them. Tennessee (1866), Virginia (1869), North Carolina (1880), Mississippi (1889), and Alabama (1902) opened segregated institutions. Segregated wards were set up in various states, including West Virginia (1864), Missouri (1865), Georgia (1866), and Arkansas (1882).[52] North or South, there was a broad consensus: racially integrated wards were inconceivable, and "virtually no member of the psychiatric profession was willing to challenge the dominant separate but equal doctrine that was the basis of public policy."[53]

Just how stark the differences in the treatment of Black Americans were has been detailed in a number of recent analyses of asylums in the South—in Georgia, Louisiana, Virginia, and Washington, DC. Collectively, they show the profound differences in the way African American patients were viewed and treated both by psychiatrists and the staff of these institutions. Martin Summers has helpfully documented the large discrepancies in the use of restraints and seclusion between the Black and white inmates at the federal hospital in Washington, DC.[54] And he and Elodie Edwards-Grossi show the equally profound differences in the labor demanded of Black inmates compared with their white counterparts. If most American asylums were massively underfunded and overcrowded, conditions in the "colored" wards and institutions were far worse. Some Black patients sought to resist what they perceived as slave labor.[55]

The heads of mental hospitals in the South were unapologetic about the differential treatment of the races, insisting that it was "essential" and beneficial to both races. Under slavery, T. O. Powell proclaimed, "there were, comparatively speaking, few negro lunatics. Following their sudden emancipation their number of insane began to multiply, and, as accumulating statistics show, the number is now alarmingly large and on the increase."[56] He and his colleagues knew why: as primitive people barely removed from savagery, they were simply not equipped to cope with the pressures of a free society.[57] In reality, slaveholders had been reluctant to assume the cost of institutionalizing the mentally ill; some even resorted to "freeing" them rather than providing for them.[58]

The "necessity" of segregating Black patients was adopted as an article of faith by the first generation of asylum physicians, and it exercised a continuing hold on the profession.[59] In 1914, Mary O'Malley, a junior psychiatrist on the staff of St. Elizabeths in Washington, DC, explained the origins of "psychoses in the colored race" in the pages of the profession's journal, the *American Journal of Insanity*: "300 years ago the negro ancestors of this race were naked dwellers on the west coast of Africa . . . in the depths of savagery and suddenly transplanted to an environment of the highest civilization, and 250 years later had all the responsibilities of this higher race thrust upon them."[60] Small wonder that so many of them became psychotic.

Seven years later, her colleague W. M. Bevis echoed these views. "The colored race," he contended, were simply too primitive to cope with the stresses of a free society. They were, after all, the descendants of "savages or cannibals in the jungles of central Africa." Consequently, their "biological development" left them "poorly prepared" for emancipation, though "their talent for mimicry . . . is remarkable . . . sometimes sufficiently exact to delude the uninitiated into the belief that the mental level of the negro is only slightly inferior to that of the Caucasian." Nothing, of course, could be further from the truth.[61] Persuaded that their happy-go-lucky nature and primitive emotional equipment meant that few suffered from depression, psychiatrists believed they disproportionally exhibited the symptoms of dementia praecox and mania.

The early twentieth-century anti-immigrant sentiment thus built on decades of prejudice and unequal treatment that were most deeply embedded in "scientific" beliefs about the biological inadequacies of African Americans. If the Grants, the McKims, and the Davenports directed their ire at Jews, Italians, Greeks, Poles, and Russians, that did not mean that they were not even more convinced of the inferiority of Black Americans. It was simply that those prejudices were so deeply rooted and widely shared in the dominant culture as to be not worth mentioning. Segregation and neglect were seemingly immutable facts of life throughout the asylum era. For the most part, psychiatrists shared the profound racial prejudices of the larger society and even helped give a scientific gloss to those beliefs. Those who did not share the broad consensus were largely complaisant, reluctant to make waves or simply oblivious to the racial inequities of the mental health system.

ASYLUM DOCTORS and their allies had by the early decades of the twentieth century put in place yet another set of policies that derived from their belief in the heritability of mental illness. Sexual surgery to "cure" female patients of their insanity had enjoyed something of a vogue in the last three decades of the nineteenth century. By and large, the operations had been performed on "nervous" cases outside asylum walls, mostly by those associated with the newly emerging field of gynecology. Pioneered by a Georgia physician, the aptly named Robert Battey, some thousands of "normal ovariotomies"—the surgical removal of apparently healthy ovaries for the relief of mental distress—were carried out all across the United States and on a smaller scale in Europe.[62]

Asylum superintendents, resentful of the encroachments of another group of medical men on their turf and perhaps worried that embarking on a novel, potentially controversial, and highly experimental therapeutics would be perilous, for the most part shunned the operation. By the turn of the century, however, a handful of asylums in the United States and Canada had dispensed with these scruples and embarked on ovariotomies—on occasion (as at Norristown State Hospital in Pennsylvania) through the efforts of female physicians in charge of patients of their own sex.[63]

Their interventions were short-lived. There had been much in nineteenth-century medicine's views of "the female animal" that lent seeming substance to the claim that an operation on a woman's reproductive organs might influence her mental state for the better, but by the 1890s, as elite physicians began for the first time to gain some understanding of the role of the endocrine system in the regulation of bodily functions, a growing disquiet emerged about the long-term impact of such drastic and mutilating surgery.[64] The use of normal ovariotomies on asylum inmates was soon halted, and experiments with castration, attempted in some homes for the feeble-minded and institutions for juvenile delinquents, were met with fierce criticism, in at least one case costing the physician who had performed them his job.[65]

But less drastic surgeries—severing the vas deferens of the male and tying the fallopian tubes of the female—were now available (the first vasectomy was performed in 1897), and these spared the testes and ovaries, preserving their role in the body's hormonal system. In short order, the availability of these alternatives led states to adopt statutes permitting involuntary sterilization of the unfit. Indiana passed the first such law in 1907, after several failed attempts to pass similar legislation in New York and

Pennsylvania, and soon many other states followed suit. Though seven of the first sixteen laws were subsequently struck down by the courts, others passed constitutional muster and were enthusiastically implemented, nowhere more so than in the progressive state of California.[66]

Legislation permitting such operations was passed in Sacramento in 1909. Its eugenic aims were made explicit in a 1917 amendment, which spoke of the need to prevent reproduction by those with "mental disease which may have been inherited and is likely to be transmitted to descendants."[67] By 1921, California had performed more than 2,000 sterilizations, 80 percent of all such operations in the country, and thousands more were performed over ensuing decades—11,491 by 1950, out of a total for the United States of some 23,466.[68] Margaret Smyth, the superintendent of California's Stockton State Hospital, was a particular enthusiast. She noted with pride in 1938 that California's aggressive moves to prevent the reproduction of the unfit had served as a salutary example and been emulated in Nazi Germany:

> California adopted its first sterilization law April 26, 1909. This law has attracted attention from countries all over the world. The German government applies sterilization systematically in accordance with its law, the total number of operations to date having reached something like 250,000. Investigators agree that the policy there is being enforced in a scientific spirit without racial or political implications and with a minimum of difficulty. The leaders in the German sterilization movement state repeatedly that their legislation was formulated only after careful study of the California experiment. . . . It would have been impossible they say, to undertake such a venture involving one million people, without drawing heavily upon previous experience elsewhere.[69]

A decade before Smyth spoke so proudly of her state's place in the vanguard of psychiatric progress, a test case of the constitutionality of involuntarily sterilizing the mentally defective had reached the United States Supreme Court. Carrie Buck was a twenty-one-year-old woman who had been a resident of the Virginia Colony for the Epileptic and Feeble-minded in Lynchburg for three years before, on the morning of October 19, 1927, she was wheeled into the operating theater to have her fallopian tubes severed, cauterized, and tied. She had been pregnant on admission, apparently the aftermath of being raped by her cousin, in whose parents' house she had lived after her mother had been institutionalized as feeble-minded.

Carrie's out-of-wedlock pregnancy had been the primary reason for her own subsequent confinement in the same Colony as her mother. (Her baby, incidentally, was "disposed of," as the Colony's official records put it, by being handed over to her aunt and uncle, who took their grandchild with the understanding that the little girl "will be committed later on if it is found to be feeble-minded also.")[70]

Buck was unaware of why she was being operated on, just as she did not know that, months previously, after Virginia had passed a statute permitting involuntary sterilization, she had been carefully selected by the law's defenders to serve as the test case of its constitutionality. The state had provided her with a lawyer who proceeded to sue the Colony's superintendent, Dr. John Bell, for operating on her. The original trial was little more than a formality. Buck's lawyer made little effort to challenge the factual basis of the state's version of the case, and it passed quickly through Virginia's state courts before being appealed to the United States Supreme Court. Here the lawyer did finally speak eloquently about the underlying constitutional issue, whether the 14th Amendment provided a guarantee of the individual's right to bodily integrity. But as the measure's backers had hoped, the Justices were unmoved. The majority ruled 8–1 that there was no constitutional obstacle to the involuntary sterilization of an American citizen. Oliver Wendell Holmes, Jr., widely regarded as one of the most eminent jurists in the nation's history, was assigned the task of writing the opinion. He ringingly endorsed the state's position: "It is better for all the world," he wrote, "if instead of waiting to execute degenerate offspring for crime, or to let them starve for their imbecility, society can prevent those who are manifestly unfit from continuing their kind. The principle that sustains compulsory vaccination is broad enough to cover cutting Fallopian tubes. Three generations of imbeciles are enough."[71] In a rich bit of historical irony, the parents of the politician who wrote the Virginia statute, and who subsequently successfully argued its constitutionality before the United States Supreme Court, had both died of insanity in Virginia asylums. Circumstantial evidence strongly suggests that they were driven mad by what was then seen as the quintessential disease of the degenerate classes, tertiary syphilis.

Holmes's decision had a broad impact. Indiana, which had passed the first law allowing for involuntary sterilization two decades earlier, had seen that statute invalidated in the courts. It now became the first state to pass a new statute created to pass constitutional muster by employing the reasoning behind the *Buck v. Bell* decision. North Dakota soon followed suit, as did Mississippi, and by 1931, four years after the Supreme Court ruling, twenty-eight states had enacted legislation that would pass the test established in *Buck v. Bell.* That same brief period saw as many involuntary sterilizations undertaken in the United States as the total number performed in the preceding two decades.[72]

A year later, on July 3, 1932, Buck's daughter, Vivian, the third-generation "imbecile" Holmes had denounced, died at eight of complications from measles—but not before she had been placed on the academic honor roll at the elementary school she attended in Charlottesville.

CHAPTER THREE

Psychobiology

THE ERA OF COMPULSORY MASS STERILIZATION CREATED, for many psychiatrists, a clear ethical dilemma. Nominally, their professional duty was to act in the best interests of their patients. But when psychiatrists made the decision to sever the vas deferens or the fallopian tubes of a patient, the rationale enshrined in the law was that the state had a compelling interest in preventing the multiplication of biologically inferior specimens of humanity. The implication was that the operations benefited society as a whole, but not the individuals these physicians were charged with treating. On its face, that might seem to violate the Hippocratic oath, but fortunately an ideological solution came readily to hand. Sterilization, these doctors argued in increasing numbers, provided hidden health benefits for the patients themselves. Both physically and psychologically, patients were seen as improving post-surgically, thus providing a therapeutic rationale for the operation. It was at best a tenuous fig leaf, but its widespread employment in clinical records is a testament to psychiatrists' desperate need to see themselves as something other than jailers.[1]

These mental gymnastics point to a larger problem that the theory of degeneration created for psychiatrists and for patients and their families. For the former, biological determinism provided an excuse for therapeutic failure and a new rationale for the institutions over which they presided, at the price of their claims to be part of a therapeutic profession. For patients and their families, it added a layer of shame and disgrace on top of the stigma that traditionally clung to the mad. The poor folk who contributed the bulk of those confined in what were now called "mental hospitals" were largely powerless to contest these realities and would not have been listened to anyway. Their inferiority was manifest in their lowly social status, and their lack of education or social capital precluded their mounting any sustained challenge to the opinions of certified experts.

Their rich counterparts, however, cannot have appreciated being dismissed as tainted creatures. Nor were wealthy families pleased to be told that their mentally ill relations had no prospect of a cure. Some, at least, were now threatening to take their custom elsewhere. The last thing such respectable citizens wanted was to draw attention to mental disturbance in their midst. Still, a hopeless prognosis was not one they readily embraced. Some families began to distance themselves from orthodox psychiatry and to look to religious or psychologically based explanations and treatments of mental pathology. Mary Baker Eddy's Christian Science offered one alternative that acquired great popularity in the 1880s. By 1891, her *Science and Health* had appeared in fifty editions. The credulous followed her teaching that illness was all in the mind, an illusion. That was claimed to be as true of mental suffering as of cancer, tuberculosis, or heart disease. The healing power of prayer "results now, as in Jesus' time, from the operation of divine Principle, before which sin and disease lose the reality in human consciousness and disappear."[2] For those who found these doctrines a step too far, the Emmanuel Movement that emerged in Boston in the early twentieth century offered another version of religious consolation and psychotherapy.[3]

Partly out of concern that patients were voting with their feet for mental therapeutics, and partly in response to frustration with the limitations of the treatments at their disposal, some nerve doctors catering to private patients began to argue that "there ought to be some definite form of psychotherapeutics approved by the profession so that people would not go after 'soul massage' or other faked forms of psychotherapeutics. What are we going to do with the large number who won't come to us and will go to anyone who will raise his psychic standard?"[4] It was a highly controversial stance. Some argued that following this path risked psychiatry's "dignity and scientific integrity."[5] "Let those who want to go to Christian Science go, we are not seeking patients. A certain number of them will go. There will be plenty left. We cannot keep people from consulting quacks of every description."[6] Undissuaded, those who wanted to experiment with psychotherapy were increasingly drawn to European theorists: the Swiss Paul Dubois, the Frenchman Pierre Janet, and the Viennese Sigmund Freud. Of these, it was Freud and Freudian psychoanalysis that would ultimately attract the greatest number of followers, and we shall return to them later.[7] For the overwhelming majority of psychiatrists, however, madness remained rooted in the body. For the more ambitious among them, the therapeutic nihilism associated with degenerationism had little appeal. Consequently, the search for alternative biological accounts for mental illness, and for new kinds of somatic treatment, took on a new urgency.

Mainstream medicine had benefited mightily from its embrace of the germ theory of disease and from the rise of the laboratory and new sorts of medical technologies. The work of Louis Pasteur and Joseph Lister in the 1860s and 1870s was originally greeted with skepticism, but by the last two decades of the nineteenth century, particularly with the advances of antiseptic surgery, the detection of bacteria in the laboratory, and the discovery of effective vaccines by both Pasteur in France and Robert Koch in Germany, bacteriology and the laboratory acquired a new prestige among medical men.[8] Might psychiatry acquire new luster by moving in the same direction? Perhaps a different sort of biological account of mental disorder might allow for some prospect of therapeutic progress?

The experience of German psychiatry, which had invested heavily in the postmortem study of mental patients' brains, did not at first blush offer much by way of encouragement. Wilhelm Griesinger, Germany's first professor of psychiatry, had insisted that all mental illness was disease of the brain, and his example helped establish a network of university-based clinics devoted to uncovering what those brain pathologies might be. German psychiatrists soon developed tools to section, stain, and examine microscopically the brains of patients who died in the asylums. But their virtuosity in the laboratory offered no evidence of the brain abnormalities their theories insisted must lie at the root of psychosis.[9]

In time two such discoveries *were* made, though neither led directly to therapeutic advances. The first was the discovery by a neuroanatomist working in Emil Kraepelin's clinic in Munich, who sectioned the brain of a female patient, Auguste Deter, whom he had first encountered at a Frankfurt asylum. Though only fifty-one when she was hospitalized and fifty-five when she died, Deter had shown signs of senile dementia, with increasingly profound disorientation and loss of memory. Careful examination of her brain revealed the presence of plaques and neurofibrillary tangles in her cortex—structural changes he reported in 1906, and that others soon replicated on conducting postmortem examinations of mental patients whose disease had followed a similar course. Here was one type of mental disturbance that appeared to be unambiguously rooted in pathological changes in the brain—a disorder that Kraepelin dubbed Alzheimer's disease in 1910, to honor the friend and colleague who had made the discovery.

It was a diagnosis that could only be definitively confirmed after a patient died—and was completely resistant to treatment.

The second discovery had a longer history. As early as 1822, a young French alienist, Antoine-Laurent Bayle, had called attention to a subset of patients in the Charenton Asylum where he worked. Most were men, and they exhibited a characteristic combination of symptoms: a staggering gait, difficulties of articulation and vocal tremors, an abnormal reaction to light, abnormal reflexes, and growing muscular weakness that eventually led to paralysis and death. Their increasing physical debility coincided with the most grandiose of delusions—they were Christ, or Napoleon, or the richest and most sexually potent people on the planet. Before being confined, they had often acted on these delusions, spending money they did not have and inviting all manner of legal troubles. Within a matter of months, they became demented and died, usually riddled with bedsores, and often choking on their own vomit.

Over the course of the nineteenth century, alienists had become expert at detecting the subtle signs that often signaled the onset of this terrible malady. Some thought the condition was the end state toward which all cases of insanity tended. Others blamed sinful lifestyles, excessive drinking, and sexual promiscuity, and indeed those in the throes of delirium tremens were often initially misdiagnosed with what came to be called General Paralysis of the Insane (GPI)—except that alcoholics tended to recover their wits when deprived of their tipple, while the victims of GPI most certainly did not. As the century wore on, there were growing suspicions that the cause of the malady might be infection with syphilis—a suspicion amplified by the fact that its female victims were often prostitutes—though the relationship remained ambiguous and, in some quarters, controversial.[10]

Again, German laboratory science helped settle the matter. In 1905, Fritz Schaudinn and Erich Hoffmann isolated the bacterium that caused syphilis, a corkscrew-shaped organism dubbed *Treponema pallidum*. The following year, August von Wassermann developed a blood test for syphilis.[11] Then, in 1912, working at the Rockefeller Institute in New York on brains from patients who had died from GPI at the Central Islip State Hospital on Long Island, the Japanese scientist Hideo Noguchi discovered the offending organism, essentially ending the controversy.[12] (Much of Noguchi's other scientific work—though not this paper—proved unverifiable; not only was a good deal of his research shoddy and subsequently rejected, but he was accused of experimenting with syphilis injected into children and other vulnerable populations without their consent, a charge the Rockefeller Institute attempted, with some success, to suppress.)

This was scarcely a trivial discovery, since GPI accounted for as many as 25 percent of male admissions to New York's asylums in the early twentieth century, and perhaps 5 percent of female admissions. Unfortunately, although another German scientist, Paul Ehrlich, had discovered compound 606, or Salvarsan, a partially effective treatment for early-stage syphilis, in 1909—and coined the concept of a medical "magic bullet"—Salvarsan had no comparable efficacy when used to treat tertiary syphilis, or GPI.

Others had flirted with the idea of an infectious origin of psychosis. Ira Van Gieson at the New York Pathological Institute had boasted, just before he was fired from his post, that he was on the track of "the germ of madness" and had insisted that "the majority of the diseases of the nervous system (including diseases of the mind) are to be led back to one form or another of 'poisoning.'"[13] Kraepelin had begun to hint that the origins of dementia praecox might lie in autointoxication, a theme he returned to in each edition of his increasingly influential textbook. So had the Scottish psychiatrist Lewis Bruce and a number of others.[14] For Kraepelin, not the least of the appeal of the autointoxication hypothesis was the hope it offered of heading off a disorder he considered hopeless once it had established itself.

THE ISOLATION OF AMERICAN PSYCHIATRY from the world of university medicine was almost complete. The microscope and X-ray machine symbolized the bacteriological and technical revolution now under way in American medical education, but this was a revolution that left asylum medicine essentially untouched. A few desultory lectures were offered at a handful of medical schools on the subject of insanity, but psychiatry was otherwise notably absent from the curriculum. Entry into its ranks came via a haphazard apprenticeship as an assistant physician at a barracks-asylum, a process that, as Silas Weir Mitchell acerbically pointed out, left its recruits bereft of any pretense of scientific training.

Under the circumstances, it is perhaps hardly surprising that the figure who was most responsible for altering this state of affairs had trained abroad in very different surroundings. The Swiss-German immigrant Adolf Meyer was scarcely a typical assistant asylum physician. Possessed of a splendid professional pedigree, his solid training in neuropathology and laboratory research and exposure to the advanced teachings of Hughlings Jackson and Jean-Martin Charcot left him shocked at the isolated and intellectually impoverished state of American psychiatry. He had written a thesis on the forebrain of reptiles under the supervision of the eminent neuroanatomist August Forel, a prominent proponent of degenerationist doctrines, and adopted the new conventional wisdom as his own.[15]

Privately, his acceptance of these doctrines caused Meyer considerable concern, because his family bore the "taint" of nervous disease on both the paternal and maternal sides, and Meyer himself had what he regarded as an "inherited tremor of the hands," which had precluded a career in surgery and had "pushed [him] . . . to the nerve clinic where relatively few fine manipulations are necessary." On his own account, he suffered in these early years from "chronic feelings of inadequacy" when forced to examine patients, and confessed that, as an adolescent, he suffered from "an inherited and . . . considerably developed neurasthenia."[16]

Meyer had trained at a university clinic devoted to pathological research aimed at uncovering the origins of mental diseases—a regime in which the mass of patients in asylums were little more than a source of brains for postmortem examination. He displayed neither the aptitude nor inclination for contact with live patients during his training, preferring anatomical research on brain tissue, and his formal training in psychiatry was slender indeed by the time he graduated. Still, he was bound to be disappointed by the purely administrative atmosphere that dominated the American psychiatric scene—and by the corrupt, patronage-dominated politics of Illinois.

Born in Niederweningen, in the German-speaking region of Switzerland, in 1866, Meyer was the son of a Zwinglian minister. Adolf preferred to save both body and soul. Supplementing his neurological training in Zürich with a year in Edinburgh, London, and Paris, he concluded that his career opportunities would be better in the United States and emigrated immediately after receiving his MD in 1892, leaving behind his widowed mother, whom he had persuaded to mortgage her house to support his New World venture. Struggling with debts she had trouble paying, Anna Meyer promptly fell into a deep clinical depression, a "melancholia with fancies of persecution" that led to her being committed to the Burghölzli mental hospital, where she was treated by none other than Adolf's former professor, August Forel. Whatever guilty feelings Adolf may have felt, they were insufficient to deter him from pursuing his ambitions.

Meyer initially chose to settle in Chicago, one of the major centers of neurology in the United States. Despite his impressive European credentials, he was unable to find a full-time, salaried post at the new Rockefeller-funded University of Chicago, and he struggled to generate sufficient income from private practice. This should scarcely have come as a great surprise, given his habit of decorating his consulting rooms with jars full of human brains preserved in formaldehyde. Eager to secure a post worthy of his training, he proposed to the University of Chicago that it appoint him head of a newly created "Brain Institute." For good measure, he also wrote to Rush Medical College, suggesting he be hired to preside over a subdivision of neurology, and to Clark University in Massachusetts, where he proposed to "establish an institution for the anatomy of the nervous system." All three institutions declined the brash twenty-six-year-old's offers.[17]

In a state of some desperation, with visions of his American adventure turning into a nightmare, the following year Meyer felt compelled to accept an otherwise undesirable appointment at the vast Illinois Eastern Hospital for the Insane at Kankakee, some ninety miles from Chicago. It was an inauspicious place to begin his career, and Meyer's position as the hospital pathologist placed him at the very bottom of the medical hierarchy, below the most junior assistant physicians. But it meant a small salary, plus room and board, and he had hopes that in time he could parlay his appointment into something better.

Meyer began his time at Kankakee by conducting large numbers of autopsies in an effort to correlate brain lesions with psychiatric diagnoses. Soon, he realized that the disarray of the hospital's patient records and absence of any systematic effort to record the patients' symptoms while they were alive made all such endeavors pointless. "I found the medical staff," he later recalled, "hopelessly sunk into routine and perfectly satisfied with it." As he sought a way forward, "a bewildering multiplicity of cases stared me in the face. It would have been the easiest thing for me to settle into the traditions—I had to make autopsies at random on cases without any decent clinical observation, and examinations of urine and sputum of patients whom I did not know. . . . [W]henever I tried to collect my results, I saw no safe clinical and general medical foundation to put my findings on."[18]

Crucially, this led Meyer toward a greater interest in studying the clinical course of psychiatric illness in living patients.[19] If he hoped to train the hospital staff in systematic history taking and record keeping, he had perforce to develop the necessary techniques himself. Assembling the hospital staff and employing a stenographer to take notes as he examined patients, Meyer pioneered the standardized case record. Soon, he was emphasizing the need to create a comprehensive record of all aspects of the patient's mental, physical, and developmental history. Such records became standard elements in Meyerian psychiatry and the central feature of his future teaching and mentoring of young psychiatrists.

The year in Chicago and the two and a half years at Kankakee were decisive in bringing about his transition from neurology to psychiatry. These years brought Meyer into close contact with the ideas of John Dewey and Charles Peirce, two of the three founders of American pragmatism whose philosophic doctrines exercised a considerable hold on him and helped fuel his enthusiasm for collecting endless amounts of data. Like Dewey, he thought that knowledge resulted from practical inquiry, from our attempts to solve problems in the world. He was, he claimed, a proponent of a "common-sense" psychiatry, one that eschewed metaphysics and preferred a rigorous focus on "facts." A later move to Massachusetts led Meyer to develop intellectual ties to the third major figure among American pragmatists, the Harvard psychologist William James. One of Meyer's most prominent later students, the Yale psychiatrist Theodore Lidz, remarked, "I had long been puzzled—since I first met Adolf Meyer and recognized the similarity of his teachings to those of James and Dewey—how it happened that a Swiss had embraced pragmatism, indeed, had found in it his natural voice."[20]

Kankakee gave Meyer ready access to a broad array of clinical material, allowing him to make some of his few original contributions to the neurological literature, while encouraging him to take a broader view of the problems of psychiatric illness. As he remarked many years later, "It was the work of American thinkers, especially of Charles S. Peirce, of John Dewey and of William James, which justified in us a basic sense of pluralism; that is to say, a recognition that nature is not just one smooth continuity."[21] James would write that "pragmatism is uncomfortable away from facts," and Meyer emphasized "the facts" like a mantra.[22]

Besides the overcrowding, isolation, and anemic resources that were endemic to most state hospitals in this period, the Illinois asylums were notoriously at the mercy of state politicians, who viewed them as patronage plums. Two years after Meyer had taken up his post, S. V. Clevenger, the superintendent at Kankakee, found himself under growing political pressure.

He succumbed in a matter of months. Some observers concluded that his erratic behavior and "delirious delusions" were the product of alcoholism. Meyer simply noted that he "broke down" under pressure, "so that he was considered insane even by his friends." Meyer must by now have begun to question his decision to leave Switzerland, and it was surely with great relief that, in 1895, he received an offer to move to Worcester State Hospital in Massachusetts as director of research.[23]

In a letter to George Alder Blumer, superintendent of the Utica State Hospital in New York (where psychiatry's official journal, the *American Journal of Insanity*, was still housed and printed—by the patients, no less), Meyer was vocal about his dispiriting encounter with the realities of institutional psychiatry. He was, he informed his influential correspondent,

> more and more convinced that the atmosphere of the place shows little sign of being improved to such a degree as to make life satisfactory enough to spare energy for the work I am longing for. Catering towards political effects, towards mere show and granting insufficient liberty of action, the administration discourages progress along sound principles. The library facilities are poor and the whole mechanism of medical work little promising although much better than when I came here. My courses on neurology and mental disease have certainly roused the interest of the Staff, but the ground does not promise much fruit as long as the simplest means of clinical observation and examination are absent![24]

Meyer's invitation to become research director was a signal that one of the largest state networks of mental hospitals recognized the depth of the crisis facing institutional psychiatry. That same year, the New York state asylum system set up the Pathological Institute of the New York State Hospital on premises at Madison Avenue and Twenty-Third Street in Manhattan. If the United States had no university department of psychiatry, or tradition of linking together research, clinical teaching, and patient care that had combined to make German medicine preeminent in this era, the new establishments in Worcester and New York were intended as a partial remedy.

Modeling his approach on the best German clinics, Meyer developed links to Clark University and put in place a program to bring in four or five new assistants a year, people who were recruited from the best medical schools in the country on the basis of their academic records and a competitive examination. But before this program could be launched, he faced a major problem: his own desultory background in psychiatry. Late in his time at Kankakee, Meyer had delivered lectures on brain anatomy and basic neurology to the assistant asylum physicians, but he had little to contribute on psychiatric matters. It was a deficiency he knew he would have to address if he were to succeed in his new position. After his first months at Worcester, he sought leave to return to Europe, to gather ideas and information he could apply in Massachusetts.

IN THE SPRING AND SUMMER OF 1896, Meyer visited Italy and Switzerland, where he found his mother finally on the mend, and he then spent six weeks at Kraepelin's clinic in Heidelberg. German psychiatry, like German medicine, was widely considered to lead the world. It commanded, besides its barracks-asylums, clinics attached to universities where intensive research could be undertaken and where the leading figures in the field could avoid the dreary work of mental hospitals—indeed, they could and largely did avoid living patients altogether. Convinced that mental illness was brain disease, they polished their techniques in the laboratory, staining brain cells and peering through microscopes in search of the elusive traits that distinguished mad brains from sane ones. Asylums, for them, were simply a source of pathological material and their interest in clinical care almost nonexistent.

Following an apprenticeship in a number of provincial asylums and a spell as an assistant to Wilhelm Wundt, the father of experimental psychology, Kraepelin's first appointment as a professor of psychiatry was an unpromising one: as director of the clinic at the University of Dorpat, what is now Tartu in modern Estonia, then part of the Russian empire. Kraepelin found his patients spoke languages he could not understand, the facilities were run-down, and the finances precarious.[25] After four long years of exile, in 1890 he secured an appointment as professor at the ancient and prestigious University of Heidelberg. It was here that Adolf Meyer encountered him, spending six weeks observing at Kraepelin's clinic and interacting with its staff.

Kraepelin's work and orientation were quite unlike that of other leading German psychiatrists, for his eyesight was too poor to work at a microscope. Instead, he devoted himself to following the clinical course of mental disturbance among the hundreds and eventually thousands of patients who passed through the clinic, usually the prelude to a long stay in an asylum.

Like a naturalist (his older brother Karl was a botanist who became director of the Museum of Natural History in Hamburg), he sought patterns in the apparent mental chaos that confronted him. Notes on the course of each patient's illness were recorded on cards *(Zählkarten)* and eventually, from this mass of information, Kraepelin developed a new classification of mental disorders. He looked not just at initial symptoms but also at the clinical course of the disease and its outcome, and he decided that he could divide cases of serious mental disorder into two fundamental classes: what he called dementia praecox (a category that in most respects later came to be called schizophrenia), and manic-depressive psychosis.

It was a distinction he elaborated on and refined in successive editions of a textbook he first published in 1883, at the very outset of his career. Meyer's approach constituted a sharp break with the unitary theory of psychosis most of his German colleagues embraced, the idea that all forms of mental illness were matters of degree. It also broke with conceptions of madness that dated all the way back to the ancient Greeks, who had established basic divisions between mania, melancholia, and dementia. Observing the work of the clinic, Meyer was simultaneously exposed to Kraepelin's new diagnostic system, its empirical basis, and its application.

Kraepelin was deeply skeptical of what his German colleagues' work on neuropathology and the microscopic study of insane brains had accomplished. There were, he contended, "no definitive achievements" one could point to, and for "most fundamental forms of madness . . . results of postmortems continue to leave us entirely in the lurch." As for confident assertions that psychosis was grounded solely in the brain, these were "a fantasy unfettered by the uncomfortable shackles of fact."[26] The implied contrast was with Kraepelin's own work, built from the close observation of living patients.

Kraepelin was nonetheless convinced that his diagnostic system was converging on the discovery of "real natural disease entities," *Natürliche Krankheitseinheiten,* that would prove to have distinct causes and pathologies that were ultimately biological. For him, dementia praecox was a dire diagnosis. It struck early and followed a deteriorating course. The parade of disasters included agitation, incoherence, hallucinations and delusions, an incapacity to form relationships with others, and ultimately a descent into a grossly denuded mental universe. Its roots, he later speculated, perhaps lay in an autointoxication, a poisoning of the brain by internal secretions or persistent infections. Regardless, it was a devastating disorder.

Manic-depressive psychosis, by contrast, was a remitting disorder, one that did not entail cognitive deterioration and the demise of the personality.[27]

Kraepelin was confident that as long as one modified this diagnostic system to account for subsequent experience, a classification developed through clinical observation would allow the reliable identification of different forms of mental disease. Those ideas—that mental illnesses were distinct and separable like the diseases of general medicine; that they were ultimately rooted in biology though constructed from clinical observation; and that such divisions should guide the course of future research—would resurface in the closing decades of the twentieth century, albeit in very different circumstances. Those determined to reemphasize psychiatry's medical identity would call themselves "neo-Kraepelinians" though their appropriation of his ideas was partial and distorted.[28]

THE FIFTH EDITION OF KRAEPELIN'S TEXTBOOK appeared in print during Meyer's stay in Heidelberg. He returned to Worcester with a copy in hand, along with a superficial if more extensive grasp of matters psychiatric than he had possessed before. With his intimidating personal manner, his German accent, and the credentials provided by his European training, no one was inclined to challenge his presumed expertise. "I am," he wrote to his fellow Swiss émigré, August Hoch, "strictly Kraepelinianer just now. I make my men swear by Kraepelin although sometimes I should like to do the swearing myself."[29]

The structure provided by Kraepelin's system, particularly the ominous diagnostic category of dementia praecox, with its intimations of a hopeless prognosis, was invaluable for a time. Already, though, Meyer was chafing at being someone else's "pupil," complaining that Kraepelin's "schematism hurts me considerably" and adding, "It is really hard for me to follow laws."[30] Such caveats foreshadowed the more full-blown criticism he would develop within a few years, as he invented his own rival system of "psychobiology." Meyer was not content to operate for long in someone else's shadow, and his distancing from Kraepelin was similar to his subsequent reaction to Freud, whose ideas at first he expressed interest in, only to become a critic of psychoanalysis.

Little or no formal instruction in psychiatry was provided at American medical schools at the turn of the century, a reflection of the specialty's marginality and low professional status. Meyer's new recruits (the twenty-nine appointments he made during his time at Worcester included two women) were to be the shock troops of a new scientific psychiatry, employing serious and sustained clinical and pathological research to uncover the roots of mental disorder. He insisted that they learn German and French, opening up the European literature to them and providing them with a veneer of sophistication that helped separate them from the provincial outlook of most of their generation in American psychiatry. With their aid, Meyer began to develop the eclectic approach he would soon dub "psychobiology." His young assistants went on to become leaders of the next generation of American psychiatrists; three subsequently served as presidents of the American Psychiatric Association.[31]

Meyer's counterpart at New York's Pathological Institute, Ira Van Gieson, enjoyed no such success, despite occupying a very influential bureaucratic niche. A pathologist by training, with no experience in asylum work, Van Gieson was contemptuous of institutional psychiatry, viewing its practitioners as poorly trained and "ignorant" of medical science.[32] Though he was in charge of an institute that was meant to improve psychiatry's knowledge and clinical applications, he made no effort to hide his disdain for his colleagues. "As a science psychiatry, at present, is dead," he proclaimed in 1899, "and a mummy may be its symbol."[33]

However accurate this perception may have been, expressing it made him powerful enemies. Research into the origins of mental disease ought to be "unshackled from the narrow-circumscribed connections which have so long governed it," he insisted, and it was time for psychiatry "to escape from the confines of the asylum walls."[34] Van Gieson may have been echoing his German models in making these claims, but he only succeeded in arousing the fury of asylum superintendents, who sharply resented the slanders of someone they dismissed as an ignorant and arrogant outsider.

In an era when the bacteriological revolution was all the rage and the medical profession was tempted to exaggerate the reach and scope of germ theory as an explanation for disease, Van Gieson seized on the notion that mental illness might be connected to autointoxication, a poisoning of the brain, and devoted his energies in this direction.[35] He worked spasmodically, and for long periods the institute seemed almost somnolent. One visiting psychiatrist, Edward Brush, reported that he had found the premises empty, and he expressed doubt that "anything was being done or had been done for some days . . . I judge entirely by the amount of dust on the majority of tables."[36] Van Gieson's enemies had by then been circling for some time. The program of basic research he had sought to establish had little obvious therapeutic or intellectual payoff, and when a press report circulated of an interview in which he had boasted that his institute was on the brink of discovering "the germ of insanity"—a claim met with ridicule—political maneuvering was under way to secure his dismissal. A committee was established to assess his work, stacked with his enemies. It reached its foreordained conclusion, aided by a damning secret report from Meyer on the institute's operations. By the autumn of 1901, Van Gieson was out of a job.[37]

His replacement, after some characteristic vacillation from the man himself, was none other than Adolf Meyer, by now embroiled in arguments with his superior at Worcester and anxious to move to a larger and more visible stage. Certainly, the move to New York was another step up the professional ladder. Meyer had promised to give the institute a new direction, one far more congenial to the asylum superintendents who had orchestrated his predecessor's departure. As a token of his intentions, the Pathological Institute was renamed the Psychiatric Institute and moved from its separate downtown premises to Wards Island in the East River between Manhattan and Queens, where it became an adjunct to a particularly noisome state hospital. So far from being contemptuous of clinical psychiatry, Meyer embraced it, and this made for better relations with the superintendents.

Meyer's work at Worcester had begun to win him a national reputation, and he now set about remaking the institute. He established an outpatient clinic, standardized record-keeping and statistics statewide, and sought to provide research training for mental hospital staff throughout the system. His cooperative stance toward the hospitals, expressed desire to "raise the standards of medical work in the State institutions," and frequent visits to the various mental hospitals throughout the state helped to smooth ruffled feathers. But they did not, in the end, provoke a reciprocal engagement on the part of the heads of most institutions. Meyer's insatiable appetite for collecting "facts" was not matched by any comparable creativity in organizing or making sense of this mass of material.[38] As with his time at Kankakee and Worcester, little of his efforts in New York were directed at the question of therapy, and none produced advances in the actual treatment of mental patients.

A VARIETY OF ATTRIBUTES contributed to making Meyer the dominant figure in American psychiatry from the turn of the century until the outbreak of the Second World War: his European training and grasp of

German psychiatric and neurological literature; his icy and intimidating bearing; the obscurity of his prose, which many mistook for profundity; and his eclectic willingness to consider anything and everything as potentially relevant to the understanding of mental illness. His position in New York gave him considerable visibility and helped him acquire a lucrative consultative practice, advising on the treatment of the psychotic Stanley McCormick—a sinecure that would extend over two decades and bring in tens of thousands of dollars in fees.[39] Of great importance, too, was his appointment in 1908 as the first professor of psychiatry at Johns Hopkins and as founding director of its Phipps Clinic. The number of his former students who subsequently went on to head departments of psychiatry across the United States helped cement his influence and authority.

Yet the decision to appoint Meyer to the first chair at Hopkins was by no means preordained. From 1904, he had held a much less prestigious post as a part-time clinical professor of medicine at Cornell's medical college. When William Henry Welch, the dean of the medical school, decided he should take the plunge and venture into murky psychiatric waters, he first approached Stewart "Felix" Paton, who had studied in Europe and written one of the most widely used psychiatric texts of the time. But Paton had no interest in being tied down by a full-time clinical appointment, and he urged Welch to appoint Meyer instead.

Hopkins opened its doors in 1893 and was the first American medical school to adopt the German model of combining medical education and research. By almost any measure, it quickly became America's premier medical school. Apart from its institutional innovations, the stature of its first four chairs—Welch, William Osler, William Stewart Halstead, and Howard Kelly—had done much to cement its dominance. In an era when most medical schools ignored psychiatry, Meyer's arrival in Baltimore in 1910, two years after he first accepted the appointment, represented a coup for the specialty, as well as for him personally. Hopkins provided Meyer with a prominent base to build his influence in the discipline, and he used it to great effect.

Meyer believed that mental illness represented a failure of functional adaptation to the demands of modern life—dementia praecox, for example, was a progressive disorganization in an individual's habits—and virtually anything might explain that adaptive failure. Hence his emphasis on the rigorous recording of everything that might have affected the patient's mental state—biology, medical history, social and familial context, psychological factors, environment (physical and social)—the list was almost infinite. Without any criteria of relevance or way to assign weight or significance, assembling the patient's "life chart," as he called it, might therefore proceed indefinitely, a splendid form of make-work that became an end in itself.[40]

Faced by the bewildering complexities of an array of disorders whose origins and treatment remained a matter of guesswork and improvisation, Meyer's psychobiology provided an elastic overarching framework within which a whole array of hypotheses and interventions could be accommodated. His stress on the meticulous collection of detailed case histories created a host of tasks with which his students and the profession at large could engage, consoling themselves with the illusion that piling up fact upon fact would somehow lead to a solution to the problems posed by mental disturbances.

Meyer's prose was notoriously impenetrable, but the word "psychobiology" seemed to gesture toward a scientific basis for psychiatry that brought together all the myriad features of human existence: social context, consciousness, and the complexities of human biology. In the more straightforward words of one of his disciples, Franklin Ebaugh, "mental disorders are considered to result from the gradual accumulation of unhealthy reaction tendencies . . . [and their treatment requires] the study of all previous experiences of the patient, the total biography of the individual and the forces he may be reacting to, whether physical, organic, psychogenic or constitutional."[41]

For Meyer, nothing was too trivial or inconsequential to record and to enter on a patient's life chart. Even the color of the wallpaper in a nursery might have some bearing on the distress of the adult patient in front of him. "A fact," he proclaimed, "is anything which makes a difference."[42] But his system advanced no clear criteria for determining what *did* make a difference. In later years, Phyllis Greenacre, who served as one of his assistants from 1916 to 1927, recalled that this approach rested on "an obsessional and probably futile search for accuracy," one in which "the emphasis on recording all possible phenomenological details" about the patient's life history had "sometimes reached fantastic proportions." Meyer's intimidating effect on his students, she recalled, led most of them "to drive recording observation to a stage of the infinite and the absurd in the effort to cover everything."[43] Nurses, for example, were told to keep hourly logs of patient behavior: what patients said, signs of disturbance and emotional displays, and any overt or covert actions they undertook. When Meyer walked through the wards, it was with a stenographer in tow, and even impromptu encounters with patients in the garden or elevator were carefully noted down and placed in their case histories.[44] The oft-repeated emphasis on the value of "facts," coupled with the teaching of more systematic ways of recording this mass of data, helped structure the practice of psychiatry and created mountains of busy-work for the professionals, filling their time and creating the illusion that they were doing something for their patients. But whatever Meyer's disciples were doing, it was not leading to greater understanding of the etiology of mental disorder or to more effective treatment.

The stream of students passing through Hopkins between 1910 and 1940 went on to prominent positions in these new departments, declaring their allegiance to Meyerian psychobiology. By one estimate, in 1937, as many as 10 percent of all academic psychiatrists in the United States had trained at Hopkins under Meyer.[45] His influence extended abroad, too, particularly to Britain, where Aubrey Lewis in London and David Henderson in Edinburgh came to dominate the discipline. Both had received extensive training at Hopkins under Meyer.[46]

Many of his pupils remained in his shadow. Smith Ely Jelliffe, editor of the *Journal of Nervous and Mental Disease,* once commented waspishly that Meyer had put "partly castrated pupils in professional chairs" all across the country.[47] As we shall see, Meyer came to play a vital role in legitimizing the enthusiasm for some of the most far-reaching of the somatic therapies that were embraced by American psychiatrists in the 1920s and 1930s. He repeatedly used his authority to quell any questioning of the legitimacy of these approaches and encourage their spread, though his stance coexisted uneasily with his own invocation of the role of social and psychological factors in the genesis and persistence of mental illness.

MEYER CLAIMED THAT HIS APPROACH was a "common-sense psychiatry." If mental illness was a problem of ingrained bad habits, psychotherapy should seek to modify them. "Habit-training," Meyer pronounced, "is the backbone of psychotherapy; suggestion is merely a step to the end and only of use to one who knows what that end can and must be attained. Action with flesh and bone is the only safe criterion of efficient mental activity; and actions and attitude and their adaptation is the issue in psychotherapy."[48] Mental illnesses, however slight or serious their manifestations, represented a failure of adjustment. Acute psychosis, mania or depression, obsessions and chronic unease, even "the trivial fears . . . and petty foibles of everyday life" were but variants on a theme, the product of an individual's psychobiological makeup and of the environmental challenges he or she confronted. Those at less disturbed points on the continuum might more readily be induced to abandon their pathological habits.

The difficulty, of course, was that, by the age of thirty, as William James had succinctly observed, most people's character and habits had "set like plaster, and will never soften again."[49] Perhaps this explained Meyer's poor therapeutic results with the seriously mentally ill, whose defective habits were simply too deeply entrenched. If noxious habits of the mind, once set, were hard to remold, early intervention might have better results. "Every kind of training," as James had pointed out in an essay on habit, quoting from the English physiologist William Carpenter, "is both far more effective, and leaves a more permanent impress, on the growing organism, than when brought to bear on an adult."[50]

CHAPTER FOUR

Freud Visits America

ON A SUNDAY EVENING IN SEPTEMBER OF 1909, a fifty-three-year-old neurologist from Vienna, accompanied by two of his close intellectual companions, stepped off the train in Worcester, Massachusetts. It was the last leg of a long journey that left him temporarily exhausted. The overnight train he had taken from Munich to Bremen after leaving his holiday residence in the Alps meant that he only had a day to explore the old city, and then there had been a tedious transatlantic journey on the German-American liner the *George Washington*. The weather had been unkind—cold, wet, and foggy—and the seas rough, "not friendly," as our weary traveler confided in his diary, and accordingly "the uplift in mood through the sea air did not occur as I had surmised." The presence among his fellow passengers of a fierce rival and critic, Professor William Stern of Breslau, did nothing to improve his mood. Stern, he wrote to his wife, is "a repulsive person" and "he will not find us hospitable."[1] Ignoring Stern for the most part, Sigmund Freud and his companions, Carl Jung and Sándor Ferenczi, passed much of the time analyzing one another's dreams.

On August 29, Freud's party docked at Hoboken, New Jersey. A few days of sightseeing in New York, the high point of which was a visit to the Metropolitan Museum to view Greek antiquities, had been marred by gastric upsets and an embarrassing episode when he publicly wet himself. It was a measure of Freud's deep-seated dislike of the United States that he would ever afterward blame his years of digestive problems on his encounter with American cooking. There was a lengthy trip north, first by an overnight boat trip as far as New Haven, and then by train via Boston—an exhausting itinerary made worse by Freud's growing prostate problems. The weary travelers welcomed a few hours of rest at the Standish Hotel before they repaired to their host's house for dinner.

Waiting to greet them was the president of Clark University, the psychologist G. Stanley Hall. He was, Jung wrote to his wife, "a refined, distinguished old gentleman close on seventy who received us with the kindest hospitality. He has a plump, jolly, good-natured, and extremely ugly wife, who, however, serves wonderful food. She promptly took over Freud and me as her 'boys' and plied us with delicious nourishment and noble wine, so we began visibly to recover."[2] On Monday morning, Freud, Jung, and Ferenczi moved in to stay with the Halls for the balance of the week.

Over the next few days, Freud delivered five lectures on psychoanalysis in his native German. In return, besides a fee of $750, he received an honorary doctorate. It was Freud's first and last academic honor of this sort, and his first and last visit to the country where, in the middle years of the twentieth century, his intellectual brainchild would achieve its greatest influence, albeit in a bowdlerized and simplified form. Yet the famous occasion nearly failed to materialize. Freud was undoubtedly eager for some public acknowledgment of his genius, but he had little regard for America or Americans even before setting foot in the United States, and his first instinct had been to refuse the invitation. Clark's first president, Hall had sought to make his university a center of graduate training, modeled to some extent on Johns Hopkins University, the first American institution to embrace the German principle of a research university. The year 1909 marked the twentieth anniversary of Clark's opening, and he conceived the plan of inviting a host of eminent scientists and scholars for a week of lectures to celebrate the occasion. Many were far better known than Freud, and the plan had originally been for the distinguished group to assemble in early July.

For Freud, the dates simply wouldn't do. He wrote back, "I am a practicing physician and because of the summer habits of my countrymen, I am obliged to discontinue work from July 15 to the end of September. If I were to lecture in America in the first week of July, I should have to suspend my medical work three weeks earlier than usual, which would mean a significant and irretrievable loss for me. This consideration makes it impossible for me to accept your proposal."[3] Privately, he complained to Ferenczi, "America should bring in money, not cost money."[4]

Learning of Freud's decision, Carl Jung promptly wrote to his mentor, urging him "to speak in America if only because of the echo it would arouse in Europe." Far more attuned to the indirect rewards the trip might bring in its wake, he returned to the subject at the end of his letter: "About America I would like to remark that [Pierre] Janet's travel expenses were amply compensated by his subsequent American clientele. Recently Kraepelin gave one [consultation] in California for the tidy sum of 50,000 marks. I think this side of things should also be taken into account."[5]

As Jung well knew, the mention of Emil Kraepelin was bound to attract Freud's attention. In successive editions of Kraepelin's highly influential textbook, the German psychiatrist had developed a new diagnostic system that distinguished two major classes of psychosis, what he termed "dementia praecox" (relabeled as "schizophrenia" in 1910 by the Swiss psychiatrist Eugen Bleuler, who saw it as not necessarily involving dementia or restricted to the young), and manic-depressive psychosis, each characterized by particular clusters of symptoms and a differing prognosis. Derived inductively from some of the thousands of cases that filled Germany's barracks-asylums, Kraepelin's nosology made him the most famous German psychiatrist of his generation. He was Freud's bête noire, and Freud referred to him sarcastically as "the Great Pope of Psychiatry."

KRAEPELIN'S STANDING as the leading psychiatrist of the age had led to his invitation to perform a highly lucrative examination of perhaps the wealthiest psychiatric patient in the United States. Stanley McCormick was the youngest son of Silas McCormick, a man who had parlayed his invention of a reaping machine into one of the great nineteenth-century fortunes. The family company, International Harvester, controlled 85 percent of the market for agricultural equipment from its headquarters in Chicago, and by the turn of the century the McCormicks' wealth rivaled that of Carnegie, the Rockefellers, and the Vanderbilts. Stanley had married Katharine Dexter, the first woman to graduate with a science degree from the Massachusetts Institute of Technology, in September 1904. After a nine-month honeymoon in Paris, partly spent in the company of Stanley's domineering mother, Nettie, and Katharine's mother, Josephine, the couple returned to the United States. Much to Katharine's dismay, Stanley insisted at first that they live in Chicago. Nettie had bombarded Stanley with telegrams during their European sojourn, telling him what to eat and wear, where to visit, and what to avoid (with special concern for that "den of evil," Monte Carlo). The prospect of living in a city where Nettie's proximity would allow still more attempts to control her son, whose behavior was notably unstable, scarcely boded well for the marriage.

Within weeks, Katharine relocated to Boston, where she had designs on attending medical school. Stanley, nominally occupying an executive position in the family firm, stayed behind in Chicago and moved back in with his mother. His behavior grew more erratic, and after some weeks his brothers, in the face of Nettie's vociferous opposition, persuaded him that he ought to join his wife in Boston. Katharine sought to keep a low profile, to minimize the stress that social situations evidently caused her husband, but Stanley's mental state was fragile. The marriage, it subsequently materialized, had never been consummated, and as the months passed, Stanley began to hallucinate and behave increasingly erratically and violently. In June 1906, his decline forced his resignation from International Harvester. By October, as his irrational outbursts worsened, scandal threatened to erupt, and at Katharine's behest he was confined in the McLean Asylum—to this day the institution favored by the New England plutocracy. Pugnacious and delusional by turns, given to shouting obscenities and to unpredictable violence, Stanley seemed particularly prone to paroxysms in the presence of women, including his wife, and soon all female attendants were removed. Ice-cold baths were used to calm him whenever he refused to behave, and in therapy Stanley confessed his impotence.

Months of confinement produced no improvement in his condition. Adolf Meyer, brought from New York to examine him, pronounced him to be suffering from "psychasthenia" overlain with psychotic and manic-depressive features. Meyer's report, filled with his usual prevarications and obscure prose, suggested that, while Stanley might improve over time, the prospects of a full recovery were fading. In a subsequent letter to Stanley's mother, he indicated that the case was "one of the saddest" he had encountered, since earlier opportunities to head off his now-serious problems had been missed.[6] Months passed, with many behind-the-scenes squabbles between Katharine and the McCormicks over who would control Stanley's treatment. The deeply religious Nettie prayed for his recovery, and other family members enlisted a pliant Meyer in their campaign to persuade Katharine to accept a lucrative divorce—a course of action she summarily rejected.

Fortified by the promise of a substantial fee for his trouble, Meyer subsequently renewed his attempt to secure Katharine's agreement to a divorce, arguing that it would be therapeutic for her husband. Again, he was rebuffed. Ever afterward, Katharine would view his involvement in the case with suspicion, seeing him as the ally of her enemies, Stanley's mother and siblings. Not that Meyer was in the least deterred by her hostility, clinging for nearly two decades to the stream of fees that Stanley's relations continued to send his way.[7]

At length, with the utmost precautions against his escape, Stanley was moved by private train across the continent to Santa Barbara. Here he was locked up in a family mansion, Riven Rock, with a full-time medical attendant and staff. (In his saner days, Stanley had helped design and build his new place of imprisonment as somewhere to confine his mad older sister, Mary Elizabeth—though she had been relocated to an Arkansas sanatorium by this point.) Insistent on securing the services of the most eminent psychiatrist in the world, with money no object, Nettie summoned Kraepelin to her son's bedside, offering him so princely a sum that he could not refuse the weeks of travel the consultation would necessitate.

Kraepelin examined Stanley and pronounced him a case of catatonic dementia praecox, endorsing Meyer's previous diagnosis. His report made no mention of the fact that Stanley was kept in virtually complete physical restraints most of the time. Like most of the other eminent psychiatrists who had preceded and would follow him to Riven Rock, Kraepelin tried hard to find some sliver of optimism to proffer the family in return for their largesse. "These conditions," he reported, "have always to be regarded as very grave disorders. In a very large proportion of the cases the disease progresses to a more or less pronounced degree of mental enfeeblement. There are, however, patients in whom marked improvement takes place; indeed, in isolated instances, a complete recovery has been observed. . . . [W]e would conclude that at present we are still dealing with an acute illness in which a more or less favourable turn may still occur, although this can by no means be predicted with any certainty."[8]

This was all rather disingenuous. In reality, Stanley would remain psychotic till his death in 1947, the arrangements for his care complicated by a long-running power struggle between his mother and siblings on the one hand and his equally strong-willed spouse on the other. Between them, the two sides would engage a veritable galaxy of eminent psychiatrists (and their not-so-eminent assistants) to examine and treat a patient who was mostly violent and incoherent.

Stanley McCormick's unfathomable wealth made him a one-man psychiatric gravy train. One psychiatrist, a Dr. Edward Kemp who had once briefly met Freud, would be paid $10,000 a month—almost $2 million a year in today's dollars—to live with Stanley and "treat" him psychoanalytically between 1927 and 1930, when Katharine finally obtained a court order to terminate his services: all this two years after Kraepelin had found it "impossible to have any continued conversation" with someone scarcely able "to carry out movements or utter words."[9]

WHEN JUNG URGED FREUD to consider the rewards an American visit might bring, he could draw on more than just the gossip about the fabulous fee Kraepelin had secured for his services. For Jung himself had already begun treating two other members of the McCormick clan at the Burghölzli, the famous mental hospital in Zürich. The first was Stanley's cousin Medill McCormick, who sought help supposedly for his alcoholism.[10] Jung's treatment would extend intermittently over almost two decades before the then Senator McCormick committed suicide in a Washington hotel in 1925.[11] In March 1909, Jung would be consulted by Stanley's brother Harold and his wife, Edith Rockefeller McCormick.[12] Surely there were other rich, nervous Americans whose patronage the nascent science of psychoanalysis might secure?

Perhaps Jung's invocation of these future possibilities carried weight. In any event, when Hall renewed his invitation with more accommodating dates, Freud's response was very different. The major celebration had been moved to early September. Perhaps just as significantly, Hall had raised the honorarium from $400 to $750 and dangled the prospect of an honorary doctor of laws degree. Freud accepted with alacrity, thanking his host for "a very happy surprise."[13]

Freud was only one of a long roster of distinguished speakers during a busy week at Clark. Given the prominent place the visit has come to occupy in the annals of psychoanalytic history, it is important to bear in mind that, despite Hall's own academic interests in psychology, the lectures given that week spanned a wide spectrum of subjects: mathematics, physics, chemistry and biology, as well as anthropology, pedagogy, psychology, and psychiatry.[14] The speakers included two Nobel Prize winners (the physicists Albert Michelson and Ernest Rutherford) and such eminent figures as Franz Boas of Columbia.[15] Within the field of psychiatry and psychology, Freud and Jung were far from having the platform to themselves. Some speakers were actively hostile to psychoanalysis, most especially Professor William Stern, who had traveled on the same ship as Freud and was harshly critical of his work, and Edward Bradford Titchener, the British-born experimental psychologist who wanted nothing to do with psychotherapy and loathed what he termed the perversion of psychological science in the service of "saving souls." Still others, like Adolf Meyer, displayed a nominal openness to Freud's ideas that would soon dissolve into hostility.[16]

Meyer's ambivalence, and subsequent rejection, helped ensure that, whatever following psychoanalysis developed in some intellectual circles and among wealthy patients over the next three decades, it would attract very few American psychiatrists. Yet his stance was far from the only reason Freudian psychoanalysis remained a distinctly minority taste in those years. Freud's focus on neurotic, fee-paying patients who were treated in his private consulting room had little relevance for psychiatrists whose experience was almost wholly confined to seriously disturbed patients involuntarily confined in institutions. Sensitive to their marginal status, they clung tightly to the vestiges of their medical identity, insisting on madness's biological roots and repeatedly opting for somatic treatments for their patients' troubles. Freud's embrace of the "talking cure" would have had little appeal to them, had they been aware of his work. Still, as psychiatry began to broaden its remit, particularly after the Great War, and as an office-based practice became possible, Meyer's open hostility to psychoanalysis added to the obstacles it faced, for by then his dominance of American psychiatry was entrenched.

Throughout his career, even when he rose to defend the most extreme physical interventions for mental disorders, Meyer continued to invoke the role of social and psychological factors in the genesis and persistence of mental illness. Yet if that prompted him to voice some initial sympathy for Freud's views, it was one that would not—indeed, could not—last. Just as he had initially expressed a loyalty to Kraepelin's doctrines, only to subsequently spurn them, Meyer was not content to operate for long in anyone's shadow. Besides, the pragmatism and eclecticism of his psychobiology, where interpretation was loose and unstructured, put him firmly at odds with the highly deterministic intellectual system Freud had constructed. Psychoanalysis was, in that sense, psychobiology's polar opposite. Until Meyer's advancing age and the transformative effects of the Second World War weakened the hold of his teachings on America's psychiatry, Freud's ideas would gain only limited traction.

Twenty-eight of the twenty-nine participants delivering lectures at Clark later received honorary doctorates of law. (Titchener was awarded the somewhat more prestigious DLitt.) For Freud, the ceremony had to be a gratifying occasion, given the lack of recognition he had received so far in

Europe. It was, as he commented when the degree was awarded, "the first official recognition of our labors." Whatever pleasure the occasion may have provided was diluted by his underlying contempt for those offering him their applause. "America," he would later remark, "is gigantic—a gigantic mistake."[17] The informality of Americans rankled him, and he was convinced that their hypocritical prudery would preclude their acceptance of his theories. Before he had even visited, he confided to Jung, "I . . . think that once they discover the sexual core of our psychological theories they will drop us."[18] Nothing changed once he had seen the country. It was, he informed Arnold Zweig, an "anti-Paradise" populated by "savages" and swindlers, and it ought to be renamed "Dollaria."[19] "What is the use of Americans," he belligerently asked Ernest Jones in 1924, "if they bring no money? They are not good for anything else."[20]

SADLY FOR FREUD, Americans never seemed to offer him enough of that particular commodity. An invitation in 1920 to spend six months in New York in return for a payment of $10,000 was scornfully rejected: "In other times," he wrote to his daughter Anna, "no American would have dared to make me such a proposition. But they're counting on our poverty [in the aftermath of the First World War] to buy us cheap."[21] Freud's ambivalence or outright hostility thus compounds the irony that it was only in the United States that psychoanalysis would enjoy an unbridled period of success.

That success came some decades later, and it is easy to exaggerate the importance of the visit to Worcester, set against what was to come. And yet the occasion *was* an important one. To be sure, by no means was all his audience converted. Experimental psychologists like Edward Titchener greeted Freud's talk of the unconscious with scorn. William James, already suffering from the heart disease that would soon end his life, might have been expected to be more receptive to Freud's theories, but he attended only the lecture on the interpretation of dreams, and found the Viennese physician pretentious, "a man obsessed with fixed ideas. I can make nothing in my own case with his dream theories, and obviously 'symbolism' is a most dangerous method."[22]

But Freud also made a handful of enthusiastic converts. Stanley Hall became his champion, though that was in some ways a mixed blessing, given Hall's controversial career and propensity for all sorts of intellectual enthusiasms. More important, given his intellectual and social standing among the New England aristocracy, was James Jackson Putnam, the impeccably connected professor of neurology at Harvard, whose benediction did much to lend legitimacy to psychoanalysis in its early years. Along with transplanted European disciples like Ernest Jones, who had fled to Toronto in the face of a looming sexual scandal, these men gave Freudian psychoanalysis an important early bridgehead in North America, one that would be amplified and extended when a flood of analysts arrived in the 1930s to escape persecution at the hands of the Nazis.

Then there were the lectures themselves, a fluent and engaging outline of Freud's ideas as they were then constituted, beginning with the patient Anna O. and her famous (and mythical) cure. Bertha Pappenheim, the real Anna O., had been treated not by Freud but by his colleague Josef Breuer, for symptoms that included a chronic cough; paralysis of the extremities on the right side of her body; and disturbances of vision, hearing, and speech.[23] As recorded in the monograph the two men published in 1895, *Studies on Hysteria,* her "cure" came about as she recalled repressed events and emotions. For Freud and his followers, she thus became the patient who launched psychoanalysis. Her treatment, Freud emphasized, was but one example of the force and power of recovering memories via the talking cure.

At the very time Freud was delivering these lectures, Pappenheim was visiting North America to give a series of lectures of her own on white slavery and prostitution. She had become one of the foremost feminist activists of her generation. Many decades later, we learned that, far from being cured by Breuer's ministrations, on leaving his care she was immediately hospitalized in the Belleview Sanitorium in Kreuzlingen, where her symptoms continued unabated. Her subsequent recovery owed nothing to psychoanalytic therapy, of which she was fiercely critical.[24]

The Worcester lectures were delivered in German and apparently extemporaneously, though in reality Freud had carefully rehearsed what he would say while walking with Ferenczi just before delivering each lecture. German was then the language of science and particularly of medicine, so for the scholars in the audience Freud's decision to speak in his own idiom posed no great difficulty, but it is striking that the Massachusetts newspapers managed to provide extensive and largely accurate summaries of Freud's arguments. To be sure, they glossed over most of the Saturday lecture, devoted as it was to the sexual aspects of Freud's theories, for such subjects were hardly suitable for family newspapers. But the essential elements of what purported to be a radically new approach to the understanding and treatment of mental disorder were laid out in a brief and appealing compass.[25]

Freud began with a becoming modesty—one he would regret when his audience took his self-effacement at face value—granting to his early collaborator, Josef Breuer, the primary credit for the invention of psychoanalysis. He announced that he intended, over five days, "to give you in a very brief compass a historical survey of the origin and further development of this new method of investigation and treatment."[26] He proceeded to do so in a clever fashion, one that recapitulated for his audience the slow, halting, and uneven process by which he had moved from Breuer's initial emphasis on recovering repressed traumatic memories via hypnotism to his own completely different approach. He sought to make plausible the notion of "a purely psychological theory . . . in which we assign the affective processes the chief place." Breuer's faltering first attempt, it turned out, "was able to give only a very incomplete and unsatisfying explanation of the observed phenomena." Psychoanalysis had advanced markedly in the succeeding decade and a half, but it took much time and patient effort, Freud emphasized, to construct "a well-rounded theory without any gaps," one that was not "a child of speculation" but rather "the product of an unprejudiced and objective investigation."

Struggling with this great task, and with the simultaneous creation of effective therapeutic techniques, Freud gradually revealed how, moving from a stress on repressed traumatic memories and resistances to acknowledging them, he had developed the notion that the libido, or unconscious sexual drive, was the central psychological underpinning for all human beings. All sorts of discomforts and disturbances flowed from that fundamental reality, and from civilization's demands that these forces be channeled in "acceptable" directions—a fraught process, and one that remained incomplete and unsatisfactory. The nuclear family was the scene of frightful and dangerous psychodramas that populated the unconscious, fomented its repressions, and created its psychopathologies. As the infant struggled to grow up, and the child to mature, the perils of Oedipal conflicts awaited—the unacknowledged and unacknowledgeable desire for an erotic relationship with the parent of the opposite sex—only to be suppressed in ways that wreaked havoc on the adult personality.

Forced to deny their fantasies, or drive them underground, children were riven with psychical conflict. Cravings and suppressions, a search for substitute satisfactions, false forgetting, the constraints of "civilized" morality—in all these respects and more, the conflict between Eros and Psyche created a minefield from which few emerged unscathed. Madness was not a condition unique to the degraded and degenerate, Freud argued, but something that lurked to some degree within all of us. Here, Freud emphasized, was what differentiated his theory from that of biological reductionists like the French psychotherapist Pierre Janet, who had spoken at Harvard some years previously. Sublimated in some with greater success, the same forces that led one to mental invalidism allowed another to produce accomplishments of surpassing cultural importance. Civilization and its discontents were thus locked in an indissoluble embrace. Where resistance to the repressed was strong, that which had been pushed into the unconscious surfaced in disguised form as symptoms. Rendering the unconscious conscious was necessarily a complex and drawn-out process. Here, dreams could play a vital role in uncovering what we hid from ourselves, giving Freud the opening for an extended treatment of his ideas on this subject, in his fourth lecture, attended by William James.

In *Studies on Hysteria,* the urtext of psychoanalysis, Freud acknowledged that the case studies he had contributed, a series of psychologically charged vignettes, read "like short stories." As such, he lamented, they lacked "the serious stamp of science." This was a thought that rankled, and he immediately sought to blunt the charge with the assertion that "the nature of the subject is responsible for this, rather than any preference of my own."[27] It was an insightful remark, however painful he found it. By the time of his American visit, Freud could offer a new theory of mind, one that presented human psychology as deterministic and just as rulebound as the functioning of the human organism.

It was a project that rested unabashedly on transforming the neurotic symptoms of his patients and seemingly chaotic landscape of our dreams into the raw materials of psychoanalytic science. Rendering the meaningless meaningful was Freud's great accomplishment. If (and it was and remains a very large if) one bought into Freud's increasingly elaborate set of assumptions about how human minds are constructed, then one possessed the key to unlock all sorts of puzzles about our psychic functioning. So it was that Freud's intellectual edifice came to embrace the stranger-than-fiction case history as its intellectual foundation—Sherlock Holmes–like narratives that are the source of so much of psychoanalysis's enduring popular appeal.

Who can resist the charms of a story-driven science that purports to unveil our innermost secrets, ones we keep resolutely hidden even from ourselves? Certainly, many of those in attendance and in the larger audience who learned of his ideas via the press found Freud's ideas seductive. Delivered with wit and élan, here was a presentation that was straightforward without seeming overly simple, constructed in a narrative form that invited the audience to feel they were part of the process of discovery. Perhaps as a concession to the American culture he professed to abhor, Freud's performance gave the impression that therapeutic success, if not easy, was assured.

There was one more feature of Freud's lectures that proved attractive to some of the medical men in his audience: his resolute rejection of the then-popular mania for religiously based mind cures. New England had witnessed the birth of many of the most notable of these, from the effusions of Mary Baker Eddy and the Church of Christ, Scientist, to the initially more "respectable" interventions of the Reverend Elmer Southard at the Church of Christ Emmanuel in Boston. (Southard's effort to combine religious consolation and psychotherapy with a veneer of medical blessing initially attracted men like William James and James Jackson Putnam, until they recoiled from the Frankenstein they belatedly realized they may have helped create.)[28] Freud's fierce rejection of these affronts to "science and reason" was welcome indeed among the medics in his audience. It was a message he sought to convey to a wider audience in an interview with the *Boston Evening Transcript*. "The instrument of the soul," he noted solemnly, "is not so easy to play, and my technique is very painstaking and tedious. Any amateur attempt may have the most evil consequence."[29]

After the conference, Freud, Jung, and Ferenczi traveled west to see Niagara Falls, and then south to the Adirondacks, where they spent a week in James Jackson Putnam's camp near Lake Placid. As it happened, their time there coincided with a visit from a nineteen-year-old Harvard student, Alan Gregg, who had come to join Putnam's niece and nephew. Gregg participated in conversations and learned much about psychoanalysis. Just over two decades later, he became medical director of the Rockefeller Foundation, a vital position where his adolescent interest in psychoanalysis would resurface. He played a significant role in underwriting its growth.

Their week in the country behind them, Freud's party headed for New York, where they boarded the *Kaiser Wilhelm der Grosse*. A stormy crossing ensued, and they passed a somewhat-uncomfortable eight days before docking in Bremen. Freud proclaimed the visit a success, and in a limited sense he was right. To be sure, he was delighted to have escaped a country that held little charm for him. As he wrote to his daughter Mathilde, "I am very glad I am away from it, and even more that I don't have to live there."[30] Still, his ideas had received a respectful hearing, and he had won a handful of important converts. Psychoanalysis would remain a distinctly minority taste among physicians and the public at large, but a bridgehead had been secured upon which, some three decades later, it would anchor its dominance in American psychiatry.

THE HOPED-FOR FINANCIAL WINDFALL was slow to materialize, at least for Freud. A handful of American patients and would-be analysts came his way, but none with the fabled wealth of the McCormicks. For Jung, however, the trip had a far more pleasant aftermath. His earlier connections with the McCormick clan now bore fruit. Edith Rockefeller McCormick was, in Jung's own words, a woman who "thought she could buy anything," and Jung was someone she decided she would buy. Suffering from depression and what she claimed was severe agoraphobia, she sought help from a variety of quacks and society doctors, to no avail. As early as 1893, Edith Rockefeller had been sent to Weir Mitchell's Hospital for Orthopedic and Nervous Diseases in Philadelphia, where she was "treated" to his famous rest cure: bed rest, lots of food (to build up "fat and blood"), and massage and electrical stimulation of the muscles. The treatment had continued until just weeks before her marriage to Harold McCormick in November 1895. With the assistance of her husband's cousin Medill, a sometime patient, Jung took the occasion of another visit to New York in 1912 to examine her, diagnosing her as a "latent sch[izophrenic]." Edith imperiously demanded that he relocate to Chicago where she would buy him a large house near hers and set him up in practice, bearing the entire cost of relocating his family and reestablishing him professionally in America. Jung toyed with the idea, and Edith even arranged a meeting with her father, John D. senior, notwithstanding the old man's disapproval of her extravagant and pleasure-seeking lifestyle. But in the end, that particular scheme came to nothing.

By now, Edith was determined to have her analysis with Jung, and if the man would not come to her, she would go to him. Pleading agoraphobia, she insisted that Jung come to New York and travel with her to Zürich. Accompanied by some of her children, a huge retinue of servants and retainers, and the plethora of personal effects she simply could not do without, the grand dame made her way to a city where she would live for the next eight years, undergoing her own treatment with Jung, encouraging her children and her increasingly estranged husband to follow suit, and eventually offering herself for training as a lay Jungian analyst—for by 1914 the man Freud had once called his "Crown Prince and successor" had broken with his onetime mentor and struck out on his own.[31]

Besides the money that flowed from the analytic sessions themselves, Jung soon had other reasons to bless his American benefactor. In 1916, Edith gave him $200,000 to found a Psychological Club to promulgate his ideas, going so far as to take out $80,000 in bank loans for the purpose when her cash balances were temporarily inadequate.[32] It was a fabulous sum.[33] A quarter century later, Jung would receive similar bounty from a woman wedded to another great American fortune, Mary Mellon, wife of the spectacularly wealthy Paul Mellon and daughter-in-law of the ruthless Andrew Mellon. The Mellons' money would fund the Bollingen Foundation, whose first great aim was to translate and publish the complete works of Carl Jung.

Vain, hedonistic, and a spendthrift, Edith Rockefeller McCormick had exhibited all the qualities her stern Baptist father abhorred. In Zürich, she inhabited a palatial suite at the Hotel Baur au Lac. Periodically, in an effort to cure her travel phobia, she would board a train, remaining on it as long as she could bear. Her chauffeur drove ahead in the Rolls Royce, stopping at each station lest his employer found herself unable to continue and felt compelled to return to the safety of the car. For a while, Harold joined her in Switzerland, resigning his position as secretary of International Harvester to do so. But in 1918, he was summoned back to Chicago to serve as the corporation's president. Even while nominally together, the McCormicks had followed Jung's lead and engaged in a multitude of affairs. Edith's were now becoming increasingly public, threatening scandal. She was also piling up debts, many from massive investments in a Ponzi scheme that purported to have discovered a novel technique for hardening wood.

At last, in August 1921, she pronounced herself sufficiently recovered to return to the United States, her latest lover in tow. Embroiled in an affair of his own, Harold was less than delighted to see her, and within a month had filed for divorce. He had been more discreet than she and, faced with testimony about her serial infidelities, she was forced to sign a settlement that not only left her without alimony, but obliged to pay Harold $2.7 million as compensation, in return for retaining title to their houses. Harold promptly married his mistress, a young Polish "opera singer," Ganna Walska, who couldn't sing but whose career he besottedly kept subsidizing. (Walska was successively married to a half-dozen wealthy men; Harold was her fourth husband and contrived to stay married to her for nine years, but he invited ridicule, in a pre-Viagra age, when he had monkey testicles implanted by Serge Voronov in a desperate effort to improve his potency.)

Meanwhile, in 1923, Edith had pronounced herself to be the reincarnated wife of Tutankhamen, whose tomb had just been discovered. She set herself up in Chicago as a Jungian analyst, and before long she attracted as many as a hundred socialites as her patients. Simultaneously, she was investing yet more vast sums of money in fraudulent real estate schemes set up by her lover, losing all in the aftermath of the 1929 stock market crash. Though her father continued reluctantly to rescue her from her financial foolishness, she never saw him, pleading her travel phobia whenever a meeting was proposed. Through it all, her devotion to Jung and Jungian analysis remained unshaken. Diagnosed with a recurrence of cancer in 1932 (she had been operated on for breast cancer two years earlier), she elected to treat it with psychotherapy. By late August she was dead.[34]

By any measure, Edith and Harold and the McCormick extended clan were a rum lot, if a great blessing to psychiatrists, Carl Gustav Jung foremost among them. Freud himself managed to attract no such American benefactor, though a handful of his disciples eventually began to tap into the wealth of the American superrich in the 1930s. He had to content himself with repeated subventions from Princess Marie Bonaparte, the great-grandniece of Napoleon, who became his patient in September 1925 and supported him financially and emotionally ever after.[35]

Support from American heiresses notwithstanding, Jung's analytic psychology developed no mass appeal in the United States, nor any sizable following among American psychiatrists. His talk of not just an individual unconscious but of a collective unconscious populated by archetypes, a universal and objective part of the psyche made up of everything that has been inherited from human evolution, with its mystical overtones, was off-putting to many, as was his dabbling in the occult and his embrace of the irrational. Freud's psychoanalysis, by contrast, after a fitful start, would for a time become the dominant belief system embraced by American psychiatry, enjoying an extraordinarily wide influence in American popular culture, in Hollywood, in fiction, advertising, and even the socialization of American babies. For decades after Freud's famous voyage to America, though, psychoanalysis remained a distinctly minority taste, and the talking cure was the butt of ridicule and hostility from most psychiatrists and the medical profession at large.

CHAPTER FIVE

The Germ of Madness

THE BACTERIOLOGICAL REVOLUTION in medicine that took place in the closing decades of the nineteenth century, like most intellectual revolutions, did not succeed overnight. For years, conservative medical men had resisted the implications of the work of Louis Pasteur and Robert Koch, and the warnings about the perils of pus that emanated from that apostle of antiseptic surgery, Joseph Lister. But by the dawn of the twentieth century, the gospel of "germs" was sweeping all before it. Medicine embraced the laboratory as the source of cultural authority. Bacteriological models of disease brought gains in etiological understanding and, to a more limited degree, in therapeutic efficacy. The upshot was that physicians and surgeons, donning the mantle of the new science, found their prestige and prospects soaring. And yet there were diseases and disorders that remained recalcitrant, resistant to the new paradigm and frustratingly beyond the reach of modern therapeutics: rheumatism and arthritis, for example, and atherosclerosis and nephritis, not to mention serious mental illness.

The triumph of the germ theory of disease had heightened the sense that the future of medicine lay in discoveries in the laboratory, rather than at the bedside.[1] Bacteriological models of disease very often invoked the idea that it was not necessarily the bacteria themselves that caused disease, but the damage the toxins they unleashed wrought upon vulnerable bodies. Many who embraced the new alliance with science saw focal sepsis—the presence of unobserved low-grade infections lurking in the corners and crevices of the human body, pumping out poisons via the bloodstream and the lymphatic system—as the likely cause of a host of chronic disorders whose etiology remained baffling and mysterious. Among the most prominent supporters of this view were the Philadelphia neurologist Francis X. Dercum, and Frank Billings, dean of the Rush Medical School from 1901 until 1924, and professor of medicine at the University of Chicago.[2]

It was Henry Cotton who took the lead in applying ideas about focal sepsis to the treatment of psychosis. Cotton had an impressive pedigree for a young psychiatrist. He had trained under Adolf Meyer at Worcester State Hospital, one of those Meyer hoped would become the leaders of a new psychiatry. Meyer continued to promote his career in the following years, sponsoring his move to Germany, where he worked at Emil Kraepelin's Munich clinic. German training was de rigueur in these years for those who aspired to join the elite circles in American medicine, but it was still relatively rare among psychiatrists. Cotton's training included work on sectioning the brains of patients who had died from General Paralysis of the Insane (GPI), and he may have been exposed to Kraepelin's speculations on the possible infectious origins of psychosis. Most certainly, the experience reinforced his commitment to discovering the biological roots of mental disorder.

On returning to the United States, Cotton's glittering résumé and the enthusiastic support of Adolf Meyer secured him the position of superintendent of the New Jersey State Hospital at Trenton, an unusual accomplishment for a young man barely in his thirties. He inherited an overcrowded institution that had had only two superintendents since its foundation in 1848, a backward establishment that routinely employed chains and other forms of mechanical restraint to impose some semblance of order. The superintendent Cotton replaced, John Wesley Ward, preferred tending to his collection of seashells to caring for the patients he was nominally charged with treating. Trained long before the bacteriological revolution in medicine, he openly disdained the modern notion that microscopic organisms caused disease.

That skepticism proved Ward's downfall. Lack of attention to elementary sanitation—the dairy attached to the asylum farm was filthy, swarming with flies and encrusted with excrement, and the water supply contaminated with human waste—meant that dysentery was endemic, and then the hospital suffered an outbreak of typhoid fever. Patients developed prolonged high fevers and chills, agonizing muscular pains and seizures, and soon the epidemic spread to the staff. Deaths mounted. The asylum discharged its untreated sewage into the Delaware River, from which the city of Trenton drew its water. Making matters worse, rumors began to circulate of a cover-up of murders of patients by hospital attendants. Ward was forced to resign.

Cotton moved swiftly to banish all traces of the ancien régime. Mechanical restraint was abolished, the institution's water-supply and waste-disposal systems improved, and occupational therapy was introduced to rouse both patients and staff from their torpor. Above all, Cotton sought to connect the asylum to the world of modern medicine. A new operating theater was opened, together with laboratories and a professional library that gave the medical staff access to contemporary medical research. The physicians met frequently to discuss cases and were encouraged to attend medical congresses and contribute to the professional literature. Meyer's detailed case notes were introduced, and two full-time social workers were employed. Cotton even succeeded in luring one of Meyer's chief assistants, Clarence Farrar, to forsake his Hopkins appointment and join the Trenton staff.[3] Modern psychiatry was on the march.

Except that the expected progress failed to materialize. Cotton, undaunted, was determined to make therapeutic progress. He had attempted to circumvent the failure of Salvarsan to produce therapeutic results in cases of GPI by drilling holes in his patients' skulls and injecting the drug directly into the cranium. Beginning in 1916, he took things a step further. Others had speculated that insanity might have an infectious origin, and the demonstration that GPI was the product of syphilitic infection had shown that some asylum inmates' illnesses were the product of bacterial disease. It was time to move beyond speculation and to launch an assault on the reservoirs of chronic infections that Cotton was certain must be poisoning his patients' brains.[4]

Billings's lectures on focal sepsis at Stanford had just appeared in print and may have provided not just the inspiration to move forward, but specific targets to pursue.[5] Billings had mentioned the teeth and tonsils as likely sources of overlooked infections. "Deplorable as the loss of teeth may be," he wrote, "that misfortune is justified if it is necessary to obliterate the infectious focus which is a continued menace to the general health. . . . Too often the tonsillar tissue in children and also in some adults is a culture medium of pathogenic bacteria and as such is a constant source of danger as a portal of entry of infectious bacteria through the lymph and blood streams to the tissues of the body. . . . [E]ntire removal is the only safe procedure."[6] Physicians at such august institutions as the Mayo Clinic in Minnesota and Johns Hopkins (in the person of the great William Osler's successor, Llewellys Barker) endorsed these conclusions.[7] So Cotton had respectable company as he launched his assault on what he was convinced were the sources of all forms of madness, from the mildest to the most severe.

COTTON'S 1916 *Annual Report* on the Trenton Asylum contained a brief passage in which he commented that "we have found that focal infections and the absorption of toxins may appear in the etiology of certain groups primarily held to be purely psychogenic in origin." The first fifty patients whose teeth had been removed had not, he confessed, seemed much better, but he was undeterred. The following year, he reported on twenty-five patients whose teeth and tonsils had been removed, claiming that twenty-four of the twenty-five had subsequently been discharged as recovered. Unerupted and impacted teeth, teeth with infected roots and abscesses, decayed or carious teeth, apparently healthy teeth with periodontitis, poorly filled teeth, sclerotic teeth, teeth with crowns—all should be regarded with deep suspicion, and, if possible, removed. Modern cosmetic dentistry was a menace. It left in place what were in reality still-decaying teeth, and subterranean pathology continued to undermine health. "It seems incredible," Cotton remarked, "but it is nonetheless a fact, that the dental schools of today are teaching the installation of gold crowns, fixed bridge work, pivot teeth or Richmond crowns, all of which have been definitively proven to be a serious menace to the individual's health. To paraphrase an old proverb, 'Unhealthy is the tooth that wears a crown.'"[8]

Suitably encouraged by what he claimed were remarkable therapeutic results, he announced proudly, "We started literally to 'clean up' our patients of all foci of chronic sepsis."[9] X-rays, laboratory analyses, and the regular resort to surgical interventions in the operating theater marked Trenton as a mental hospital where the most advanced medical technology of the era—technology most general hospitals had not yet begun to employ—was mobilized to cure what others deemed hopeless conditions. The key, Cotton argued, was germs. Germs and pus. When many of his patients stubbornly refused to recover, he was undeterred. Other hidden sites of infection had to be tracked down and eliminated. Harmful bacteria had been swallowed and disseminated to other sites in the body. So tonsils and teeth were soon joined by spleens and stomachs, colons and cervixes, as he ruthlessly pursued the goal of a thorough cleansing of the patients' bodies. Colons were the most prominent target, but it was increasingly clear that there were simply too many lesions in too many sites for a single surgical procedure to suffice.[10]

Cotton's program of surgical bacteriology even extended to organs others saw as untouchable. Experience had taught him, for example, that the stomach "is one of the least important organs of the body. . . . The principal function of the stomach is the storage and motility, each easily dispensed with. . . . The stomach is for all the world like a cement mixer often used in the erection of large buildings and just about as necessary. The large bowel is, similarly, for storage and we can dispense with it just as freely as with the stomach."[11] By 1921, Cotton had removed a half dozen thyroid glands. For reasons he could not fathom, he found that colons were infected twice as often in female patients, and the cervix, he reported, was also a potent site of low-grade infections. It was these factors that explained the disproportionate number of women who underwent operations.[12] Patients often needed several rounds of surgery, since a failure to recover often indicated other hidden reservoirs of infection. One patient admitted for depression and anxiety set some sort of record, for she successively underwent a gastroenterostomy for a stomach ulcer, a thyroidectomy, a complete colectomy, a removal of both ovaries and fallopian tubes, enucleation of her cervix, and a series of "vaccine" treatments designed to address any lingering and overlooked problems. Cotton was nothing if not persistent, and his persistence paid off. Mrs. Llewellyn was at length discharged as cured.[13]

In laying the foundation for all these interventions, Cotton emphasized not just the virtue of using the most advanced medical technology to track down lurking sepsis, but also the importance of an array of specialists—four surgeons, three gynecologists, a laryngologist, a rhinologist, two ophthalmologists, a dentist, a genito-urinary surgeon, an oral surgeon, a pathologist, and a bacteriologist, not to mention six assistant physicians and a roentgenologist (or specialist in X-rays). Small wonder that when Cotton opened two new wards in 1921, the ceremony was attended by the presidents of both the American Medical Association (Hubert Work) and the American Psychiatric Association (Albert Barrett of Michigan). Work pronounced Trenton one of the "great institutions" in the country: "This is a general hospital, really the first one I ever saw. It excludes nothing. It regards mental alienation as a symptom, as most physicians regard delirium in a fever. . . . It does not make a bit of difference what the name for a condition is, provided the cause of that condition is found and eliminated." Here, at last, was a place where "the treatment of the psychoses is surrounded by medical science and not set apart from any part of it."[14]

Work's endorsement was highly significant, and not just because of his prominence in American medicine. President Harding would soon appoint him to his cabinet, and later Work successfully managed Herbert Hoover's candidacy for the presidency in 1928. Others were equally enthusiastic, including another physician-turned-politician, Senator Royal Copeland, who had previously served as the head of New York City's Department of Health. Copeland wrote a widely syndicated health advice column, "Your Health," and often used it to warn of "the perils of pus infection." Cotton's work drew particular praise from him on several occasions, an opinion that, he claimed, followed on ward inspections, conversations with patients, and observations of surgical operations: "I have never seen an institution conducted in a better way. There is every consideration given to the latest medical methods, and we should commend its work in every way possible."[15]

John Harvey Kellogg, the physician who headed the famous Battle Creek Sanitarium in Michigan, would have preferred that Cotton's campaign against focal sepsis extend to the exclusion of meat from his patients' diet, since he saw the putrefaction of dead-animal matter in the intestines as a major source of focal sepsis. But still he offered his support for Cotton's pioneering assault on the roots of mental disorder.[16] And while both Copeland and Kellogg were viewed with some suspicion by many medical men as publicity hounds and (in the case of Kellogg) a food faddist, their influence with the public was considerable.[17]

There were murmurs of discontent from some of Cotton's fellow psychiatrists about his mono-causal approach to madness, though the force of their objections was lessened by their simultaneous acknowledgment, as Richard Hutchings, superintendent of the Utica State Hospital in New York, put it, that "there cannot be two opinions as to the advisability of removing sources of infection, whether located in the tonsils, at the roots of the teeth, or elsewhere."[18] Other asylum superintendents eagerly endorsed Cotton's findings, their opinions solicited by the prominent journalist Albert Shaw, a close friend and confidant of Woodrow Wilson during his years as president and subsequently editor of Wilson's papers and speeches. Charles Page of Danvers State Hospital in Massachusetts, where Cotton had worked earlier in his career, wrote of his pride in his former assistant and the way he had accumulated "a mass of indisputable facts to . . . establish . . . cause and effect, between bacterial toxins and functional insanity." His encomiums were echoed by the superintendents of King's Park State Hospital on Long Island, the Arkansas State Hospital, and the Veteran's Hospital Number 24 in Palo Alto, California.[19] The governor of New York was so impressed by Cotton's work that he sent a three-man commission to examine and report on the work at Trenton, resulting in a special state appropriation to fund a resident dentist at all of New York's mental hospitals.[20]

In the early 1920s, Cotton secured two prestigious platforms from which to proclaim his message that "the insane are physically sick."[21] For a number of years, he had lectured on psychopathology to Princeton undergraduates, and two of his most prominent supporters (and sometime trustees of the Trenton State Hospital) were the eminent neurologist and psychiatrist Stewart Paton and the chair of Princeton's Biology Department, Edwin Conklin. The two men had nominated him for a signal honor, the invitation to deliver the prestigious Louis Clark Vanuxem Foundation Lecture Series at the university.

ON JANUARY 11, 1921, at 4:30 p.m., Cotton rose to deliver the first of four lectures. Nearly 400 people had assembled to hear him. They learned that at the state hospital, not far from the Princeton campus, a full-fledged surgical assault on sepsis was now the order of the day. Each year, thousands of teeth and tonsils were extracted, and scores of colons and other internal organs were removed. The payoff, Cotton proclaimed, was a massive increase in the number of cures, and equally major savings for the state's treasury. When the *New York Times* reviewed the published version of the lectures in June 1922, its reviewer, Thomas Quinn Beesley, had no doubt of their importance. "At the State Hospital at Trenton, New Jersey, under the brilliant leadership of the medical director, Dr. Henry A. Cotton, there is on foot the most searching, aggressive, and profound scientific investigation that has yet been made of the whole field of mental and nervous disorders." Across the country, others had given way to despair, as rates of mental illness grew four times as fast as the general population increase, the *Times* noted. But thanks to Cotton, "there is hope, high hope . . . for the future."[22]

Desperate for relief from the demons that tormented them (or their nearest and dearest) and dazzled by the seemingly authoritative reports emanating from Trenton, patients and their families urgently sought to share in the new miracle cures. Affluent madmen and madwomen flocked to Trenton—where their willingness to pay premium rates for the attention of Cotton and his consultants made them a highly desirable commodity. Meanwhile, across the country psychiatrists found themselves besieged by supplicants seeking the new wonder cure. Frantic families urged that teeth, tonsils, and guts be ransacked for the source of the germs that prompted hallucinations and delusions, ranting and raving, and depression. For so long, madness had seemed a condition beyond help, a source of stigma and shame. If modern biological science had revealed that it was just another physical affliction, no more than the effects of bacterial poisoning of the brain, then deliverance might be at hand.

Across the Atlantic, news of Cotton's work had attracted much attention, and in the late spring of 1923 he sailed for Britain, invited to address the Medico-Psychological Association, the professional organization of Britain's psychiatrists. A number of British physicians and surgeons had by then embraced the doctrine of focal sepsis, as had the head of the city of Birmingham's mental hospitals, Thomas Chivers Graves. Cotton was greeted as psychiatry's savior. Illustrating his lecture with a chart of the various locations where chronic infections might lurk, and with X-ray photographs of infected teeth and colons, Cotton spoke at length of the challenges he had faced in overcoming skeptics, genuflecting toward British figures who had previously acknowledged the systemic problems low-grade infections could bring in their train. His own ruthless pursuit and elimination of sepsis had had the happiest of outcomes, he assured his audience. Relying on the most recent advances of scientific medicine—gastric analyses and serology, bacteriological work and X-rays, serums and vaccines, and above all the miracles of modern aseptic surgery—and employing the surgeon's scalpel and the dentist's forceps, he claimed to have increased "our recoveries . . . from 37 percent to 85 percent."[23]

Rather than considering the various forms of psychosis as different types of illness, madness ought rather to be seen as "a symptom, and often a terminal symptom of a long-continued chronic sepsis or masked infection, the accumulating toxaemia of which acts directly or indirectly on the brain-cells." What appeared to be different diseases were in fact simply the reflections of the ways "the psychosis is modified by several factors: first, the duration of the sepsis, the severity of the toxaemia produced, plus the patient's resistance, or lack of resistance, to septic processes." One could thus reject competing accounts of the origins of mental illness. Blaming insanity on defective heredity had been "a cloak to mask our ignorance of other factors," he claimed, which "has had the most unhappy result of stifling investigation, and retarding constructive work." As for psychoanalysis (and here his audience was united in its approval of his critique), "the extravagant claims made by its advocates are without foundation or justification. Freudism has proven to be a tremendous handicap to psychiatry."[24]

A parade of eminent British physicians hastened to praise Cotton. "Wholly admirable," said the Edinburgh physician Chalmers Watson, except that he thought the proportion of mental patients with diseased colons was considerably higher than the 20 percent Cotton had estimated. In Watson's view, more rather than less abdominal surgery would be needed to reach the 80 or 90 percent cure rate Cotton's work made possible. Sir Frederick Mott FRS, the head of the pathological laboratories for all of London's mental hospitals, was equally complimentary. Cotton's specimens of bowel disease "closely resembled those which he, Sir Frederick, had seen in his own experience" and the "beautiful pictures and photographs . . . were most convincing. He referred especially to the beautiful radiograms of teeth and of the bowel conditions." William Hunter, a prominent surgeon, added his own encomiums: "The striking individual and statistical results described by Dr. Cotton placed the matter beyond all reasonable doubt. It only remained to put measures against sepsis into routine operation not merely in isolated cases, but in all cases of insanity."[25]

Summing up the day's proceedings, the newly elected head of British psychiatry, Edwin Goodall, enthusiastically endorsed Cotton's remarkable work. His American guest's new therapeutic ideas "should have served to draw members from the alluring and tempting pastures of psychogenesis back to the narrower, steeper, more rugged and arduous, yet straighter paths, of general medicine. . . . Before seeking to summon spirits from the vasty deep and one's subliminal consciousness, let members remember that they were brought up as materialists and biologists; let them, before plunging into those depths, exhaust every material means for dealing with and curing their mental patients." After all, "here presented today, were results which no-one could deny; seeing was believing."[26]

Responding to these tributes, Henry Cotton graciously acknowledged the role British physicians had played in drawing attention to the importance of focal sepsis. His was an approach, like theirs, which insisted on the importance of "real science" as opposed to "the metaphysical, fantastical and otherwise objectionable theory of psychoanalysis." Speculation must give way before "the facts." "In our own institution we have a recovery rate of 37 percent up to 1918; but when we put into our work the clearing up of focal sepsis we have, in the last five years, increased that to 85 percent recovery rate. . . . These figures are based on very conservative facts, and are not due to enthusiasm."[27] It appeared to virtually all in attendance that a new era in psychiatry had dawned.

HENRY COTTON WAS A MASTER at securing publicity for his work. At times, this brought trouble in its wake. One of his earliest publications on focal sepsis was meant to appear in the *Psychiatric Bulletin*, the official organ of the New York Psychiatric Institute, but when Cotton boasted about his findings to a journalist, not even Adolf Meyer's intervention could save him from the wrath of its editors.[28] He was forced to take the paper elsewhere but seemed undeterred by the experience. Public education came first, or so he rationalized. The longer focal sepsis lingered, the more likely the damage to the brain would become so extensive as to preclude the cure that was otherwise possible. His claims to cure large fractions of those he treated circulated widely, drawing growing numbers of patients to Trenton, many coming from great distances.

The case of Margaret Fisher, one of the first private patients to be transferred to Trenton, illustrates the eagerness with which even highly educated and well-connected people arrived, bringing with them relations in desperate need of miraculous new remedies. Margaret was the daughter of Irving Fisher, a Yale professor lionized by Joseph Schumpeter as "the greatest economist the United States has produced."[29] Fisher was an arrogant, ruthless, and domineering man who made and eventually lost a fortune exceeding $10 million (more than $150 million in today's money). He enjoyed access to the highest circles of American society, and he embraced a host of causes, including Prohibition, eugenics, dietary reform, and the extension of the human life span.

Fisher developed close ties with John Harvey Kellogg, with whom he shared interests in "race betterment" and in the value of a healthy, fiber-filled diet. Beginning in the early 1900s, Fisher began to take his family to Kellogg's establishment each year to partake of the cure. It was an increasingly popular pilgrimage undertaken by many politicians and captains of industry. Hydrotherapy, exercise, a vegetarian diet, close attention to the working of the bowels—all these central elements of Kellogg's regimen became a regular part of the Fisher family routine. As a dutiful daughter, Margaret embraced such healthy practices at her father's urging. Still living at home as she entered her twenties and serving as an unpaid office assistant to her father, Margaret seems to have undergone a slow mental deterioration beginning about 1916. The changes were subtle at first, the onset of her symptoms insidious and easy to overlook or rationalize. Only in retrospect did her parents come to see them as signs of incipient pathology.

On April 27, 1918, three days before her twenty-fourth birthday, Margaret became engaged to be married. Her parents were delighted, and Margaret's father, having checked the young man's pedigree for a family history of insanity or criminal activity with one of his oldest friends, urged her to marry as soon as possible. But the prospect seems to have unhinged her. Within days, as Cotton later noted in the last of his Vanuxem lectures (without mentioning Margaret by name), she began to babble "queer things about portents and was afraid her fiancé would not come back [from the war]." "She soon began to talk at random about 'God, Christ, and immortality,'" and began to have auditory hallucinations. Her conduct was peculiar in many ways. Her condition gradually worsened, and on June 1 she was sent to a private hospital.[30]

Thus far, the Fishers had defined her condition as a temporary nervous prostration and had kept her out of any sort of psychiatric facility. Once hospitalized, as Cotton's case notes recorded, "she became much worse, and could not be controlled." Her parents' hands were forced. Fisher and his wife concluded that "it was necessary to send her to the Bloomingdale Asylum" in White Plains, long regarded as a suitable institution for those of their social class. Admitted on June 27, Margaret was "pensive and preoccupied, and at times depressed. She responded slowly to questions and when aroused was irrelevant."[31]

Her psychiatrists soon despaired of her prospects. Noting the "acute distortion of the patient's personality with marked distortion in thinking, peculiar behavior, and disharmony between mood and thought content," they concluded that her psychosis "seems more nearly related to the schizophrenic disorders than to the exhaustive or manic-depressive disorders." These were important and potentially devastating diagnostic distinctions. Schizophrenia in this era was widely considered to be an incurable condition, and the Fishers were informed that "a recovery without defect symptoms seems improbable."[32] It was not a verdict Irving Fisher was willing to accept. He promptly arranged for Margaret to be released from Bloomingdale on March 29, 1919. Later that day, she was spirited out of state and admitted as a private patient to Trenton State Hospital.

Over the years, Fisher had maintained close contact with John Harvey Kellogg. In August 1914, the two men had jointly organized the First International Congress on Racial Betterment in Battle Creek, and Fisher had written for Kellogg's magazine, *Good Health.* Kellogg, like Cotton, emphasized the nefarious influences of decayed teeth and also the poisons that lurked in the bowels. So when Fisher learned of Cotton's assertions about the etiological connections between focal sepsis and insanity, and the possibility of intervening to cure the apparently hopeless mental patient through a program of surgical bacteriology, he was primed to accept those claims.

Neurologically, Cotton reported, Margaret Fisher appeared to be normal. But there was ominous evidence of "marked retention of fecal matter in the colon with marked enlargement of the colon in this area." "Because of her resistiveness, X-ray studies of the intestinal tract could not be made," but Cotton was convinced that the source of a substantial portion of her problems had been uncovered. Proceeding further, he found evidence that her "cervix was eroded." Deeply suspect as well were two unerupted molars, which Cotton immediately insisted must be extracted. He next approached the Fishers for permission to perform "an exploratory laparotomy"—a surgical opening of the abdomen to examine the internal organs—"based upon the physical examination and the fact of long-continued constipation."[33]

Irving Fisher and his wife were eager to embrace this physical account of their daughter's disorder. It was in close accord with their own beliefs about human health, and a far more hopeful prognosis than the one the doctors at the Bloomingdale Asylum had delivered. Still, they hesitated to endorse so drastic a remedy as surgery on Margaret's bowels. They announced that they "preferred to wait till other means such as vaccines and serum should be exhausted."[34]

In August, however, they did consent to the removal of a portion of Margaret's cervix, after being advised of the presence of "pure colon bacillus" in her tissues. The operation was performed by Cotton's assistant, Dr. Robert Stone, on August 15, 1919. The following day, Fisher and Cotton took the train to Battle Creek to consult Kellogg on how next to proceed. Cotton was clearly doing all he could to overcome Fisher's hesitations about further surgery for his daughter. Fisher wrote to his wife, "Dr. C. doesn't think M. will suffer any pain. The uterus, like the intestines and other internal organs, has few nerves. . . . I suspect that the colon bacillus is the little demon most to blame."[35] Two days later, another letter hinted that Mrs. Fisher had harbored such suspicions long before Margaret had been referred to Trenton for treatment: "I imagine," Irving Fisher wrote, "that you have been as right as anyone that constipation is the key."[36]

Yet still Margaret's parents hesitated. Back at the hospital, Cotton acknowledged that "the family preferred to wait. So in September, another course of antistreptococous [*sic*] treatment was given."[37] Again, he urged surgery on the bowels. Again, Fisher temporized. "As to operating on M.," he wrote to his wife in early October, "we'll talk it over with Dr. C. and each other."[38] And then events took the decision out of their hands.

Cotton obtained another batch of "vaccine" prepared for him by the nearby Squibb Company. In all probability, it was this intervention that produced a fatal crisis, the result of a failure to kill the streptococci before injecting them into poor Margaret's body. In any event, in late October Margaret exhibited symptoms of inflammation of the lungs, and a deep-seated abscess developed over the ribs on her left side—an abscess that, when lanced and cultured, Cotton recorded, "gave pure streptococcus—the same type found in the teeth and stomach. The condition of the patient did not improve and her temperature continued to be high. She failed rapidly and died on Nov. 7, 1919," a little more than seven months after she had been admitted to Cotton's care.[39]

Despite Margaret's death, Cotton believed that her case demonstrated the septic origins of psychosis. Fisher, though devastated by his daughter's death, continued to believe in Cotton's theories and to insist that there had been a physical cause of her mental breakdown. "Even in later years," according to one of his biographers, Robert Loring Allen, "he wrote his friend Will Eliot that some form of toxemia causes a nervous breakdown."[40] Of course, such sustained faith was a natural defense mechanism in the face of the choices he had made and the treatments he had authorized.

FISHER WAS NOT ALONE. Legions of other well-to-do Americans followed in his footsteps, so many that the number of private patients showing up in Trenton to receive treatment began to exceed the capacity of the state hospital to receive them. As their ranks swelled, Henry Cotton seized the opportunity to open a private hospital in Trenton. Henceforth, the bulk of his paying patients were referred to this facility for their treatment.

In one important respect, Henry Cotton's approach to Margaret Fisher's treatment was at variance with his usual procedures. In her case, he had taken great pains to consult with her family and to accede to their wishes when they sought to delay or avoid certain forms of treatment for their daughter. As a general rule, however, Cotton ignored objections from patients and their families and was quite open about doing so. Such protests were, he claimed, short-sighted and ignorant. If they were voiced by patients, their madness had rendered them incapable of rendering a valid choice, and if by their families, they could be dismissed as the product of a lack of medical knowledge: "If we wish to eradicate focal infections, we must bear in mind that it is only by being persistent, often against the wishes of the patient . . . [that we can] expect our efforts to be successful. Failure in these cases at once casts discredit upon the theory, when the reason lies in the fact that we have not been radical enough."[41] He was blunter still in one of his *Annual Reports*. Psychoanalysts, he pointed out, often sought to excuse their therapeutic failures by blaming patients' resistances and their refusal to cooperate with the talking cure. His approach suffered from no such obstacles: "We offer no such excuse for our work because patients who are resistive and non-cooperative can be given an anesthetic and the work of deseptization thoroughly carried out."[42]

The outcome in Margaret Fisher's case was sadly not in the least unusual. From the very first reports Cotton made in the professional literature about abdominal operations to cure psychosis, he had acknowledged large numbers of deaths. When he and John Draper (the surgeon Cotton had brought in to New York to perform the surgery) reported on the seventy-nine cases they had operated on between mid-1919 and mid-1920, they acknowledged that twenty-three had "died as a result of the operation," generally from peritonitis.[43] Elsewhere, when Cotton reported on a series of fifty cases of what he called "developmental reconstruction of the colon," he noted that fourteen more (28 percent of the total) had died.[44] A year later, he recorded some improvement of the mortality rates, down to "about twenty-five percent" (again with "many of the deaths being due to "peritonitis").[45] But the improvement, if any, was short-lived. A 1922 paper given to the American Psychiatric Association reported on a larger series of 250 colon operations, with 30 percent of these patients having died.[46] (Cotton had by now decided to perform his own surgery.) At his extraordinarily well-received lecture in London to the assembled British psychiatrists, Cotton provided two more sets of statistics: the first for total colectomies (133 cases with 44 deaths); the second for a less extensive operation, resection of the right side of the colon (148 cases with 59 deaths).[47]

The mortality statistics might seem high, but Cotton assured his audience that they were tolerable, since they were "largely due to the very poor physical condition of most of the patients."[48] It was an explanation the diverse audiences he addressed seem to have accepted without demur. On no occasion was there any comment or criticism directed at the proportion of patients who had died from the surgery. Indeed, in London, more than one speaker complained that Cotton had perhaps been too conservative in performing abdominal surgery on only 20 percent of his patients, and suggested a more extensive use of colectomies was in order.[49]

Cotton's claims about the success of his treatment had, however, begun to attract criticism in some quarters on other grounds. At the annual meeting of the American Psychiatric Association in Montreal in 1922, a Columbia bacteriologist, Nicholas Kopeloff, and the assistant director of the New York Psychiatric Institute, Clarence Cheyney, delivered a paper reporting on their attempt to replicate Cotton's work. Eschewing abdominal surgery as too risky, they had subjected a group of patients to defocalization. Teeth and tonsils had been extracted, and where there was some evidence of infection, gynecological surgery performed. Alongside those treated to eliminate focal sepsis, they had constructed a control group matched as to age, sex, and psychiatric diagnosis and prognosis. They pronounced themselves unable to discover any evidence that removal of sepsis had any positive effects.[50]

Cotton spoke afterward and simply dismissed their findings as the product of their conservative approach and unwillingness to search out sepsis wherever it lay. Besides, he noted, Kopeloff lacked the necessary medical training to evaluate the evidence, being a mere PhD. A fierce discussion ensued. Some clearly saw Cotton as an enthusiast, though even they did not raise doubts about the wisdom of removing sepsis. Many others rallied to his defense, and a burst of applause greeted one of the final speakers, who urged: "We want this matter to go on. We want Dr. Cotton to proceed with his investigation; to present facts and not mere opinions. If there is anything in it we want to help him. (Applause.) We do not want to put ourselves in a position of opposition to anything that promises benefit or good to our patients. . . . We need to do more of the work Dr. Cotton is doing."[51] Recognizing that most were lining up behind Cotton, Bernard Glueck, who had suggested the formation of a committee to investigate the work at Trenton, rose to withdraw his motion, protesting that he had never meant his proposal to be seen as "putting a check on [Dr. Cotton's] work."[52]

There was evidence, however, of mounting concern among some senior figures in the field. Their objections were not to his experimentation on vulnerable human beings, or his boasts about ignoring the objections of patients and their families, or the alarming rate of death among his patients. Rather, what upset them was "the rather remarkable, and in the minds of many, unethical, exploitation of the methods and results claimed at Trenton in a lay periodical . . . in such a manner as to minimize the value of Dr. Cotton's fellow members. . . . The value of Dr. Cotton's work and the soundness of his conclusions, cannot be measured or discussed, with any good to the public or the profession in lay journals or the daily press."[53]

Sensing that a hostile outside investigation of Cotton's work might be forthcoming, the Trenton Hospital's board, proud of their superintendent's accomplishments and the attention they had drawn, decided to forestall that eventuality by commissioning an outside investigation of their own. They approached Adolf Meyer to conduct it—a curious choice if they wanted an independent review, since Meyer was Cotton's mentor, had written an approving foreword to the published version of the Vanuxem lectures, and remained on close terms with his protégé throughout his career.[54] Meyer at first wrote that he was too busy but then agreed to second one of his chief assistants, Phyllis Greenacre, to the day-to-day work and to oversee the project. For eighteen months, Greenacre labored to reexamine Cotton's claims, traveling around New Jersey, Pennsylvania, and New York to check on the condition of former patients, and constructing detailed summaries of treatments and outcome.

A meeting between Greenacre, Cotton, and Meyer to discuss her findings had to be postponed for some months when Cotton suffered a breakdown. Packed off to Hot Springs, Arkansas, to recover, he arranged for some of his teeth to be pulled to eliminate the focal sepsis that must surely be at the root of his mental troubles. When he at last arrived in Baltimore and came to Meyer's office, the meeting proved tense. Greenacre's detailed report was uncompromising and unambiguous in its findings. And quite devastating.

The three principals met over three uncomfortable days. Meyer mostly sat silent while Greenacre reviewed what she had found. Cotton bullied and blustered and, when he found he could not silence or intimidate a woman many years his junior, stormed out and took the train back to Trenton.

Greenacre reported that recoveries, contrary to Cotton's claims of an 85 percent cure rate, were few and far between, and her findings were "somewhat paradoxical in that . . . the least treatment was found in the recovered cases and the most thorough treatment among the unimproved and dead groups." Beyond this, death was all too often the direct sequel of the surgery, whether from shock, from peritonitis, or from postoperative complications that included uncontrollable diarrhea. "There is," she concluded, "practically no evidence of positive results obtained by detoxication methods."[55]

One might have expected that this would have marked the end of the affair. Nothing of the kind. Meyer suppressed Greenacre's report, and it never surfaced. The hospital board at Trenton, which also possessed a copy of her detailed findings, simply sided with their superintendent. And Cotton resumed his defocalization, pursuing infected teeth, tonsils, and colons with the same enthusiasm as before.

In mid-1927, Cotton sailed for Britain, invited to appear at a joint meeting of the British Medical Association and the Royal Medico-Psychological Association in Edinburgh. Once more, he was lionized by the great and good of British medicine. The year 1927 marked the centenary of the birth of Lord Lister, the man who had introduced antiseptic surgery, and Cotton was hailed as the new Lister of psychiatry, the man who had launched "a desperate frontal attack with horse, foot, and artillery . . . on the whole field of the sepsis . . . in the teeth, the tonsils, nasal sinuses, stomach, intestine and colon, and the genito-urinary tract, with the result of doubling the number of his discharges, and reducing the average stay in hospital from ten months to three months."[56]

What was needed now, as Cotton himself had urged, was a campaign of prophylaxis, eliminating suspicious teeth and tonsils and operating on children suffering from constipation to head off future mental troubles. Demonstrating how completely he was convinced by his own argument, Cotton arranged to have his wife's and his two children's teeth extracted, and when his younger son suffered from constipation, to have his son's colon resected. Sadly, the prophylaxis did not succeed. Both Adolph Cotton and his brother Henry later committed suicide.

Back in the United States, where his British triumph had made the papers, Cotton was honored by the State of New Jersey for his twenty years as superintendent of their flagship state hospital. The dinner was a splendid occasion, attended by the governor and over 400 invited guests. The rich continued to flock to Trenton in search of Cotton's miracle cures, and they were mostly operated on at his private facility. Meanwhile, he attacked the problems of the state-hospital patients with renewed ferocity. He was increasingly convinced that he had not been "radical enough." X-rays and inspections of patients' mouths were not sufficient. Previously, he had lacked the courage of his convictions, but no such inhibitions restrained him now:

> Many dentists would hesitate to extract vital [healthy] teeth and for some years we followed this practice. But we have found within the last year that many of our failures were due to the fact that we allowed vital teeth to remain in the mouth although we had extracted a large number of devitalized teeth, which were infected. It has become the rule now that if a patient is found to have a considerable number of infected teeth complete extraction must be done in order to eradicate all infection in the mouth. . . . [W]e have by necessity become more radical.[57]

The same logic applied to other regions of the body, "for there can be no question that infection originating in the mouth, especially in the teeth and tonsils, migrates to other parts of the body setting up secondary foci of infection."[58]

"Our success," he noted, "has been due to a combination of methods, all directed to the elimination of all sources of infection." Closer inspection had revealed that 76 percent of his female patients had infected cervixes. Problems involving the "lower intestinal tract" were even more common, identifiable by "sluggishness and delay due to the toxic effects of chronic infections on the musculature of the colon. . . . [A]t least 86 percent [of both male and female patients] will be found to have internal stasis and toxemia." Perhaps fortunately, he had discovered that in many cases these putrefactive processes and the mental illnesses they spawned could be forestalled by a new treatment he had discovered, "massive colonic irrigations, 15–20 gallons in a treatment." So successful had these proved that "each case right after admission [is] being given colonic irrigations as soon as possible." And if they didn't recover, resection of their colons was the obvious next step.[59]

As the mutilations and deaths continued to accumulate, Meyer kept a tight lid on Greenacre's findings. Another of his assistants, Solomon Katzenelbogen, traveled to Trenton to view Cotton's work for himself. On his return, Katzenelbogen reported on what he had seen to his chief. He had been startled to see how careless the diagnostic processes at the hospital were. The staff's response was that "the accurate discrimination between different types of psychosis matters very little, for the reason that *any psychosis would be of septic origin.*"[60] He found patients were being given typhoid vaccines for dubious reasons. "The patients have fever, chills, and they quiet down," he reported. "Those who have insight and had once the injection [*sic*] are threatened with being given a second one if they do not behave. The menace works well." Colonic surgery was performed on the least excuse, with the "certainty of finding abnormalities in the removed colons and adhesions on laparotomy."

Walking the wards, Katzenelbogen reported, was a sobering experience:

> I felt sad, seeing hundreds of people without teeth. Only a very few have sets of false teeth. The hospital takes care as to the pulling out of teeth, but does not provide false teeth. . . . The extraction of teeth does great harm to those who cannot afford to pay for a set of false teeth, and these patients are numerous. While in the hospital they suffer from indigestion . . . not being able to masticate their food. At home, recovered, these poor people have the same troubles, not being in a position to choose food which they would be able to eat without teeth. In addition, they are ashamed of being without teeth, since in their communities it is known to be a token of a previous sojourn in the State Hospital. They abstain from mixing with other people, refuse to go out and look for a job. . . . Thus, many of those recovered develop a reactive depression.[61]

Meyer read Katzenelbogen's report, filed it, and sat on his hands. Though the psychiatric profession, with some exceptions, had grown increasingly skeptical of Cotton's claims, the most influential psychiatrist in the country remained determined to avoid scandal.[62]

On May 8, 1933, the *Trenton Evening Times* reported some sad and unexpected news. While lunching at his club, Henry Cotton had suffered a heart attack and was dead. The following day, its obituary paid tribute to a psychiatrist who had brought so much attention to the state: "Thousands of people who have suffered from mental affliction owe him an enduring debt of gratitude for . . . displacing confusion and despair with hope and confidence." The paper further suggested that all must lament the loss of "this great pioneer whose humanitarian influence was, and will continue to be, of such monumental proportions."[63]

His successors at Trenton, all former assistants, continued to profess their fealty to the doctrine of focal sepsis. Abdominal surgery was quietly abandoned almost immediately, but the use of vaccines, colonic irrigations, and the extraction of teeth and tonsils continued to be an extensive part of the therapeutic armamentarium employed by the hospital's staff well into the 1950s, alongside a host of other somatic treatments that entered American psychiatry in the 1930s. Absent Henry Cotton's loud proselytization, it was these more up-to-date remedies the profession came to embrace.

But Adolf Meyer was not content to let his protégé's efforts simply fade from view. He volunteered to write Cotton's obituary for the *American Journal of Psychiatry.* Here the formidable Swiss-German psychiatrist who dominated American psychiatry for the first half of the twentieth century delivered his verdict. The profession, he wrote, had suffered "an outstanding and premature [loss]." Henry Cotton had been "a man of action and results. . . . [H]e made an extraordinary record of achievement. His views and practices were a vigorous challenge which stood non-compromisingly for an almost unitary explanation by focal infection, supported by the testimony of a number of patients and a number of colleagues." Meyer acknowledged that these claims had been controversial. But what Cotton had achieved "at Trenton State Hospital is a most remarkable achievement of the pioneer spirit" and one could only lament the premature departure of "one of the most stimulating figures of our generation" with his work "only partially fulfilled."[64]

Thus was the trail of maimed and dead bodies and the record of thousands of patients treated against their wills summarized for the profession at large. An independent study of the results of Cotton's colonectomies conducted just before his death (and never published) concluded that the extraordinary mortality rates he acknowledged significantly underestimated the carnage he left in his wake, putting the actual death rate at more than 44 percent.[65]

CHAPTER SIX

Body and Mind

HENRY COTTON'S EXPERIMENTS were part of a broader pattern of new treatments aimed at mental patients' bodies that marked the first four decades of the twentieth century. Some might argue that the psychopharmacological revolution that began in the early 1950s represented an extension of that pattern of cavalier experimentation. Be that as it may, a remarkable array of somatic treatments emerged in the 1920s and 1930s and, with a single notable exception, subsequently vanished from the scene.

Once biology had come to seem a route to cures, rather than a reason for pessimism, the search was on for novel approaches. One of the discoveries that helped persuade Henry Cotton of the infectious origins of mental illness was the series of studies that had shown the syphilitic origins of General Paralysis of the Insane (GPI). Almost contemporaneously with Cotton's early experiments with "surgical bacteriology," a Viennese physician, Julius Wagner-Jauregg, had begun treating patients with GPI by infecting them with malaria.

Wagner-Jauregg entered psychiatry as a young man after failing to secure a post in internal medicine. As early as the 1880s, he had begun to speculate that fever might prove therapeutic in some cases of psychosis, having observed temporary improvement in febrile patients. At first, his junior status stood in the way of implementing his hypothesis, but following his appointment to a chair in psychiatry at Graz in 1889, he began experimenting with Robert Koch's newly discovered tuberculin to induce fever, choosing to inject those with GPI, "because a favourable result cannot be so easily regarded as fortuitous as in other psychoses."[1] Adverse reactions forced him to stop, but following his appointment as the professor of psychiatry at the University of Vienna in 1902, he experimented with a variety of other means of inducing fevers. The most common of these were injections of streptococcus pyogenes, the cause of the disease erysipelas, popularly known as St. Anthony's Fire, which was characterized by a rapidly developing skin disorder accompanied by shivering, pain, fever, and chills. But this was a dangerous technique in a pre-antibiotic era. It could leave patients with lymphatic damage or even result in death. Besides, patients stubbornly failed to improve.

Then, on June 14, 1917, Wagner-Jauregg came upon an Italian prisoner of war suffering from tertian malaria.[2] Here was a novel febrile agent, and Wagner-Jauregg at once extracted some of the man's blood, which he then injected into two psychiatric patients afflicted with GPI. These patients in turn supplied malarial blood to treat another small group of patients. Within about a week, all developed malaria, and they proceeded to have between seven and twelve attacks each. Three of those he had treated, by his account, soon showed signs of improvement and were subsequently able to return to work. Unlike erysipelas, for which no treatment existed, the high spiking fever associated with malaria could (usually) be curtailed by administering quinine. Here, perhaps, was the febrile cure that Wagner-Jauregg had long sought. If so, it would surely secure his fame.

Within a matter of months, the Great War was over. Wagner-Jauregg found himself on the losing side and was soon arraigned on charges of war crimes. Industrialized warfare had created hundreds of thousands of psychiatric casualties, those whom the British and Americans called victims of shell shock, and the Central Powers termed sufferers from *die Kriegneurose*. On both sides, treatment was often brutal. British or French, German or Austrian, it made little difference: psychiatrists often resorted to using powerful electrical currents to inflict great pain on soldiers who were mute or claimed to be paralyzed. The French neuropsychiatrist Clovis Vincent called the process *torpillage*. Attaching electrodes to tongues and genitals, he administered powerful jolts of faradic electricity, persisting until the soldier spoke (if mute) or moved arms and legs (if paralyzed). As one of his assistants, Andre Gilles, put it, "These pseudo-impotents of the voice, the arms, or the legs, are really only impotents of the will; it is the doctor's job to will on their behalf."[3] At Queen Square Hospital in London, Lewis Yealland and Edgar Adrian (who would later win a Nobel Prize) pursued almost identical tactics. In Germany, this approach was nicknamed the "Kaufmann cure," and Fritz Kaufmann added shouted military commands to the electrical "treatment."[4]

Wagner-Jauregg had supervised and directed similar tactics on Austrian troops in Vienna. After the defeat, angry veterans demanded that he be held accountable for visiting torture on them, and forced a trial. Aided in part by testimony from Sigmund Freud and bolstered by his own testimony that he had been motivated only by a genuine desire to cure, he secured an acquittal.[5] The cloud apparently lifted, he returned to his chair at the university.

News of his malaria treatment spread rapidly. He presented a paper on his approach at the 1920 meeting of the German Psychiatric Society in Hamburg, and two years later, a translated version appeared in the *Journal of Nervous and Mental Disease.* American psychiatrists, like their European counterparts, responded enthusiastically to his work. Trials were soon under way at the New York State Psychiatric Institute and at the Manhattan State Hospital. George Kirby, head of the institute, reported a 34.9 percent remission rate among the 106 cases he had treated, and by 1926, mental hospitals throughout the New York State system were employing fever therapy.[6] Wagner-Jauregg had pointed out that "when one has once inoculated a paretic one can inoculate other paretics from him and so establish a malarial stock and further cultivate it."[7] It was an approach the American psychiatrist William Alanson White, superintendent of the St. Elizabeths Mental Hospital in Washington, DC, was fiercely critical of, for, as he pointed out, the diagnostic tests for syphilis were far from foolproof, and a misdiagnosed patient receiving malarial blood might simultaneously contract syphilis.[8]

Most mental hospitals ignored this concern, and thermos flasks of malarial blood soon passed through the mails. Some psychiatrists preferred a more "natural" method of transmission, encouraging mosquitos to bite infected patients, and then placing those they wanted to "treat," confined in straitjackets, into rooms occupied by the insects and waiting for nature to take its course.[9] The treatment, when it was not fatal, produced savage symptoms: high spiking fevers and chills that resembled a near-death experience. Nor were doses of quinine always sufficient to terminate the process. Still, since the alternative was a lingering death, bedridden, afflicted with "gangrenous trophic ulcers, total incontinence, and profound dementia," and victims of syphilis were destined in many cases to die choking on their own vomit, both patients and their families were driven to seek the new cure.[10] Curiously, as Joel Braslow has shown using California clinical records, one unintended side effect was to reduce some of the double stigma that GPI patients had traditionally faced, as both mad and damned for engaging in illicit sex.[11] Psychiatrists who had previously viewed their patients as "hopeless," "immoral," and "stupid"—objects to be acted upon and people who were sinful and depraved—now reacted much more empathically and positively toward their patients, who were allowed to become active participants in their treatment. Patients with syphilis voluntarily sought admissions to asylums they had previously shunned, and they began to see asylums as places of treatment rather than confinement. Whatever else it accomplished, malaria therapy did succeed in producing powerful shifts in the perceptions of both doctors and patients.

The 1920s and 1930s were a period of experimentation with other mechanisms for inducing fever. Some tried injecting horse serum into patients' spinal canals, thereby producing meningitis.[12] Injections of the organisms that caused rat-bite fever were tried, as were injections of killed typhoid bacilli and colloidal sulphur, a technique that by design led to the formation of abscesses. Alternatively, efforts were made to employ sweat boxes (or diathermy machines, to use the preferred term) to break down the body's ability to maintain a steady temperature.[13] Like malaria therapy, diathermy machines were not without their hazards. After his first attempt to use the technique, the superintendent of Arizona State Hospital reported to Winfred Overholser, the influential superintendent of St. Elizabeths Hospital in Washington, DC, that he had been forced to resume using malaria to induce fever because "as sometimes happens, the first patient treated in the new-fangled [machine] had died. Since that time the few graduate nurses we have have refused to operate the cabinet and I am in no position to force them to do so."[14] Some psychiatrists experimented with fever as a treatment for other forms of serious mental disorder, schizophrenia in particular.[15]

The consensus was clear: none of the alternative febrile agents were as convenient or effective as malaria. And other forms of psychosis were unaffected by fever. That the original malarial treatment produced a substantial number of remissions in cases of GPI was widely, though not universally, believed, and as the progenitor of this breakthrough, Wagner-Jauregg was rewarded with the 1927 Nobel Prize in medicine (one of only two psychiatric interventions to achieve this distinction).[16]

Did the treatment work? Certainly, many psychiatrists at the time were convinced that it did. What had been a uniformly fatal disease now seemed somewhat treatable, with one 1926 examination of the published data on a collection of 2,336 cases claiming that 27.5 percent of them were greatly improved and another quarter moderately improved.[17] Some historians seem inclined to accept these data, but others have been more skeptical, perhaps remembering that for many centuries, physicians were convinced by their clinical experience that blood-letting, purges, and vomits were sovereign remedies in the treatment of a host of diseases. In the words of Gerald Grob, "The evidence to support the alleged effectiveness of fever therapy was extraordinarily weak. The criteria to judge either remission or improvement was vague, and the fact that the psychiatric therapist was also the evaluator vitiated the results still further."[18]

In this connection, it is worth recalling that Wagner-Jauregg's initial claims about the efficacy of his approach rested on nine cases he had treated in 1917. On his account, six of the nine improved after undergoing a series of malaria treatments and were subsequently discharged. If these statistics are taken at face value, this would seem a remarkable turn of events, even though one of the patients died in the course of his treatment. Those initial optimistic statistics, however, belie what we now know about the actual fate of the remaining eight patients. Two of them, a tram conductor and a clerical worker, returned to work and we know no more of their fates, so being generous one may presume (though we cannot be certain) that they were successes. Four others are a different matter. We know that in two cases, a sergeant-major and a woman cleaner, the remission lasted barely a few months—and it was not uncommon for there to be some temporary remission of symptoms in cases of GPI. A third patient, a railway worker, committed suicide following his discharge. Finally, there was a thirty-seven-year-old actor who had been admitted "suffering from loss of memory and epileptic fits due to *suspected* paralysis."[19]

Audiences found this case to be particularly telling evidence of malaria's efficacy, for on Wagner-Jauregg's account, within two months of his treatment he was able to give readings for his fellow patients, and after his discharge, he was able to resume his career on the stage. Well and good, except for the inconvenient postscript Wagner-Jauregg neglected to add until 1936: within months he had learned from another psychiatrist, Dr. Raphael Weichbrodt of the Frankfurt-am-Main Asylum, that the patient had relapsed and been rehospitalized.[20] To these four we can add the remaining three cases, who were acknowledged at the time not to have responded to the malarial treatment. So, to review what we now know of the results in these nine cases: two appear to have recovered, though we cannot know for certain as they disappeared after their release; four had brief remissions; three did not even briefly improve; and one died as a direct result of the treatment.

There are many troubling issues surrounding the statistics on treatment for GPI. For example, the notion that all these patients were definitively suffering from tertiary syphilis rests upon the twin assumptions that all had serological tests for the disease and that those Wasserman tests never produced false positives. Both assumptions are false, and at least some of the cases diagnosed with GPI may in fact have not been suffering from this disease. Just how careless Wagner-Jauregg was with his serological technique is demonstrated by what happened a few months after he had treated his first group of patients. Seeking to treat another patient but not having malarial blood available, he approached an internal medicine specialist at a nearby hospital for some blood he could use. Not pausing to run a serological test to ensure that the sample was of the relatively benign tertian form of malaria, he injected the blood into one patient and, when that patient exhibited symptoms of the disease, drew blood from him and injected it into three more patients. When he attempted to short-circuit the fever with large doses of quinine, following his usual procedure, the intervention was a failure. He had, it turned out, infected these poor creatures with the malignant variant of malaria tropica. Three of the four patients were dead within a matter of weeks, their white blood cells having disappeared, and their red corpuscles declined massively. It was an incident that gave Wagner-Jauregg pause, and it was a year before he resumed malaria therapy. This incident remained hidden until the posthumous publication of his memoirs.[21]

Perhaps a Scottish verdict of "not proven" on malaria therapy might be in order. The advent of penicillin in the 1940s (which was a real magic bullet when used to treat syphilis) eventually led to the demise of the malaria treatment without its ever being put to a controlled test. But in certain respects, to debate this question is to miss the point: what mattered at the time was the widespread acceptance of the claim that a biological treatment for a major form of mental illness had been discovered—one that, as crude as it was, seemed to improve the fate of at least a fraction of the afflicted. Still more crucially, these claims had been validated by the award of a Nobel Prize—an accolade no other psychiatric therapy would achieve until Egas Moniz won the same award in 1949 for inventing the lobotomy. Perhaps at last psychiatry's claims to be more than the custodian of the crazy would have some substance, and its isolation from the rest of medicine might begin to diminish. And perhaps other treatments directed at the body might now emerge.

Wagner-Jauregg had speculated that "the brutalities of war" may have allowed him to avoid "possible censure for experimentally infecting sick people with a new disease."[22] The example of Henry Cotton and his "surgical bacteriology" raises some doubts on this score. Shut up in every sense of the term, deprived of all legal rights, stigmatized as barely human, and seen as an immense burden on the public purse, the mentally ill were uniquely vulnerable to therapeutic experimentation, and experiments there most certainly were.

IN FACT, within three years of Wagner-Jauregg's first experiments in Vienna, a Swiss psychiatrist at the famous Burghölzli Asylum in Zurich, Jakob Kläsi, suggested a possible way to intervene in the apparently hopeless course of another form of psychosis: schizophrenia. Nineteenth-century psychiatrists had long relied on a variety of hypnotics to keep many patients sedated. From opium and morphine, they had gravitated to new classes of drugs, including paraldehyde, chloral hydrate, and the bromides. There were no claims that these drugs were therapeutic, but they did make the management of asylums (and the calming of manic patients) somewhat easier. The bromides, in particular, were highly toxic, and these medications often gave rise to psychiatric side effects, but in a general climate of therapeutic hopelessness, their sedative properties were too useful to ignore. From 1904 onward, when the Bayer company in Germany began to market the first barbiturate, these drugs too came to be widely employed in asylums.[23]

In 1920, Kläsi proposed that barbiturates might be used not just as a symptomatic treatment, but as a curative therapy. He termed the approach *Dauernarkose,* or, as it became known in the English-speaking world, prolonged sleep therapy. Kläsi believed that his preferred barbiturate, Somnifen, was completely harmless, especially with respect to heart and lung functions.[24] He had found that using it to produce prolonged sleep could reduce manic excitation, both psychic and physical, and, by increasing the patient's dependence on the physician, it might create the possibility of increased contact with his doctor.[25] Patients undergoing the treatment had to be periodically awakened to eat and drink and to excrete. Otherwise, intravenous doses of barbiturates were used to keep them unconscious, sometimes for as long as eleven days.

Three of Kläsi's first twenty-six patients died in the course of treatment, something he neglected to mention in his first report on his work.[26] Instead, he emphasized that between a quarter and a third of those he had treated had recovered sufficiently to be discharged from the asylum. (The deaths had resulted from bronchial pneumonia and hemorrhages in the heart muscle.) As with Cotton's reports of his surgical interventions and Wagner-Jauregg's of fever therapy, these results seem to have been accepted completely uncritically, and the treatment spread rapidly, with many others reporting on their use of this novel approach. A number of these reports mentioned the heightened risk of death among those being treated, as well as a host of other complications: high fever, vomiting, lowered blood pressure, rapid breathing, cyanosis, and symptoms of the general collapse of various critical bodily functions.[27] Recoveries, however defined, were few and far between, though the treating psychiatrists often concluded that there was some degree of symptomatic improvement. Even at Kläsi's own institution, Burghölzli, increasing doubts were expressed about the dangerous side effects of Somnifen, and efforts were made to find an alternative compound, to which drugs designed to prevent collapse of the circulatory system were added. To little avail: life-threatening complications persisted, often forcing the early termination of the treatment. And yet, as one scholar put it, "in the face of equivocal evidence in regard to its effectiveness, but with ample evidence of complications and mortality," the Burghölzli and other institutions persisted in using deep sleep therapy into the mid-1930s.[28]

ONE OF THE GREAT DISCOVERIES of Western medicine in the 1920s was insulin. Until then, what at the time was called juvenile diabetes had had a uniformly fatal course. Doses of the newly discovered hormone suddenly transformed that situation. Excessive doses of insulin, however, could produce hypoglycemia and, in extreme cases, unconsciousness or even death. It was this property that led the Polish physician Manfred Sakel to suggest that insulin might have a role in the treatment of schizophrenia and, having developed a technique for inducing comas, to proclaim that he had discovered a remarkably effective new therapy.

Sakel had first encountered insulin when he worked at a private clinic, the Lichterfelder Hospital just outside Berlin, which specialized in treating morphine addicts. One of the techniques used to manage patients' withdrawal symptoms had been to administer insulin to put them into a semi-comatose state. At some point in the early 1930s, Sakel began to play with the idea of extending the use of insulin to the treatment of schizophrenia.

Convinced he had discovered a valuable new approach, he moved to Vienna in 1933 and continued his experiments at the University Clinic.[29]

If Kläsi's deep-sleep therapy had rendered his patients unconscious for hours at a time, Sakel's approach went further still. By injecting his patients with steadily increasing doses of insulin, he eventually produced a sufficiently severe state of hypoglycemia that they lapsed into a coma. The consequences could be dire, and those undergoing the treatment required constant nursing and medical attention, lest they slip into irreversible unconsciousness or simply expire. Even after the comas had been terminated by the administration of glucose, there were risks of hypoglycemic "after-shocks," so close medical supervision remained vital. The comas were induced five or six times a week, and a course of treatment often consisted of sixty or more such comas before it ceased.

As one observer recorded, "Shortly after an injection, the patients became quiet as the insulin began to lower blood sugar and deprive the brain of energy. . . . By the end of the first hour, the patients who had higher insulin doses had lapsed into a coma; many were tossing, rolling, and moaning, their muscles starting to twitch; and some had tremors and spasms. Here and there an arm would shoot up uncontrollably. Some of the patients began to grasp the air, reflexively . . . [and there were] other 'primitive movements,' including rapid licking of the lips."[30]

Isabel Wilson, who visited Sakel's clinic on behalf of the Board of Control, the body that supervised England's mental hospitals, added her own observations of the treatment. She noted that before lapsing into a coma, the patient "may indeed look very ill, groan or shriek, grind his teeth, twitch or splutter or make snorting or rasping noises. . . . There can be no doubt that the treatment is unpleasant for the patient . . . the feeling of anxiety, confusion, and distress in the process of waking are obvious to the onlooker." She attempted to minimize these problems in her report with the claim that, "fortunately, the patient himself is usually left with an amnesia for most of this period." But this contention was somewhat undermined by her simultaneous acknowledgment that "insulin therapy, with all its consequences, was carried out in full sight and hearing of every patient within range in any given dormitory."[31]

During the comas, patients sweated profusely. Their reflex responses declined nearly to the vanishing point. Not infrequently, their stupor was interrupted by a series of convulsions. As two experienced observers noted, during this phase, "the face is now pale and sunken. The pupils gradually contract to pinpoints and light reaction is lost. . . . [T]he greatest depth of the insulin effect consistent with safety has been reached. Beyond this point, the changes are likely to be irreversible, and protracted coma with central pulse and respiratory difficulties may be expected."[32] In the immediate aftermath of their comas, patients were often described as calmer and briefly less delusional. The influential Philadelphia psychiatrist Earl Bond rhapsodized about the lucid period that could occur: "From an aloof, withdrawn, bizarre and suspicious individual," he and a colleague noted, "the schizophrenic is momentarily transformed into a warm, friendly, responsive, and lucid person whose symptoms are either absent or greatly diminished in intensity."[33] Unfortunately, such improvements usually proved evanescent, but the hope was that with time, the lucid periods would become more extensive and perhaps permanent. Hence the fifty or sixty comas or more to which many patients were subjected.

Sakel referred to the preconvulsive comas as "wet shock" and the convulsions, should they occur, as "dry shock." He concluded that these seizures were beneficial and, in later years, deliberately added intravenous doses of a drug that produced convulsions to ensure the proper therapeutic effect. As he put it, "The mode of action of the epileptic seizure is on the one hand like a battering ram which breaks through the barriers in resistant cases, so that the 'regular troops' of hypoglycemia can march through." Secondarily, "The epileptic seizure also affects the psychosis . . . by means of the retrograde amnesia it produces." Only such risky and radical measures could hope "to break through the fixated and petrified psychotic processes and to devitalize them."[34]

Between 1934 and 1935, Sakel published thirteen reports on his work in the *Wiener Medizinische Wochenschrift.* Henry Cotton had claimed his treatment of focal sepsis had cured 80 percent of those he operated on. Sakel was bolder yet: he maintained that 88 percent of his schizophrenic patients had improved.[35] An American researcher who visited his clinic wrote back to his colleagues at Bellevue, in New York, that the therapy was one of the most remarkable things he had seen, and in an article in the *Journal of the American Medical Association* in 1936, his fellow New York psychiatrist Bernard Glueck informed his readers that if the improvements he had witnessed proved permanent, Sakel's work would constitute "one of the greatest achievements of medicine."[36]

News of these remarkable results was sufficient to induce the commissioner of mental health for New York State to issue an invitation for Sakel to come to America at the department's expense to demonstrate his approach. Sakel's lecture at the Harlem River State Hospital in 1936 was attended by twenty-five psychiatrists from throughout the state, and it led to the rapid spread of insulin shock units, first in New York and then in many other states. As a Jew, Sakel had endured vicious antisemitism in Vienna, and he gratefully accepted a visiting appointment at Harlem River State Hospital. He spent the rest of his career in the United States. At the invitation of the New York Academy of Medicine, he delivered a lecture on his work in January 1937, based on what he claimed were 300 cases of schizophrenia. In his first one hundred recent cases (ill for less than six months), he claimed that he had completely cured 70 percent, leaving them symptom-free, and had greatly improved the condition of 19 percent more.[37] Soon he was being feted in the popular press. The *Reader's Digest* spoke of a "Bedside Miracle," and that other oracle of middle-class America, *Time*, informed its readers that "Sakel has cured hundreds of cases of schizophrenia at his Vienna clinic by means of insulin injections."[38] With powerful endorsements like these, Sakel proceeded to make a fortune in private practice.[39]

The psychiatric profession was almost equally enthusiastic. The *American Journal of Psychiatry* devoted an entire issue in 1937 to insulin shock. Subsequently the journal published a special supplement reprinting papers from the Swiss Psychiatric Society devoted to the new shock therapies—a highly unusual move for a profession that was generally insular and unconcerned with developments elsewhere in the world. Professional enthusiasm for insulin coma therapy rested on more than just anecdote. Benjamin Malzberg, a statistician employed by the New York State mental hospital system, wrote an influential study of 1,039 patients treated with insulin comas, and compared them with a "historical" control group of nontreated patients admitted to the same hospitals in 1935 and 1936. The results, though far from the "cures" Sakel had claimed for his treatment, seemed promising to most psychiatrists who encountered them. Malzberg reported that some degree of improvement was evident in 65 percent of those given insulin, as compared with 22 percent among the control group, and that 37.6 percent of the treated group were much improved compared with only 14.7 percent of the control group.[40]

Aaron Rosenoff, overseeing California's sizable state hospital system, used the enthusiasm for insulin coma therapy to extract $2 million from a previously dubious state legislature in 1939 to construct an acute care psychiatric facility, the Langley Porter Clinic in San Francisco. He estimated that employing insulin treatment there would save the state $2 million each year.[41] And where New York and California led, the rest of the country soon followed. A federal survey undertaken just before the United States entered the Second World War found that 72 percent of the 305 public and private mental hospitals included in the study were using insulin coma treatment.[42]

In reality, Malzberg's "controls" were nothing of the sort. He relied on the assessments of the psychiatrists employing the treatment, and there was no indication (or likelihood) that the same standards were used to measure the two groups, since there was no systematic guideline establishing how to proceed, and different psychiatrists made the assessments. Nor did the two groups form a meaningful comparison in any case. Those subjected to the insulin treatment were "carefully selected with regard to their physical status" and likely for their potential treatability, whereas many of the "controls" were in far poorer physical and mental condition—a discrepancy that showed up in the dramatic differences in their death rates, which were almost four times higher among the so-called controls than among those being treated.[43] For the most part, however, these difficulties were ignored at the time by a profession unused to the notion of controls and impressed by any sort of comparison between treated and untreated patients.

Even a handful of studies reporting negative results had no discernible effects on enthusiasm for the treatment.[44] Just as Cotton had dismissed those who attempted to replicate his findings as having failed because they had not been ruthless enough, so Sakel and his allies contended that the poor results some reported simply reflected poor technique. Managing the comas and knowing when to terminate them was, they contended, a delicate art, not reducible to a formula, and deeply dependent on clinical skill and experience. Those who reported failures simply were incompetent practitioners or had been insufficiently aggressive in their treatment.[45]

Those employing insulin could see themselves as using one of the great advances of twentieth-century medicine in order to rescue an otherwise hopeless victim of schizophrenia. The sense of intoxication, power, and psychiatric heroism was nicely captured by Charles Burlingame, whose Institute for Living attracted some of the most affluent patients in America:

> There are few scenes that can match the dramatic intensity of the insulin treatment room in a modern psychiatric center. Here, each day, the patient is brought to the very fringe of life and allowed to hover there for several

> hours; sometimes he even starts to slide across the border, and only heroic measures, applied without delay, bring him back. Only the immeasurable reward of a mind restored could possibly justify the extremity of the method. From the moment the insulin is injected into the muscle of the patient he is under the most careful scrutiny. A special nurse is assigned to him. Pulse and temperature are taken every few minutes, and the physician is constantly nearby, ready to go into action at any sign of danger. The nurse looks up from her patient: "Doctor, I can detect no respiration." The doctor moves to the side of the patient and speaks sharply in an attempt to rouse him. There is no response, nor is there any visible sign of breathing; a few minutes ago the breathing was stertorous, but now the patient is silent and limp. Meanwhile, artificial respiration is being given. It is a matter of seconds before the sterilized needle is injected. Breathing follows almost instantly. Then follows the administration of sugar. The effect is quick. In five minutes the patient is sitting up in bed, smiling and, fortunately, remembers nothing of what has taken place. So it goes, this skillful sparring with death, where a few moments of neglect, inattention or inadvertence may cost a life.[46]

What could be more at odds with the conventional image of the psychiatrist as a curator of a museum of madness, a mere boarding housekeeper? Here was active treatment with a vengeance.

Not everyone was as sanguine about insulin coma therapy. Its dramatic effects led some to recoil and a number of doctors to raise questions about employing so savage a treatment. "The neurological and psychiatric phenomena suggest an extensive disintegration of cerebral function," one University of Minnesota physician noted in 1939. "The profound clinical manifestations make one wonder how these patients are able to withstand such severe reactions without some ill effect upon the nervous system."[47]

Stanley Cobb, professor of psychiatry at Harvard, was one of the principal skeptics. In 1937 and 1938 he cited a variety of animal experiments to show that brain damage followed the administration of insulin (and metrazol, a treatment we shall turn to next). "Such evidence," he complained, "makes me believe that the therapeutic effect of insulin and metrazol may be due to the destruction of great numbers of nerve cells in the cerebral cortex. The destruction is irreparable. . . . [T]he physician recommending these radical measures should do so with his eyes open to the fact that he may be removing the symptoms by practically destroying the most highly organized part of the brain. The use of these measures in the treatment of psychoses and neuroses from which recovery may occur seems to me entirely unjustifiable."[48] A letter in the *New England Journal of Medicine* elaborated on his concerns. Two recent "authoritative" studies reporting on autopsies of patients who had died during the course of insulin coma treatment had shown "petechial hemorrhages in the brains of [those] who died of hypoglycemia" and "widespread devastation" of their brains. "In certain devastated areas no nerve cells remained." Animal experiments had shown similar results. In the face of the extraordinary claims, "one is obliged to be skeptical. One suspects that the treatment is merely palliative and is carried out at the risk of permanent damage to the brain."[49]

Cobb's leading position in the still-tiny field of academic psychiatry ensured that he had a prominent platform from which to promulgate his views. Most notably, he wrote an annual assessment of the state of neurology and psychiatry for the *Archives of Internal Medicine,* and these essays provided a forum for his firmly held doubts about the shock therapies, insulin prominent among them. He complained repeatedly that the public was being led astray by misleading press reports, pointing out that the patients most likely to recover were being chosen for the treatment and that "remission is not cure." Citing Cotton's colectomies, he pointed out that "it has long been known that any situation that brings a schizophrenic patient near to death may rid him temporarily of his symptoms." And he returned again to the physical consequences of the comas: "It is now demonstrated beyond reasonable doubt that repeated insulin shock destroys nerve cells in the brain. It is probably that the mechanism is asphyxia, resulting from hypoglycemia." Optimistically, he proclaimed that "the tide of enthusiasm [that rose to a flood . . . now has ebbed."[50]

Cobb's concerns about the damaging effects of hypoglycemia on the brain were echoed in a whole series of studies published in the late 1930s in *Science,* the *American Journal of Psychiatry,* the *Archives of Neurology and Psychiatry,* and other publications.[51] Contrary to Cobb's expectations, however, these findings did not seem to diminish most psychiatrists' enthusiasm for the procedure, as Lawrence Kolb's survey for the American Public Health Service showed.[52] The press, which routinely seems to be willing to hype the latest medical breakthrough, took this as their cue to continue to praise Sakel's treatment, and the public, eager for some solution to the plague of mental illness, had little reason to dissent.

Sakel had developed insulin coma therapy purely serendipitously, without any theory to explain why it might work. Under some pressure to develop an explanation after the fact, he had first proffered the notion that neurons were like engines suffering from an oversupply of fuel and that insulin diminished this oversupply. Alternatively, he suggested that somehow overactive adrenal glands caused schizophrenia, and that insulin "opposes the action of the products of the adrenal system, so that excessive stimuli are muffled, and the cells kept relatively quiescent, to the benefit of the individual."[53] Rather than being discomforted by the evidence that hypoglycemia was producing brain damage, some suggested that "these very alterations may be responsible for the favorable transformation of the morbid psychic picture of schizophrenia."[54]

Sakel was increasingly inclining to the view that brain damage was central to insulin's therapeutic effect. Hans Hoff, who worked closely with him, noted that "Dr. Sakel's working theory held that under the influence of shock treatments, sick and defective cell connections in the brain would be separated."[55] The resulting brain damage selectively killed "psychotically functioning nerve cells which, because of their hyperactivity, were more sensitive to the hypoglycemic blockade than normal ones."[56] Accepting the notion that insulin (and its contemporary rival, metrazol) worked by producing anoxia and thus selective cell death in the brain, Harold Himwich and his colleagues experimented with a different approach that they thought might produce the same result: they argued that nitrogen might be used as a "safer and more controllable" way to starve the brain of oxygen. They boasted that "unlike previous attempts of other workers, the anoxia was intense. At the height of the bout the patient was respiring almost pure nitrogen." They observed "respiratory stimulation, tachycardia, cyanosis, spasms, and convulsive jerkings"—and no cures.[57]

Many were unimpressed by these attempts to provide a rationale for the comas and accentuate the effects. As one Chicago psychiatrist commented, "Obviously this 'theory' does not deserve serious physiologic consideration, since it is based on concepts having no scientific foundation."[58] This criticism seems not to have bothered Sakel. "The mistakes in theory," he said, "should not be counted against the treatment itself, which seems to be accomplishing more than the theory."[59] Pragmatically, American psychiatrists seem to have concurred.

The fact that insulin coma treatment required such an enormous and ongoing commitment of medical and nursing resources and entailed considerable costs to set up and run the specialized units used for treatment meant that its practical application was always limited. In some respects, it functioned more as a symbolic gesture toward a curative psychiatry than an intervention routinely employed on large numbers of patients. During the Second World War, the shortages of trained personnel added to these problems. The British, short of sugar as well, improvised and tried to use a slurry of potatoes administered through a gastric tube to bring patients out of their comas. Yet once the war ended, the treatment was readopted widely in many mental hospitals, and leading psychiatric texts continued to proclaim that it was the single most effective treatment for schizophrenia. Lothar Kalinowsky and Paul Hoch's textbook, *Shock Treatments and Other Somatic Procedures in Psychiatry*, which first appeared in 1946 and was published in successive editions through 1961, endorsed the procedure. Similarly, Arthur Noyes's *Modern Clinical Psychiatry*, the standard American textbook that passed through a multitude of editions, first published in 1935 and continuously in print for nearly forty years, continued to give prominent place to insulin coma therapy. Leading figures in British medicine had dubbed Henry Cotton "the Lister of psychiatry." As late as 1943, the science reporter for the *New York Times* thought the inventor of insulin coma therapy stood comparison to an even greater figure in the history of medicine: Manfred Sakel, he announced, was "the Pasteur of Psychiatry."[60]

INSULIN COMA THERAPY'S DEATH came slowly and stealthily. In 1953, a young British psychiatrist, Harold Bourne, published a paper in the *Lancet* decrying "The Insulin Myth." He pointed out that patients given the treatment "are given, at a conservative estimate, 50–100 times as much medical and nursing care, measured by the clock, as non-insulin treated patients." Beyond this, they were carefully selected recent cases, judged particularly likely to recover anyway. Surveying the existing literature, he found that the reported success rates were inflated and based on research so riddled with methodological flaws as to be useless, with no common diagnostic standards, and no clear criteria of what constituted "improvements." "It is clear," he concluded, "that . . . insulin offers the schizophrenic no long term benefits [and that] insulin has no place in [the treatment of] chronic schizophrenia."[61]

The response from senior members of the psychiatric establishment was swift and scathing. Bourne was an inexperienced junior doctor who had selectively examined the literature, and he was an example of clinical ignorance and youthful intemperance. Such prominent figures as Willy Mayer-Gross, Richard Hunter, Linford Rees, and William Sargant marshalled their years of clinical experience and professional standing to squash the young upstart.[62]

The following year, the American psychiatrist Henry David echoed Bourne's complaints in the pages of the *American Journal of Psychiatry.* Notwithstanding the ubiquity of insulin coma units in mental hospitals, and the evident conviction among his fellow psychiatrists of the value of the treatment, the fact remained that "in twenty years of work with IST [insulin shock treatment], there has been only one controlled study reported in the literature"—and that controlled study had found no significant differences in outcome between cases of schizophrenia treated with insulin, brief psychotherapy, or electroshock treatment. Moreover, the range and extent of the flaws in the research made it unsafe to draw any conclusions about the value of the treatment.[63]

Three years on, the *Lancet* published a controlled study in which patients were rendered unconscious using either insulin or barbiturates. Commenting on the existing literature, the authors noted that "despite the claims of enthusiastic supporters, and despite the vigorous protestations of Sakel himself, the evidence for the value of insulin remains far from convincing." Their study compared randomly assigned groups of first-break schizophrenics who had been ill for less than a year and showed as yet no signs or recovery. Each got daily treatment five times a week and were subjected to thirty-five or forty comas before being assessed by psychiatrists who were not aware which patients had been treated with insulin and which with barbiturates. Those assessments took place at the end of the treatment, six months later, and then annually. There were no significant differences between the two groups, prompting the authors to comment that "insulin is not the specific therapeutic agent of the coma regime as has so often been claimed."[64]

One might be tempted to conclude that science ultimately triumphed, that after two decades, carefully controlled work debunked the insulin myth. In reality, the picture was more complicated. The use of insulin comas had begun to decline three years before the study appeared, the consequence of the emergence of an alternative, far cheaper, and much more readily administered form of treatment, one that far more closely resembled the approach to disease employed by mainstream medicine: the first antipsychotic drugs, which appeared on the market in 1954.

Nor did the publication of findings that undermined the claims about insulin's therapeutic properties bring about the demise of the therapy in any simple fashion. To be sure, insulin therapy continued its slow decline in the late 1950s and early 1960s—lamented though it was by its supporters, who continued to claim that it constituted a "pivotal advance in the treatment of the mentally sick" and "THE ONLY SUCCESSFUL BIOLOGICAL TREATMENT OF OUR TIME."[65] Far from being deterred as they learned more about the brain damage insulin comas inflicted on mental patients, the treatment's advocates insisted that this damage constituted its singular merit: "The objective of IST is the destruction of brain cells, in other words, the production of a controlled brain lesion," O. H. Arnold proudly announced at the conference held in 1958 to honor Sakel's legacy, and "in many such cases we have to risk considerable organic brain damage."[66] Hoch, who had been head of the department of experimental psychiatry at the prestigious New York Psychiatric Institute and from 1955 was commissioner of mental health for the State of New York, hastened to add his endorsement of Arnold's claims. If the treatment of schizophrenia were to succeed, "cells which are sick have to be destroyed. Otherwise relapses will come. That means that one of the most important things is to see that really every cell which is affected is really destroyed."[67] The British psychiatrist D. N. Parfitt had previously suggested that the appropriate analogy was to psychosurgery, contending that the effects of insulin coma therapy could "most easily be understood in terms of a physiological rather than anatomical lobotomy and generally this physiological lobotomy is of better quality"—a sort of "bloodless operation."[68]

Three years later, the future Nobel Prize winner John Nash, who had been diagnosed as schizophrenic, was admitted to Trenton State Hospital. There he underwent a lengthy course of insulin coma therapy—to no avail. At what cost no one can say.[69]

CHAPTER SEVEN

Shocking the Brain

IN THE EARLY 1930S, the Hungarian psychiatrist Ladislas Meduna began to experiment with a different biological approach to the treatment of schizophrenia. Like Sakel, Meduna was of Jewish descent, and the rising tide of antisemitism eventually forced him to emigrate to the United States. His initial experiments were conducted in his native Hungary.

Trained originally in Budapest in the neuroanatomical tradition of his time by a psychiatrist who hewed closely to the orthodoxy of the age, Meduna was taught that schizophrenia was an endogenous hereditary disease that could not be cured. The very idea, according to his mentor Karl Schaffer, was "nonsensical." Working full time in the laboratory, Meduna did not encounter his first patient with schizophrenia until the late 1920s. He was put in charge of the outpatient department at the Royal National Mental and Nervous Asylum in Budapest, and those with seizures constituted a large part of his clinical load. Continuing his neuroanatomical work, in 1931 he and a colleague noticed that at autopsy there often appeared to be "tremendous changes in the brain" in cases of epilepsy, changes he did not observe in schizophrenia.[1]

Wagner-Jauregg's malaria therapy and Jakov Kläsi's experiments with deep-sleep therapy had persuaded a number of the rising generation of psychiatrists that a view of mental illness as rooted in the body was not incompatible with an active therapeutics. It was this changed intellectual context that encouraged ambitious young psychiatrists like Sakel and Meduna to launch their own experiments. In 1932, Meduna published his findings of an increased proliferation of glia cells in the brains of those suffering from epilepsy, and much later he claimed that he had simultaneously observed a deficit of glia cells in schizophrenics.[2] From the late 1920s onward, a handful of psychiatrists had claimed that epilepsy and schizophrenia rarely occurred together.[3] Some contended that if epileptics developed schizophrenia, their seizures diminished; others that epileptic seizures seemed to result in a temporary abatement of schizophrenic symptomatology.[4] Modern researchers have shown that the claimed antagonism between schizophrenia and epilepsy is completely spurious: in reality, schizophrenia is associated with a higher risk of seizures, and epilepsy goes hand in hand with an elevated risk of developing schizophrenia—a demonstration of the hazards of relying on anecdote and clinical intuition.[5] Yet the clinical observations and speculations of the late 1920s and early 1930s that claimed that one disorder precluded the other would prove to have dramatic consequences for the treatment of psychosis.[6]

Encountering these claims, Meduna reasoned that there might be a biological antagonism between epilepsy and schizophrenia. From that intellectual leap, he reasoned that if he could somehow induce seizures artificially, he might be able to cure schizophrenics. He first attacked the problem with a series of animal experiments, using a series of alkaloids, including caffeine and strychnine, before settling on injections of camphor dissolved in oil. Having tried this solution on guinea pigs and determined to his satisfaction that it was safe to inject, he moved rapidly to experiment on human subjects.[7]

ON THE MORNING OF JANUARY 23, 1934, Meduna brought in his first patient, a man diagnosed as suffering from catatonic schizophrenia. The man was injected with camphor. Nothing happened for forty-five minutes, and then, at last, the patient had a major seizure. After a short interval, during which the patient briefly stopped breathing, he recovered consciousness but seemed little changed. The episode left Meduna in considerable distress: "My body began to tremble, a profuse sweat almost drenched me, and, as I later heard, my face was ash gray."[8] He had to be supported by two nurses as he was led back to his room.

The patient subsequently underwent several more injections and convulsions, though Meduna soon switched away from camphor to induce them. Its effects were so unpredictable that clinicians were unable to anticipate the timing of the subsequent convulsions, which might occur fifteen minutes to three hours later, or not at all. It tended to produce nausea and vomiting and abscesses at the site of the injection.[9] Instead, he settled on a synthetic drug, pentathylenetetrazol, soon known in the United States as metrazol, as his treatment of choice.[10]

Perhaps Meduna's reaction should not occasion surprise, given the scene he had just witnessed and the uncertainty about whether the patient would even survive camphor's "powerful and brutal effects on the organism."[11] Camphor and metrazol had similarly savage impacts, though the latter had somewhat more predictable consequences. Solomon Katzenelbogen, a Swiss-trained psychiatrist who by 1938 occupied the post of director of research at St. Elizabeths Hospital in Washington, DC, provided a vivid account of the standard procedure in these cases. A syringe with a large-diameter needle was introduced into a vein, and its contents injected as rapidly as possible. Delivering the metrazol quickly was vital if it was to induce a seizure, and if a seizure failed to result, a still larger dose was tried. Katzenelbogen advocated being "generous with the dosage" to obviate the need for a second injection. Some recommended repeating the injections every other day until as many as thirty or more convulsions had been induced. Katzenelbogen preferred no more than two treatments a week, such was "the strain imposed on the organism by the convulsions."[12]

Yet far from being deterred by the severity of the response, Meduna was certain it was vital to the success of the treatment. To remove an entrenched psychosis, he was convinced, required even more powerful countermeasures. The use of "brute force" was inescapable: "We act with both [metrazol and insulin] as with dynamite, endeavoring to blow asunder the pathological sequences and restore the diseased organism to normal functioning. . . . [W]e are undertaking a violent onslaught with either method we choose, because at present nothing less than such a shock to the organism is powerful enough to break the chain of noxious processes that leads to schizophrenia."[13]

The treatment was worth the trauma because the results were so spectacular, or so Meduna asserted. His first paper on the new treatment reported on twenty-six patients. He claimed there had been ten long-lasting remissions, three temporary remissions, and thirteen patients who were not improved. Two years later, his monograph on convulsion therapy touted much more impressive figures: of sixty-two patients with a particularly pernicious form of schizophrenia, 80 percent had entered remission.[14]

With results like these, psychiatrists in Europe and America were quick to adopt the new metrazol treatment. Unlike insulin comas, which required special treatment wards and large amounts of scarce and expensive medical and nursing attention, metrazol injections were quick, easy to administer, and cheap. Or usually easy to administer: once patients had had one such treatment, they were often desperate to avoid a second. The British psychiatrist Henry Rollin recalled "the unseemly and tragic farce of an unwilling patient being pursued by a posse of nurses with me, a fully charged syringe in hand, bringing up the rear." Others reported that patients were so violently resistive they had to be sedated before the seizure could be induced.[15]

For the professionals administering the treatment, the procedure was deeply upsetting until it became routine. Once the injection had been given, "color drained from the patient's face, which became stiff and motionless. Onset of seizure was signaled by a cough or a cry, before tonic contractions began . . . [followed by] a sudden yawning spasm, at which an attendant inserted a gag to avert dislocated jaw." As the seizure continued, incontinence was common, and the patient then fell into a coma.[16] Katzenelbogen, who by then had administered the treatment to hundreds of patients, added that "the patient's face shows successively flushing, cyanosis, ashy-white color deepening with the period of apnea and then gradually clearing up."[17] But these events, psychiatrists were assured, were usually not remembered by the patients, who were amnesiac for the period of the seizure itself.

What patients usually *did* remember was the interval between injection and seizure. "In this pre-paroxysmal phase, the most common and striking observation is the patient's facial expression of fright, of being tortured, of extreme anxiety." A number of those given the treatment reported that they felt on the brink of dying, and "one patient was sure that embalming fluid was being injected into his veins." Patients' accounts of their terror varied, but "they contain nevertheless one common and outstanding feature, namely the feeling of being tortured, and of intense fear of imminent death." Making matters worse, "If no convulsion takes place, anxiety, restlessness, and general discomfort may continue for hours." Not to worry, the author of one of these descriptions hastened to reassure his colleagues, "this extremely drastic therapeutic procedure has proven . . . to be quite safe from danger to life, the rate of fatalities being less than 1 per cent."[18]

The sheer violence of the seizures often produced fractures of the long bones or of hip sockets. In 1939, still another complication was reported, soon confirmed by broader studies.[19] X-rays of the backs of patients subjected to metrazol seizures revealed that more than 40 percent of them had suffered compression fractures of their spines—a finding that some attempted to minimize by stressing that these spinal fractures were asymptomatic.

Then there were the murmurings, some private, some public, that the extraordinary results claimed by Meduna for his therapy (and Sakel for his) could not be reproduced. Phyllis Greenacre, who had attempted to expose the fallacies of Henry Cotton's claims, only to be stymied by the resolute refusal of her superior Adolf Meyer to publish her findings, wrote privately from Cornell to her former chief, professing herself deeply worried "by the present therapeutic enthusiasm about the use of insulin, metrazol, and camphor in the treatment not only of schizophrenia, but of almost anything else in the field of psychiatry." "The present epidemic," she continued, "seems to be following quite closely in the lines that the Cotton one went;—with similar premature claims of cures; confusion between therapeutic hopes and results;—the utilization of manic-depressive remissions in the interest of 'proving' and demonstrating 'results' in schizophrenia, etc., etc." With considerable understatement, she added that "while one does sympathize with the urge to 'do something,' it seems that the landslide which is sometimes set in motion has some dangers too."[20]

Meyer professed some sympathy with her letter, adding that "I am not at all surprised that . . . this work of 1925 looms up in the present situation." But his sympathy did not extend to sharing her critique of the new shock therapies. "I must confess," he told her, "that in general I have the impression that these non-bloody assaults upon the person have given some very interesting results." While they were not the complete answer, "where one is almost powerless with regard to the ability to get out of the day-dreaming or scattering conditions" they were certainly an improvement, however much one might regret "that it is so hard for some patients to get *en rapport* along less aggressive lines."[21]

A SHORT TIME AFTER MEDUNA'S FIRST REPORTS on his shock treatment, the Italian psychiatrist Ugo Cerletti was prompted to find an alternative to metrazol by reports of the existential terror metrazol produced in patients. (He had tried Sakel's insulin coma treatment as early as 1936 and began using metrazol the following year.) Cerletti had been experimenting with electricity and its effects on animals for several years, as part of his research on epilepsy. His trials with dogs, though, at first produced negative results. He had placed one electrode on the head and the other on the anus, and the electric currents that passed through the thoracic cavity stopped the heart and killed half of the dogs he shocked. But a serendipitous visit to the Rome slaughterhouse provided the solution to this problem.

Cerletti discovered that pigs destined for slaughter were first stunned by an electric current passed through their brains via electrodes applied to their temples. "Then the butcher, taking advantage of the unconscious state of the animal, gave its neck a deep slash, thus bleeding it to death. I at once saw that the fits were the same as those I had been producing in dogs." Further animal experiments assured him that the procedure was safe, but still he hesitated "because of the terror with which the notion of subjecting a man to high tension currents was regarded. The specter of the electric chair was in the minds of all."[22] At length, though, Cerletti and his chief assistant, Lucio Bini, decided to experiment on a human subject.

Fortunately, one soon came to hand. On April 10, 1938, the Rome police arrived with a vagrant they had picked up at the Termini train station. Enrico X was confused and incapable of normal communication. The following day, Cerletti and Bini made repeated attempts to induce a grand mal seizure, with the attendant loss of consciousness and violent muscle contractions, but afraid of a possibly fatal outcome, the doses they used failed to produce one, and after the third attempt they gave up. On April 20, they resumed their efforts. This time, Enrico X, treated with increased voltage, responded with a classic seizure, during which Bini recorded that the patient ceased breathing for 105 seconds—what must have seemed an eternity for those clustered around his bedside. "True it is," Cerletti remarked some years afterward, "that all had their hearts in their mouths and were truly oppressed during the tonic phase with apnea, ashy paleness, and cadaverous facial cyanosis . . . until at the first deep, stertorous inhalation and the first clonic shudders, the blood ran more freely in the bystanders' veins as well."[23]

Ten minutes after the convulsion, Enrico X began to return to consciousness and, five minutes later, began to speak a few words. "I asked him," Cerletti recalled, "'What has been happening to you?' He answered, 'I don't know; perhaps I have been asleep.' . . . So electroshock was born, for such was the name I forthwith gave it."[24] Cerletti and Bini subsequently gave him another ten shocks, then presented him before the Royal Academy of Medicine of Rome, where for good measure and, to demonstrate the procedure, they gave him another treatment. The demonstration that electroconvulsive therapy (ECT) could return catatonic patients to temporary contact with reality was greeted with astonishment and, when replicated, did much to advance claims about the procedure's efficacy.

Enrico X was subsequently discharged as cured, and he returned to his family in Milan. Two years later, his wife reported that he had relapsed and been rehospitalized in Milan's Mombello Psychiatric Hospital. But his case provided Cerletti's team with a demonstration that convulsions could be induced electrically without causing death, and that after a number of treatments psychiatric symptoms abated. A previously unreachable patient was once more able to communicate. Over the next six months, twenty more patients were treated with an average of more than twenty-one shocks apiece. Only then did Cerletti venture into print. His account of administering the new form of shock therapy remained the standard one throughout the war years.[25]

In general medical journals, the advent of ECT was greeted with some circumspection. The *Lancet* expressed concern that there were as yet no adequate data on either efficacy or safety, adding that psychiatrists, who had previously been accused of being "resigned . . . and torpid," now seemed to be seized by a kind of "therapeutic fury." Still, the editorial professed agnosticism about ECT's worth: "We must not let unconscious associations with what is done periodically in a room in Sing Sing prejudice us against what may turn out to be a valuable way forward."[26] The *Journal of the American Medical Association* echoed such concerns: "Sufficient data are apparently not available to determine whether irreversible changes are produced in patients after several shocks." There was obvious concern among "many physicians and physiologists" about whether "the passing of electric current through the brain is a most hazardous procedure."[27] But it swiftly became apparent that most psychiatrists did not share these doubts.

The Second World War contributed to ECT's adoption in the United States. Renato Almansi, one of Bini's assistants, was Jewish and fled to the United States in 1939, bringing with him an Italian-designed ECT machine. The following year, Lothar Kalinowsky, a German psychiatrist present at the first ECT trials who was also Jewish and forced into exile, arrived in New York, having previously stopped in Paris and London, where he had introduced colleagues to Cerletti's technique. Kalinowsky proved the better publicist of the two, and though Almansi has a stronger claim to have been the first to administer ECT in the United States, it was Kalinowsky who would be its most effective promoter throughout his career until his death in 1992.

By October 1941, ECT was being used for a wide spectrum of patients in 42 percent of American mental hospitals.[28] ECT induced seizures more reliably than metrazol, and its effects were almost instantaneous, avoiding the long delay and profound terror that followed injection with the drug.

It was cheap, easy to administer, and the technique did not take long to learn—a combination of virtues that, when attached to optimistic claims about its therapeutic efficacy, hastened its adoption.[29] By the end of the war ECT had largely, though not completely, replaced metrazol as the convulsion therapy of choice.

In some quarters, however, the use of ECT was controversial. In the quarter century after the Second World War, the most prestigious and profitable branches of American psychiatry came to be dominated by psychoanalysts and their supporters. A handful of psychoanalysts provided fanciful psychodynamic accounts of why ECT might work, but most of their number thought the treatment rested on a category mistake. Mental illness was about memories and meanings, and crude jolts of electricity were hardly the way to deal with people's psychopathology.

The analysts sneered at what one of their number, the Yale psychiatrist Fritz Redlich, called "directive-organic psychiatry." Those engaging in such practice were, by their lights, an inferior and misguided lot. In the immediate aftermath of the war, the analysts formed a group within the profession to promote their views, and one of their first acts was to publish a report condemning ECT. The Group for the Advancement of Psychiatry (GAP) issued its *Report No. 1* on shock therapy in October 1947. It condemned the "promiscuous and indiscriminate use" of ECT, contended that brain damage was the inevitable consequence of its use, and cast doubt on its therapeutic value, save for possibly shortening episodes of depression. Psychiatrists committed to the treatment immediately fought back fiercely. After three years of behind-the-scenes battles, GAP gave in to pressure from the supporters of the treatment and issued a revised report in August 1950. This time GAP report pronounced ECT safe and effective for a wide range of conditions, even when administered on an outpatient basis.[30] Freudian analysts' control over the profession had, it transpired, its limits.

Most psychoanalysts nonetheless remained vigorously opposed to ECT even as professional politics dictated a retreat from expressing their distaste openly, and their status as the elite portion of the psychiatric profession grew steadily more obvious through the 1950s and 1960s. In the future, one important consequence of this situation was that, as the procedure came under attack in the broader culture during the 1960s, prominent psychiatrists either absented themselves from the fray or expressed their own reservations about ECT.

CHAPTER EIGHT

The Checkered Career of Electroconvulsive Therapy

BEFORE AND FOR SOME YEARS AFTER the introduction of the first drugs to treat mental illness in the early 1950s, electroconvulsive therapy (ECT) was widely used in American mental hospitals. It was administered almost regardless of a patient's diagnosis and, though no systematic records were kept, was unquestionably the most frequently employed active treatment in most of these establishments, public and private alike. Most psychoanalysts disliked ECT and counseled their patients against it, but inside mental hospitals it had an unchallenged place in the psychiatric armamentarium.[1]

Then things changed dramatically. The antipsychiatry movement that emerged in the 1960s saw ECT as a symbol of psychiatric oppression. Hollywood, novelists, and journalists increasingly dismissed it as a barbaric intervention that broke bones and destroyed memories. In an era when mental hospitals began to empty, those running them increasingly relied on drugs, not ECT. Psychoanalysts, the dominant fraction among outpatient psychiatrists, renewed their own criticisms of the procedure. Medical schools stopped teaching how to perform ECT, and psychiatry for the most part abandoned it. Like the other somatic treatments developed in the 1920s and 1930s, it appeared to be moribund.[2]

But that was not the end of the story. Those convinced that shock therapy had a role to play in the treatment of depression had found ways to head off the fractures and upsetting scenes that accompanied the administration of unmodified ECT. Administering muscle relaxants and performing the procedure under general anesthetic avoided broken bones and rendered the whole process less fraught. Yet those developments did not suffice to rescue the procedure.

Decades on, however, as we shall see later in this book, while some patients described how undergoing ECT had destroyed their lives, a number of prominent figures published memoirs testifying that it had saved them from killing themselves and alleviated the miseries of their depressive state. In the early twenty-first century, with the side effects of medication more evident, interest in ECT began to revive, new claims were made about its efficacy, and, though stigma still clung to its administration, it reentered the realm of acceptable psychiatric practice.

AS WITH INSULIN AND METRAZOL, there was no generally accepted account of why ECT worked.[3] Metrazol had been launched on the basis of the fallacious claim that schizophrenia and epilepsy were mutually exclusive conditions, and ECT's rationale was initially similar. In both forms of treatment there was growing skepticism about their value in treating schizophrenia, and more and more clinicians were suggesting that their primary utility lay in their impact on affective disorders, particularly depression. This meant that biological antagonism played no role in whatever therapeutic gains ECT brought in its train; its administration was purely empirical, bereft of any clear foundation.

Despite an emerging consensus that ECT worked best in cases of depression, during the 1940s and 1950s it was deployed extensively across the board. To some extent, this reflected how unreliable and labile psychiatric diagnoses were in this era, and the fact that ECT acted to modify patients' behavior, quieting them, and rendering them more tractable, particularly when several shocks were given in rapid succession. Ugo Cerletti's assistant, Lucio Bini, had given this approach the unfortunate name of "annihilation therapy," and there is abundant if necessarily anecdotal evidence that this kind of "therapeutic discipline," as Joel Braslow calls it, was a routine feature of mental hospital life.[4]

Braslow's examination of the clinical records at Stockton State Hospital in California provides some systematic data about how widely ECT was used across diagnostic labels and how frequently patients were singled out for this treatment as a means of controlling those who acted out. Even the hopeless diagnosis of senile psychosis or general paresis (tertiary syphilis) did not preclude treatment with ECT. What kind of psychiatric "disease" one was held to have seems to have exercised little or no effect on the choice of treatment. Disordered behavior was, for these psychiatrists, symptomatic of disordered bodies, and by whatever mysterious alchemical process electroshock worked, it demonstrably acted to change the disordered behavior and thus was, as Braslow puts it, "unassailably therapeutic: the control of bodies was the control of disease."[5]

Doctors clearly hoped that in at least some cases, the therapeutic effects would result in patients being cured. In the immediate aftermath of ECT, patients were often confused and cloudy, but these effects gradually wore off, and they then seemed to be less distressed and more amenable to ward discipline.[6] All too often, these benefits proved transient, but then most patients were being shocked three times a week for several weeks, and the working assumption (or hope) was that the intervals of lessened distress and improved behavior would grow longer with time.

When clinicians chose whom to treat, clinical records suggest that disciplinary considerations were the primary driver of the interventions. Psychiatrists did not seem inclined to differentiate sharply between improvements in mental status and improvements in behavior, repeatedly conflating the two. Stockton psychiatrists described the patients they selected for treatment as "fighting," "restless," "noisy," "quarrelsome," "resistive," "combative," "stubborn," "aggressive," "obstinate," "uncooperative," "hyperactive," and "disrobes." Physicians saw incorrigible behavior as a signifier of treatable disease. Its elimination measured therapeutic success. "Quieter," "manageable," "calm," "more cheerful," "cooperative," and "not so aggressive" all described effective therapeutic outcomes.[7]

In the 1940s and 1950s, the expansive use of ECT that Braslow documented at Stockton State Hospital was clearly the norm, and the psychiatric literature is replete with evidence of its widespread employment to manage the problems posed by schizophrenics and other troublesome inmates. The magic phrase was "maintenance therapy." Active treatment offered demonstrable results, producing calmer wards valued by psychiatrists, ward attendants, and perhaps patients themselves.[8] If disorder threatened to return, that provided a rationale for a new round of interventions.

At Pilgrim State Hospital, where Lothar Kalinowsky had set up shop in October 1940, the superintendent, Harry Worthing, quickly grasped that treated patients became "quieter and more manageable. It was therefore decided to treat chronic, disturbed patients in the disturbed buildings . . . with the sole intention of making them less disturbed, less assaultive, and less destructive." Worthing made no attempt to disguise what counted as success here: "We feel that results from a purely symptomatic standpoint are worthwhile considering the little effort involved in giving three or four treatments, if the patients become more manageable even for several weeks."[9]

There were no checks on how many shocks individual patients might receive, and the numbers reported were often remarkably high. There was one published reference, for example, to a patient receiving more than 800 shocks.[10] Two British psychiatrists, Kino and Thorpe, reported on 500 patients to whom they had administered ECT. "Most" were female (this was the norm), and if they were melancholic, they might receive sixty or seventy treatments. Cases of mania and schizophrenia, they reported, required longer and more intensive treatment.[11] Thorpe separately provided an example of what this more intensive treatment could involve. In cases of mania, he recommended "several shocks daily until the excited state is suppressed" and then "frequent spacing of shocks to prevent relapse." Making use of such a regimen would ensure that "the most maniacal patient can be rapidly and dramatically be brought under control"—and, thereafter, "the quiet cooperation of the patient will be appreciated by the nursing staff."[12]

Early on, there were occasional experiments with the subconvulsive use of electricity. In 1946 at Rockland State Hospital in New York, for example, Walter Thompson and his colleagues, following earlier trials along these lines by Nathaniel Berkwitz, tried connecting one electrode to the patient's head and the other to a leg. (They had earlier tried putting both electrodes on the head and using a nonconvulsive dose of electricity but dismissed this approach since "the transmission of the current through the brain only was comparatively painless and innocuous.")[13] All 213 patients they experimented on (one as young as thirteen) were female. Turning on the current produced a tetanuslike convulsion with forcible contraction of the face and eye muscles and obvious signs of pain. During the contractions, patients stopped breathing temporarily but remained fully conscious.

Initially, the women were given thirty of these subconvulsive shocks, which "had no beneficial effects as far as the psychosis was concerned." In other respects, the treatment seemed useful, proving "of definite help in improving antisocial behavior." For some odd reason, both patients and attendants considered this to be "a form of punishment . . . but every attempt was made to prevent this attitude." Metrazol, Thompson noted, elicited similar responses. By contrast, subconvulsive shocks carried fewer risks of fractures or other untoward effects, and they were a useful means of reducing "destructiveness, combativeness, untidiness, self-mutilation and at times, although rarely, hallucinations and delusions." All these "beneficial effects appear to be on a reality basis and due to the associated discomfort."[14]

Overt reliance on aversive therapy of this sort was uncommon, though the use of ECT to achieve similar results was widespread.[15] Kalinowsky and

Paul Hoch's 1946 textbook on shock treatment advised that "maintenance treatment should be available on every chronic ward" and lauded "the great satisfaction that active treatment gives to the personnel of a chronic ward." Repeated shocks produced docile patients and instead of inmates being left to rot on "continuous treatment wards," the staff could see themselves as dispensing therapy. Such interventions thus served multiple goals, and their use "inevitably improves morale and, consequently, the interest of the nursing and medical staff in the individual patient." Some patients might even become suitable for discharge, returning for repeat ECT if and when their condition deteriorated.[16] The authors of the major textbook on shock therapy assured their colleagues that "such treatment can be continued for years without damage."[17]

Maintenance treatment on the wards of understaffed state hospitals was more extreme than this sanitized account suggests. At Milledgeville in Georgia, for example, one of the largest state hospitals in the country, "maintenance therapy" had a different name: "the Georgia Power Cocktail."[18] By 1942, the hospital census had reached 10,000 patients whom a staff of fifteen physicians were charged with treating. ECT proved to be an essential tool for managing the massively overcrowded wards. One of the staff psychologists described the routine:

> The matter of who was to receive electric shock treatment on the various wards was largely based on the reports of nurses and attendants. The words "punish" and "shock treatment" were often synonymous to the disturbed. Which electric shocks were given for treatment, which for punishment, and which for both presented confusing problems to the patients, many of whom were paranoid to begin with and felt they were being punished for their "guilty" deeds prior to their illness. . . . The attendant himself was confused when he was criticized for using force to subdue a patient who might have attacked him when he had heard a physician say that force was unnecessary "because shock treatment left no marks." The physician could have added that the patient would not even be able to remember the circumstances surrounding the behavior leading to the punitive shock treatment.[19]

Edward Shorter and David Healy imply that this abusive and "vengeful" use of ECT at Milledgeville was exceptional.[20] But considerable evidence casts doubt on this contention. At Traverse City State Hospital in Michigan, Paul Wilcox reported on 500 chronic patients who had been given an extensive series of electroshocks. "Many were selected," he noted, "because they were extremely difficult ward problems." Wilcox stressed that others had failed to achieve results with patients like these because they had not treated them adequately. "Adequate treatment means intensive treatment until the expected improvement has occurred." Diagnosis was irrelevant. Patients drawn from thirty-six different diagnostic categories were treated, and women on the average were given twice as many shocks as the men. No justification was offered for this discrepancy. As for what all this was intended to accomplish: "No attempt was made to judge basic improvement in the psychosis other than in the ward behavior." Perhaps that was just as well, since "only a small number of these chronic patients improved enough to leave the hospital, even with the help of the treatments." But the patients were no longer so difficult to manage.[21]

The role of ECT in enforcing order on mental hospital wards and punishing patients whom the staff deemed troublesome was likewise a central theme of Ivan Belknap's ethnography of a Texas state hospital in the early 1950s.[22] Lists of patients to be given ECT were compiled daily by the ward attendants, singling out those who were violent, had proved uncooperative or troublesome, or had annoyed the staff: hallucinatory and delusional "worry warts" were particularly likely to be put on the shock lists.

Sometimes the mere threat of being referred for ECT was sufficient to rein in the patients. They had witnessed the effects firsthand, since they were often called on to hold down a fellow patient while the shock was administered. The procedure occurred in front of them, with no attempt to hide the results: "The patient's convulsions often resemble those of an accident victim in death agony and are accompanied by choking gasps and at times by foaming overflow of saliva from the mouth. . . . Moreover, in the early disorientation and vacuity of his recovery he is obviously upsetting to the other patients." For those who refused to be deterred, or were too disturbed to respond to these sights, "the amnesia and disorientation produced . . . by the shock treatment keeps them quiet and prevents their disturbing or hurting the other patients and upsetting the ward routine."[23]

THE WORKING ASSUMPTION of many of those advocating ECT was that failure to improve was either the result of the patient suffering from a particularly malignant psychosis, or because the patient's treatment had not proceeded far enough.[24] There were no markers to indicate when the limits of treatment had been reached, and there were complaints that "the vital question remains unanswered at what number of treatment should one stop."[25] There was always the hope that additional shocks would produce the improvement psychiatrists sought. Quite early on, however, there were suggestions that patients who failed to respond to the conventional sequence of three or four shocks a week might need a more drastic approach.

In 1941 at St. James Hospital in Portsmouth, England, the superintendent, Thomas Beaton, and his assistant, Liddell Milligan, began giving as many as four shocks a day to a series of "psychoneurotic" patients, an approach that deliberately served "to reduce the patient to the infantile level, in which he is completely helpless and doubly incontinent" and to create "confusion, amnesia, and complete disorientation." Milligan did not report the results of the hundred patients they treated in this fashion till after the war.[26] In 1944, Bini tried a similar approach, one he dubbed *metodo dell'annichilimento,* or "annihilation therapy." Soon others followed, some calling it BLITZ ECT.

Clarence Neymann of Northwestern University, mostly famous for his efforts to develop artificial fever therapy, had suggested in 1945 that if a psychotic patient became "violent and unmanageable," ECT should be employed to the point of "disorientation. . . . Even beyond this, his psychic state or mentality should be reduced to a merely vegetative level. . . . No matter how great the excitement, this, of necessity, will cease under continued electric shock therapy."[27] Two years later, the New York State Psychiatric Institute, regarded as one of the foremost centers of therapeutic research in the country, began experimenting with using ECT four times a day, as Neymann had recommended, but soon found that it required too much nursing care and abandoned it.

At the same time, at the nearby Kings Park State Hospital, Cyril Kennedy and David Anchel began their own trial with intensive electroshock and gave the procedure the name most generally used thereafter, REST (for regressive electroshock therapy). Their patients were a group of twenty-five schizophrenics who had previously received "adequate" courses of treatment with insulin coma, metrazol, or conventional ECT, with little or no sign of improvement. A number of these patients had been recommended for another drastic therapy—prefrontal lobotomy—and Kennedy and Anchel, citing Milligan's paper, preferred to try a more aggressive form of electroshock. Each patient was shocked two to four times a day, "until the desired degree of regression was obtained. . . . We considered a patient sufficiently regressed when he wet and soiled, or acted and talked like a child of four." At that point, patients were thoroughly confused and unable to care for themselves, and "liable to fall and injure themselves," but "their minds seem like clean slates upon which we can write. They are usually cooperative and very suggestible, and thus amenable to psychotherapy." There were, they claimed, no lasting ill effects from the shocks, and within ten days to two weeks after treatment, in all but one case, the patients improved—though whether the improvement would last, they cautioned, "time alone will tell." What improvement meant was ambiguous, since they simultaneously conceded that, of the twenty-five patients, "15 regressed to wetting and soiling."[28]

Others soon were reporting their own experiments with regressive electroshock.[29] In 1950, at Stockton State Hospital, Mervyn Shoor and Freeman Adams, claiming that "intensive electric shock therapy . . . has become a validated procedure in acute excitement states," decided to try the approach, once again selecting 123 female patients in "the most disturbed of the chronic women's wards," selecting their patients "on the basis of their ward behavior." The goal was quite explicitly "not curative; they were limited to the level of improved ward behavior." And this they claimed to have achieved: "Within two weeks from the beginning of our intensive electric shock treatment the character of the ward changed radically from that of a chronic disturbed ward to that of a quiet chronic ward." With some patients, they were forced to give as many as a hundred shocks before they achieved their goal, though others started to conform much sooner. They were struck, they reported, by the close resemblance those "who receive large numbers of electric shock treatments daily for many weeks bear to the lobotomized patients." The treatment, they concluded, had promise.[30]

A decade later, Sylvia Cheng and Sinclair Tait of the Weston State Hospital in West Virginia reported that "electroshock therapy has been used almost universally in the control of disturbed psychotic patients."[31] With chronic patients, by the end of the 1940s, the use of the regressive form of ECT was "not uncommon."[32] Still the hoped-for cure rarely materialized. Koenig and Feldman lamented that despite "insulin, electric shock, metrazol, or combinations of these[,] great numbers of patients remain within the walls of our state hospitals permanently. . . . [W]ith the passing of time, their regression and deterioration come more and more to the surface." These patients become great behavior problems. They could be kept in isolation, or immobilized in wet packs, but then they tended to "develop decubitus ulcers and secondary infections."[33] What was to be done?

At Arkansas State Hospital, Feldman had experimented as early as 1942 with giving such patients up to four ECTs a day twice a week, something he dubbed the "block method." But "though marked improvements were achieved, these were only of short duration."[34] In December 1949, now relocated to Manhattan State Hospital, he and a colleague implemented a similar program. All the patients they chose were assaultive and destructive, and the shocks were used "purely as a palliative treatment and the main aim was to improve the condition of the disturbed wards." The thirty-one patients they reported on had all previously received insulin comas and metrazol or ECT but had remained hard to manage. All patients received an ECT at 9 a.m. and then a second shock an hour and a half later, with sandbags placed under their backs to minimize the chances of fractured vertebrae. Each received ten to fifteen double treatments, and, thereafter, "the patients quieted down to a point where restraint and sedation were done away with practically altogether."[35] Staff morale improved, the hospital realized considerable savings, and even patients' relatives commented on how much calmer the patients were.

At the Chillicothe Veterans Hospital in Ohio in 1952, thirty of the most disturbed and hopeless patients, this time men, were chosen for regressive ECT. All of the patients had received at least twenty prior ECT treatments and fifty or more insulin comas. They were now given three shocks a day five days a week, till they had received a total of sixty to seventy-two ECTs each. "No permanent ill-effects have been observed," E. S. Garrett and G. W. Mockbee informed their colleagues. Unfortunately, there had also been no positive effects so far as the psychosis was concerned: "The present psychotic pattern of each patient follows that of his preshock picture." This was "disappointing." On the other hand, twenty-seven of the thirty were better behaved on the ward, so there was "a definite place [for regressive electroshock] where all other methods of treatment have failed."[36]

WORCESTER STATE HOSPITAL in Massachusetts had long been regarded as one of the best in the country. It was where Adolf Meyer had trained his cadre of young psychiatrists, and its programs for treating schizophrenia had garnered both approval and substantial subventions from the Rockefeller Foundation, then the leading source of support for psychiatric and medical research. Here, too, were experiments with regressive ECT. In 1951, fifty-two patients (all young and in good physical condition) who had shown no lasting improvement with other physical therapies were given four shocks a day for seven days: "By the end of this intensive course of treatment practically all patients showed profound disturbances. . . . [They were] dazed, out of contact and for the most part helpless . . . prostrated and apathetic. . . . [M]ost of them whined, whimpered, and cried readily, and some were resistant and petulant in a childish way." D. Rothschild and his colleagues commented that "regressive" was an appropriate term to apply, as these patients, while they could walk only with assistance and had to be spoon fed, would happily suck on a baby's bottle. The procedure might appear "drastic" but "clinical psychiatric observations did not reveal any [brain damage]." Five of the fifty-two patients were said to be "much improved," and "the fact that this improvement may be only temporary should stimulate further research in the field of therapy for such patients."[37]

Others were eagerly experimenting along similar lines. At the Willard State Hospital in New York, J. A. Brussel and J. Schneider asked the ward attendants on the most disturbed female ward, where "the problem of the chronically disturbed patient . . . reached its zenith," to select their most troublesome fifty patients, all of whom had already received full courses of regular ECT and were currently on "maintenance shock." The problem was urgent, they claimed, because no one on the ward was improving, and "new admissions were constantly arriving."[38]

A series of intensive shocks produced remissions that lasted from six to forty-two days, and "so uniformly gratifying have been the results that it has been most difficult to restrain enthusiasm." The short duration of improved behavior might at first sight be of concern, but close observation could pave the way for another course of shocks, "enabling patients to enjoy unbroken remissions and maintain improvement." There was even the prospect that some might be released, returning every two or three weeks for a series of shocks to "maintain improvement." Not incidentally, employee morale had soared, and the "monetary savings on otherwise smashed windows, destroyed clothing, blankets and bed linen, destroyed furniture and supplies, and the savings in outlay for sedatives and restraint apparatus is an important budgetary benefit."[39]

WHEN ECT WAS INTRODUCED, there was considerable concern about whether it was safe. The research on this question was unsatisfactory and usually poorly designed. It was obvious, in the immediate aftermath of regular ECT, let alone its regressive variant, that patients' memories were severely affected. There were bland assurances in the literature that this amnesia was of limited duration. There was, however, little systematic attention to this issue, and where extended memory loss ensued, there was a propensity to blame it on the depression or schizophrenia the patient was suffering from.[40] Follow-up studies were infrequent and as carelessly conducted as the original decision to administer ECT, but concerns about the degree and extent of memory loss and anecdotal complaints from patients cast a long shadow.[41] Those dispensing ECT in the 1940s and early 1950s fiercely objected to these criticisms. Kalinowsky spoke dismissively of those who complained as the "hysterical" and "neurotic."[42] He insisted ECT was safe and that memory problems "disappear within the one or two weeks after the last treatment."[43]

Many feared that the passage of electricity through the brain might produce brain damage. Those promoting it emphatically declared that there was none, or that the changes involved were temporary.[44] Critics professed doubts, pointing to animal studies and autopsy reports that found brain hemorrhages and other signs of damage. Most but not all of the animal experiments "did not closely enough duplicate the conditions pertaining in human electroshock to be important in answering the question." The better studies of this sort did seem to document damage, but the clinical significance was unclear.[45] Thus the position psychiatrists ending up adopting on the question of brain damage depended on their existing view of the treatment, and both sides ended up drawing opposite conclusions from anecdotal evidence and from essentially the same data.

Then there was the question of broken bones. The emphasis in many quarters on repeated convulsions immediately raised questions about whether ECT shared with metrazol a tendency to cause fractured spines and femurs. Once again, Kalinowsky sought to minimize the problem, arguing that X-ray studies that showed evidence of fractures "have limited or no clinical significance at all."[46]

His efforts, and those of other enthusiasts for ECT, to devise ways to avoid or limit fractures suggest that behind the scenes they were more concerned than these public protestations would indicate. In his 1946 textbook, Kalinowsky acknowledged that "it is difficult to describe any special holding techniques because none has proven to be absolutely safe."[47] Even four or five nurses holding the patient down did not always suffice.[48] He advocated placing sandbags under the middle of the back. "The shoulders and hips are then manually applied to the table with some force."[49] The procedure "looked terrible."[50] Worse still, it was apparently ineffectual. In 1950, I. Meschan and his colleagues conducted a careful study of 212 successive cases treated at a veterans hospital in Arkansas. Despite using a specially designed hyperextension table, pre- and posttreatment X-rays showed vertebral fractures in 35.4 percent of the patients, with, on average, 2.56 vertebrae broken per patient, and they concluded that "the incidence of electric shock convulsive fractures and metrazol fractures is approximately the same."[51]

Others, less complacent, sought a better way to perform ECT. Abram Bennett experimented with using curare as a muscle relaxant as early as 1940, in an attempt to mitigate the explosive force of metrazol seizures. Since curare temporarily paralyzed respiratory muscles, its use introduced an additional source of terror unless it was combined with anesthesia, which added still another element to the cocktail.[52] The margin between a therapeutic and a fatal dose was slight and uncertain, as the potency of curare, a natural substance, varied considerably.[53] Bennett was soon campaigning to extend the use of curare to ECT, but there was fierce resistance among advocates of ECT, Kalinowsky prominent among them.

Discussing the question in his widely used textbook, Kalinowsky wrongly asserted that "fractures in ECT are rare." "Curarization," he complained, "adds to the possibility of complications which are more dangerous than those it is designed to prevent." While ECT rarely led to patient death, curare all too easily could do so. "Another disadvantage of curarization is that most patients dislike the feeling of being paralyzed," with the result that "the number of our patients who refuse to continue treatment is unduly high among those treated with curare." One of the great advantages of ECT over metrazol was that it dispensed with the need for injections and minimized the terror metrazol brought in its train. The use of curare with ECT squandered that advantage.[54]

Bennett attempted to dismiss these objections, attributing deaths among curarized patients to clumsiness or inexperience on the part of the practitioner. Only an effective muscle relaxant could adequately protect patients from fractures, and this, he claimed, was a far more serious problem than Kalinowsky and Hoch were willing to acknowledge. He cited one recent study by Lingley and Roberts that had systematically underestimated the problem by only X-raying patients who complained of severe pain, and yet had found a 23 percent fracture rate.[55] Bennett attracted some followers, especially among those practicing in private hospitals.[56] But most shied away from the risks, and state mental hospitals, grossly unstaffed and overcrowded, could in any event scarcely afford the extra expense of the drugs and an anesthesiologist.

Succinylcholine, commonly known as "sux," or by its trade name, Anectine, was a synthetic and much safer and more predictable muscle relaxant that began to be marketed in the early 1950s by Burroughs Wellcome. In May 1952, two Swedish psychiatrists suggested that it could eliminate both the problem of fractures and the brutal sight of patients thrashing about as if in their death throes. If patients were first given a dose of a barbiturate as an anesthetic and were given oxygen to support respiration, ECT could be administered without provoking a grand mal seizure.[57] In time, this modified ECT would become the standard way of administering shock, but the parlous situation of the state hospitals drastically slowed its adoption in the institutions where hundreds of thousands of patients still languished. Thus, even in the late 1950s, many psychiatrists continued to administer unmodified ECT. No one challenged their professional authority to do so, and ECT remained in their eyes a valuable means of securing patient compliance. Patients, their minds disordered, had no rights over their bodies and perforce submitted to the orders of those who controlled their fate. Though Anectine, anesthesia, and oxygen made ECT visually and physically a much less taxing procedure, not long after modified ECT became standard, shock treatment fell into disrepute.[58]

THE ARRIVAL OF A NEW CLASS of psychiatric drugs, the phenothiazines, in 1954, had dramatic effects on the practice of psychiatry. At first this did not dampen enthusiasm for ECT, which was often given alongside the new drug therapy. At the private Stony Lodge Hospital in Ossining, New York, and in the psychiatric unit at the nearby Sing Sing Prison where he was on staff, Bernard Glueck, Jr. continued to experiment with regressive ECT, lamenting in 1957 that others seemed to have abandoned a potentially powerful therapy.[59]

Not everyone had, however. At the Allan Memorial Institute in Montreal, the psychiatrist in charge, D. Ewen Cameron, remained convinced that regressive ECT had promise, not just as a means of controlling and disciplining the most disturbed of patients, but as a form of therapy. Cameron was no marginal figure. He had trained under Adolf Meyer and Eugen

Bleuler and served as president of the American Psychiatric Association, Canadian Psychiatric Association, and the Society of Biological Psychiatry and, for five years in the 1960s, as the first president of the World Psychiatric Association. Ironically, given what many regard as his criminal abuse of his patients, Cameron had also been an expert at the Nuremberg trials of Nazi war criminals.

As chair of the department of psychiatry at McGill University, Cameron had been showered with money from the Rockefeller Foundation and also (secretly) from the CIA.[60] The CIA involvement was conducted under the code name MKUltra, or Project Artichoke, and involved other prominent psychiatrists, including Jolyon West, chair of the psychiatry department at UCLA. Cameron (and the CIA) saw in regressive ECT the possibility of erasing patients' existing memories and then reprogramming them. He experimented with prolonged sleep for up to two months, massive doses of neuroleptic drugs, and what he called "psychic driving" to his program of "de-patterning." Where others had used regressive ECT primarily on chronic schizophrenics, Cameron embraced few such limits. Even patients who arrived at the Allan Memorial Institute with relatively minor psychiatric diagnoses might find themselves part of his experiments.

Though Cameron's CIA funding was hidden, what was happening to his patients was not. In 1957, at the Second World Congress of Psychiatry, he announced that his intervention was designed to produce temporary disturbance of brain function. Five of the patients in the series he was reporting on were men, and twenty-one were women; ten had been "ill" for fewer than two years. They were kept asleep for twenty to twenty-two hours a day using barbiturates and chlorpromazine and then subjected to four or five electric shocks in the space of two or three minutes. This combination, Cameron concluded, generally sufficed to de-pattern the patients "somewhere between the 30th and 60th day of sleep and after about 30 electroshocks," leaving them—allegedly temporarily—with "severe recent memory deficit, disorientation and impairment of judgment." The patient during this phase was "smiling and unconcerned . . . does not recognize anyone, had no idea where he [*sic*] is and is not troubled by that fact . . . , incontinen[t] and has difficulty performing quite simple physical skills." The upshot, by his account, was "some blunting of affect, some loss of drive" compared to that shown in earlier years, but this was probably the result of "schizophrenic damage." Most of the women, he claimed, were then able to resume their careers as housewives, though "some have had relapses . . .

[which] we are always able to terminate within two or three days and often within 24 hours by intensive electroshock therapy."[61]

In two later papers, published in 1960 and 1962, Cameron elaborated on aspects of his de-patterning therapy. He freely admitted that the treatment produced massive amnesia. Total amnesia sometimes extended to five years, while "the longest period of differential amnesia has been for ten years prior to treatment." Where patients relapsed, more de-patterning was in order, and "in rare instances, we have had to repeat this procedure several times for a given patient."[62] By now, the treatment had been extended from schizophrenics to psychoneurotics, some of whom were treated only three months after the onset of symptoms. In 1962, he reported that six ECTs in rapid succession were being given twice a day, with some patients receiving up to sixty such clusters.[63] Once again, the patients were disproportionately female: twenty-one out of the thirty reported cases. Depatterning therapy constituted, Cameron claimed, "a noteworthy advance" over insulin comas, chemical therapies, and even earlier forms of regressive electroshock.[64]

In 1964, Cameron abruptly left Montreal under mysterious circumstances. Despite his professional prominence and twenty years as head of McGill's psychiatry department, he departed for a new post in Albany, New York, without notice and without any ceremony to mark his exit.[65] Depatterning continued for some months at the Allan Memorial Institute after he left, and then, without fanfare, was abolished.[66] Three years after his departure, he suffered a heart attack while mountain climbing and died, eulogized as a giant of twentieth-century psychiatry. The *Canadian Medical Association Journal* mourned the passing of "a man who was vitally concerned with the well-being of men everywhere"—someone whose extraordinary abilities had transformed Montreal into "one of the leading psychiatric centres in the world."[67]

Years later, the CIA and the Canadian government were sued for the human damage Cameron had left in his wake: patients depressed and crippled, tortured by what had happened to them, who never recovered their memories and in some cases control over bodily functions. Many despaired, and there were a number of suicides. Several hundred patients belatedly secured a small measure of financial compensation.[68]

As with virtually all published research on ECT through the 1960s, Cameron's findings rested upon research conducted without control groups against whom to compare the treated population. Assessments of "improvement" were unsystematic and anecdotal. The body of research on ECT consisted of little more than strings of case reports and judgments about outcomes with no provisions to guard against observer bias or independent yardsticks to measure changes in patients' mental states. In many instances, follow-up data were missing altogether, or so brief as to be worthless. In the 1940s, these methodological shortcomings were characteristic of medicine as a whole. Most historians regard the trial of streptomycin in the treatment of tuberculosis, published in 1948, as the research that led to the establishment of the double-blind randomized trial as the gold standard in medical research.[69] Even in medicine, the new approach was not adopted overnight, but psychiatry's status as the poor relation of the profession allowed such unreliable research to pass muster for still longer.

THE HORRORS ASSOCIATED with Ewen Cameron's "psychic driving" were not exposed until 1975. Tales of the destruction he had inflicted on the lives of the patients he experimented on provided critics with a dramatic illustration of just how damaging ECT could be. But the attacks on the procedure had begun many years earlier and had built to a crescendo by the 1970s. Much of the criticism came from nonscientific sources. Writers as different as Ernest Hemingway and Sylvia Plath, one almost a caricature of the macho man and the other (at least after her suicide) a feminist icon, were both seen as victims of a failed treatment. Before Hemingway blew his brains out with a shotgun, he denounced his doctors at the Mayo Clinic: "What these shock doctors don't know is about writers . . . and what they do to them. . . . What is the sense of ruining my head and erasing my memory, which is my capital, and putting me out of business? It was a brilliant cure but we lost the patient."[70]

Ken Kesey's novel *One Flew over the Cuckoo's Nest* and Janet Frame's *Faces in the Water* were both based on personal experience—Kesey worked as an attendant in a mental hospital in Menlo Park, California, in 1960, and Frame was a patient who underwent hundreds of ECTs. (Frame's psychiatrists judged those treatments a failure and were about to lobotomize her when news that she had won a major literary prize led them to abort the surgery.) Both novels portrayed ECT in harsh terms, and both were subsequently made into highly influential films. Jack Nicholson's scene in the movie *One Flew over the Cuckoo's Nest* seems to have been particularly effective in demonizing ECT. But it fed into a narrative already launched by renegade psychiatrists like Thomas Szasz and R. D. Laing that portrayed psychiatrists as little more than agents of social control, and ECT as one of the tools used to secure conformity.

Symptomatic of the changed attitudes toward ECT was the passage of legislation in California (soon followed by other states) that sharply constrained the use of ECT, virtually abolishing its use except on patients who volunteered for it.[71] ECT's sharply diminished profile was reflected in the fact that the National Institute of Mental Health awarded $9.9 million in research grants on somatic therapies in psychiatry in 1972 and 1973, less than $5,000 of which was for research on ECT.[72] More subtly, ECT simply disappeared from the psychiatric journals and the academic conference circuit.

The advent of the first modern drug therapies for schizophrenia in the early 1950s had proved an unexpected bonanza for the pharmaceutical industry, and as the companies realized the massive profits that lay within reach, the volume of drug advertising in the psychiatric journals exploded. In the 1950s, these advertisements often included references to ECT. After 1965, such references simply vanished. The upshot was that "ECT disappeared from the awareness of most doctors, except insofar as they read negative and stereotyped references to it in the popular press."[73] A contemporary survey of ECT practice, completed in 1976, confirmed that "public and professional opinion of ECT was generally unaccepting, negative, and somewhat hostile."[74] It appeared to be on its deathbed.

CHAPTER NINE

Brain Surgery

THE SCENE IS AN OPERATING THEATER at the George Washington University Medical School in Washington, DC, in the mid-1940s. A number of support staff are present in the room, but there are three main actors: a masked figure, who is performing the surgery; another medical man, who sits holding the hand of the patient, talking in measured tones, asking the questions; and the patient himself, a twenty-four-year-old railroad brakeman named Frank, who has been given a local anesthetic to dull pain at the site of the initial surgical incision, and a dose of morphine to calm him, but not so strong as to prevent him from taking part in the conversation that follows. The site of the incision has been shaved, wiped down with soap and water, cleansed with ether, and marked with gentian violet. The patient's head is supported by a sandbag, and his hands and feet are strapped to the operating table. A rubber dam has been placed to direct the flow of blood that accompanies the start of the operation, and towels screen off the site of the incision.

What follows is a transcript provided by one of the principals, who made a point of recording many of the operations he took part in. The conversation takes place against the background of the usual sound effects of surgery: the rattling of instruments, the hissing of the suction used to clean the wound, the crunch of the chisel, and the grinding of the drilling machine used to cut through bone:

Doctor: Are you scared?

Frank: Yeh.

Doctor: What of?

Frank: I don't know, doctor.

Doctor: What do you want?

Frank: Not a lot. I just want friends. That's all. How long's this going on?

Doctor: Two hours.

Frank: Two hours? I can't last that long. (Squeezes hand)

Doctor: How do you feel?

Frank: I don't feel anything, but they're cutting me now.

Doctor: You wanted it?

Frank: Yes, but I didn't think you would do it awake. Oh. Gee whiz, I'm dying. Oh, doctor. Please stop. Oh, God, I'm goin' again. Oh, oh, oh. Ow. (Chisel) Oh, this is awful. Ow. (he grabs my hand and sinks his nails into it) Oh, God, I'm goin', please stop.

Doctor: Frank?

Frank: Yeh?

Doctor: What work have you done?

Frank: A little bit of everything.

Doctor: Such as what?

Frank: Brakeman on a railroad. That was a good job. Ow . . . and a material checker. . . . Ow . . . stop, unh, unh, uhn. [The doctor records that at this point the patient is scarcely controllable, even though fastened down to the operating table.] I liked that one, too. Hey, listen, cut it out for God's sake. Oh, quit, I'm goin'. What's goin' on? . . . Hey, give me some air. (The towels have slipped a bit) Hey, what's goin' on? Oh, please stop.

Doctor: Relax!

Frank: I can't relax. Oh, what's going on here? (Rongeur [a device for removing bone]) (Admits he feels no pain) Hey this is . . . oh, you know I can't go on. Oh, I'm having trouble breathing. Oh, stop experimenting.

Doctor: Stop what?

Frank: I don't know. How long's this goin' on? Fix it up. I'm having trouble breathing.

Doctor: Feel better now?

Frank: No, I'm getting worse. I'm goin'. Oh, come on, will you?

Doctor: How much is a hundred minus seven?

Frank: Ninety-three, unh, unh, ow! (Tapping) eighty-six, seventy-nine, seventy-two, sixty five (Drilling) Ow! I don't know. Give me some air. Air, Air. Ow! Hey, Cut it out. Cut it out! (Trembling hands still cold. He is quick to grab my hand when I try to take it away.)

Doctor: How do you feel?

Frank: Yes, sir. Click.

Doctor: What's it like?

Frank: Oh, a pickle puffle phi, hey, stop it, will ya?

Doctor: You're grabbing me awful tight.

Frank: Am I? I can't help it. How long does this go on?

(Right lower cuts)

Doctor: Glad you're being operated?

Frank: Yes, it makes me feel better.

Doctor: Why all the fuss?

Frank: Oh, I can't help it. I can't breathe. Hey, what are you doing there?

(Right upper cuts)

Doctor: Feel all right now?

Frank: Yeh, I can't breathe. Hey, when is this thing over?

Doctor: What will you do when you are well?

Frank: Oh, go back to work. Oh, I can't stand it.

Doctor: What job?

Frank: Oh, it's a good job, brakeman with a railroad.

Doctor: Scared?

Frank: Yeh.

Doctor: Sing God Bless America.

Frank: (He starts rather high and does a couple of lines, then grunts and continues his chatter) Ow! That's hot. What's going on here? (Warm saline) (Left lower cut) (Left upper cut) (Stabs left)

Doctor: Was that hot?

Frank: No, it wasn't hot.

Doctor: How do you feel?

Frank: Yes, yes.

(10.15 a.m. He is moving his head about during the stabs) (Stabs right) (Voice suddenly becomes muffled)

Doctor: Who's operating?

Frank: I dunno.

Doctor: Are you uncomfortable?

Frank: No.

Doctor: Why do you jerk around?

Frank: I don't know.

Doctor: Can you breathe?

Frank: Yes. (He thumps with his hands which are now quite warm and pink.)[1]

The silent surgeon is Dr. James Watts. The figure holding the patient's hand and talking him through his ordeal is Walter Freeman, and the operation is a standard lobotomy, one that would be performed by the tens of thousands in the 1940s and would win for its inventor, the Portuguese neurologist Egas Moniz, the 1949 Nobel Prize for Medicine.

Freeman and Watts initially performed this surgery under local anesthetic, though Freeman adopted a different approach after the war, using a series of electroshocks to induce a coma before inserting an ice pick through each eye socket into the brain. They used a series of stock questions to elicit conversation from those whose frontal lobes they were severing. By a process of trial and error, they had concluded that the signal to stop cutting was when the patient began to be confused. Cease before that, and the patient would most likely remain as psychotic as before. Cut further and the brain damage would be too severe. It was a rough-and-ready standard but sufficient in their eyes to term this a "precision" lobotomy. On occasion, the conversations were more macabre than the one recorded here, as when Patient 53 responded, after a long pause, to Freeman's stock question "What's going through your mind?" with what under other circumstances might be regarded as remarkable sang-froid, "A knife." (The outcome of the operations, in both cases, is unrecorded.)

Freeman was quite aware of the existential agonies his patients suffered during the surgery but the need to know when to cease severing brain tissue meant that this unfortunate side effect was unavoidable. As he explained in an article he co-published with Watts in 1950: "An operation under local anesthesia is always a somewhat trying experience to the patient. This must be doubly so when the patient knows his brain is being operated upon. . . . [A] number of patients have informed us, both before and after the operation, that they accepted the operation in the hope that it would kill them. . . . Apprehension becomes a little more marked when the holes are drilled, probably because of the actual pressure on the skull and the grinding sound that is as distressing, or more so, than the drilling of a tooth."[2]

Once the skull had been opened, the cutting of the frontal lobes proceeded quite quickly; whatever protests the patient mounted generally died away as the operation drew toward a close.

IN JULY 1935, five years or so before he sat talking Frank through his lobotomy, Freeman sailed for England to attend the Second International Neurological Congress. By a stroke of good fortune (for Freeman, if not for his future patients), he made the acquaintance there of an eminent Portuguese neurologist, Egas Moniz.

Freeman had by then established himself as one of the up-and-coming neurologists of his generation in America, helped by the fact that he was the grandson of one of the founders of the field, William W. Keen, who, along with Silas Weir Mitchell and George Morehouse, had written a classic text, *Gunshot Wounds of the Nervous System*, based on their experiences in the Civil War. Later, Keen had attempted some pioneering brain surgery, and in the opening decades of the twentieth century he was widely recognized as one of the grand old men of American medicine. His eminence was something Freeman aspired to emulate, and his grandfather's influence had helped Walter secure training in Paris in the 1920s under the eminent neurologist Pierre Marie, eventually paving the way for a post at the George Washington School of Medicine, a position he took up in 1926. His professorship at George Washington was less impressive than it sounded. Neither that medical school nor its rival Georgetown had much in the way of resources, and neither had been very successful in meeting the heightened standards for medical education that the Rockefeller Foundation had been heavily involved in promoting in the first three decades of the twentieth century. Indeed, in 1931, both medical schools briefly faced the threat of loss of accreditation and closure.[3]

Two years before he joined the faculty at George Washington, his grandfather's influence had won Walter a well-paid job at the federal mental hospital St. Elizabeths. He was seeking (like many before him) to uncover a physical cause for schizophrenia, and this position, which lasted nine years, had given him the chance to perform hundreds of autopsies. In 1934 he had parlayed his professional connections into an appointment as secretary to the newly formed American Board of Psychiatry and Neurology, a powerful position he would occupy for more than a decade. The publication of a well-received text on neuropathology further burnished his credentials, though the process of writing the book almost derailed his career.

Prone to insomnia, Freeman slept little while completing the manuscript, working at a manic pace. He collapsed on completion, suffering from what he himself called a "nervous breakdown," which was in reality a full-blown manic episode, accompanied by the delusion that he was dying of cancer. Travel to Europe brought some relief, as did taking the barbiturate Nembutal to secure regular sleep. From this point forward, Freeman was addicted to (or as he preferred to put it, dependent on) the drug to function, using a cold shower in the morning to restore some semblance of normal mental functioning if his senses were still numbed.

Freeman arrived in London eager to connect with European neurologists and psychiatrists. One of the features of these medical meetings was the setting up of booths outside the meeting rooms where the participants could present versions of their work. Moniz and Freeman found themselves next to one another and struck up a conversation. Surgery on the brain had long been hampered by the absence of imaging technology that could capture soft tissues. X-rays, which had revolutionized orthopedic surgery, were of little use to neurosurgeons, and this was a problem Moniz had tackled in the late 1920s. He experimented with the injections of strontium and lithium bromide, hoping to visually capture the arterial circulation of the brain to help locate brain tumors, hematomas, and aneurysms. The technique was a failure, and one of the three patients he injected died. But he persisted, injecting another three patients with a 25 percent solution of sodium iodide, and this time got positive results.

Freeman's own exhibit featured a modified version of this technique using thorotrast, a radioactive contrast medium that was opaque to X-rays with few apparent side effects. (Unfortunately, it lingered in the body and was later found to have powerful carcinogenic effects.) Freeman found the Portuguese "a kindly old gentleman, who had made a great discovery and could now rest on his laurels."[4]

Both men subsequently attended a full day of presentations on the frontal lobes of the brain. Wilder Penfield of McGill University, renowned for his surgical treatment of epilepsy, reported that he had operated to remove an enormous brain tumor from a patient (who was, in fact, his sister) and had been forced, in the process, to remove much of the frontal lobes. Richard Brickner, a New York neurologist, spoke about a stockbroker, Patient A, with a large frontal meningioma, who had been operated on by the neurosurgeon Walter Dandy at Johns Hopkins.[5] A's intellect had survived the operation, but his personality had been transformed. Formerly shy and introverted, he had become vivacious and boastful, lost all sense of self-restraint, threw tantrums, and grew uninterested in his appearance or personal hygiene. Brickner noted that in a casual setting, the patient could appear to be normal, but his peculiarities "rapidly became manifest" when closer attention was paid. The patient showed impaired judgment and restraint, pathetic attempts to assert superiority, emotional lability, and generally infantile behavior. His capacity for complex thinking and his initiative seemed to have disappeared.

Following Brickner, Spafford Ackerly, a neurosurgeon at the University of Louisville, reported a similar set of findings in 1935 with respect to a female patient he had operated on for a brain tumor in Louisville in 1933. Putting the best possible face on the results, he nonetheless found she exhibited slowness and perseverance postoperatively. Initially her husband said that she would not do as she was told, but he reported later that she was more docile and cooperative.[6]

These presentations reflected an emerging consensus on the importance and role of the frontal lobes among the nascent neurosurgical community. The battlefield casualties of the First World War had presented the medical establishment with a number of cases of patients with brain damage, often to the frontal lobes, and they were known to exhibit a pattern of euphoria and childlike behavior, particularly evident when the brain damage extended bilaterally. To the study of these military casualties, neurosurgeons could add reports of cases involving tumors or accidental damage to the frontal lobes. Taken together, there was now a growing and consistent set of findings about the effects of lesions of this sort.

The Chicago neurosurgeon Percival Bailey, who had served at the front in the First World War, worked at Harvard on return under the tutelage of the leading neurosurgeon of the age, Harvey Cushing. Cushing, who trained at Johns Hopkins and in Switzerland and England and spent most of his career at Harvard, was the first surgeon to have even a modicum of success operating on the brain. To be sure, his patients with malignant brain tumors generally died within a year or fifteen months, as such patients still do today. But those with benign tumors or traumatic brain injuries for the first time had some chance of survival, and the flamboyant Cushing became a famous figure.[7]

In the decade that followed Bailey's time at Harvard, he and Cushing pioneered the classification of brain tumors, and he developed more effective techniques to operate on them. Their work stood in stark contrast with earlier surgery of this sort. Lacking imaging technology to locate the tumors, and without adequate lighting in the operating theater, successful interventions were previously rare. The sessions on the frontal lobes at the 1935 Congress reflected recent advances in surgical techniques, and the emergence of a still-small but growing number of specialists willing to operate on the brain.

With far more experience than most of his fellow neurosurgeons, Bailey spoke with authority about his hesitations about "amputating a frontal lobe. This procedure is always followed by a more or less great alteration in character and defects in judgment," he warned, "when the patient is a professional businessman, who must make decisions affecting many people, the results may be disastrous." With housewives and "washerwomen" he had fewer qualms about operating, reasoning that they could manage with less need to exercise judgment and initiative.[8] Men's brains mattered more.

Seen in this context, Penfield's, Brickner's, and Ackerly's papers could be read either as a sobering commentary on the deficits that followed frontal lobe surgery or, alternatively, as documenting the ability to survive and function even following such trauma.[9] Their reports added further weight to the remarks of the distinguished French neurologist Henri Claude, whose wide-ranging survey of the field had opened the session. All available evidence, Claude pointed out, suggested that "altering the frontal lobes profoundly modifies the personality of subjects."[10]

Later in the day, John Fulton moderated a separate session on the physiology of the frontal lobes. Fulton had studied under the eminent Charles Sherrington at Oxford and then took a position at Harvard, where, like Percival Bailey, he befriended Harvey Cushing. In 1927, at thirty, he had become the youngest-ever Sterling Professor of Neurophysiology at Yale.

At the Congress, one of his young associates, Carlyle Jacobsen, reported on work on memory he had been doing with primates, most notably with two chimpanzees, Becky and Lucy. If the chimpanzees remembered the tasks they had been presented with, they were rewarded. As the experiments grew more complicated, Becky failed the tests. She grew increasingly frustrated, rolling on the floor, raging, defecating, and throwing her feces at her tormentors. Jacobsen arranged for Fulton to remove the frontal lobes of both chimpanzees, and the change in Becky was startling. It was as if, he commented, she had joined a "happiness cult" or "placed [her] burdens on the Lord." Her failures no longer concerned her. She was tamer and easier to manage, but also, as he would subsequently reveal, "severely deteriorated."[11] The full extent of her deterioration was only revealed some twelve years later, when Jacobsen and other members of the laboratory confirmed how badly damaged Becky and Lucy had been postoperatively, exhibiting "erratic" performance across a wide variety of tasks and "profound behavioral deficit." Becky had lost the ability to groom or care for herself, something that would have been a fatal defect in the wild.[12]

Fulton would later claim that it was listening to his and Jacobsen's paper that prompted the Portuguese neurologist to start operating on psychiatric patients, though Moniz strenuously denied it. More than a decade later, Fulton went so far as to claim that Moniz had approached him at the end of the talk to ask if he had thought about operating on human beings. It is difficult, after all this time, to know whom to believe, but the most thorough historian of lobotomy, Jack Pressman, was skeptical of Fulton's claims, going so far as to call them "fabrications," a conclusion my own work in the archives supports.[13]

WHATEVER THE CASE, on November 12, 1935, Moniz began to perform operations on the frontal lobes of patients at the Santa Marta Hospital in Lisbon. To be more precise, since Moniz was crippled by gout and arthritis, he had the Oxford-trained neurosurgeon Almeida Lima undertake the surgery for him. At first, Moniz had Lima drill holes into the skull and inject alcohol into the brain, but he soon decided the destructive effects of this technique were too unpredictable. Instead, he had Lima "crush white matter" and cut a half-dozen cores out of the frontal lobes with a device he dubbed a leucotome. Moniz instantly pronounced the procedures, which he referred to as leucotomies, a grand success, "simple, [and] always safe." His initial announcement, in a French medical periodical, was rapidly followed by a monograph, *Tentatives operatoires dans le traitement de certaines psychoses,* and several more papers trumpeting his discovery and asserting that "it's by adopting an organic orientation that [psychiatry] will make real progress."[14]

Remarkably, Moniz's longest postsurgical follow-up was a mere eleven days. Later, he would concede that "deteriorated patients obtain slight or no benefit from the treatment"—a concession that accords awkwardly with his claimed success in his first twenty patients, since these were allegedly chronic patients from the back wards of Portuguese mental hospitals.[15] We know that Moniz's first patient, whom he proclaimed "a great success," was an agitated depressive who spent the rest of her days confined at the Bombarda mental hospital, and she was followed by a series of patients described as suffering from agitated depression or paranoid schizophrenia.[16] Sobral Cid, the superintendent of the Bombarda mental hospital in Lisbon, who had supplied the first dozen patients Moniz experimented on, refused to supply any more (presumably having observed the results), and Moniz was forced to scramble to secure his last eight patients from other hospitals.

The extravagant claims Moniz made for his new operation rested on the slenderest of empirical foundations. Their theoretical justification was flimsier yet. Moniz asserted that in compulsive psychoses and melancholia, the mental life of patients was "constricted to a very small circle of thoughts, which master all others, recurring again and again in the sick brain." The "anatomico-pathological explanation" of the psychoses, he deduced, must be that the connections between the neurons making up the brain had become stuck, and "after two years' deliberation, I determined to sever the connecting fibers of the neurons in question."[17]

The Harvard neuropsychiatrist Stanley Cobb pronounced himself singularly unimpressed. "The reports are so meager," he wrote, "that one cannot judge of the work."[18] But Walter Freeman was of a different mind. He immediately wrote to Moniz, congratulating him on his daring, and indicated that he planned to introduce psychosurgery to the United States.

There was a small obstacle, however: Freeman was a neurologist, with no training as a surgeon. Thus, like Moniz (though for different reasons), he was forced to find a partner who could perform the operations.

Fortunately, George Washington Medical School had just hired a young neurosurgeon eager to build his practice, James Watts. Though the field of neurosurgery was relatively novel and thinly populated, Watts had an impressive résumé. He had trained under Harvey Cushing at the Peter Brigham Hospital and at Harvard, and he had subsequently worked with John Fulton at Yale (encountering Becky the chimpanzee) before taking up an appointment in Freeman's department. Watts had accepted the position even though "my salary will be negligible, a few hundred a year" and "the budget for neurology and neurosurgery will be small"—a measure of how little demand there was for neurosurgeons in the mid-1930s. His hope was that over time his university connections would help him to build up a lucrative practice.[19] He was just the partner that Freeman needed.[20]

Though the heat of a Washington summer in the years before air-conditioning delayed their first surgery, Freeman and Watts managed to perform their first lobotomy in September 1936. Their patient, sixty-three-year-old Alice Hammatt, had not been hospitalized prior to the procedure, despite her husband's report of suicidal tendencies, years of agitation and depression, and episodes where she periodically exposed herself and urinated on the floor. He (and perhaps she) had had enough of this terrible situation. "The patient," Freeman and Watts later wrote, "was a past master at bitching and really led her husband a dog's life."[21]

Guided by Freeman, Watts cut out several conical sections from her brain. When she recovered from the surgery, they noted that she was calmer and less agitated in the hospital. "In a month," they reported, "[she] was managing the essentials, although her husband and her maid did most of the work." A success, they intimated, though they proceeded to describe her as shrewish, unselfconscious, lacking in initiative, indolent, and abusive.[22] Both her husband and the patient herself pronounced themselves satisfied, though Mrs. Hammatt was prone to repeated epileptic seizures brought on by the scar tissue from the operation, and she died from pneumonia five years later.

Three weeks later, Freeman and Watts operated on a fifty-nine-year-old bookkeeper, also a victim of agitated depression, again using Moniz's leucotome. Other operations followed more rapidly, most of which they claimed were highly successful. Six weeks after they started to operate, however, Freeman sent a note to John Fulton that hints at considerable doubts: "This matter of prefrontal lobotomy has me all hot and bothered and I want a little conversation with you and Jacobsen on the subject." He indicated that he planned to be in New Haven on November 10 and asked Fulton to set aside "a couple of hours to discuss things with me."[23] A decade and a half later came an indirect admission that outcomes had been rather different from the ones they were publicly proclaiming. They buried these poor results when they published the first edition of their textbook on psychosurgery in 1942, but in the second edition, they acknowledged that "the number of failures was considerable." Indeed, "we had so many failures with the original Egas Moniz technique that we tried to obtain better results."[24] To do so, they developed an alternative technique.

JUST AS HENRY COTTON had interpreted his patients' failure to recover their wits not as evidence that his theory was mistaken but that he had been insufficiently bold in his incisions, so Freeman and Watts saw themselves as failing because they had not done enough. "It seemed to us that [an] insufficient number of fibers had been sectioned," and that more brain tissue needed to be removed.[25] Accordingly, they developed a new technique that they referred to, with no sense of irony, as a standard or "precision" lobotomy. Instead of taking out conical cores of brain tissue, they now used an instrument akin to a butter knife to make sweeping cuts in the frontal lobes bilaterally. When patients failed to recover, they often resorted to a second or even a third lobotomy before giving up the case as hopeless. These decisions were often taken within days of the first operation. As they explained, "Since we carry out our work at a general hospital, and disturbed patients require private rooms and special day and night nurses, prolonged hospitalization is out of the question. The family cannot bear the expense and the hospital will not tolerate the noise. Therefore, as soon as it becomes apparent the prefrontal lobotomy is a failure, we immediately consider a second operation."[26]

Other interventions were sometimes employed before a second operation was performed. At a session on lobotomy's "successes" put on for student physicians at George Washington University in 1949, Freeman recounted how he dealt with one of his lobotomized cases who remained difficult postoperatively. The woman was given forty electroshock treatments in the space of two and a half days. When she still hallucinated, he performed a second transorbital lobotomy, after which he reported she was "alert and cheerful and dignified." Unfortunately, two weeks after she was presented as an advertisement for psychosurgery to the medical students, she relapsed and had to be institutionalized.[27]

Of their first twenty cases, one died within days of the operation from a massive brain hemorrhage, eight were submitted to a second lobotomy, and two were subjected to a third operation, with uniformly disastrous results. Among these patients, as Freeman and Watts revealed in a brief appendix to a paper published a decade later, were the following outcomes:

- A suicide, an attempted suicide, and two other deaths within three months of the operation
- A patient rendered "tactless and disagreeable"
- Another who was left with "extreme flattening of emotional life . . . extreme indolence, petulance and puerility . . . [and] a sterile intellectual life"
- A seventh who was subject to "frequent convulsions and incontinence"
- A patient who "emerged permanently relieved of his depression, but with a boisterous, arrogant and extravagant nature that required institutionalization"
- A woman who became "fat, jolly and outspoken"
- Another characterized as "indolent and sarcastic and was subject to outbursts of anger, which made it necessary to confine her in an institution for eighteen months"
- A woman who, postsurgery, was "greatly deteriorated, fat and inaccessible" (Freeman and Watts lamented that "her family refused permission for further operation")
- And another female patient who was "indolent and talkative in a silly, vapid way"

More positive results were reported in a handful of cases:

- Their second case, a bookkeeper, was reported to have returned to work within three months of the operation and had worked for eight years before retiring and living comfortably at home

- A cement finisher was "euphoric but soon relapsed" after surgery, but some years later it was reported that he had "been steadily employed as a janitor at a school, where he is highly thought of."
- A housewife who initially "made an erratic adjustment and was in and out of hospitals for four years" after which she divorced and remarried "and writes enthusiastically of her new life."
- Finally, a bookbinder who had been a lifetime hypochondriac "has been employed for the past seven years at her old job. She still complains when asked about her symptoms but never mentions them otherwise."

Though Freeman and Watts passed over the matter in silence, seventeen of these twenty patients were women, eleven of them housewives.[28] The overrepresentation of women was rarely so extreme, but virtually all the later reports on lobotomies noted that female patients were greatly in the majority.[29] Morton Kramer of the National Institute of Health's Biometry Branch found that 12,296 lobotomies were conducted in the United States in the eighteen months ending on June 30, 1951. A majority of the patients in the wards were male, yet nearly 60 percent of the psychosurgery had been performed on women.[30] A study conducted by Charles Limburg in 1949 had found an even larger gender discrepancy: women were lobotomized twice as often as males.[31]

The predominance of female patients was also notable in the surgical treatment for focal sepsis, and in the administration of electroshock therapy right down to the present. Why were female bowels and brains uniquely attractive to those bent on discovering a physical remedy for mental disorders? Gender biases of this sort can be traced back to the vogue for clitoridectomies and surgical excision of the ovaries as a remedy for psychosis, reflective of male physicians' attitudes to female bodies that were widely embraced in the nineteenth century, with roots in Western medicine extending back to Hippocrates and Galen.[32] The persistence of these gendered prejudices in what was still a heavily male-dominated profession unquestionably played a major role. And the existence within the larger culture of assumptions about women's roles and capacities fed into this narrative in some obvious ways. Women's minds, so medical men continued to believe, were much more closely linked to their peculiar biological natures.

Freeman quite openly spoke of how these concerns fed into the decisions he and Watts made about whom to operate on: "We have based our work more on social than on the psychiatric findings," he explained, and, in deciding on whether to perform a lobotomy, "we have performed operations when the patient faced prolonged or permanent disability and where the type of occupation did not require much constructive imagination."[33] Being a housewife (or, in the case of richer women, a decorative fixture in a gilded domestic environment) required in the eyes of their doctors much less intelligence than was required of most men, and readjustment to these roles was more readily accomplished than returning to the workaday world.

There are other possible explanations for the greater use of these physical therapies on women. Perhaps psychotic women on the wards acted out more frequently than their male counterparts, or perhaps (and this strikes me as more likely) disobedient and violent female patients were simply perceived as more deviant and problematic. Either way, these women were perceived as more eligible candidates for surgical solutions. At Pilgrim State Hospital on Long Island, the largest mental hospital in the country, the neurosurgeon H. S. Barahal laid out the basic calculus that many followed: "One of the criteria for surgery on chronically ill patients has been disturbed behavior; and female patients are generally more disturbed on a behavioral level."[34]

Joel Braslow has examined the records of all 241 lobotomies performed at the Stockton State Hospital in California to see whether the marked discrepancy along gender lines (85 percent of the psychosurgeries performed there were on women) can be explained by other factors. Was it that women were disproportionately clustered in particular diagnostic categories thought to warrant lobotomies? No. Nor did they outnumber men on the wards during the period he examined. To the contrary, male patients were consistently in the majority. "The inescapable conclusion," Braslow writes, "is that Stockton physicians, to a highly significant degree, preferred to lobotomize women." In this decision they were at one with their counterparts elsewhere. Braslow also found a marked propensity to discipline women more severely. "Men," he discovered, "were nearly *thirty times* less likely than women to be bound in camisoles, straitjackets, belts and cuffs or mittens, or lashed to chairs or beds."[35]

The great majority of those whom Freeman and Watts operated on in the early months and years of psychosurgery were ambulatory patients whose mental illnesses had not been so severe as to require hospitalization.

Only a dozen of the eighty patients they had operated on by the time they published the first edition of *Psychosurgery* in 1942 had been diagnosed as schizophrenic.[36] The assertion that lobotomy was a treatment of "last resort," as some have claimed, was frequently a fiction.[37] Moniz had found that deteriorated patients were generally left worse off after psychosurgery, and some American psychiatrists urged that lobotomies be performed on much less disturbed patients. As early as 1942, the prominent British psychiatrist T. P. Rees claimed that "it is unfair both to the patient and the surgeon to defer the operation. . . . [N]o-one should be in a mental hospital for longer than twelve months without the possibility of his being relieved by prefrontal leucotomy being considered."[38] Toward the end of the decade, more and more psychiatrists began to act on such views.

At the outset, seeking patients to experiment on, Freeman had approached William Alanson White, superintendent of the enormous St. Elizabeths Hospital in Washington, DC, asking if he could operate on some of its inmates. White dismissed the suggestion out of hand, saying, as Freeman later recalled, "It will be a hell of a long while before I let you operate on any of my patients."[39] White wrote about the encounter to his friend Smith Ely Jelliffe, with whom he edited the *Journal of Nervous and Mental Disease,* "I am asked to subject my patients to this operation as a legitimate experiment in therapy. I do not very often trouble you with the various propositions that are handed up to me, but here is one which I would like to have you tell me what you think of in as few words as possible. I could express the whole matter in one word, but I do not want to do that because it would be unmailable."[40]

Undeterred by this forcible rejection and by reports of postoperative complications and deterioration, Freeman and Watts relied on outpatient referrals. By November 1936, they had accumulated enough experience to present their results to their colleagues. Freeman, more comfortable in front of an audience, took the lead. On the whole, it did not go well.

IN BALTIMORE, MARYLAND, before a meeting of the Southern Medical Society, Freeman's presentation was met with astonishment and anger. Dexter Bullard, the proprietor of Chestnut Lodge, a private asylum for mental patients run along psychoanalytic lines, tried to shout him down. He was scarcely alone. The *Baltimore Sun* reported that Freeman's presentation was greeted with "cries of alarm" and for much of the discussion period "one man after another . . . joined in the chorus of hostile cross-examiners." Benjamin Wortis, a psychiatrist on the faculty at New York University, pointedly demanded to know "where Dr. Freeman obtained his evidence that obsessions are located three or four centimeters deep in the frontal lobes." Dismissively alluding to Cotton's surgical interventions on the digestive tract, he added, "I can only hope this report will not start an epidemic of progressive evisceration experiments."[41]

The official publication of the paper itself in the *Southern Medical Journal* was accompanied by a heavily sanitized version of the discussion, giving no hint of the disorder following its delivery. Mercifully for Freeman, Adolf Meyer was in the audience, and just as things threatened to get out of hand, the Johns Hopkins professor intervened. "I am not antagonistic to this work," he informed his assembled colleagues, "but find it very interesting." Though it was important not to promise miracles, he was convinced that experiments along these lines should continue. Certainly, he concluded, "the available facts are sufficient to justify the new procedures in the hands of responsible persons."[42]

Three decades later, in an unpublished autobiography written for his children, Freeman vividly recalled that "the reaction of most of the discussors was . . . unfavorable." His gratitude for the intervention that saved the day was palpable: "Adolf Meyer, bless him, wrote out a judicious statement indicating that this method had possibilities, that it was based on some of the information we were gaining about the frontal lobes, and he adjured us 'to follow each case'"—a classic Meyerian formulation—"with a view to determining the eventual results. Had it not been for his sympathetic and helpful discussion, the advance of lobotomy would probably have been much slower than it was."[43] Absent Meyer's intervention, the whole program might have come to a halt.

Now past seventy, and clinging to his post at Hopkins long after the official retirement age, Meyer continued to lend his considerable authority to the support of lobotomy until he stepped down in 1941. He provided private encouragement to Freeman and met with him on a number of occasions to examine the brains of lobotomized patients who had died during or shortly after an operation.[44]

Still, Freeman continued to encounter sharp criticism from both psychiatrists and neurologists whenever he ventured to talk to them. In February 1937, he traveled to Chicago, where he reported on twenty lobotomy cases to the Chicago Neurological Society. In the discussion period, his methodology was roundly critiqued. Why, one participant asked, had

Freeman not used a control group, opening up the skulls of a second group of patients without proceeding to damage their brains? Harry Paskin went further, raising the question of "why the operation had been done on the brain?" After all, it was well known that psychiatric patients had been found to show temporary signs of improvement in the aftermath of "trauma, intercurrent disease or surgical operation, only to relapse later." Lewis Pollock of Northwestern University challenged the "anatomic basis" of lobotomy and the locationalist model of the brain it rested on, and he called lobotomy "not an operation but a mutilation." He was outdone by the outspoken neurosurgeon Loyal Davis, who had trained under Harvey Cushing. "I had hoped that the reports were grossly exaggerated," he began. Instead, Freeman had openly confessed to the wanton destruction of brain tissue. "The offhand manner in which this surgical procedure is described and discussed is no credit to the essayist as a surgeon, a pathologist, or one who is searching for scientific truth." Freeman was unabashed. The brain, he insisted, "can stand a good deal of manhandling" and "patients with these psychoses are in a serious condition and do not have much chance of recovery otherwise."[45] Later, he confessed that in the midst of the exchange, "I nearly bit the stem of my pipe off trying to regain control of myself."[46]

Nor did the clamor show signs of dying down.[47] Beginning in 1935, Cobb had begun to write an influential annual review of developments in neuropsychiatry for the *Archives of Internal Medicine*. Though he had referred a handful of patients at the Massachusetts General Hospital for lobotomies, his doubts about the procedure led him in 1940 to attack its scientific basis, arguing that far too little was known about the frontal lobes, and the results of similar procedures performed on animals had been quite ominous, producing creatures "who no longer showed any restraint or resourcefulness. . . . In my opinion," he said, "[lobotomy] is a justifiable procedure only when the patient is old and the prognosis hopeless." As for the claims Freeman and others had made for its positive value, one could place no trust in them. Not one, he wrote, was "convincing as scientific evidence."[48]

The lead editorial in the May 1940 issue of the *Medical Record* was harsher still. It denounced "the lobotomy delusion" and attacked "the mutilating surgeons who performed the operation."[49] Four years after Freeman and Watts had performed their first lobotomy, they must have wondered what its fate (and theirs) would be. Perhaps the decision to make a direct assault on the frontal lobes of the brain had been a step too far?

CHAPTER TEN

Selling Psychosurgery

IF THE MEDICAL COMMUNITY'S INITIAL REACTION to lobotomy was often harsh and condemnatory, it was quite otherwise in the popular press. Walter Freeman was a master at manipulating the media, and he left little to chance.[1] In the days before his Baltimore announcement, he had invited Thomas Henry, the science reporter from the *Washington Evening Star,* to witness an operation, and fed him an exclusive. Henry obliged by dubbing lobotomy "one of the greatest surgical innovations of this generation." He also went on to say that the two local heroes had established "that there is an actual, tangible, physical basis in the brain for various mechanisms of both normal and abnormal personality which can be attacked with the surgeon's knife as easily as can an inflamed appendix or diseased tonsils."[2] The *New York Times,* only marginally less hyperbolic, quoted a neurosurgeon, Spafford Ackerly of Kentucky, who called lobotomy a "shining example of therapeutic courage."[3]

Six months later, Freeman managed to secure even more prominent favorable coverage in the *New York Times.* He had set up his booth at the American Medical Association convention a day early to tout the merits of lobotomy, bringing a lobotomized monkey with him to attract attention. The monkey promptly died, but he buttonholed the right reporter anyway. The next day, William Laurence posted a breathless piece on "surgery of the soul." Freeman and James Watts, the paper's readers learned, had "changed the apprehensive, anxious and hostile creatures of the jungle into creatures as gentle as the organ grinder's monkey." Their new surgery "cuts away the sick parts of the human personality" and restores the mad to sanity.[4]

Over the next five years, Freeman succeeded in getting a succession of newspapers to echo these claims. In 1941, the *Houston Post* informed its readership that a lobotomy was "a personality rejuvenator" that severed "the 'worry nerves' of the brain," and that "Dr. Watts and other brain surgeons consider it only a little more dangerous than an operation to remove an infected tooth."[5] Waldemar Kaempffert, one of the most influential science writers of the age, wrote enthusiastically of the operation's miraculous qualities in the *Saturday Evening Post,* which included photographs of Freeman and Watts in the operating theater.[6] Recycled in the regional press, such coverage worked wonders in attracting potential patients.[7]

One of the strictest "ethical" rules physicians of the age were enjoined to obey, on pain of losing their license, was a prohibition on any form of advertising, an injunction rooted in the desire to substantiate the claim that medicine was a profession, not a trade. The ban served, not coincidentally, to keep competition in check. Freeman and Watts were on the brink of publishing *Psychosurgery,* a book they had been forced to subsidize to the tune of over $3,000, a substantial sum for that period, and these newspaper and magazine stories, naming both of them, were perceived by many as a form of promotional advertising for their services. Angry doctors brought pressure on Charles C. Thomas, the publisher, to abandon the project, and more strident voices sought to suspend their medical licenses. Only frantic behind-the-scenes lobbying by John Fulton staved off the effort, and eventually the storm subsided.[8] Freeman's showmanship, which had done so much to sustain the practice of lobotomy, had nearly brought both men to professional ruin.

EVEN BEFORE HIS CRUCIAL INTERVENTION, John Fulton had played a vital role behind the scenes in vouching for the validity and value of lobotomy, asserting that it had a scientific basis in neurophysiology and persuading other neurosurgeons to use the procedure. With the prestige that had accrued from his training at Oxford under the future Nobel Prize winner Charles Sherrington, his subsequent work at Harvard under Harvey Cushing, followed by his appointment to an endowed chair at Yale, where he gathered a large and highly talented team to work at his direction, Fulton was regarded as the preeminent American neurophysiologist of his era. He had secured substantial funding from the Rockefeller Foundation almost from the moment he arrived at Yale, where he established the first primate laboratory and became a powerful academic figure. Many of those he contacted about lobotomy were his former students. Unquestionably, his endorsement of Watts and Freeman's work went a long way to allaying doubts among the still-small circle of neurosurgeons in the United States.[9]

In November 1936, accompanied by his close friend Henry Victs, one of the editors of the *New England Journal of Medicine,* he had visited Washington and examined some of the first handful of patients to have been lobotomized. One previously "impossible" patient now seemed calm and rational. "I felt somehow," he wrote in his diary, "that we were in the presence of one of the milestones of modern medicine."[10] He promptly set about trying to persuade other neurosurgeons of the merits of the operation, sending around his own copy of Egas Moniz's monograph, reminding them of the disappearance of "experimental neurosis" in the monkeys and chimpanzees on whom he and Jacobsen had performed frontal lobectomies, and reiterating his belief that "when the last word has been said about catatonic dementias and schizophrenia, we will find that the frontal association areas are the parts of the brain principally involved, and when they get tied up in knots, I can see no reason for not surgically untying them."[11]

In many respects, Fulton's energetic work behind the scenes proved even more important than Adolf Meyer's initial endorsement.[12] In the face of strong critiques from psychiatrists and neurologists, it provided valuable support for the claim that lobotomies were rooted in the best laboratory science. It was Fulton who helped to ensure that the operation began to seem more than the idée fixe of two or three professional mavericks.

HAD LOBOTOMY REDUCED all who were operated on to the status of human vegetables, its early critics would swiftly have carried the day. Neurosurgery was barely two decades old as a serious enterprise, and its practitioners hardly wanted to be perceived as careless mutilators. But Freeman's cavalier assertion that brains could withstand a good deal of manhandling had some small kernel of truth. Despite the devastating effects of his procedure, damaging the frontal lobes was seldom fatal (death rates for most surgeons ranged from 3 to 10 percent), and unless the cuts were unusually extensive or the incisions made too far back, patients retained some degree of cognitive competence.[13] Crude measures like IQ tests appeared to reveal little diminution of mental capacity.[14]

When brains are damaged by strokes, new neural pathways are built, bypassing the damaged areas of the brain and restoring some degree of function. Something similar may have occurred in some lobotomy cases. Despite Freeman's talk of "precision," the damage inflicted in each case varied enormously. Some saw this wide variation in the location of the damage as "an unqualified boon" since it unintentionally provided great experimental data on the functions of different areas of the brain.[15] Postoperative hemorrhaging often led to far more extensive extirpation of brain tissue than had been inflicted by the surgeon's knife, and in any event the massive complexity of the brain and its enormous variability meant that each operation was quite literally a stab in the dark.[16] The scar tissue left behind (the source of the seizures that plagued a large fraction of the lobotomized postsurgically) created still another layer of uncertainty. Contemporary accounts show that patients' postoperative mental states could vary widely. That variation could be read as improvement. This was particularly the case if the upshot of the surgery was that patients could be managed successfully at home and spared the horrors of a mental hospital—while seeming to be less consumed by obsessions and compulsions, or delusions and hallucinations.

Evaluating what is to count as improvement in cases of serious mental illness, and how to weigh costs and benefits, remains a fraught issue to this day. The failure of the few standardized tests that were administered to show much evidence of postoperative defect was testimony to the shortcomings of those tests, as contemporaries sometimes recognized.[17] Beyond this, it was rare for pre- and postoperative tests to be undertaken or reported. In the absence of any systematic way of evaluating patients, the neurosurgeons fell back on clinical impressions, easily biased by what they wanted to see. Follow-ups were often of surprisingly short duration. As one experimental psychologist complained at the 1946 meeting of the American Psychiatric Association, "Not a single patient had been adequately studied. For the moral and social responsibility to do this, there has been substituted a phenomenal array of case statistics. Unfortunately, the pyramiding of unknowns is scarcely the pathway to knowledge. . . . In no instance has the psychological test or battery of tests employed ever been shown to be sensitive to frontal lobe function."[18]

Postoperative shock and sedation added to the blunting effect of frontal lobe damage to persuade many neurosurgeons that they were doing therapeutic work. Many who traveled to Washington, DC, to observe Freeman and Watts's patients during and immediately after the surgery left them convinced that this was an intervention they should try. Over the next decade, university-based lobotomy programs at Harvard, Yale, Columbia, and the University of Pennsylvania were set in motion. Not for the first time in the history of psychiatry, statistics of the most dubious provenance were endowed with spurious facticity.

BEGINNING IN THE EARLY 1940S, and at an accelerating pace after the end of the Second World War, lobotomy moved from an experiment viewed in many quarters with alarm to an established part of psychiatry's therapeutic armamentarium, and not just in the United States. Two English doctors brought to America in the late 1930s under the auspices of the Rockefeller Foundation to watch Freeman and Watts at work, William Sargant and Wylie McKissock, returned to Britain just before the Second World War broke out as enthusiastic converts to psychosurgery's merits. By 1946, McKissock had performed 500 lobotomies, a total that temporarily exceeded that of Freeman and Watts.

The publication of Freeman and Watts's *Psychosurgery* in 1942 did much to legitimate and spread the gospel of psychosurgery. Though no formal statistics were kept, a handful of psychiatrists in state hospitals had begun to extend the operation to the involuntarily confined. Freeman was delighted to report that J. S. Walen had managed to perform 200 lobotomies at the Wyoming State Hospital in Evanston, working from a set of written instructions he had supplied. Meanwhile, at the ominously named State Hospital Number 4 in Missouri, Paul Schrader had "all but solved the problems of the disturbed wards at that particular hospital" by performing a lengthy series of lobotomies.[19]

Edward Strecker, a psychiatrist on the faculty of the University of Pennsylvania and clinical director of the psychiatric division of the Pennsylvania Hospital (the oldest private hospital in the United States, and the first to accept insane inmates), was among the first to extend lobotomy to the treatment of chronic schizophrenics. The first of Strecker's cases had received ninety-one insulin comas and eighteen metrazol convulsions before being lobotomized.[20] The selection criteria for his interventions were clear: these were patients who were "well-nigh intolerable and quite disruptive of hospital morale."[21] Four of the five were women. Strecker's prominence and prestige eased the way for others to follow in his footsteps.[22] The precarious position lobotomy had occupied in 1940 had been resolved. The operation was no longer dependent on the enthusiasms of a single Washington team.

Many neurosurgeons who were tempted to try lobotomies were disturbed even about Freeman and Watts's original operative technique, let alone Freeman's transorbital approach. The fact that the surgery was done "blind," without the operator being able to see the portions of the brain that were being severed, troubled them. As early as the spring of 1937, J. G. Lyerly, a neurosurgeon based in Jacksonville, Florida, had developed a new procedure that allowed him to view the brain directly as he cut into it, and eighteen months later, he reported on twenty-one operations he had performed in this fashion.[23]

A decade later, a number of alternative approaches had been developed. They involved more extensive opening of the skull to allow direct visual observation when cutting, and also experimentation with damaging different regions of the frontal lobes, ranging from J. L. Poppen's modification of Lyerly's approach in Boston, through Wilder Penfield's experiments in Montreal with an operation he called a "gyrectomy," to the "orbital undercutting" preferred by William Scoville of Yale and the "topectomy" introduced at Columbia by Lawrence Pool. It was hoped that attacking different sections of the brain would provide the claimed salutary effects of the operation while reducing its drawbacks. In many cases the alternatives were abandoned because they produced even more severe side effects, such as the greatly increased incidence of seizures cited by Penfield and his associates in Montreal following their use of gyrectomies (smaller, more targeted incisions). Not only was this a "difficult, long, and somewhat dangerous procedure," but its efficacy was so poor that four of the seven people subjected to it were subsequently given a standard lobotomy.[24]

Freeman's sardonic comment on these exercises was savage and not far from the mark: "We can agree that the surgeon sees what he cuts but are inclined to think that he does not know what he sees."[25] But the search for alternatives provided another justification for continuing psychosurgery. If the operation failed, the neurosurgeon could always sustain the hope that cutting more deeply, or in a different portion of the brain, might produce more acceptable results.

FREEMAN AND WATTS had begun operating on patients with intractable pain as early as 1943.[26] Somewhat later, they lobotomized patients prone to hyperventilation, claiming that this was "a radical yet rational method of therapy [since it] does away with the anxiety that leads to hyperventilation." Besides, they suggested, such cases were often "larval schizophrenics who were saved from a complete break with reality by operation while the somatic symptoms were still in the foreground."[27] Lothar Kalinowsky, who pioneered the use of electroconvulsive therapy (ECT) following his arrival from Italy, was equally enthusiastic about the possibilities for lobotomy. Together with his colleague Paul Hoch, he recommended psychosurgery for "severe and intractable cases of psychoneurosis . . . if they have failed under psychotherapy."[28] As for hospitalized patients, lobotomy "should become a routine procedure . . . even if the diseases cannot be cured . . . as the operation frequently means the difference between suffering and danger for the patient and his environment and a pleasant and useful life though perhaps on a lower level."[29]

Lobotomy's enthusiasts increasingly argued that rather than waiting and treating only "hopelessly deteriorated cases," intervention should begin earlier, "as soon as we can be sure that further shock treatments will be of no avail" and "at a time when the disorder has not fixed itself too profoundly." Given improvements in operative technique, "we feel it is time to make an effort to re-evaluate the indications for surgery." Neurotics were often treated with an extensive course of shock therapy, but one could legitimately question whether patients should "be subjected to this procedure when surgical results have been so promising." Certainly, "early psychosurgery is fully justified even before a year of continuous illness has elapsed."[30]

"As lobotomy has spread to more and more centers," the Harvard psychiatrists Milton Greenblatt and Harry Solomon reported in the pages of the *New England Journal of Medicine*, "the general conviction is strengthened that this method is capable of modifying the disturbed behavior of psychotic patients. Small as well as large series carry the same message. Indeed, authors who express overall disappointment are few and far between."[31] When the National Institute of Mental Health assembled psychiatrists in 1954 to develop guidelines as to when lobotomy was justified, the upshot was a resolution that "the conference feels that psychosurgery should not be restricted to committed incompetents."[32] Chances of a relatively successful outcome, all were agreed, were considerably increased in milder and more recently diagnosed cases of psychosis and neurosis.

Psychosurgery was now respectable and widely endorsed. Accordingly, the number of lobotomies performed in the United States increased steadily in the years after the war. Statistics were never compiled systematically, and official numbers almost certainly underestimate the tally, but, still, the trajectory is clear: fewer than 300 operations in 1945; more than double that the following year; 1,200 or more in 1947; between 2,600 and 3,000 in 1948; and more than 5,000 in 1949—at this point the numbers then began to plateau. Veterans hospitals ministering to psychiatric casualties of the war, university medical schools, and psychiatric research centers all participated enthusiastically in the psychosurgical revolution, as did state hospitals, some two-thirds of which had active lobotomy programs. In 1947 Milton Greenblatt, a Harvard psychiatrist and director of research at the Boston Psychopathic Hospital, endorsed lobotomy in the pages of the *American Journal of Psychiatry* as "a good method, perhaps the foremost method, for the treatment of the chronically mentally ill."[33]

HENRY COTTON'S WELL-PUBLICIZED FIGHT against focal infection had brought him private patients from all across the country, and his colleagues complained that they were inundated with requests from affluent families to try surgical remedies on their mad relations.[34] Later interventions—malaria fever therapy, deep sleep, the various shock therapies, and lobotomies—were likewise enthusiastically embraced by the well-to-do and by the institutions that serviced their disturbed relations. At the Hartford Retreat, Charles Burlingame moved rapidly to adopt these novel forms of treatment and became one of their most forceful advocates.

Burlingame's appointment as the retreat's superintendent in 1931 coincided with a change of name to the Neuro-Psychiatric Institute of the Hartford Retreat in an attempt to emphasize its claims to provide the most modern therapeutic approaches. By the 1940s, he had succeeded in attracting more than twice as many inmates—or "guests" as he insisted on calling them—to facilities that superficially resembled a country club: indoor and outdoor swimming pools, tennis courts, fashion shows, music lessons, sculpture and art studios, a street of boutiques constructed in mock-Tudor half-timbered style: Ye Royale Booke, Ye Silver Smithy, Ye Bindery, and so forth. Charming cottages were scattered over the elegant grounds, and at day's end, the patients (such as could sufficiently control themselves) gathered for formal dinners. To further the disguise, in 1943, the trustees once more changed its name to the Institute of Living, a euphemistic exercise somewhat undercut by its postal address: Asylum Avenue, Hartford.

The institute's guests could expect a rapid acquaintance with the full panoply of modern psychiatric therapeutics. Burlingame's announced goal was "the complete intellectual unification of psychiatry with all other branches of medicine."[35] Traditional forms of hydrotherapy (confining patients for hours or days in tubs of cold or warm water, bodies confined by a stout canvas cover, with only the head sticking out) were combined with pyrotherapy (fevers first induced first with malaria, and later by placing patients in a copper, coffin-shaped device where infrared lamps broke down the body's homeostatic regulatory system and induced temperatures as high as 105 or 106 degrees Fahrenheit). Burlingame swiftly added insulin comas, metrazol-induced seizures, and electroshock treatment to the mix, and when his "guests" exhibited fear of these therapies, he proposed to address the issue with another resort to word magic. If an asylum could be relabeled an institute for living, the shock treatments could be renamed insulin sleep therapy, metrazol sleep therapy, and electric sleep therapy.

Burlingame was equally enthusiastic about psychosurgery. The first lobotomies at his hospital were performed at Hartford as early as 1939. In 1941, he invited the young neurosurgeon William Scoville of Yale to offer the treatment to his guests on a more systematic basis—though his guests, who included Scoville's wife, generally had little say in the matter.[36] Burlingame's commitment to the operation persuaded the hospital's trustees to construct a suite of operating rooms on the northern portion of the grounds, devoted solely to psychosurgery.[37] Burlingame was a major proponent of the Connecticut Co-operative Lobotomy project, which drew the nearby state hospital at Middletown into the arena.[38]

In one respect, the regime at Hartford differed sharply from the routine at other resorts for rich mental patients. Increasing numbers of these establishments were proclaiming an interest in psychological approaches to the treatment of mental illness. Burlingame was content to see those seeking such treatment go elsewhere. In a widely noticed speech aimed at his medical colleagues and published in the *Journal of the American Medical Association,* he addressed the issue of "psychiatric sense and nonsense." Psychoanalysis, with its emphasis on sex and primal fantasies, was, he asserted, just "bunk" and not a medical procedure besides. To be sure, patients could benefit from "personal tutoring," such as his own hospital provided, but the talk of sex and unconscious primal drives was so much metaphysical nonsense. Instead, as he announced in his annual report the following year, psychiatry ought rather "to consolidate our professional interests with the other branches of medicine."[39]

In the years leading up to the Second World War, Burlingame's contention that psychiatry's future, "beyond any shadow of a doubt, must be cast with medicine" would have been completely uncontroversial. The war, however, had attracted large numbers of new recruits to the profession, who were men (and it was almost exclusively men) who had learned psychiatry outside the walls of the traditional mental hospitals, treating the mental casualties of total war. Soldiers breaking down under the stress of combat were an entirely different class of patient, and their troubles seemed more rooted in the chaos and frightfulness of battle than in the defective heredity to which prewar psychiatry attributed mental breakdowns. Military psychiatrists had depended largely on some sort of psychotherapy to patch soldiers up. The shape and the orientation of the profession was shifting, and shifting more quickly and suddenly, as we shall see, than Burlingame and his allies among traditional state hospital psychiatrists had yet fully grasped.

Their complacency was understandable. State hospital psychiatrists still dominated the profession and treated the great majority of the mentally ill. Even those better-paid and more prominent psychiatrists who worked in private asylums shared with their public-sector colleagues both similar experiences in managing institutionalized mental patients and a community of interests and ideas. Some of the heads of the more traditional private asylums had shown more disposition than Burlingame to provide space for psychodynamic ideas and approaches, but the somatic therapies employed so vigorously at Hartford usually found a place in their establishments as well. Desperate as their patients' families were for some purchase on psychosis, they felt they could ill afford to discard any procedure that might offer some glimmer of hope.

Wealthy families who read press accounts of the new treatment's successes eagerly sought them for their relatives. Strecker at the Institute of Pennsylvania Hospital had been an early convert to the virtues of lobotomy in making chronic schizophrenics more manageable. The McLean Hospital showed similar enthusiasm, performing its first lobotomy in March 1938 and a second eleven months later. Kenneth Tillotson, the clinical director at the McLean, acknowledged that his decision to begin using the shock therapies in 1937 reflected "the demand . . . by professional colleagues as well as by the relatives of patients."[40] Manfred Sakel personally oversaw the introduction of insulin coma therapy at the hospital. Both mental hospitals adopted an eclectic approach that combined the various physical therapies with psychotherapeutics. Strecker, indeed, was one of a number of psychiatrists who attempted to provide a psychoanalytic justification and explanation for lobotomy.

If the Hartford Institute was a country club, the McLean was more luxurious still. Here, as the trustees boasted, "a class of educated people of means" could enjoy life on a sylvan campus, as "a large family living in different houses."[41] Staff were instructed to treat their patients with the utmost deference, and physicians had to interact with patients as equals. Rank and fortune had their privileges. But only to a point, for less disturbed people of wealth opted for sanitariums or homes for the nervous; McLean was for those whose mental illness was so severe as to prompt their removal for years or even decades at a time. Prominent families even took to building individual "cottages" on McLean's extensive grounds. The first of these, built in 1898, set a pattern for those that followed: a large five-bedroom house (with the smaller bedrooms on the top floor reserved for the fleet of servants who waited on the single patient who occupied the premises). Subsequently, families desiring the comforts of home for their loved ones built similar cottages and deeded them over to the hospital in return for lifetime care of their disturbed relation.[42]

On the back wards, beginning in 1945, lobotomies were performed at an increasing rate, particularly on patients "whose behavior and aggressive psychotic symptoms rendered them incapable and unresponsive to all forms of treatment and constructive nursing and medical care."[43] In singling out the most difficult long-stay patients (of whom there were a great many), the McLean was clearly following the path pioneered by Strecker in Philadelphia. Tillotson, gratified by the greater docility of those who had been operated upon, soon began to ask, "Why should not prefrontal lobotomy be done early?"[44] The numbers of lobotomies performed increased markedly as patients were operated on much sooner after they had been admitted. Women always constituted the overwhelming majority of those operated on. In 1948, for example, all twelve of those who were lobotomized were women.

The outcomes of these operations were highly variable and hard to predict. Some patients seem to have improved considerably by most measures, and a number of them were eventually discharged—an outcome doubtless greatly assisted by the resources their families could mobilize in their support. It was these positive outcomes that the staff emphasized in the late 1940s, and that helped sustain the lobotomy program. The reality, however, is that these positive outcomes were, as always, the exception, not the rule. Many, male and female alike, were left worse, sometimes catastrophically so. Of one woman, the nursing notes indicate that she had to be "cared for as a small child would be, except that she is uncooperative, often resistant." Apparently, "the only thing she does well is eat."[45] Some patients became tamer and more easily managed, a situation the hospital gratefully accepted as improvement.

DURING THE SECOND HALF OF THE 1940S, lobotomy began to be widely endorsed by American psychiatrists. Still, Freeman felt frustrated. Despite his best efforts, and despite the increased legitimacy lobotomy had acquired, the numbers of operations performed each year numbered only in the low hundreds. He was convinced that a large fraction of a mental hospital population rapidly approaching half a million were potential beneficiaries of psychosurgery, but the number of neurosurgeons willing and capable of performing the operation remained tiny. The operation required the presence of several medical personnel, was very costly, and slow—each surgery often taking up to two hours to perform. The recovery period also required the type of skilled nursing care that was simply unavailable in massively overcrowded state hospitals.

When he came across an obscure paper by an Italian psychiatrist, Amarro Fiamberti, written in 1937, Freeman thought he had found a way to solve the problem. Fiamberti had found that it was possible to reach the frontal lobes of the brain from below by driving an instrument through the orbit of the eye, having first given the patient a local anesthetic. On his account, he then injected alcohol or formalin to destroy adjacent brain cells. Fiamberti anticipated that he might meet with professional reproof for this procedure and responded in advance: "In the present state of affairs, if some are critical about lack of caution in therapy, it is on the other hand deplorable and inexcusable to remain apathetic with folded hands content with learned lucubrations upon symptomologic minutiae or upon psychopathic curiosities."[46] This was a sentiment with which Freeman fully concurred.

Freeman chose to modify Fiamberti's procedure in one important respect: concerned that the damage from injecting fluids was unpredictable, he preferred to make sweeping cuts in the brain. He experimented first in the morgue, using corpses to determine how to proceed, till finally he was sure of his ground. Searching for a suitable instrument to break through the bone, he had settled on an ice pick from his kitchen, and though he would subsequently substitute a specially designed tool, complete with markings to measure how far into the brain he had penetrated, the image of an ice-pick lobotomy would remain firmly entrenched in the public mind.

Starting in January 1946 with a patient he referred to in print as Ellen I., Freeman performed his first transorbital lobotomies in his office on weekends, to keep them a secret from his partner, Watts. To render the patient unconscious so that he could proceed, he used one or two electroshock treatments. This was an intervention he quite routinely used in his office practice, though occasionally things went awry. On one occasion, a woman he had previously treated arrived suffering from another bout of depression. Alone in the office, Freeman administered ECT. In the absence of any muscle relaxant (still not a routine part of ECT), her convulsions were so severe that she fractured both thighs. Since she was still unconscious, he gave her morphine and left to see other patients in the hospital, returning to find her screaming in agony. The upshot was a lawsuit his insurers settled out of court. Freeman commented that his mistake had been in failing to recommend and perform a lobotomy on her, since "thus far I have never been sued on behalf of a patient who had undergone lobotomy."[47]

Such complications seldom ensued when he rendered patients unconscious prior to inserting ice picks in their brains (usually pausing to photograph the scene).[48] Generally, Freeman later indicated, he preferred to administer three electric shocks spaced two or three minutes apart. "Stated succinctly and much too simply, I believe that shock treatment disorganizes the cortical patterns that underlie the psychotic behavior, and the lobotomy . . . prevents the pattern from reforming." Needless to say, this was pure speculation on his part, ungrounded in evidence, as was his claim that "it would seem that transorbital lobotomy is an effective means of treating all but the most severe and chronic cases of schizophrenia."[49]

Some weeks later, Watts came in unexpectedly and witnessed Freeman at work. An almighty row ensued. Watts was incensed that Freeman, who was not certified as a surgeon and who routinely took no antiseptic precautions when performing transorbital lobotomies, would behave in such a fashion. It was an argument the two men could not patch up. The preface to the second edition of their text indicated as much, and thereafter, Freeman struck out on his own.[50] The two men seldom operated together again, but they continued to coexist within the same department, and Watts collaborated on the second edition of their textbook, which appeared in 1950.

The operation he had refined, Freeman proclaimed, was so simple that any damned fool could perform it, even a psychiatrist, and he could teach them to do it in twenty minutes or so. As for the surgery itself, it "can be completed in a few minutes." It was "a minor operation" and could readily be performed on an outpatient basis, with the patient able to leave under his or her own power within an hour, only needing to wear dark glasses to cover the black eyes that routinely followed the procedure.[51] Freeman conceded that those encountering the operation for the first time were often repelled. The procedure was attended with certain unfortunate sound effects: as the leucotome, the modified ice pick he had devised, was driven through the bone, the cracking sound tended to disturb the naive observer.

"The fracture of a thick orbital plate is not infrequently accompanied by a rather horrifying snap and sometimes by a gush of blood and extreme swelling of the eyelids that is definitely disturbing to both psychiatrists and surgeons until they have become accustomed to it."[52] But, fortunately, the patient was unconscious, shocked into a state of suspended animation, and his professional colleagues would soon learn to treat these phenomena as no more than a temporary aesthetic annoyance, particularly once they realized that "the transorbital method brings the possibilities of psychosurgery right into the psychiatric hospitals, where it is obviously of greatest value."[53] Freeman now set out, as his daughter Lorne had suggested he do, to become "the Henry Ford of lobotomy."[54]

IN THE SUMMER OF 1946, Freeman drove west, having arranged with the superintendent of the state hospital in Yankton, South Dakota, to demonstrate his new operation on several patients. The following summer, he drove his camper, christened the "lobotomobile," west again. This time Freeman moved on after Yankton to state hospitals in Washington State and California, teaching psychiatrists at each hospital how to perform the surgery, first on corpses in the deadhouse, and then on live patients with Freeman watching and coaching them.

Newspaper coverage, especially in Washington State, was extensive and glowing: the *Seattle Times Herald* ran a banner headline on August 20, 1947, that proclaimed "Brain Operations on 13 Patients Are a Success." Its rival, the *Seattle Post-Intelligencer*, took things a step further. Its story included two graphic photographs, one of a staff psychiatrist, James Shanklin, giving a female patient ECT to render her unconscious, and the second of a bearded Freeman using an ice pick and a mallet to break through the orbital bone and lobotomize her.[55] Following a further series of transorbitals performed at Stockton State Hospital, Freeman privately conceded that these forays had "poor results," an outcome he blamed on the bad "material" with which he had been provided.[56]

News of these activities reached John Fulton at Yale, and he wrote immediately to Freeman. Beginning in January 1947, Fulton had toured the country promoting lobotomy and speaking to large and enthusiastic crowds. "The world," he noted privately, "seems to have become interested in the frontal lobes." To his satisfaction, he found that newspapers regularly credited him as the world authority on the operation and as the man in whose laboratories the basic science behind the miracle operation had been performed. All of that promise was now threatened by the news of Freeman's grandstanding. Fulton had spoken forthrightly about his distaste for the "blind" operation and preference for "open" surgery, which he believed resulted in more precise targeting of particular brain regions. This new operation, placed into the hands of untrained psychiatrists, threatened the future of lobotomy. He and Freeman had been corresponding since 1931 and had generally been on good terms. Now he wrote an angry and bitter reproof: "What are these terrible things I hear about you doing lobotomies in your office with an ice pick? I have just been to California and Minnesota and heard about it in both places. Why not use a shotgun? It would be quicker."[57]

Determined to put lobotomy on a more satisfactory footing, Fulton had spent most of 1947 orchestrating the annual meeting of the Association for Research in Nervous and Mental Disease, which was held in New York in December, attracting an audience of almost a thousand. One of those in the audience was Robert S. Morison of the Rockefeller Foundation, who had thus far been circumspect about supporting research on lobotomy. Having listened to the proceedings, Morison concluded, "One would guess on the basis of this two-day meeting that there will be a rather rapid increase in the number of operations on the frontal lobes on incapacitated patients in our state hospitals during the next ten years."[58] Thematically organized around research on the frontal lobes, much of the meeting revolved around a series of papers on lobotomy—papers Fulton had deliberately chosen to highlight the potential of psychosurgery to begin emptying the state hospitals. He also sought to stress the need for a more circumspect and cautious approach—one that emphasized "open" operations and careful efforts to determine which regions of the brain might bring about the most improvement and the least negative effects.[59]

Fulton touted a paper by Arthur Ward, a recent alumnus from his laboratory, who reported on operations on the cingulate gyrus of monkeys, and Fulton suggested that damaging this more limited, if difficult to access, part of the brain might mitigate some of the personality defects following conventional lobotomies. This was exactly the sort of "evidence" concerning "functional localization in the frontal lobe" he had hoped would be forthcoming.[60] Fulton directly sought a confrontation with Freeman (having denied his request to present work on transorbital lobotomies), repeatedly calling on him to respond whenever criticisms of standard lobotomies were voiced.[61] He had invited, as a special guest, a Swedish psychiatrist, Gösta Rylander, who was known to be skeptical of lobotomy.

As professor of forensic psychiatry at the Royal Caroline Institute in Stockholm, Rylander had performed a small series of lobotomies. His talk focused on eight of the thirty-two cases he had operated on. He began by alluding to the great successes many had reported in the published literature. "Most authors declare that they have found no intellectual deterioration. Cases have also been published which show better results in intelligence tests after lobotomy than before. Isn't that a rather marvelous result," he commented sarcastically, "of a mutilating brain operation!"[62]

Rylander now put forward an alternative explanation for this apparent success: "If a schizophrenic or melancholic person . . . is examined before the operation, the results of the tests are greatly lowered by splitting or inhibition, faulty concentration, etc." After the operation, when these handicaps are reduced or removed, "the patient can use his intellectual tools to better advantage. The results may be better than the preoperative ones, even when the intellectual faculties have been impaired." In other respects, it was easy for naive observers to be led astray when examining lobotomized patients. "The patients themselves behave in the most perfect way at the examinations. They are smiling, polite, answer the questions rapidly and openly and say that they are very pleased with the results, that everything is all right and that they have not altered. But questioned in detail, they will explain that they forget things and that they have lost many of their interests. The more introspective of them may allege that they are unable to feel as before. They can feel neither real happiness nor deep sorrow. Something has died within them." More extended observation had demonstrated, he continued, that "all . . . have shown changes in behavior of the well-known type, namely, tactlessness, emotional lability with tendencies to outbursts, extrovertness and slight euphoric traits." The changes, he concluded, were "rather grave. . . . [S]omething is taken away from the mind of the patient, something that is more valuable than the therapeutic effects of the operation." The families of his patients had been dismayed, a typical comment being "She is with me in body but her soul is in some way lost."[63]

A lively discussion ensued. David Rioch, on the staff at the psychoanalytically oriented Chestnut Lodge in Maryland, had previously described lobotomy as a form of "partial euthanasia" and spoke dismissively of these "ablation experiments in human beings." "It is interesting," he commented, acerbically, "that at the present period of social development excision of large parts of patients' brains is socially unacceptable." Another analyst, Roy Grinker of Chicago, congratulated Rylander for pointing out the existence of "a great deal of deficit after lobotomy which laboratory tests and examinations in hospital wards completely miss."[64]

Freeman rose to object, and a bitter exchange ensued. As Fulton goaded him to respond, Freeman angrily denied that the indications for lobotomy were in any way limited. Rylander's critique obviously stung. He granted that in "superior" patients something was lost in the aftermath of psychosurgery, but he noted that even these patients were freed from "their harassing doubts, fears and distress" and so overall were better off.[65] His aggressive defense even seemed to sway Rylander, who conceded that, in certain cases, lobotomy could be justified.

THE ARGUMENT SCARCELY RESTED THERE. Freeman's incursions into the world of state mental hospitals were just getting under way, but the rich had already made clear their interest in what was touted as a new miracle cure. Early on, Joseph Kennedy, previously the American ambassador to the United Kingdom, had proffered his daughter Rosemary, whose rebelliousness and slight developmental problems he feared might lead her to pregnancy and scandal, as a suitable case for treatment. Freeman and Watts had operated on her in 1941, with disastrous results. From 1941 till her death in 2005, Rosemary Kennedy was severely mentally handicapped, unable to speak, incontinent, barely able to walk, and hidden from public view (as well as ignored by her parents while they were alive).[66] But as well as ambulatory patients like Rosemary, there were rich patients who languished in the ritzier private asylums like the McLean Hospital in Boston, the Institute of Pennsylvania Hospital in Philadelphia, and the Institute of Living (previously the Hartford Retreat) in Connecticut. All of these asylums had developed active lobotomy programs for their "guests," as the Hartford superintendent preferred to call them, and all had affiliated with major medical schools to provide the necessary surgeons: McLean with Harvard, the Pennsylvania Institute with the University of Pennsylvania, and Hartford with Yale.

In advance of the December 1947 meeting, Freeman had written to Fulton suggesting that he be allowed to demonstrate the efficacy of his new approach by operating on some patients at the nearby Institute of Living. Fulton declined, so Freeman made arrangements directly with the institute's superintendent, Charles Burlingame.

Burlingame was already presiding over a lobotomy study involving his own patients and those from the nearby Connecticut State Hospital at Middletown, Fairfield State Hospital, and Norwich State Hospital in collaboration with Yale Medical School.[67] So it was that on November 19, 1948, the two most enthusiastic and prolific lobotomists in the Western world faced off against each other in the operating theater at the Hartford establishment. They performed before an audience of more than two dozen neurosurgeons, neurologists, and psychiatrists. Each had developed a different technique for mutilating the brains of the patients they operated upon, and each man had his turn on the stage.

William Beecher Scoville, professor of neurosurgery at Yale, went first. His patient was conscious. The administration of a local anesthetic allowed the surgeon to slice through the scalp and peel down the skin from the patient's forehead, exposing her skull. Quick work with a drill opened two holes, one over each eye. Now Scoville could see her frontal lobes. He levered each side up with a flat blade so that he could perform what he called "orbital undercutting." What followed was not quite cutting: instead, Scoville inserted a suction catheter—a small electrical vacuum cleaner—and sucked out a portion of the patient's frontal lobes.

The patient was wheeled out, and a replacement was secured to the operating table. Walter Freeman, professor of neurology from George Washington University, was next. He had no surgical training and no Connecticut medical license, so he was operating illegally—not that such a minor matter seemed to bother anyone present. Freeman's swift, assembly-line technique allowed him on occasion to perform twenty or more operations in a single day. He proceeded to use shocks from an ECT machine to render his female patient unconscious, and he then inserted a modified ice pick beneath the eyelid till the point rested on the thin bony structure in the orbit. A few quick taps with a hammer broke through the bone and allowed him to sever portions of the frontal lobes, using a sweeping motion with the ice pick. The instrument was withdrawn and inserted into the other orbit, and within minutes, the process was over.

Scoville was quick to denounce the crudity of Freeman's procedure, a position the neurosurgeons in the audience were happy to endorse.[68] Freeman, in turn, was scornful of the notion that his rival's suctioning away of portions of the brain was "precise," as Scoville and his supporters contended. As Freeman put it on a later occasion, with characteristic tact, the technique employed by neurosurgeons like Scoville and James Poppen was as precise as "running a vacuum cleaner over a bathtub of spaghetti." "These brains," he added, "at least the one or two that I have seen following [these operations], are horrifying."[69]

It took a lot to horrify Freeman.

CHAPTER ELEVEN

The End of the Affair

A FEW MONTHS AFTER THIS FACE-OFF, Walter Freeman attended the First International Congress of Psychosurgery. The gathering was held in Lisbon, Portugal, in part to honor Egas Moniz for his pioneering efforts in the field. The following year, lobotomy's author would receive a much greater honor for his innovation, the Nobel Prize in Physiology or Medicine. Freeman had originally nominated Moniz in 1943, submitting a copy of his own monograph on psychosurgery along with his letter, possibly hoping to share the prize. After the Congress, he renominated Moniz, this time finding a more sympathetic audience.[1]

Freeman was not the only American hoping to share in the prize. John Fulton's active proselytizing on behalf of lobotomy between 1946 and 1948 and his steadily more elaborate and exaggerated accounts of the connections between the chimpanzee experiments in his laboratory and Moniz's decision to embark on psychosurgery were almost certainly motivated by the hope that his purported role in laying the scientific groundwork for the surgery would be sufficient justification for him to win a share of the prize. In the event, all this effort produced only a misleading reference in the selection committee's report on how the operation had come about, a claim that Fulton and his collaborators had proved that "neuroses caused experimentally disappeared if the frontal lobes [of chimpanzees] were removed and that it was impossible to cause experimental neuroses in animals deprived of their frontal lobes."[2] No prize for Fulton. Coincidentally or not, within two years he ceased working as a neurophysiologist, retreating into the presumably less demanding role of Yale's first professor of medical history. At the age of fifty-one, the wunderkind subsided into a bitter and alcoholic senescence.

For most of the operation's supporters, the award of the Nobel Prize to Moniz was greeted with éclat. Who could doubt, said Charles Burlingame, that "psychosurgery has been a milestone. A masterpiece of inductive reasoning . . . [part of] the great march toward scientific psychiatry."[3] By 1950, Freeman and Watts could accurately boast that "psychosurgery has come of age. The surgical treatment of mental disorder has been adopted on a progressively larger scale as its benefits have been recognized, its failures accounted for and its pitfalls avoided."[4]

THERE WERE DISSENTING VOICES, even in these years of the operation's greatest success. Some of the criticism came from psychoanalysts, who were becoming increasingly influential, but there were also complaints from other unexpected quarters. All thirty-seven of the new veterans hospitals established to cope with the large numbers of psychiatric casualties of the war had rapidly begun to make use of psychosurgery, and John Fulton had taken advantage of the Veterans Administration's interest and the new willingness of the federal government to support medical research to obtain substantial funding for lobotomy studies based in his laboratory.[5] But in 1949, Jay Hoffman, chief of the Veterans Administration Neuropsychiatric Service, published an article in the *New England Journal of Medicine* expressing serious doubts about lobotomy's usefulness in the treatment of schizophrenia. "The evaluation of the results after prefrontal leucotomy will be greatly influenced by the frame of reference one uses," he wrote. "If the condition of the patient is compared with his condition prior to the onset of his psychosis, all the results must be considered failures. . . . I think it should be re-emphasized that by psychosurgery an organic brain-defect syndrome has been substituted for the psychoses."[6]

Nolan Lewis, director of the New York State Psychiatric Institute and professor of psychiatry at Columbia, was even more scathing. He dismissed the notion that lobotomy was only used as a last resort and accused unnamed doctors of "an utter lack of respect for the human brain." What, he continued, were the criteria of success?

> Is the quieting of the patient a cure? Perhaps all it accomplishes is to make things more convenient for the people who have to nurse them. . . . The patients become rather childlike. . . . They act like they have been hit over the head with a club and are as dull as blazes. . . . It disturbs me to see the number of zombies that these operations turn out. I would guess that lobotomies going on all over the world have caused more mental invalids than they have

cured. . . . I think it should be stopped before we dement too large a section of the population.[7]

These fierce criticisms anticipated the negative reaction to psychosurgery that would become the norm by the mid-1960s, when the whole episode came to be seen as an embarrassing mistake the profession would prefer to forget. At the time, however, the critics had little effect in either stemming the general enthusiasm for lobotomy or reducing the numbers of operations performed. Indeed, Lewis's institute would continue to provide major backing for the Columbia-Greystone Project on psychosurgery, one of a number of collaborations between mental hospitals and academic psychiatrists, well into the 1950s. Hoffman published yet another critique of lobotomy in 1950, this time expressing concern about the lack of initiative, inertia, and disinhibition among patients after the operation, and openly questioning whether patients' families could validly consent to surgery that left patients with permanent, irreversible brain damage.[8] Yet even he conceded that on occasion the operation might be legitimately performed, and veterans hospitals continued to perform them.

So, too, did many other mental hospitals closely affiliated with major universities. Harry Solomon served simultaneously as chair of Harvard's department of psychiatry and as superintendent of the Boston Psychopathic Hospital, and in 1951 he reported on 550 lobotomies he had performed over the preceding seven years.[9] Two years later, he and his colleague Milton Greenblatt wrote a largely favorable piece on lobotomy for the *New England Journal of Medicine:* "As lobotomy has spread to more and more centers, the general conviction is strengthened that this method is capable of modifying disturbed behavior of psychotic patients. . . . [A]uthors who express over-all disappointment are few and far between."[10]

Farther south, the Connecticut Cooperative Lobotomy project led by William Scoville of Yale was operating on patients drawn both from the Institute of Living and the Connecticut State Mental Hospital in Middletown. Columbia psychiatrists and neurosurgeons were engaged in a similar study in collaboration with the Greystone State Hospital in northern New Jersey. Edward Strecker, who chaired the department of psychiatry at the University of Pennsylvania, continued to tout the operation's merits till his retirement in 1952, declaring it valuable even if it only succeeded in transforming destructive patients into complacent and less troublesome inmates. Meanwhile, at Duke University in North Carolina, a team operated on nearly 300 patients from the Raleigh and Butler State Hospitals, and at the Mayo Clinic in Minnesota, 250 lobotomies were undertaken in association with the Rochester State Hospital between 1940 and 1948. Apart from six deaths, it was reported that no one was "worse after the operation."[11]

No single individual or team, however, could match Freeman's efforts. Between January 1946 and December 1949, he personally performed "some 400 transorbital lobotomies."[12] This was but the prelude to a vast expansion of his efforts. In July of 1950, he was invited to deliver the McGhie Lecture at the University of Western Ontario and was delighted when his hosts asked for "an actual demonstration of a transorbital lobotomy." This was exactly the sort of opportunity Freeman relished, for it allowed him to demonstrate the simplicity of his innovation. He wrote back urging a more extensive program to enlighten the audience:

> I hope you have a sufficient number of patients to make things interesting for your staff. Ten operations is a fairly good number and I have done as many as twenty-one in an afternoon with time out for lunch. This requires a fair amount of teamwork. I shall bring instruments so that the two sets can be sterilized alternating, and will require a hammer. . . . I trust one or more of the capable staff will be interested in doing a few operations under my direction. Since transorbital lobotomy is a closed operation and has no exposed operative wound, you may reassure the operating room nurse that gowns, masks, and gloves are not necessary for the operation or for the audience. Some of the operations might be done under local anesthetic if the patients are cooperative . . . but I have recently found an additional reason for using electroshock: namely, that it increases the speed of coagulation of the blood.[13]

In a previous letter, he had specified the sorts of patients he had in mind: "I would prefer the most violent schizophrenic and the most agitated involutional, but if you do have a completely helpless obsessional or anxiety state of long duration, that patient might be good for the demonstration of transorbital lobotomy under local anesthesia. Otherwise, I expect to employ electroshock to produce the few moments of coma necessary for the operation."[14]

In the event, Freeman performed seven lobotomies on this occasion. Some months later, Stevenson wrote to him about the outcomes: "I am afraid there is very little to report about the patients on whom you operated when you were here. Our four schizophrenic patients are much the same. The coloured girl, X, has relapsed to her former mute condition. Y is rather acutely disturbed much of the time. Z continues on our convalescent ward and I think she has shown definite somatic improvement." Perhaps unsurprisingly, in the face of these results, Stevenson added, "Our Head Office does not give us any encouragement to do transorbital lobotomies so long as Dr. McKenzie can do a reasonable number of standard lobotomies for us." Some months later, he wrote that the state of Freeman's patients was, if anything, worse than before.[15]

Freeman once again blamed the poor outcomes on his hosts' inappropriate selection of "material" to be operated upon. His general reaction to failure was to double down. In 1949, at a conference at George Washington University, he had reported on a woman he had lobotomized transorbitally. Finding her little improved postoperatively, he had proceeded to give her forty electroconvulsive therapy (ECT) treatments in the space of two and a half days.[16] She continued to hallucinate after this, so he gave her a second, more extensive lobotomy, supposedly with success. But within two weeks she had relapsed and was permanently rehospitalized.

The resort to postoperative ECT was a rather routine part of Freeman's practice. Within a week after the operation, he noted that "a few electroshocks may alter the behavior in the most gratifying manner. . . . [W]hen employed, it should be rather vigorous—two or four grand mal seizures a day for the first two days, depending upon the result."[17]

AGE WAS NO BARRIER TO TREATMENT, and by now Freeman was operating on children as young as four. (That particular four-year-old, he reported, died three weeks postoperatively from meningitis.) Juvenile brains, Freeman contended, could withstand far more damage than their adult equivalents. These were fairly extensive operations, since "a great deal of frontal lobe tissue has to be sacrificed in children in order to obtain any more than temporary improvement. The younger the age at which the psychosis has begun the more posteriorly the incisions have to be made, with resulting greater disability." "Chiefly," Freeman continued, "the aim has been to smash the world of fantasy in which these children are becoming more and more submerged."[18] In that respect, he counted the operations a success. Freeman conceded that his interventions on children were not uniformly successful: "Two of the patients died shortly after the operation and one remains in a state hospital after three operations had been undertaken in the hope of relieving the disturbed behavior. Only one patient has been able to assume responsibilities commensurate with his age or intellectual attainments. . . . All of the patients remain at a rather childish level of dependency and in the absence of satisfactory home environment would inevitably have to be cared for in institutions."[19] But one should not dwell on the negative. After all, "they have definitely abandoned their fantasies [and] perhaps the greatest change in these individuals lies in their facial expression. . . . Lines of depression and exhaustion . . . make way for pleasant smiling countenances, a little vapid at times."[20] Jonathan Williams, Freeman's chief assistant in these years, was blunt about what these operations accomplished: "Lobotomy in children is primarily helpful in making the child more easily controlled." That said, he opined, "It has a limited but definite field of usefulness."[21]

Remarkably, the publication of these cases seems to have drawn no reproof from fellow psychiatrists. Perhaps equally remarkable is the willingness of parents to submit their offspring to the surgery. What motivated them is unclear. Was it desperation? A misplaced reliance on the authority of the therapist? In one case, Howard Dully, we have some answers. As a preadolescent he reacted badly when his father divorced and remarried. Tiring of his tantrums, his new stepmother decided that a lobotomy would quiet him down and arranged for Freeman to perform it. In other cases, we can only speculate. With the exception of the minority of psychiatrists who objected to all lobotomies, Freeman's colleagues appear to have seen it as acceptable to inflict such irreversible damage on young children. As late as December 16, 1960, Freeman performed a transorbital lobotomy on twelve-year-old Howard Dully, who has written about the experience and its effect on his life.[22] Not until later in the 1960s, when Freeman had become a pariah, was there a concerted effort to put a stop to child lobotomies.

TWO YEARS BEFORE HIS DISASTROUS Canadian venture, Freeman embarked on a much more extensive campaign to spread the gospel of transorbital lobotomy. Beginning in February 1949, he traveled to state hospitals in Texas, Arkansas, Nebraska, Minnesota, and Ohio, lobotomy tools in hand. At each stop, he demonstrated his technique and taught others how it was done. The next year, a slight stroke curtailed his missionary zeal, though he did manage a visit to the huge state hospital at Milledgeville, Georgia. One of the psychiatrists he had taught to operate subsequently was heard to quip, "You can change your mind, but not like I can change it."[23] In 1951, he visited Texas, Arkansas, and Nebraska again, but also Iowa, Missouri, South Dakota, Washington State, and California. He even went to Puerto Rico to proselytize on behalf of his operation.[24] Virginia and West Virginia closely adjoined Washington, DC, and in 1952 he attempted to launch large-scale programs in both states. The West Virginia authorities welcomed him with open arms, and until June 1954, when he retired from George Washington University and moved to California, he regularly visited that state to operate on large numbers of patients.

By now, Freeman could perform these operations as on an assembly line. A letter he wrote to Moniz in September 1952 gives a vivid portrait of what transpired. The three state hospitals in West Virginia had gathered patients for him at a single site:

> I began operating by the transorbital route on July 18, and when the study was closed on August 7 the total number of operations was 228. Only twelve days were devoted to operating. . . . I mention this matter to reveal how mass surgery can be carried out against a background of shortages of everything but patients. . . . I followed the technic developed over the past year and more, first making the patient unconscious with electric convulsive shocks, usually three at intervals of two minutes. On one occasion when the team was working to perfection, I routinely administered two shocks and then proceeded immediately with operation. Upon this day I operated upon 22 patients in a total of 135 minutes, about six minutes per operation. . . . When the patient had stopped convulsing I had the nurse hold a towel firmly over the nose and mouth of the patient. I then elevated the eyelid and inserted the point of the instrument into the superior conjunctive sac, being careful not to touch the skin with the point. I felt for the vault of the orbit, about 3cm. from the midline and then aimed the shaft of the instrument parallel with the bony ridge of the nose. I drove the point of the instrument through the orbital plate to a depth of 5cm. . . . When the instrument was thus in place on one side I inserted the other instrument in the same fashion on the other side. This maneuver not only makes for greater symmetry in the lesions, but also prevents the frontal lobes from moving laterally with the moving of the instruments. It also hastens the operation.[25]

In two years, Freeman succeeded in lobotomizing a tenth of all the patients at the Lakin State Hospital in West Virginia. He claimed half of these patients were later discharged from the hospital, an obvious boon to the financially strapped state treasury. At the hospital in Spencer, West Virginia, he lobotomized twenty-five women in a single day. Back at Lakin State Hospital, a segregated facility for African Americans, he reported that on one visit, he operated on "twenty very dangerous Negroes." A week later, he was gratified to find "fifteen of them sitting under the trees with only one guard in sight. It was the first time they had been out of the seclusion rooms for anywhere from six months to seven years."[26]

The plan was for him to launch a similar program in Virginia, but when he was demonstrating how he intended to proceed, things went awry. The superintendents of both the Eastern and Western State Hospitals had endorsed the project, as had the state's mental health commissioner. Freeman left the premises before the psychiatrist he had been instructing had finished operating. After his departure, his protégé had broken the tip of the leucotome, which had slipped back into the brain, necessitating an emergency visit from a neurosurgeon to open the skull and remove the piece of metal. For once, it turned out that not just any damned fool could perform a transorbital lobotomy. Freeman tried to make light of the incident, but it cast a pall over the plans, and though the Virginia state hospitals did resume the procedure, it was on a very limited basis.

By the early 1950s, lobotomy was widely embraced by American psychiatrists as one of the best tools they possessed to cope with the depredations of major mental illness. No one regarded psychosurgery as a panacea, and even enthusiasts conceded that the operations exacted a price, but it was one many were willing to inflict on their patients. Paul Hoch, head of experimental psychiatry at the New York Psychiatric Institute and previously director of shock treatment at Manhattan State Hospital, summarized some of the most common outcomes of the surgery: "Clinically the most conspicuous damage is a certain inertia or apathy [along with] careless and uninhibited behavior. . . . The highest symbolic functions like planning, foresight, and creativeness . . . [are] impaired."[27]

Freeman concurred: given "the somewhat limited imaginative powers of the lobotomized individual," he conceded, "it stands to reason that the occupational adjustment in the case of most of our patients has to be on a rather simple level."[28] Three years earlier, Freeman had quipped, "It is remarkable how little frontal lobe is required to earn a living."[29] Fortunately, he commented on another occasion, "Many patients are content with routine jobs that they would previously have thought beneath their dignity."[30]

For wealthier patients, whose families had the resources to cope with an indolent, tactless, and impulsive but more docile husband or wife, or son or daughter, and to manage such things as occasional seizures, an unfortunate tendency to masturbate in public, incontinence, or a failure to observe conventional niceties, the lobotomized patient could be cared for at home. Both families and psychiatrists saw this as a positive outcome. Some of the psychological suffering patients experienced and inflicted had been alleviated, and compared to the horrors of the wards, the return to domesticity was interpreted as a blessing.[31] Often, though, this took an enormous toll and a great deal of ideological work by the families, who had strong reasons to put a positive spin on the outcome. After all, it was they, for the most part, who had chosen this fate for their loved ones.

Patients, Freeman explained, were like children. And like children, their behavior might improve over time. Or it might not. They could and should be disciplined—by verbal barbs and prodding, and, if necessary, by more direct intervention. "Lobotomy patients sometimes need to be punished," he wrote in a paper on the nursing problems that lobotomized patients presented. One needed to use

> action of a sufficiently visible character to make the patient realize in his childlike reasoning who is boss. Spanking is as effective a measure as we know, since it seems to recall childhood experiences of a similar nature. If the patient is too big for a nurse to handle alone, family assistance may be required. We would stress the fact that spanking is good for the patient though it may distress the family. . . . Punishment does not need to be severe, but it must be prompt, vigorous, and pointed if it is to have a good effect.[32]

One should not worry unduly, since the lobotomized patient "does not feel the shame and indignity of the procedure as would a normal individual."[33] In their docile and degraded state, another analogy was pertinent: "About a quarter of the patients in our series can be considered as adjusting at the level of the domestic invalid or household pet." Some might recoil at such an outcome, but such squeamishness was misplaced: "When it comes down to the question of such domestic invalidism as against the type of raving maniac that was operated on, the results could hardly be called anything but good."[34]

These were children—or household pets—who displayed an uninhibited sexual appetite, which could produce social embarrassment or worse. This was particularly problematic for married couples. Here again, psychosurgery had an answer: "Her husband may have regressed to the cave-man level, and she owes it to him to be responsive at the cave-woman level. It may not be agreeable at first, but she will soon find it exhilarating if unconventional."[35]

IN STATE HOSPITALS, a rather different rationale for lobotomy prevailed. Joel Braslow's study of lobotomy in California state hospitals showed it was largely aimed at the most "recalcitrant" patients, those who exhibited "severely disordered behavior" and were "assaultive, destructive and unpredictable." They had been drugged, bathed, wrapped, repeatedly put into comas, and shocked and shocked again. And when those techniques failed to quell the difficulties they presented, psychosurgery was prescribed for their unrelenting madness.[36] At the Boston State Hospital, Greenblatt and his associates chose "hopeless" cases for treatment. "In not a few instances," he reported, "the selection was made on the basis of very disturbed behavior with the inherent difficulty of caring for the patient."[37] In Missouri, Schrader and Robinson conceded that psychosurgery did "produce some mental deficits" but insisted that "the efficacy of prefrontal lobotomy as regards gross behavior seems incontrovertible." What efficacy meant to them (and to other state hospital psychiatrists) is quite revealing: patients were described postoperatively as displaying "little initiative" but being "highly cooperative" and "content to be completely idle for long periods." Another model patient "seldom loses his temper."[38]

B. E. Moore and his colleagues found "frontal lobotomy to be a highly effective procedure for the treatment of intransigent mental disorders," since "the more troublesome symptoms to patient and society, such as anxiety, depression, hypochondriasis, impulsiveness, assaultiveness, destructiveness, etc., are eliminated in 75 per cent or more cases."[39] Winfred Overholser, superintendent of St. Elizabeths Hospital in Washington, DC, was even blunter. Having initially balked at using lobotomy, he began employing it in a targeted fashion, selecting patients on a series of criteria: "First, the existence of marked destructiveness (as shown, for example, by homicidal or suicidal trends) and destructiveness or marked tension or anxiety;

second, an ample trial of other means of treatment; third, institutional care for a substantial period. . . . All the patients were serious behavioral problems before the operation, so almost any change was bound to be for the better. 'Improved' may mean, therefore, that the patient is still a problem but perhaps less assaultive, or destructive or noisy."[40] Those who did not calm down could expect to receive "electric shocks and, if they still remained unresponsive, underwent another lobotomy."[41]

It was scarcely a secret that lobotomy was extensively used in state hospitals not as a means to discharge patients, but—as was often the case with electroshock treatments—to make the work of the attendants on the wards less dangerous and unpleasant. Still, when participants at a gathering sponsored by the National Institute of Mental Health issued their report on the uses of psychosurgery, the majority balked at the statement that a small group of psychiatrists proposed, namely "that psychosurgery is justifiable for the sole purpose of rendering an individual more tractable." Many felt uncomfortable with the word "tractable" and preferred to substitute "comfortable."[42] For patients, the embrace of the euphemism made no practical difference.[43] Troublesome patients, wherever they were confined, were at heightened risk of having their frontal lobes severed.[44]

SCHOLARS HAVE SUGGESTED that the advent of a new generation of antipsychotic drugs, the phenothiazines, which came to market in 1954, brought about the swift demise of psychosurgery. This is not altogether true. All through the 1950s and into the early 1960s, lobotomies continued to be performed in substantial numbers—a pattern that is scarcely surprising when one recalls how committed prominent psychiatrists were to the procedure.[45] A decade later, what earlier adopters had hailed as a "chemical lobotomy" did indeed largely replace psychosurgery, but lobotomy's demise was slow and uncertain, and in most quarters it was not till the 1960s that the practice began to seem beyond the pale, an episode best forgotten or treated as a bizarre aberration.[46]

Before that, there were a few outstanding exceptions to the general reluctance to acknowledge that a grievous error had been made. Percival Bailey, the distinguished professor of neurology and neurosurgery at the University of Chicago, had performed a handful of lobotomies in the late 1930s, influenced in part by John Fulton's recommendation of the procedure. He was disenchanted by what he saw, however, and by 1956, he had had enough:

> Now, after the wave of psychosurgery has swept across the country, leaving hecatombs of mutilated frontal lobes behind it, I read many accounts of the results and am frankly appalled. Abusive and obscene language, disability in long-term planning, uninhibited sexual drive, obnoxious mannerisms, stealing, suggestibility, laughing spells, convulsions and other untoward symptoms are frequent and discouragingly persistent. The most favorable result seems to be that the disturbed patients after lobotomy are easier to manage in the hospital. But . . . they would be still less trouble if taken to a gas chamber.[47]

In time, this would become the orthodox view, but such harsh critiques were far from the norm. Across the Atlantic, the *Lancet* was calling for a controlled randomized trial as late as 1962 to determine whether lobotomy was useful, and though the numbers performed in the United Kingdom had fallen from the annual average of 1,100 between 1948 and 1954, the total for the year ending on June 30, 1961, was still 525.[48] There are no comparable national statistics for the United States, but there too lobotomy continued to be seen by many psychiatrists as a legitimate treatment for at least a decade after the new drugs arrived on the scene.

For decades, Arthur Noyes's *Modern Clinical Psychiatry* was America's leading psychiatric textbook, used in virtually every training program in the country and going through a multitude of editions after its first appearance in 1934. Its pronouncements were enormously influential, and as late as the sixth edition, which appeared in 1963 (the year of Noyes's death), its coverage of lobotomy remained predominantly positive. Noyes and his by-then co-author Lawrence Kolb insisted that psychosurgery "contributes greatly to the control of pathological emotional states" and concluded that properly targeted "incisions leave sufficient quantities of the frontal lobes to permit the patient to retain the capacity for productive work and to make a satisfactory adaptation in a social environment." Despite the trail of broken lives, they continued to tout lobotomy's virtues for the treatment of severe mental illness. "In schizophrenia prefrontal lobotomy has no effect on the fundamental disease process, but in well selected cases may be followed by a great improvement in adjustment. . . . [T]he behavior of the over-active, resistive, excitable, destructive, restless, unmanageable schizophrenic may be greatly improved. Such a patient may become quiet and somewhat sociable and perhaps later he may be able to leave the hospital and even become self-sustaining."[49]

One of the key factors in bringing about the procedure's demise was a process of generational change.[50] As psychiatrists who had performed lobotomies or acquiesced in their performance retired, a younger generation proved steadily more reluctant to embrace the practice. In part, this may have reflected their contact with patients on the back wards who represented the worse examples of what psychosurgery could produce: deteriorated human vegetables. And in part, the growing dominance of psychopharmacology had provided this generation of psychiatrists with forms of treatment that more closely resembled the kinds of remedies employed elsewhere in contemporary medicine. Doubtless, the hugely negative image of lobotomy (and electroshock) by the late 1960s can only have hastened its abandonment.

Yet it was not just generational change among psychiatrists that helped to consign lobotomy to the status of a failed therapy, best forgotten as soon as possible. Neurosurgery was maturing as a profession, and the attractions of operating on patients in the back wards of the local asylum were waning as a result. When lobotomy arrived on the scene in the mid-1930s, neurosurgeons were few and far between, and members of the young specialty faced grave difficulties securing a sufficiently large practice to earn a living. Surgery on brain tumors remained fraught with danger and had poor clinical outcomes, even after Harvey Cushing pioneered more effective approaches to the technical difficulties this form of brain surgery entailed.

Thirty years on, the range of interventions neurosurgeons could perform had expanded greatly, with work on back injuries and pathologies of the nervous system forming a large part of their work, and brain operations outside the psychiatric realm became easier to perform (and, in the case of nonmalignant tumors, more routinely successful). Neurosurgery had established itself as a high-status specialism. If lobotomy came under attack, there were fewer and fewer reasons for the neurosurgical fraternity to perform it. Major psychoses remained as an enormous social and clinical challenge, but few were tempted to resort to the brutal remedy embraced by their predecessors.[51]

LOBOTOMY WAS ONLY the most extreme example of the orgy of experimentation that marked the period between the 1910s and 1950s. Talk of degeneration and defective heredity provided a eugenic rationale for sequestering the mad in the vast bins that grew ever-larger in the new century. But the price the profession paid for embracing these ideas was a heavy one.

If the role of psychiatry was to keep the mentally ill out of circulation and unable to reproduce, then its practitioners were little better than boarding-house keepers. While some might be satisfied with the existence that asylum practice offered, the more ambitious sought alternative ways forward.

The desperation of patients and their families found its counterpart, in some quarters, in the quiet frustration of doctors who wanted to find some way to establish themselves as members of a healing profession. The heavy costs that mental disturbance imposed on families and society were inescapable, and the potential rewards for those who could manage a therapeutic leap forward were correspondingly large. The desire to find a more satisfactory solution to the grave burdens that psychosis imposed had persuaded the Rockefeller Foundation to devote substantial sums to underwriting the new enterprise of academic psychiatry. Within the ranks of the profession, it bolstered ambitions to look for more promising avenues that might link the biological sources of mental illness to therapeutic interventions.

This constellation of aspirations might seem to point psychiatrists in largely positive directions. In practice, it fueled a set of interventions that proclaimed themselves to be dramatic therapeutic advances and for the most part turned out to be nothing of the sort.

MANY OF THESE THERAPIES inflicted grave harm on those subjected to them. One may presume that those seeking to discover new forms of treatment were no more venal or reckless than their counterparts elsewhere in medicine. Yet it is not just hindsight, in my view, that leads us to recoil at what was done in the name of psychiatric science in those years—nowhere more obvious than in the surgical mutilation of thousands performed by advocates of focal sepsis, or the psychosurgery justified by speculations about the connections between the frontal lobes and psychosis. What licensed such cavalier approaches to disability, allowing them to proceed unchecked for so long?

What stands out here is the singularly vulnerable position of mental patients. By virtue of their illness, psychotic patients were deemed to be (and often were) incapable of making rational choices about their fate. Worse still, their condition was highly stigmatized, as it has been almost everywhere for centuries—a stigma that all the talk of biological inferiority and degeneracy had exacerbated.[52] Those beliefs made it easier to treat them as objects, not sentient beings, and to act upon them accordingly. Legally, involuntarily confined patients were stripped of all civil rights and had little ability to resist whatever was proposed. Such protests as they offered were easily dismissed as the product of their pathology.

Family members were poorly placed to dissent from the advice proffered by the credentialed professionals to whom their loved ones had been consigned, and they were often desperate for some intervention that might relieve the disturbance and distress that accompanied chronic mental illness.[53] Families often volunteered their loved ones for lobotomy, and even those with the resources to consult medical luminaries such as John Fulton of Yale, or psychiatrists at the most expensive and prestigious mental hospitals in the land—the McLean Hospital, Bloomingdale, the Institute of Living, or the Pennsylvania Hospital's psychiatric branch—would as likely as not have been counseled to proceed. Medical journalists fed fantasies of miracle cures, but even those who might have been skeptical of such stories would have been hard-pressed to find contrary views of the procedure.

Internally, too, the medical profession had neither the means nor the inclination to police new therapies. Not just in psychiatry, but in medicine more broadly, the "gentlemanly" ethos of the profession placed obstacles in the way of challenging fellow practitioners. Not until the 1960s would any agreed-upon systematic techniques develop that purported to provide a check on clinical claims and enthusiasms.[54] Case reports of clinical success remained the standard way that medical innovations were introduced to the profession, and the accumulation of such clinical observations was the basis on which the novel became the routine. Some sociologists define professions as entities that exercise internal discipline over their members. In reality, evidence of such interventions is scarce, and few established mechanisms existed even for suppressing techniques professional elites had lost faith in. If that were true of medicine in general, it was even more so of psychiatry, where superintendents ruled asylums with almost unchallenged authority, largely hidden from view.

Finally, the cost of caring for the growing numbers of patients confined in public facilities fell squarely on the state. However hard politicians tried to rein in mental hospital budgets, the burden these institutions represented remained enormous. State authorities sought earnestly for some relief from the fiscal burden of mental illness, for while medical care in the United States was (and remains) entrusted to the mercies of the marketplace, from the middle decades of the nineteenth century, madness had been the exception to this rule. Serious mental disturbance was an almost insurmountable barrier to earning a living, and its depredations threatened to reduce all but the richest families to penury. That reality, and the threat some of the disturbed posed to the social order, had proved sufficient to persuade public authorities to intervene. Given the common desire to cut costs, credulous politicians were readily seduced by claims that a cure was at last at hand. States would remain disturbingly susceptible to such arguments in the years to come.

WALTER FREEMAN RETIRED from his position at George Washington University in 1954 and moved to California, where he continued to perform lobotomies in private practice on both adults and children. He lost his last hospital privileges in 1967, after a series of clinical disasters, but continued to travel and promote the value of lobotomy until his death from cancer in 1972. Scoville founded the International Society for Psychiatric Surgery in 1970, and when Yale sought to limit his practice in his late 1970s, he shifted to a local hospital where he could continue lobotomizing patients. He remained a vociferous advocate for psychosurgery until his death in 1984 in a car accident.

PART TWO

DISTURBED MINDS

CHAPTER TWELVE

Creating a New Psychiatry

NINETEENTH-CENTURY ALIENISM WAS BORN IN THE ASYLUM, and throughout the first half of the twentieth century, most psychiatrists still practiced in institutional settings and dealt with those legally certified as insane. The years immediately before the First World War had seen the birth of the National Committee for Mental Hygiene (NCMH), the outgrowth of a campaign by a former mental patient, Clifford Beers, whose own mistreatment in a series of institutions, public and private, inspired him to launch an effort to reform the treatment of the mentally ill. Beers's best-selling autobiography, *A Mind That Found Itself*, written in part while he was a patient at the Hartford Retreat and published in 1908, had made him a national celebrity and brought him into contact with Adolf Meyer.[1] Meyer sought, without much success, to rein in Beers's grandiosity, ultimately prompting a break between the two men. He suggested Beers should go to work as an attendant at the mental hospital run by his protégé Henry Cotton. Then, when Beers sought to launch a national campaign, he tried gently to persuade him to limit his ambitions to his native state of Connecticut. He succeeded in arousing Beers's suspicion of his motives but not before he had planted the idea that, instead of trying to reform mental hospitals, Beers should focus on "mental hygiene," the attempt to promote mental health and head off outbreaks of insanity.[2]

The estrangement deepened when Beers successfully sought funds for the NCMH from Henry Phipps, the industrial magnate who had funded Meyer's clinic at Johns Hopkins. Beers saw his organization as acting on three fronts: campaigning to improve the care and treatment of those in mental hospitals; spreading the word that it was possible to recover from mental illness; and setting up programs to forestall mental illness and the need for hospitalization. For three years after the NCMH's founding in 1909, financial difficulties limited its scope, but with the help of the Phipps money, eventually further funds were forthcoming from the Rockefeller Foundation. In 1912, an office was established in New York, and a new director, the psychiatrist Thomas Salmon, was recruited to run the organization. Two years later, the Rockefeller Foundation began to pay Salmon's salary and to provide funds for national surveys of mental health issues.

The late entrance of the United States into the First World War had curtailed, but not prevented, the emergence of shell shock among the troops. Army medics found themselves treating both traumatic injuries to the brain and the central nervous system, and war neuroses of a far more mysterious sort. Though some continued to proclaim that the shell shocked were malingerers or degenerates whose underlying defects had surfaced under combat conditions, there was a growing acceptance that the horrors of combat were to blame. Madness, it seemed, was not just a condition found among the biologically inferior who thronged the wards of the asylum. It existed along a continuum, from psychoneurosis to psychosis, and early intervention in a noninstitutional setting might perhaps head off more serious forms of disturbance. Salmon's brief period as an army psychiatrist in the closing months of the war had brought him into direct contact with victims of shell shock, and this helped cement his commitment to a psychiatry practicing outside the traditional asylums and to a program of mental hygiene, an effort to prevent mental illness.

If one adopted the Meyerian view that mental disorder was maladjustment, that meant that normal functioning, neuroses, and psychoses were simply variations along a continuum. It was a stance that opened up a far wider range of problems to psychiatric intervention. Juvenile delinquency, marital disharmony, crime, alcoholism, and so forth could all be viewed as psychological and thus medical problems.

To Salmon and his organization, the most promising way to head off insanity was to treat the young, an idea he soon sold to the first major foundation created by a woman, the Commonwealth Fund. Preventing juvenile delinquency became a major goal of the fund, conceived and financed by Mary Harkness, the widow of Stephen Harkness, John D. Rockefeller's silent partner in Standard Oil. Child guidance clinics and marriage guidance clinics were established as new institutions that were to be the weapons in that fight. By providing advice on child-rearing and using suasion rather than harsh and punitive interventions, it was hoped that delinquency would be headed off.

With foundation money, demonstration clinics were opened, and some psychiatrists now found work outside the walls of traditional institutions.[3]

A handful of others found employment in industry, and a few more in the outpatient clinics that progressive states like New York, Pennsylvania, Michigan, and Massachusetts had begun to establish in the second decade of the twentieth century. The plan was to treat incipient cases of insanity among children and adults with advice and brief psychotherapy in an effort to slow or reverse the remorseless increase in the number of patients confined in state mental hospitals. In that regard, these programs were an abysmal failure, but they did mark a tentative effort by a small minority of psychiatrists to tackle what we now think of as "disordered thinking" before it manifested itself as overt pathology.[4]

Privately, some leading psychiatrists were scornful about the very notion of mental hygiene. Maxwell Gitelson, an early convert to psychoanalysis, commented that "its capacity for creating a need has developed far in advance of its capacity for meeting the developed need."[5] Even Adolf Meyer, whose ideas underpinned the movement, grew skeptical, complaining in a letter to a colleague of "the unfortunate noise and propaganda that has become necessary to maintain the salaries and professionalism of so many half-doctors and new 'professions' under the name of mental hygiene, and under the guise of unattainable panaceas."[6]

Efforts to establish outpatient clinics in general hospitals were met with resistance or, more often, simply ignored. Hospitals had little interest in treating those exhibiting only mild symptoms of mental disorders, and also had little confidence that psychiatrists possessed the capacity to do so. A 1930 survey by the NCMH found that only 3 percent of the general hospitals responding made provision for psychiatric patients, and even that was likely an overestimate, given the tendency to conflate psychiatric and neurological conditions.[7]

Psychiatry's efforts to break out of its institutional straitjacket thus enjoyed only limited success. Thomas Salmon left the NCMH in 1922 for a post at Columbia University. The following year, he became the first practitioner without a background in institutional psychiatry to assume the presidency of the American Psychiatric Association. But his death in a boating accident on Long Island Sound in 1927 silenced one of the main advocates for noninstitutional psychiatry.[8] After his death, the Rockefeller Foundation began to distance itself from the NCMH, and privately Rockefeller staff voiced increasing mistrust and disdain for the whole mental hygiene enterprise. Alan Gregg, who had become director of the Division of Medical Sciences in 1930, spoke repeatedly in his official diary of his sense that "mental hygiene has been much oversold and expectations excited beyond likelihood of gratification"—a feeling that he found was "widespread" as he traveled around the country.[9]

THE YEARS OF THE GREAT DEPRESSION, however, witnessed developments that presaged the great changes that would overtake the profession in the aftermath of the Second World War. The first of these was the creation for the first time in America of a sizable presence of psychiatry in medical schools. Though Johns Hopkins had created some small space for academic psychiatry by appointing Adolf Meyer to head its new department, there were few signs of that experiment being replicated elsewhere until an enormously influential private foundation decided to devote its considerable resources to that task.

In the first decades of the twentieth century, the General Education Board of the Rockefeller Foundation had sought to transform American medical education, bribing and cajoling medical schools to reform their practices, and investing vast sums in universities that embraced its vision.[10] It hoped to drive out of business the proprietary and substandard medical schools that were so notable a feature of the American scene in the nineteenth century—those whose admissions standards, laboratory facilities, and clinical instruction were so poor as to be beyond rescue. Ultimately, as many as a third of the existing schools closed their doors. The Rockefeller Foundation had considerable though mixed success in creating medical schools in the mold it sought, emphasizing biomedicine, the laboratory, and the production of new medical knowledge—to the detriment, some have argued, of the clinical care of patients.[11]

Unquestionably, the financial resources the Rockefeller Foundation mobilized had a transformational impact on American medical education. Since pharmaceutical companies had yet to enter the picture in any major way and government funding of universities and of medicine was essentially nonexistent—as it would remain until the Second World War ushered in the era of the imperial state—alternative sources of funding were few and far between. In the absence of the federal largesse that would soon underwrite research on disease, it was funding from the great private foundations that provided for such research equipment and support as universities could muster, and they underwrote the establishment of new specialized academic departments. In the medical arena, the Rockefeller Foundation was the preeminent actor. It was thus of great consequence when, in the late 1920s, the foundation shifted its focus away from the reform of American medical schools toward a much greater emphasis on research and the generation of new knowledge.

Two divisions were soon established to implement this new policy: a Division of Natural Sciences, headed by Warren Weaver from 1932 onward, and a Division of Medical Sciences, run from 1927 to 1930 by Richard Pearce and then by his deputy, Alan Gregg. In the early 1930s, Weaver and Gregg faced the task of setting the agenda for their respective divisions, subject to the approval of the Rockefeller trustees. For Weaver, a mathematician impressed by the advances in physics and chemistry that quantitative work had brought in its train, these priorities quickly became genetics and molecular biology. Gregg's surprising choice was the highly unfashionable field of psychiatry.

The parlous state of psychiatry had not gone unnoticed even before the Rockefeller Foundation's decision to focus on the field. In 1930, the foundation had solicited a memorandum from David Edsall, the dean of the Harvard Medical School and a trustee, to assist their deliberations. Edsall was hardly encouraging, noting that "traditionally, psychiatry has been distinctly separated from general medical interests and thought to such a degree that, to very many medical men it seems a wholly distinct thing with which they have no relation." Nor was this state of affairs surprising: "In most places psychiatry now is dominated by elusive and inexact methods of study and by speculative thought. Any efforts to employ the more precise methods that are available have been slight and sporadic." Edsall dismissed psychoanalysis as "speculative" and argued that any assistance to the field "would seem to have an element of real danger. . . . [I]t has a strong emotional appeal to many able young men, and I have known a number of men highly trained in science who began activities in psychiatry but, through the fascination of psychoanalysis, gave up their scientific training practically entirely for the more immediate returns of psycho-analysis." He was equally dubious about the value of "the psychological or sociological aspects of psychiatry . . . romantic and appealing" as they might be.[12]

On taking up his position as head of the Division of Medical Sciences, Gregg had spent several months in conversation with a wide variety of psychiatrists: Meyer of Hopkins; his own brother Donald Gregg, who ran the exclusive Channing Sanitarium in the Boston suburbs (having married the founder's daughter); W. G. Hoskins of the Worcester State Hospital in Massachusetts (whose work on dementia praecox was supported by a sizable grant from Mrs. Stanley McCormick); and Franklin McLean of the

University of Chicago, among others.[13] Edsall's memorandum accurately reflected the skepticism most physicians had for psychiatry, but Gregg seems to have become convinced that this status as medicine's ugly stepchild provided precisely the opportunity he was looking for. Psychiatry was, he granted, "one of the most backward fields of science. In some particulars, it was an island rather than an integral part of the mainland of scientific medicine. . . . [T]eaching was poor, research was fragmentary and application was feeble and incomplete."[14]

Prudence might have argued for directing the foundation's resources toward other fields with better prospects for advances. But Gregg decided to roll the dice and urged the trustees to make psychiatry the foundation's top priority in medicine. It was a bold decision. His chances of success were greatly heightened by the fact that at least two leading members of the board had personal experience of serious mental illness. Max Mason, who was nominally president of the foundation from 1929 to 1936 (though his unsuitability led to his being largely sidelined), had been forced to institutionalize his wife for schizophrenia in the 1920s.[15] Raymond Fosdick, who succeeded him, grappled with even more devastating circumstances: on April 4, 1932, his mentally ill wife had shot and killed their two children and then taken her own life—a tragedy that haunted him for the rest of his life.

The double murder and suicide were front-page news in the *New York Times* the following day. Fosdick traveled a great deal in the course of his work, and the *Times* reported that he generally stayed in a New York hotel when in town. The family had gathered together only for the Easter holidays. Fosdick slept through the shootings, only to discover the bodies on the morning of April 4. Diagnosed as manic-depressive, Winifred Fosdick reportedly "had been deranged for ten years" (the age of her young son), but lived next door to her parents and was being treated on an outpatient basis by a local Montclair doctor, Victor Seidler. Fosdick's autobiography is largely bereft of references to his private life. The exception is a brief aside in which he acknowledges losing his wife and two children "in a moment of manic violence. The letters which my wife left behind her showed a mind completely out of touch with reality; she was far more ill than we realized even in the anxious years that preceded her death. . . . It takes time to recover from such a blow—if indeed one ever recovers."[16] It was perhaps no coincidence that, by the end of the year, the trustees endorsed Gregg's recommendation.[17]

When Gregg spoke to the Rockefeller Foundation trustees in April 1933, he outlined the rationale for the priority he proposed to establish. The major reason to throw the foundation's support behind the development of psychiatry and neurological science, he explained, was "because it is the most backward, the most needed, and probably the most fruitful field in medicine." A few years later, justifying the first of several grants to Worcester State Hospital in Massachusetts, there was a note that of each dollar spent "for all state purposes twenty cents is consumed by the institutions for the mentally defective and diseased." Worse still, "The present increase in Massachusetts of committed cases of mental disease is at a rate of 600 a year, requiring a new 2,000 bed hospital every four years."[18] Beyond this, the population of America's mental hospitals was rapidly approaching 400,000 souls on any given day, and the mental health sector was the largest single element in many states' budgets.[19]

A decade later, in a confidential memorandum to the trustees, designed to justify the fact that "approximately three fourths of the Foundation's allotment for work in the medical [arena] is devoted to projects in psychiatry and related or contributory fields," Gregg returned to these themes, emphasizing that the costs associated with mental illness were "tremendous and oppressive. In New York, for example, more than a third of the state budget (apart from debt service) is being spent for the care of the mentally defective and diseased." This was clearly untenable. "Because teaching was poor, research was fragmentary and application was feeble and incomplete," he suggested that "the first problem was to strengthen the teaching of psychiatry."[20]

GREGG WAS RELATIVELY CLEAR-EYED about the difficulties associated with his choice. Though intrigued by psychoanalysis, he was not at first disposed to provide funding. He initially proposed to concentrate the foundation's resources on the "sciences underlying psychiatry," which he enumerated as including "the functions of the nervous system, the role of internal secretions, the factors of heredity, the diseases affecting the mental and psychic phenomena of the entity we have been accustomed erroneously to divide into mind and body." The way forward was complicated by the fact that these were not medical specialties "in which the finest minds are now at work, nor in the field intrinsically easiest for the application of the scientific method."[21]

Some historians have suggested that Gregg simply sought to draw on Meyer's work at Johns Hopkins, having it serve as the basis for his new program.[22] Certainly, Hopkins had by far the largest academic department of psychiatry in the country—indeed, some might argue that it was the only one that came close to matching the intellectual range and staffing that could be found in other areas of medicine at first-rate medical schools. It is also true that the term "psychobiology" was frequently bandied about when the Rockefeller officers discussed their support for psychiatry. But a useful turn of phrase by itself has little significance. There are ample grounds to doubt the claims that Meyer's work underpinned the foundation's support of psychiatry. Gregg and his colleagues provided funds to Horsley Gantt for work on Pavlovian ideas in relation to mental disturbances (financing his attempts, over a four-year period, to create neurotic dogs), and to Curt Richter for his pathbreaking research on such topics as the existence of an internal clock in humans and other organisms, as well as smaller sums to support Leo Kanner's work on child psychiatry, but Meyer himself was not given much support, nor was his department.[23] That remained the case even though Gregg recognized that Meyer was having to subsidize much of his operation from his clinical income and some rather odd donors, and was constantly short of funds.

The lion's share of the Rockefeller money went toward underwriting a massive expansion of small and inadequate academic programs in psychiatry at major medical schools, or creating new departments from scratch. The support for Hopkins was dwarfed by the resources directed to Stanley Cobb's work at Harvard (Cobb received five times as much as the Hopkins researchers put together). Psychiatry at Yale was funded still more munificently. In 1942, Gregg noted that the Rockefeller Foundation "has maintained the department since 1929 to the tune of $1,600,000."[24] Beginning in 1938, the foundation offered a further endowment of $1,500,000 to Yale on condition that it fund a fifty-bed psychopathic hospital for use as a teaching hospital. Yale temporized for years, and the foundation eventually lost patience and withdrew the offer.

Hopkins also received much less than the amounts directed to McGill, Rochester, Illinois, Duke, Tulane, Washington University in St. Louis, and Chicago, where entire departments of psychiatry were founded with Rockefeller money. Though some funds continued to be provided to individual researchers after Meyer's much-postponed retirement in 1941, and Alan Gregg voiced initial support for the appointment of John C. Whitehorn as his successor, the foundation's expectation that Whitehorn's background in biochemistry and physiology would foster a closer engagement between psychiatry and the basic sciences proved misplaced.[25]

Setting research priorities was extremely difficult. The foundation's solution was to fund an extraordinarily heterogeneous array of projects. Throw enough money at the problem, and one or more lines of inquiry would surely yield results. As early as the 1920s, Rockefeller money had flowed to Emil Kraepelin's Munich institute and elsewhere, to fund research on genetics and mental illness. The accession of Hitler to power, and the growing racial dimension of this line of research brought about no change of heart, and money continued to flow to German researchers with a commitment to Nazi racial policies until the outbreak of war.

The recipients included the laboratory of Otmar von Verschuer—who counted Josef Mengele among his employees—and Ernst Rüdin, one of the architects of Hitler's plans for mass sterilization and later extermination of the mentally ill. Rüdin's activities were largely funded by the Rockefeller Foundation until 1939. Even after the war, Gregg awarded "a substantial series of grants to Franz Kallman for research on the genetics of schizophrenia." Kallman, half-Jewish, had been forced to flee Germany in 1936, joining New York State Psychiatric Institute. While in Germany, he had proposed to extend the 1933 compulsory sterilization law, testing all relatives of schizophrenics for even minor anomalies and sterilizing any found to be "defective." "The testing program would have been so massive, and would have involved the consequent sterilization of so many people, that it was considered impracticable even by the Nazis."[26]

At Montreal, Gregg put large amounts of money behind a young neurologist and neurosurgeon, Wilder Penfield, and sought to bring together neurology, neurophysiology, and neurosurgery. At Harvard, large sums were mobilized to create a psychiatric service within a general hospital, while at Yale funds were found for everything from psychoanalysis to primate neurophysiology.

Neurotic disorders, seen as a less extreme form of mental disturbance than psychoses, were treated as a function of the autonomic nervous system on the one hand, and as examples of psychosocial maladjustment on the other. Gantt's research on neurotic dogs at Hopkins was matched by Cornell's program to study neurotic pigs and work on conditioned reflexes in sheep. (The latter proved particularly problematic because sheep "are so markedly gregarious that they cannot endure the loneliness of the laboratory alone and only perform satisfactorily when another sheep is tethered in the corner.")[27] George Draper of Columbia University was given money to examine the relationship between personality and body types. In other words, a thoroughgoing eclecticism characterized the way money was allocated, and the term "psychobiology" simply provided a convenient umbrella that lent some sort of spurious coherence to the whole.

ALTHOUGH THE TEENAGED ALAN GREGG had met Freud in the aftermath of his visit to Clark University, and Donald Gregg, his older brother, dabbled in psychoanalytic techniques at the Channing Sanitarium, the foundation at first shied away from Freud's creation. A staff conference held on October 7, 1930, concluded that "psychoanalysis is in a stage of development where it cannot be attacked philosophically and can be left to its own devices—does not need money but needs maturity and needs defeat in places where it does not stand up. . . . Psychoanalysts are fighting enough among themselves to winnow out a great deal of chaff—nothing for us to do; but may not be dismissed as non-existent."[28]

At just that moment, one of Freud's closest disciples, Franz Alexander, had arrived from Berlin to lecture on psychoanalysis at the invitation of Robert Hutchings, the president of the University of Chicago. The lectures went very badly—Alexander called the visit "a fiasco"—and at the end of the year, he retreated to Boston to lick his wounds.[29] By chance, during his time in Chicago, he had analyzed Alfred K. Stern, who had been suffering from a stomach ulcer that his sessions with Alexander apparently cured. Stern had inherited a banking fortune and had the good fortune to marry the daughter of Julius Rosenwald, the driving force behind the emergence of Sears Roebuck and one of the richest of Chicago's plutocrats. He now became the sort of Dollar Onkel for Alexander that Freud had long fantasized of finding.

By 1932, Alexander was back in Chicago as head of the newly established Chicago Psychoanalytic Institute, a position he would occupy until 1956. Stern was installed as chair of its lay board of trustees, and with his assistance Alexander was soon raising funds from other wealthy Chicagoans and adding more former patients to his board. The presence of Alexander's analysands as trustees ensured that he exercised almost complete power over the institute, which now became a center for psychoanalytic training in America. It was here that Karl Menninger and (more briefly) his younger brother Will gained their acquaintance with psychoanalysis.

Alexander thought he deserved the income and standing of a German Herr Doktor Professor, and paid himself accordingly. When he brought Karen Horney, another prominent refugee analyst from Berlin, to his staff in 1934, she too was paid handsomely as his deputy before they fell out, and she moved to Boston. Before the fallout, Alexander and Stern had succeeded in gaining an audience with Gregg. Gregg initially declined to fund the institute, indicating "he thinks it unwise to back a non-university Institute of Psychoanalysis at Chicago, where there is as yet not even a department of psychiatry."[30] Alexander and Stern were persistent, and finally, in early 1934, Gregg agreed to recommend that the Rockefeller Foundation provide assistance to the institute. In the end, the foundation awarded a total of $220,000 to the institute over an eight-year period. Some of the money was earmarked for psychoanalytic training, but the bulk was intended to support work on psychosomatic disorders.[31] Alexander was shrewd enough to grasp that this emphasis was key to obtaining Rockefeller money.[32] Gregg hoped that the study of psychosomatic illnesses would help create close linkages between psychoanalysis and mainstream medicine.[33]

As early as 1937, there were signs that Gregg's confidence in Alexander was beginning to fray. He acknowledged that the training program seemed to be functioning, but he complained that "the physiological correlations they are attempting are not in competent hands" and the place seemed to be "a one-man show or at least rather too much dominated by Alexander." With some hesitation, he recommended the renewal of funding for three to five years but insisted that any further support "should be based on a clear understanding of termination at the end of the period."[34] Alexander's lifestyle and salary had begun to rankle. Gregg's diary entry for December 8, 1938, records a conversation with Stern, where he expressed dismay with unnamed psychoanalysts who ought to be "prepared to further what they regard as the cause by a larger measure of personal sacrifice in point particularly of salaries." The foundation would not provide "permanent maintenance" for such people. (Stern by now was no longer closely associated with the Chicago Institute and indicated that he concurred with Gregg's assessment.) Gregg's junior associates were even blunter. Robert Lambert noted: "I still don't think much of the Chicago Institute crowd. Maybe Alexander has contributed a little something towards making psychoanalysis respectable but he certainly has not brought it into the scientific fold. I shall feel relief when the [foundation] award terminates—and is not renewed." His colleague Daniel O'Brien concurred: "I have the same general hesitation as you about Alexander and some of the other people at Chicago. . . . Frankly, I would like to see the directorship of any institute of psychoanalysis turned over to, say, a sound physiologist or a good internist in medicine."[35]

Over the next four years, Alexander's lavish salary was the source of repeated negative commentary. Stern distanced himself from Alexander following his divorce from Marion Rosenwald and his marriage to the flamboyant Martha Dodd, daughter of the US ambassador in Berlin.[36] Finally, at a tense meeting on October 31, 1941, Gregg rejected Alexander's overtures. The psychosomatic research seemed to have led nowhere, and the salaries paid at the institute, partly with Rockefeller money, were "too large." There was, he informed Alexander, "no chance" that he would recommend further support from the foundation.[37]

The only significant involvement of the Rockefeller Foundation with psychoanalysis in America thus came to an abrupt end. There were a few crumbs in grants to universities like Harvard, Yale, Chicago, and Washington University in St. Louis, but not much more than that. Even the flirtation with Alexander's institute had cost less than 2 percent of the money the foundation used to underwrite American psychiatry. In 1934, when Gregg had received a proposal to provide support for an international institute of psychoanalysis in Vienna headed by Freud, he wanted nothing to do with it. The amount was small—$30,000 to $60,000—but the project struck him as worthless "in view of the present status of psychoanalysis," in particular its "Cinderella position in point of academic status" and the fact that "Vienna as a locus has not optimum prospects from the standpoint of racial liberties."[38]

And yet, without intending to do so, another branch of the Rockefeller Foundation had helped create the preconditions for the flourishing state of psychoanalysis in the United States. Beginning in 1933, and subject to much internal debate and hand-wringing, the foundation had started a program of "Special Aid to Displaced Scholars," designed to help persecuted European scholars escape fascist Europe. It was this program, among others, that allowed almost 200 psychoanalysts and psychiatrists sympathetic to analysis to relocate to the United States. Refugee analysts would dominate psychoanalysis along the East Coast, and most especially in New York, and for a time constituted more than half the psychoanalytic community in the New World.

CHAPTER THIRTEEN

Talk Therapy

WHEN SIGMUND FREUD DIED IN SEPTEMBER 1939, with Europe on the brink of war, W. H. Auden, who had recently arrived in America, wrote a poem lamenting his passing:

> To us, he is no more a person
> now but a whole climate of opinion
> under whom we conduct our different lives.[1]

It was an acute comment, one that captured how influential Freud's ideas had become. Ultimately, psychoanalysis would capture the imagination of American psychiatrists, but this postwar success threatens to distort our understanding of history, for before 1940, it was only among intellectuals and the urban elites—and emphatically *not* among those working in asylums who claimed expertise in the treatment of mental illness—that psychoanalysis achieved its greatest influence. This divergence of perspectives is one of the most striking features of the interwar years. One is tempted to speak of a cultural schizophrenia.

To the chagrin of most American psychiatrists, Freud's ideas about the sources and treatment of mental illness drew considerable public interest in the years immediately following the First World War. America's belated entrance had spared its troops some of the horrors of trench warfare, but the United States army and its medical corps was nonetheless confronted, like other combatants, with a raft of psychiatric casualties, the victims of shell shock or what the Germans called *Schreckneurosen*. To some observers, these breakdowns under the stress of combat emphasized the role of psychosocial factors in the genesis of serious forms of mental illness, and the limited success of crude forms of psychotherapy in treating shell shock's victims, gave some credence to the idea that talk therapy might prove useful.

One American psychiatrist who had experienced the horrors of trench warfare wrote movingly of what he had seen: "Streams have been choked with the dead and rivers have run red with the blood of the wounded. . . . The terrible pictures of the war have been pictures of death *en masse:* death of men, women and children; death of trees and shrubs, even the grass; death of hope and everywhere the despair of death."[2] Emotions seemed to be converted into physical symptoms: nightmares haunted the traumatized and caused them to relive their trauma. It would be some decades before anyone spoke of post-traumatic stress disorder, but the catastrophes of war appeared to be inextricably connected to the flight into illness.

Intellectuals exhibited a growing fascination with Freud's ideas. Even before the Great War, those who frequented the salon in Greenwich Village presided over by the wealthy banking heiress Mabel Dodge couldn't resist the chance to debate sex and psychoanalysis. Lincoln Steffens, the leading muckraking journalist, recalled an occasion when the discussion was led by Walter Lippmann, then just beginning his distinguished career: "There were no warmer, quieter, more intensely thoughtful conversations at Mabel Dodge's than those on Freud and his implications."[3] Here were doctrines that promised emancipation from American puritanism and provincialism. Mabel Dodge was a living exemplar of the liberation that these ideas could bring in their train, embarking as she was on the third of a string of marriages, adorned by a series of affairs with lovers of both sexes. Her money and charm, and willingness to serve as a patron of the arts, had attracted people such as John Reed, Margaret Sanger, Emma Goldman, and Max Eastman to her soirées, where they could discuss art, revolution, and psychoanalysis, confident that they constituted the new cosmopolitan avant-garde.[4]

Nor were they mistaken. By the 1920s and 1930s, among cultural critics, novelists, and visual artists—even in the venal new world of Hollywood—simplified versions of psychoanalytic ideas and speculations came to enjoy a remarkable influence. Freud's ideas were popularized and spread to wide swaths of the literate middle class. Here was a set of novel notions about human psychology that almost became part of the ether, surfacing in the pages of the *New York Times* or the most widely circulated and influential magazine of the age, the *Saturday Evening Post*, as matter-of-fact, unexceptionable revelations with which their readers were assumed to be familiar.

Dodge herself had become a nationally syndicated columnist for the Hearst newspapers in 1916, and her postwar circle of acquaintances now extended to Gertrude Stein and Pablo Picasso, D. H. Lawrence and Willa Cather, Georgia O'Keefe and Aldous Huxley, among many others. Freudian and Jungian ideas were a frequent topic of conversation. Although Jung, the Crown Prince (as Freud once dubbed him), had long since been excommunicated from the ranks of orthodox psychoanalysis, to Freud's dismay Americans still gave him their attention and the richest among them—Edith Rockefeller McCormick and the Mellons—patronized Jung, not Freud.

In April 1915, five months after he had helped to found the *New Republic*, Walter Lippmann provided its readers with a laudatory piece, "Freud and the Layman."[5] Not to be outdone, Max Eastman and Floyd Dell regaled readers of *The Masses* with essays titled "The Science of the Soul."[6] Theodore Dreiser, whose sexual life was nearly as colorful as Mabel Dodge's (albeit confined to members of the opposite sex), soon published his own pamphlet, "Neurotic America and the Sex Impulse."[7] Mass-market magazines such as *Vanity Fair* and *Ladies Home Journal* rapidly followed suit. A nodding acquaintance with Freud was de rigueur in educated circles in 1920s America. Among fashionable intellectuals, and those who followed them, to profess ignorance of psychoanalysis was to reveal oneself to be hopelessly provincial and unsophisticated.

Freud had spoken wistfully about the fact that "the case histories that I write should read like short stories and that, as one might say, they lack the serious stamp of science."[8] This was, of course, precisely what attracted many readers to his work: the narratives that showed the analyst as detective, painstakingly decoding behavior, uncovering the hidden wellsprings of action, stripping away surface disguises and finally revealing the secrets of the soul. In the hands of popularizers, these features were played up as readers were assured that all sorts of troubles could now be smoothed away.

Writing in *Everybody's Magazine*, Max Eastman extolled Freud's discoveries as "a kind of 'magic' that is rapidly winning the attention of scientific minds in the world of medicine." So convinced was he of its value that he set aside any pretense of objectivity and embraced, as he put it, "the language of the patent medicine advertisement": "Are you worried? Are you worried when there is nothing to worry about? Have you lost confidence in yourself? Are you afraid? Are you nervous, irritable, unable to be decent-tempered around the house? . . . Do you suffer from headaches, nausea, 'neuralgia,' paralysis, or any other mysterious disorder?"[9] Never fear, a miraculous cure for all these ailments was now at hand.

In advancing such an optimistic view of the prospects for curing mental upsets, Eastman was entirely representative of the new generation of Freudian popularizers. In the 1920s, Freud's work was moving in a darker, more pessimistic direction. He wrote of the fundamental tension between civilization and the individual and spoke openly about a death instinct.[10] Repression and perpetual feelings of discontent were, he argued, all-but-inevitable concomitants of civilized existence. Those aspects of psychoanalysis were ignored or played down in most American discussions of the new "science." The audience that encountered psychoanalysis in the popular press absorbed a version shorn of its most troublesome features, one that seemed to promise a facile resolution of life's problems.

Like their European counterparts, American artists, dramatists, and novelists were quick to grasp the possibilities of psychoanalysis. Religious and cultural conservatives reacted with hostility and dismay to the more open discussion of sexuality that Freud's ideas had encouraged, rushing to the defense of "civilized morality." Many shared these traditional views, but especially in urban settings, many did not. In the decade we know as "the roaring Twenties," authors and artists pushed back against notions of propriety and reticence, and they found in the idea that repression led to neurosis a useful prop for their determination to break with the prudery of the past. The emphasis on psychological introspection, on hidden motives, and on the complexities of language all appealed to novelists and dramatists and found echoes in their work, as did Freud's stress on the importance of sex.

Broadway in the 1920s and 1930s saw a host of plays by dramatists now forgotten that embraced and exploited the theories of Freud and Jung. Guilt, repression, Oedipal problems with overbearing mother figures, and sex (repressed and overt) became staples of the stage. The drama critic for the conservative *The Sun* pandered to its largely working-class readership by denouncing the newfangled ideas: "Keep Freud and Jung and the horrors of their psychoanalysis, their subconscious repressions, their complexes and inhibitions off the stage, and let them flourish on the printed page, where they belong."[11] But it was a forlorn protest. The audiences that frequented Broadway plays demanded more psychodrama and they got it.

Rachel Barton Butler's rather frivolous *Mamma's Affair* reached Broadway in 1920 and drew considerable critical attention. Its plot revolved around an unhealthily close relationship between mother and daughter, who is rescued when the physician who diagnoses her pathology falls in love and marries her—a happy resolution all around, as the mother consoles herself with the thought that there is now a doctor in the family. Sex surfaces more openly two years later in Ruth Woodward's *Red Geranium*, whose country-girl heroine moves to Greenwich Village, where she discovers Freud and free love and takes up with a lover who presents each of his many mistresses with a red geranium when he tires of her. More melodramatically, Mary Hoyt Wiborg's *Taboo*, which opened the same year, has a child hitherto mute recovering her speech at the crucial moment, allowing her to save a black man unjustly accused of murder. Freudian themes surface repeatedly in the plays of Susan Glaspell, from the one-act *Suppressed Desires*, through her portrait of a homicidal psychotic in *The Verge*, to her Pulitzer Prize–winning drama *Alison's House*, a fictionalized account of Emily Dickinson's life, which has Dickinson sublimating her unfulfilled sexuality into poetry. Frank Stammers and Harold Orlob produced the first Freudian musical, *Nothing but Love*, as early as 1919. Later on, comedies like Preston Sturges's *Strictly Dishonorable* and Dorrance Davis's *Apron Strings* revolved around sexual repression and the means to overcome it. Alongside these lighter riffs were many plays dealing with sexual suppression and frustration, puritanical matriarchs, and the unconscious drives lurking barely beneath the surface of civilization.

Dreiser, never one to shy away from the dramatic possibilities of sex, created the story of a deformed young man, Isadore, obsessed with young girls, whom he molested and in one case murdered, before ultimately resisting his frustrated urges, releasing his last potential victim, and killing himself. Dreiser had formed a close friendship with the Freudian Smith Ely Jelliffe, and the play drew heavily on his psychoanalytic studies. Lest the audience miss the point, *The Hand of the Potter* closes with an Irish reporter instructing them what it all means:

> I've been readin' up on these cases for some time, an' from what I can make out they're no more guilty than any other person with a disease. . . . This felly could no more help bein' what he was than a fly can help being a fly an' naht an' elephant. . . . If ye'd ever made a study ave the passion ave love in the sense that Freud an' some others have ye'd understand it will enough. It's a great force about which we know naathing as yet an' which we're just beginnin' to look into—what it manes, how it affects people.[12]

Dreiser touched on a crucial new issue—the exculpating nature of insanity—but of all the dramas of the 1920s and 1930s, it is the plays of Eugene O'Neill that have had the most lasting impact. The son of an alcoholic father and a mother addicted to morphine, packed off to a Catholic boarding school and expelled from Princeton after his freshman year, O'Neill was acquainted at first hand with the trials and tribulations of family life. His older brother Jamie drank himself to death at the age of forty-five, and O'Neill's own battles with depression and alcoholism persisted throughout his life. O'Neill was an appalling parent, who had disowned his daughter and had to come to terms with the suicides of his two sons; his dramas (with the single exception of the comedy *Ah, Wilderness!*—a fantasy of the happy childhood he had been denied—heaped tragedy on top of tragedy, culminating in the autobiographical *A Long Day's Journey into Night,* a play he sought, unsuccessfully, to hide from public view until he had been dead for twenty-five years, so raw were the intimate details it exposed to public view.[13] O'Neill was close friends with the psychoanalytically inclined psychiatrist Smith Ely Jelliffe and underwent a brief six-week analysis with G. V. Hamilton in 1927 (who informed him that he suffered from an acute Oedipal complex). Hatred of father figures, near-incest, guilt, repression, and hidden secrets surface repeatedly in his plays.

Visual artists, too, were attracted to these new theories—none more so than the surrealists, whose desire to *épater la bourgeoisie* found in Freudian dream symbolism and psychosexual imagery an ideal rationalization for their approach to art and its creation. Art could at once draw on and reveal the secrets of the unconscious. Salvador Dali, who met with Freud in London, attempted without much success to engage him in a discussion of his painting *The Metamorphosis of Narcissus.* But other surrealist artists were equally open about their debt to psychoanalysis and about its linkages to their efforts to free their imaginations from the constraints of social and psychological censorship. They dabbled in dreams, their paintings dripping with distortions and subliminal references to sex and the unconscious. Experiments proliferated with "automatic" painting and writing, blurring the boundaries between dreams and waking life.[14] And with the scandalous and subversive film *Un Chien Andalou* (1929), psychoanalysis began its entry into the cinema.

Yet any focus on the penetration of psychoanalytic ideas into highbrow and (to a more limited extent) middlebrow culture in the interwar years runs the risk of grossly overestimating the importance of Freud's ideas in

American psychiatry of the 1920s and 1930s, and even more so its impact on the treatment of the mentally ill. Mainstream psychiatry remained implacably hostile to psychoanalysis and clung desperately to its feeble claims to medical identity. The popularity of psychoanalysis in some lay circles may well have hardened psychiatrists' opposition to Freud's doctrines. Throughout the 1930s, analysis as a treatment for mental illness was distinctly a minority taste. The limited marketplace for its wares, exacerbated by the Great Depression, ensured that only a small number of psychoanalysts could successfully earn a living from their craft.

IN THE YEARS IMMEDIATELY after the First World War, the problems psychoanalysis faced were exacerbated by an almost complete lack of control over who counted as a trained practitioner. All sorts of people claimed to be analysts—some of whom had no more claim to the title than a few weeks spent in Vienna or Berlin. The deep splits in the analytic community created yet further problems, as followers of Jung and Adler set up shop, and a number of American analysts eclectically drew on the work of these apostates. Contributing to the chaos, Freud had personally selected one of his own analysands, Horace Frink, to bring the Americans to heel. It was a disastrous choice.

In the course of Frink's first round of treatment, Freud had counseled him to leave his wife and marry another former patient. (The lady in question was unfortunately already married to someone else, an older gentleman who objected strenuously to being cast aside.) The divorce was eventually secured, but the struggle precipitated a series of manic-depressive episodes in Frink, which in turn required more analytic sessions before Freud pronounced him cured. In early 1923, the great man informed the New York Psychoanalytic Society and Institute that they should elect Frink their president, and they dutifully obliged. Sadly, Frink's cure proved evanescent. Not long after his return to the United States, his second wife divorced him. He became psychotic and sought treatment at the Phipps Clinic at Johns Hopkins from Adolf Meyer.

Meyer's initial cautious welcome of psychoanalysis had by now disappeared. He found Freud's psychosexual reductionism and determinism unpalatable and sharply at odds with his own eclectic views. Freud spoke of the unconscious and relied on the couch and free association. Meyer talked of faulty habits and emphasized the use of hospital routines and a carefully orchestrated environment to accomplish habit training. Rather than seeking to make the unconscious conscious, Meyer urged the need to "build up a foundation of habits and interests of a conservative character which will crowd out the feelings which have become morbidly habitual," and whereas Freud regarded transference as a vital tool in the therapeutic process, Meyer dismissed it as trivial.[15]

If the first attempt to bring order to the American psychoanalytic scene failed miserably, others were soon under way. In Europe, with the formation of the Secret Committee in 1912, Freud created a Praetorian guard to fend off heresy and enforce orthodoxy, but there was no comparable body in the New World until the formation of psychoanalytic training institutes in the 1930s.[16] The New York Psychoanalytic Society and Institute, founded in 1911 by one of Freud's earliest American enthusiasts, Abraham Brill, who also served as the American translator of Freud's work, included a number of analysts who had been trained at psychoanalytic institutes in Vienna, Budapest, and Berlin. It was not until 1931, however, that the society created the first American training institute, modeled on these European precedents, and brought Sándor Radó from Berlin to run it. In Chicago, Franz Alexander, who had also emigrated from Berlin, soon founded another institute, which he would dominate for decades. Boston's psychoanalytic society, founded in 1914 and headed by the Harvard neurologist James Jackson Putnam, had ceased to meet on Putnam's death in 1918, though it began to reestablish itself a decade later. It, too, opted to form a training institute in the early 1930s. The Boston Institute was led by an American, Ives Hendrick, who had undergone a two-year training analysis at the Berlin Institute, so here, too, the basic training plan followed existing European precedents, with one crucial difference.

To Freud's dismay, and contrary to his own strongly held views, as these institutes consolidated their curricula they all insisted that candidates must already possess an MD degree. It was a step their leadership considered vital if they were to exclude the wilder forms of analysis that had emerged in the 1920s and secure a rapprochement with mainstream psychiatry and medicine. Equally consequential, and indicative of the marginal and embattled state of the field, the institutes were freestanding and completely independent of universities. That setup had its immediate advantages at the time, allowing the founders absolute control over training, and since universities had not yet transformed themselves into the knowledge factories they would become, the drawbacks were, for the moment, largely invisible. They would become starkly apparent with the passage of time.

These institutes and others that followed became the organized centers of psychoanalysis in America. Training was initially quite informal. At Chicago, for example, the first analyst certified by Franz Alexander was Karl Menninger. Initially drawn to psychiatry at Harvard by his work under Elmer Southard, Menninger had learned what little he knew of Freud from conversations with Smith Ely Jelliffe, the New York neurologist and editor of the *Journal of Nervous and Mental Disease*. It was on this slender basis that he had returned to his hometown of Topeka, Kansas, and decided to found a family clinic modeled on the Mayo Clinic in Minnesota, but specializing in psychiatry, not surgery.

Founded in 1919 as a partnership between Karl Menninger and his father, Charles, the Menninger Clinic's emphasis on psychiatry was driven by the interests of the son, who had returned from training at Harvard Medical School and the Boston Psychopathic Hospital.[17] By 1926, Karl's younger brother William had joined them. His presence allowed Karl to take a leave to undergo more formal training in analysis by the Freudian refugee Franz Alexander at the Chicago Psychoanalytic Institute.

Karl spent the winter of 1931 in Chicago, coupled with a few brief visits between then and February 1932.[18] His brother William, who managed the family clinic in his absence, also sought analysis with Alexander, though his encounter was briefer still.[19] Alexander had encouraged both brothers to keep a mistress, and they were quick to comply. Karl and his wife, Grace, had for a time entered into a *ménage-à-quatre* with a local judge and his wife, till the prospect of scandal prompted the two couples to separate. Karl then turned his attentions to his secretary, with whom he conducted a years-long affair. Will, meantime, had engaged in his own longtime affair with the clinic's director of nursing, which was an open secret around the hospital. Other doctors and staff were strictly warned to avoid romantic entanglements. In the puritanical environment of 1930s Kansas, it is remarkable that gossip and scandal were somehow contained.

Karl's best-selling writings, particularly *The Human Mind* (which appeared in 1930 and became the best-selling mental health book of its time, despite or more probably because of its shallowness and superficiality), helped raise the clinic's visibility on the national stage.[20] It was Karl's brother William who took charge of running the hospital and the organization, tasks for which Karl was temperamentally unfit.

Conscious of how limited his encounter with Alexander had been, Karl sought further training. In October 1938 he undertook a second analysis with Ruth Mack Brunswick in New York. Brunswick had been one of Freud's closest Viennese colleagues, but she was by this time addicted to drugs, prone to falling asleep during the analytic hour, and often more concerned with shopping and talking on the phone than focusing on the patient in front of her. Menninger stayed in New York for a little over eighteen months. How much he learned about psychoanalytic practice in this period is unclear, but the long separation from his wife did lead to a decision about his private life: he divorced and married his longtime mistress. (Neither his children nor his wife reacted well to his desertion, and his now ex-wife became acutely depressed.)

In its early days, the Menninger Clinic admitted around forty patients, the majority of whom were quite disturbed. There were alcoholics among them, as well as those the Menningers diagnosed as psychoneurotic or outright psychotic. Most were rich—essential if they were to afford the high fees the institution charged in return for generous levels of staffing and close personal attention—but they came from relatively newly wealthy midwestern families, not the East Coast old-money types who frequented places like the Channing Sanitarium or Chestnut Lodge (or the McLean or Institute of Living if the sanitaria found them impossible to manage). A handful of patients were treated with classical psychoanalysis; the rest lived in a closely supervised and regulated environment that Will Menninger called milieu therapy, including programs of occupational therapy, recreational therapy, and exercise. By the hospital's own account, only between 5 and 10 percent of those under treatment were ever discharged as fully recovered.[21]

By the late 1930s, the place had a dozen or more doctors on staff, along with an array of female nurses and male attendants. Patients certainly didn't suffer from a lack of attention. "Firm kindness" was enjoined of the staff, and the place in many ways ran like a patriarchal Victorian family. Money was tight, however, and the survival of the clinic was by no means assured. Staff lived on the premises and were basically trapped within its walls.

As in many such families, strife and turmoil lurked beneath the beneficent face the institution presented to the world. Nor were financial and romantic rifts the only sources of tension. The two brothers could scarcely stand each other, a situation temporarily solved by the war, when Will left to run psychiatry for the US army.[22]

Unlike the very brief and informal instruction the Menninger brothers received at Chicago, other recruits to psychoanalysis soon found themselves forced to undertake a lengthy and demanding apprenticeship. Instruction at the institutes became increasingly rigid, hierarchical, and time-consuming. Would-be analysts were forced to spend five years or more engaging in supervised training treating patients under the oversight of a teaching analyst before they could become full-fledged psychoanalysts. It was even longer before they could join the magic circle of those who performed the training. Training analysts alone set the curricula, determined who qualified to join their ranks, and secured the fees that would-be analysts had to pay as the price of qualification. In 1938, the American Psychoanalytic Association further lengthened the period before someone could become a full-fledged analyst by requiring candidates to complete a psychiatric residency before they could begin their training.

Lengthy training programs are typical of modern professions, of course, serving at once to transmit esoteric knowledge, socialize would-be practitioners into the worldview of the occupation, and curtail the supply of new recruits, thus creating artificial scarcity. The training offered by the institutes, however, created extreme dependency among those who sought its blessings, for the very nature of the analysis entailed the exposure of much potentially discrediting information to a supervisor who, in many cases, controlled vital referrals to future patients.[23] For all the claims of analysts to have reconstructed their own personalities and moved beyond the petty neuroses that consume the unanalyzed, one of the most remarkable features of their intraprofessional lives is how prone they have been to bitter arguments, splits, and schisms. Over the years, three institutes begat many, and even within a given institute, backbiting, scheming, and dissatisfaction were often the order of the day.[24] Still, psychoanalytic training now had a definite structure, and there was a measure of control over who could set out his or her shingle as a qualified practitioner.

But what of the market for what these therapists had to offer? With only a small handful of exceptions, the gates of the public asylums that catered to the hundreds of thousands of seriously disturbed patients were shut tight. The somatically oriented psychiatrists who ruled over these institutions wanted nothing to do with analysis, preferring their shock therapies, induced comas, and lobotomies. Besides, even setting aside the question of whether psychoanalysis had anything to offer the psychotic (which Freud believed it did not), the prospect of treating patients in these massively overcrowded and impoverished establishments with individual talk therapy for five hours each week was absurd on its face. True, William Alanson

White, who had read Freud, allowed one or two junior staff members at the massive St. Elizabeths to experiment on a handful of the inmates. But White's token hires were mere window-dressing (the St. Elizabeths patient population was between 6,000 and 7,000).[25] The possibility of employment in state mental hospitals simply didn't exist for most analysts, even supposing that an institutional career would have interested them.

A handful of the private hospitals and homes for the nervous catering to the wealthy provided one possible alternative way of making a living. Alan Gregg's older brother offered some version of psychoanalysis at the Channing Sanitarium, the institution he had inherited from his father-in-law in 1921. Many other sanitaria followed this family-based business model. Chestnut Lodge in Maryland, for example, was founded after Ernest Bullard purchased a failing hotel in Rockville at auction in 1906. Remodeling the building, he reopened it as a sanatorium for nervous patients in 1910, passing it on to his son Dexter Bullard on his death in 1931.[26] In Dexter Bullard's hands, it employed a number of prominent psychoanalysts fleeing Nazi Germany, including Frieda Fromm-Reichmann, and became nationally prominent, for a time drawing wealthy patients from all over the country.

Other European psychoanalysts arrived in Kansas in the late 1930s, as refugees from Nazi persecution who were rescued from certain death by the offer of positions at the Menninger Clinic. They worked for what they increasingly regarded as inadequate salaries while adding to the psychoanalytic gloss of the establishment. Bernard Kamm, a Viennese analyst, hired in 1936 at a salary of $3,000 a year, generated several thousand dollars of income for the clinic each month. When he asked Karl Menninger for a raise after a year, Menninger flew into a rage. Others had the same experience. Privately, Karl Menninger complained bitterly about their "audacity" and added, "They are not in Europe now, and ought to be damned grateful."[27]

Antisemitism lurked only just below the surface. Karl wrote to his father that he was "not at all averse to Jews, but I think we must not get too many Jews in the Clinic, or it will be bad for them and us." Thereafter, he made sure only two Jews a year were allowed into his residency program.[28] The refugees found Topeka a cultural desert, a cow town, and left as soon as they could find other stable employment. Eventually, those who remained on staff collectively demanded pay more commensurate with their qualifications. By threatening to withdraw their labor, they finally succeeded.[29]

Another institution for wealthy, mildly disturbed patients, the Stockbridge Institute for the Study and Treatment of the Psychoneuroses in Massachusetts, founded in 1913, had started out as hostile to analysis. Originally a New York internist, Austen Riggs conceived of the idea for his home for nervous invalids while recuperating from tuberculosis, and the Stockbridge Institute (renamed the Austen Riggs Foundation in 1919) soon accommodated forty patients. Like many physicians of his era, Riggs was fiercely critical of Freud and what he called his "mental gymnastics." Instead, he provided his patients with a structured routine of work, play, and exercise, coupled with commonsense "psychotherapy." The sanitarium's setting, on the main street of an idyllic New England town, proved attractive to the moneyed classes, and the prosperous business easily survived the death of its founder in 1940.

Given Riggs's hostility to psychoanalysis, it is more than a little ironic that within a few years, its leadership had been taken over by Robert Knight, David Rapaport, and Margaret Brennan-Gibson, under whom it became (as it remains to this day) a facility attempting to apply psychoanalytic principles to the treatment of deeply disturbed patients.[30] Perhaps that shift was unsurprising, given the growing popularity of Freudian ideas in the circles from which its patients were drawn.

FREUD HAD LONG EXPRESSED DOUBTS about the applicability of psychoanalysis in the treatment of the psychotic, and though a minority of American analysts would eventually dispute this conclusion, most of those deciding to practice psychoanalysis in the 1930s and 1940s sought an extra-institutional clientele. Besides those who traveled to Vienna to secure Freud's ministrations at first hand—a large fraction of his paying clientele, whom he despised but whose money he gladly took—a small but growing number of wealthy Americans had begun to dabble in psychoanalysis. But these were the years of the Great Depression, many analysts struggled to find enough patients, and often these patients could afford only sharply reduced fees. Still, a handful of analysts did manage to earn more, by some accounts, than most doctors.[31]

Given the dramatic increase in the number and prominence of psychoanalysts after the war, it is important to bear in mind how small the numbers of practicing analysts were in the 1930s. Most analysts belonged to the American Psychoanalytic Association (APsA), which in 1930 had the grand total of sixty-five members.[32] In 1932, the association had reconstituted itself as a federation of the three recognized psychoanalytic societies then in existence in New York, Chicago, and the Baltimore-Washington area. Its membership in that year was ninety-two analysts.[33] (All members of its constituent societies were automatically members of the APsA and of the International Psycho-Analytic Association.) By the end of the decade, the New York Psychoanalytic Society and Institute, which was the largest and most powerful in the country, boasted a total of just over twenty analysts who, between them, were training 106 students.[34] The larger society was at that point engulfed in what has come to be called "the second psychoanalytical civil war," one of the many schisms that have been so central to the enterprise from its earliest years. The first had erupted in 1930, when attempts were made to tighten the definition of who qualified as an analyst and to exclude members of the older generation whose acquaintance with the field was minimal. The number of native-born American analysts had grown slowly, but from the mid-1930s onward, their ranks were steadily augmented by analysts fleeing Hitler and the Nazis, whose arrival doubled the numbers of analysts in the United States. (In 1940, the APsA counted 192 members.)[35] This suggests around 1,000 or 2,000 patients were receiving analytic treatment in 1940, a far cry from the more than 400,000 patients confined in state and county mental hospitals that year.

The passage to the New World for refugee analysts was far from easy. The immigration authorities placed severe obstacles in the path of Jews and political refugees, a policy that condemned most to concentration camps and death. Like other intellectuals and scientists, psychoanalysts needed sponsors, financial support, and jobs if they were to secure the vital visas necessary for resettlement, and these guarantees were in desperately short supply. The Rockefeller Foundation provided some funds, but their archives reveal how difficult it was to secure academic positions for those who looked "too Jewish" and how much antisemitism its officers faced when attempting to place refugees. Milton Winternitz, who was dean of the Yale Medical School until 1935 and himself Jewish, was particularly hesitant to accept Jewish scientists, fearing a backlash from his colleagues.[36]

Unsurprisingly, the sums the foundation devoted to this program to underwrite visas for refugee scientists and physicians were inadequate to meet the need. In desperation, American analysts were canvased for funds and positions. Some funds were forthcoming, but they were paltry. This reflected the straitened circumstances many analysts faced, but there was also a less noble reason: refugee analysts were potential competition for a finite patient population. In the words of Bertram Lewin, president of the New York Psychoanalytic Society and Institute, "What in the world would we do with all these additional analysts?"[37]

Displaced from Vienna, Budapest, Berlin, and elsewhere, the refugees arrived (mostly in New York) having "lost their birthplace and their mother tongue," not to mention the culture that had given birth to Freud's ideas.[38] They brought with them the sense of intellectual superiority and disdain for Dollaria (as Freud contemptuously called the United States) that they had inherited from Freud, and confronted an analytic community that was at once intimidated by them and less than delighted to see them. Attempts were made to persuade the new arrivals to leave New York and spread the doctrines of psychoanalysis to the American heartland, and if that seemed too daunting, as Margaret Mahler recalled, they were invited to consider "'pioneering' to Buffalo, Utica, Syracuse, or some other upstate location."[39] But these suggestions were rarely heeded. The problem was not just that the great majority of American states prohibited foreign physicians from practicing medicine until they had become citizens, necessitating an economically impossible five- or six-year wait time before being able to earn a living, or that there was neither audience nor market for their services. It was rather that, like most immigrants, they sought the comfort and security of proximity. New York had the most established psychoanalytic culture, and the state medical licensing laws were among the most liberal in the country.

Soon, the Viennese and their allies dominated the New York Psychoanalytic Society and Institute. The institute had twenty-seven teaching faculty in 1939. A year later, it had added a further nineteen, and thereafter, European analysts dominated their proceedings.[40] That change did not come about without a fight, massive ill will, and a series of schisms, with Karen Horney and four others leaving to form their own organization in 1941, and Sándor Radó (the man originally brought from Berlin to run the institute) seceding with his followers and establishing a separate institute at Columbia University at the end of the war.[41] By then, psychoanalytic institutes had begun to surface elsewhere, in Topeka (1938), Philadelphia (1939), Detroit (1940), San Francisco (1942), Los Angeles (1946), and Baltimore (1947), for the Second World War had brought about a sea change in psychoanalysis's standing and prospects.[42] The New York Psychoanalytic Society and Institute remained, however, the most prestigious and powerful, its ruling members doing their best to ensure that it secured its standing as the home of Freudian orthodoxy.

CHAPTER FOURTEEN

War

LONG BEFORE THE JAPANESE ATTACK ON PEARL HARBOR, it was evident to most informed observers that the United States would eventually be drawn into the war. In the summer of 1940, with an election looming, Franklin Delano Roosevelt moved cautiously, adding a supplementary appropriation for the army of $8 billion while avoiding full-scale mobilization. The increase was said to be necessary for hemispheric defense, though the passage of the first peacetime draft in September suggested otherwise. Once reelected, Roosevelt sharply increased the pace of preparations, with military expenditures ramping up by a further $26 billion. Massive industrial mobilization soon followed the attack on the Pacific fleet, alongside an intensified military buildup.

While America's similarly late entry into the First World War had limited the number of soldiers who had succumbed to shell shock, their effects on morale and military efficiency had not been forgotten. Leading psychiatrists argued that they could make a crucial contribution to the war effort by screening those subjected to the new draft, eliminating the psychiatrically unfit and mentally marginal. This would spare the country the great expense of training them and avoid the damaging effects of their breakdowns on army operations and morale. These arguments proved persuasive, and two Washington-based psychiatrists—Winfred Overholser, the superintendent of the federal mental hospital, St. Elizabeths, and Harry Stack Sullivan, a prominent if unorthodox psychoanalyst—were tasked with drawing up guidelines to keep out those likely to crumble.[1]

On its face, the job was impossible. The sheer number of men to be screened meant that intake interviews lasted a few minutes at best and were ludicrously superficial. Worse still was the complete absence of reliable techniques for predicting future psychiatric casualties.[2] Sullivan was a closeted gay man, and the one disqualifying condition he fought to eliminate was the exclusion of homosexuals from the draft. Not surprisingly, he failed. But the system he and Overholser devised was by one measure a great success: almost two million potential draftees were discharged as unfit for service on psychiatric grounds. Disqualifications of this sort occurred nearly eight times as often as in the First World War, and the loss of manpower dismayed the military brass. The appointment of General Lewis B. Hershey as director of the Selective Service in November 1941 reflected this hostility, and Hershey's evident disdain for psychiatry swiftly resulted in Sullivan's resignation from his post.[3]

There would be no repeat of the shell-shock epidemic of the First World War—or so it was assumed. Unfortunately, events rapidly discredited that notion. As early as 1942, months after the United States entered the war, problems had begun to manifest themselves among the troops, as if prescreening had never taken place. The horrors of the battlefield—sometimes even the *prospect* of the horrors of the battlefield—created masses of new psychiatric casualties. Inevitably, this triggered massive demands for skilled manpower to respond to the looming threat to army morale and efficiency. The term "shell shock" had been abandoned. War-related casualties were relabeled as cases of "war neurosis" or, toward the closing years of the war, "combat exhaustion."[4]

For more than a year, the military bureaucracy failed to come to grips with the scale of the crisis. The official army history of the war blandly notes that "no plans were made for meeting the problems" of the Second World War II and that until late in 1943, "the Neuropsychiatry Branch was severely handicapped by lack of personnel, information, and statistics."[5] Army policy dictated that soldiers suffering from psychiatric disorders had to be repatriated, and as fighting flared in North Africa and elsewhere, the burden of returning the high numbers of traumatized young men provoked increasing consternation. During the Tunisian campaign in early 1943, for example, a third of all battle-related disorders were diagnosed as psychiatric problems. Moving so many mentally disturbed troops back across the Atlantic (not to mention their counterparts from the Pacific) constituted a major burden. In some months, such as September 1943, more recruits were discharged from the army on psychiatric grounds than were entering the service.

The former superintendent of the Metropolitan State Hospital in Waltham, Massachusetts, Roy Halloran, was appointed at the suggestion of Winfred Overholser to provide new direction, but his attempts to nominate division psychiatrists, as requested by the European theater, were repeatedly blocked by the Army Ground Forces, who claimed that "the position was not necessary." Only the intervention of a new surgeon general, Norman Kirk, enabled him to overcome these objections, though it took months of bureaucratic maneuvering to upgrade the neuropsychiatry division to the same status as general medicine and surgery.

Just as these changes were about to take effect, Lieutenant Colonel Halloran died. Once again Overholser, an old Washington hand who was secretary of the War Committee of the American Psychiatric Association, was asked to provide a list of possible replacements. On December 17, 1943, William Menninger was appointed as Halloran's successor. Soon elevated to the rank of brigadier general, Menninger proved to be a talented empire-builder. Within a month, he had secured additional staff resources and direct access to military command. With fighting in both the Pacific and European theaters rapidly intensifying, psychiatric casualties were an ever more pressing problem. Menninger's stature and influence, and those of his division, rose accordingly.

The psychiatrists working under Menninger confronted an enormous task. During the war years, more than a million troops were admitted to hospitals for neuropsychiatric disorders. In the European theater, overall admission rates were around 70 per 1,000 men per year, but this already substantial figure paled alongside the figures for troops in combat, where admissions rose to as high as 250 per 1,000 men.[6] In the Pacific, during the fierce fighting at Guadalcanal, "of the casualties severe enough to require evacuation," according to one scholar, "40 percent were psychiatric."[7] Nor did the surge in the ranks of the psychiatrically impaired show any signs of slowing as combat came to an end. In 1945, there were 50,662 neuropsychiatric casualties crowding the wards of military hospitals, and 475,397 discharged servicemen were still receiving Veterans Administration (VA) pensions for psychiatric disabilities in 1947.[8]

One of the great challenges facing Menninger was where to find adequate numbers of physicians to treat all these patients. In 1940, the American Psychiatric Association had a total of only 2,295 members. By 1945, the military alone had some 2,400 physicians assigned to psychiatric duties, only a small fraction of whom had any prior experience in the field.[9] As with the need for battlefield surgeons, demand had far exceeded supply, and brief training and orientation to the field—no more than six months and often less—had to suffice, a sharp departure from prewar norms.[10]

In securing rapid training for those under his command, Menninger had an obvious bias toward psychotherapy. His principal deputies—men such as Roy Grinker, John Spiegel, Ralph Kauffman, Norman Rieder, and John Murray—shared his psychoanalytic outlook and played a key role in training the new recruits. Theirs was an Americanized psychoanalysis, one that saw mental illness as the product of stress and unconscious inner conflicts, but they played down its psychosexual origins. Rather than stressing their distance from the rest of psychiatry, Menninger's lieutenants sought to minimize their differences from it. It was a stance that placed them at odds with the émigré analysts who dominated psychoanalytic institutes on the East Coast, whom they dismissed as preoccupied with abstruse theoretical debates and sectarian squabbles.[11] In the hands of these pragmatists, most army psychiatrists were indoctrinated with a thin veneer of knowledge to allow them to play their expected roles. Once placed in the field, these men rapidly acquired extensive experience with a much broader range of psychiatric disturbances than the serious psychoses that had been the focus of prewar institutional psychiatry.

The massive number of breakdowns among previously mentally sound young men cast doubt on mainstream psychiatry's claim that mental illness could be attributed to biological or hereditary defect and helped reinforce links between overwhelming stress and mental pathology. The experience of military psychiatrists suggested that anyone might experience mental collapse if placed under sufficient duress and that the territory of psychiatry might expand far beyond the sorts of patients who crowded the wards of state hospitals. Increasingly, military psychiatrists concluded (like Meyer) that mental health and illness constituted a continuum, rather than being discrete and radically different states of being, and that environmental factors could play a critical role in creating, and perhaps in curing, mental disturbances. If their field experiences encouraged such conclusions, the ideas they absorbed during their training powerfully reinforced them. When the war came to a close, this changed perspective would become the mantra of psychiatrists who built careers outside the walls of the asylum.

Three kinds of treatment regimes emerged in an effort to deal with the profound threats to morale and military efficiency of soldiers who had broken down: brief interventions lasting a day or two as close to the front lines as possible; removal to a more formal psychiatric facility containing a few hundred beds further up the supply chain for up to two weeks of treatment; or removal from the battlefield entirely to something that more closely approximated a traditional mental hospital, where more elaborate interventions could be attempted.[12] The last two restored only a very small fraction to a combat role. Many patients treated in them became permanent invalids. "From a long-term point of view," according to Moses Kaufman, a consultant psychiatrist in the Pacific theater, "one did not always do a soldier a favor by evacuating him from the combat zones."[13] By comparison, treatment at the front lines generally consisted of little more than warm food and a sedative to secure a good night's sleep, and the mobilization of guilt, manipulating the soldiers' feelings of solidarity with their units and their desire not to let down their fellow fighting men. In North Africa, Captain Fred Hanson, a US Army psychiatrist, claimed a success rate of more than 70 percent "after 48 hours of treatment, which basically consisted of resting the soldier and indicating to him that he would soon be rejoining his unit."[14] Other elements were sometimes added to the mix—barbiturate narcosis, hypnosis, chemically induced recall of traumatic events, subcoma doses of insulin, and group therapy—seemingly to reassure the professionals that something medical was taking place.[15] An American version of tea and (not too much) sympathy, it was scarcely a secure foundation on which to erect claims that psychiatrists possessed powerful therapeutic weapons in the fight against mental illness.

The actual techniques in use, while dignified by fancy labels, were little more than common sense. Lawrence Kubie, a prominent American psychoanalyst, claimed that "psychotherapy embraces any effort to influence human thought or feeling or conduct, by precept or by example, by wit or humor, by exhortation or appeals to reason, by distraction or diversion, by rewards or punishments, by charity or social service, by education or by the contagion of another's spirit."[16] Statistics were manipulated to demonstrate purported "cure" rates of 80 or 90 percent. As one American psychiatrist conceded, "The higher command [was] apparently composed of men impressed by figures and much too busy to look behind them and inquire into their meaning."[17] Yet another set of statistics, less attended to at the time, told a more sobering story. Among the American ground forces alone, as John Keegan pointed out in *The Face of Battle,* "504,000 men were permanently lost to the fighting effort for psychiatric reasons—enough manpower to outfit fifty combat divisions."[18] For reasons that were not fully understood, "the incidence of psychiatric breakdown in the United States Army was two to three times higher in World War Two than in World War One."[19]

With typical understatement, the medical historian Ben Shephard remarks that "it might seem paradoxical that American psychiatry should have emerged from the war with its reputation enhanced when its wartime record was so poor." We do know that the further soldiers were removed from the front, and the more extensive their psychiatric treatment, the worse the outcome, and the more likely they were to become permanent casualties. In the Sicilian campaign in 1943, for example, of the American psychiatric casualties evacuated to North Africa for treatment, only 3 percent returned to the fight.[20]

It might be tempting to think that these differential outcomes were largely the product of selection bias. After all, the more disturbed people were, the more likely it was that they would be evacuated from the front, with the most disturbed of all hospitalized. Other evidence, however, suggests that the location and extent of the treatment acted most powerfully to produce these results. "In the North African campaign," as one scholar has observed, "psychiatric casualties were sent to base hospitals located as much as three to five hundred miles away." Distance from the front and from the soldier's platoon seems to have exacerbated symptoms and created more chronicity.[21]

Herbert Spiegel, one of the first psychiatrists involved with treating breakdowns among American soldiers in North Africa, suggested that group morale and loyalty to their buddies played a primary role in determining the incidence of combat neurosis.[22] It was an insight reinforced by the research of social scientists led by the Columbia sociologist Samuel Stouffer and published after the war. They concluded that there was an inverse relationship between the incidence of breakdowns and unit morale, including the strength of soldiers' trust in their commanders and links to their fellow combatants.[23] These factors help explain why treating cases of psychoneurosis close to the front lines and mobilizing feelings of guilt about letting down one's buddies produced better results than moving victims away from the fighting. They also help us account for the higher incidence of psychiatric disorders among Black troops. Segregated from the rest of the armed forces and the victims of systematic discrimination and ill-treatment, they proved particularly vulnerable to breakdowns.[24]

Military psychiatrists recognized that assigning psychiatric labels had profoundly deleterious effects. Hence the decision to call these breakdowns "combat exhaustion," words that suggested that a little rest and recuperation was all that was needed for recovery. Their "treatment" reflected the need to keep soldiers close to the front and to their comrades. "Treatment in the battle zone," as one psychiatrist interviewed for the Army's "Lessons Learned" retrospective offered, articulating the consensus, "was of crucial importance in providing the atmosphere of expectancy for recovery and return to combat duty. Forward, brief treatment clearly communicated to patients, treatment personnel, and the combat reference group that psychiatric casualties were only temporarily unable to function. Conversely, evacuation to distant medical facilities weakened relationships with the combat group and implied failure in battle for which a continuation in the sick role was the only honorable explanation."[25] Giving the soldier a psychiatric diagnosis made that move more likely. If a mere label had such negative effects, sustained psychiatric treatment appeared to make matters even worse. Yet on this curious foundation, these "trick cyclists"—the favorite term of abuse for psychiatrists in the ranks—proclaimed the success of their contributions to the war effort.

Those treating soldiers with shell shock in the First World War had made exactly the same discovery.[26] The upshot was a decision to create special treatment units only ten miles from the front. The French and Americans came to the same conclusion: in the words of the French neurologist Georges Guillain, speaking as early as 1915, these "disorders are perfectly curable at the outset. . . . [S]uch patients must not be evacuated behind the lines, they must be kept in the militarized zone."[27] It was this shared conclusion that led the chief American psychiatrist in France, Thomas Salmon, to coin the descriptive acronym PIE: proximity, immediacy, and expectancy were key to minimizing the creation of chronic patients. As their successors relearned what an earlier generation of military psychiatrists had discovered, PIE became the ruling ideology of the day, and the army brass were reassured that the problem was controllable.

Statistics were produced that appeared to suggest a remarkable success was being achieved.[28] But leading historians of military psychiatry have concluded there are strong reasons to be skeptical of these numbers. The military physicians who produced them had ample incentives to exaggerate their prowess. It would scarcely have advanced their careers to suggest that their treatments did little good. Nor would their superiors have welcomed the news that cures were few and far between, for that would hardly have improved the morale of the troops. It was better to concoct optimistic statistics, and that was easy to do, as the military brass, who otherwise held psychiatrists in low regard, were content to receive these reports without looking too closely into their accuracy.[29]

Internal records kept by military psychiatrists painted a far less rosy picture. A series of records examined by Edgar Jones and Simon Wessely included a Canadian estimate of 22 percent returning to active duty (a number that fell to only 15 percent after relapses), and two further studies that showed 20 percent and 16 percent initially returning to their units, falling within weeks to 12 percent and 5 percent, respectively.[30] Even these figures overstate the success rate, since many of these men were incapable of taking part in combat duties and were reduced to providing support behind the lines. We know that in at least some instances, the gap between what was reported to the military authorities and the actual results was extraordinarily wide. (One British psychiatrist, Thomas Main, reported that "some 65% of [psychiatric] casualties have been returned to active duty" when the actual number was around 10 percent.)[31]

William Menninger put his own spin on what his new recruits had accomplished. Brimming with confidence, armed with their radically simplified version of psychoanalysis, America's military psychiatrists persuaded the politicians, the public, and, just as important, themselves, that their techniques were immensely powerful, and, if implemented early enough, offered a revolutionary new approach to the cure of mental disorder.[32] As Menninger put it, "As a result of our experience in the Army, it is vividly apparent that psychiatry can and must play a much more important role in the health problems of the civilian."[33] Large numbers of those who had served under Menninger decided they wanted to continue in the field, and many sought further training to bolster their credentials.

The shock troops of this new generation of psychoanalytically oriented psychiatrists moved quickly into positions of leadership within the profession in the aftermath of demobilization. William Menninger returned to Kansas where, with crucial financial support from the newly expanded VA, he and his brother Karl began to run a massively expanded training program for new psychiatrists. An extraordinary fraction of the psychiatrists trained in the United States in the aftermath of the war were educated at the Menninger Clinic and the associated veterans hospital, which opened in Topeka in a cosmetically renovated "temporary" wartime facility that had previously served as a general army hospital. Within the psychiatric profession, the new generation swamped the institutionally based old guard, forming the Group for the Advancement of Psychiatry (GAP) to advance their aims. By 1948 they had succeeded in electing William Menninger as president of the American Psychiatric Association.[34]

By the 1960s, the chairs of the great majority of university departments were analysts by training and persuasion, and the discipline's major textbooks heavily emphasized psychoanalytic perspectives. Psychiatry attracted growing numbers of applicants for its internships and residencies, and the best of these supplemented their university training with their own training analyses at powerful analytic institutes that remained at a distance from medical schools. Psychoanalytic training was the ticket, if not quite the sine qua non, for a successful career as an academic psychiatrist. And high-status practice largely consisted of office-based psychotherapy. Patients with severe and chronic forms of mental disorder were for the most part marginalized or ignored by the professional elite, who much preferred their outpatient clientele.

Yet psychoanalytic treatment was seen as potentially relevant even in the treatment of psychosis, and at some private establishments—places like the Menninger Clinic, Chestnut Lodge, Austen Riggs, and the McLean Hospital—efforts were made to treat schizophrenics with the talking cure.[35] As early as 1947, in a remarkable break from prewar precedents, more than half of all American psychiatrists worked in private practice or at outpatient clinics; by 1958, as few as 16 percent practiced their trade in traditional state hospitals. This rapid shift in the profession's center of gravity occurred in the context of an extraordinary expansion in the absolute size of the profession. From fewer than 5,000 members in 1948, the American Psychiatric Association's ranks had risen to more than 27,000 by 1976.[36]

This transformation of the size and composition of American psychiatry was financed to an overwhelming extent by the influx of new federal money. Much postwar education was subsidized by the grants provided to returning veterans under the GI Bill. Under General Omar Bradley, appointed as veterans administrator in April 1945, sixty-nine veterans hospitals were constructed, each with a psychiatric unit.[37] The National Institute of Mental Health (NIMH), formally established on April 1, 1949, devoted as much as 70 percent of its budget in its early years to training grants, and the VA was similarly generous. The Menninger Foundation was the largest single recipient of NIMH funds for this purpose, though Yale and UCLA, whose psychiatry departments were among the first to be led by psychoanalysts, also figured prominently among those receiving support.[38]

In 1946, the newly established Menninger School of Psychiatry had been pressured to undertake a massive VA-funded training of residents. Karl

Menninger had suggested they could train a dozen. General Hawley insisted that, given the staff shortages the VA faced, he must take 100, and 108 showed up in the first year. By the following year, the school was training "roughly half the psychiatrists in the V.A. system, or one third of all the psychiatric residents in the United States." By way of comparison, between 1931 and 1945, the Menninger Clinic had trained a total of thirty-six residents. The plan to embrace this remarkable expansion at the VA proceeded despite much staff resistance and the understandable inability to endow it with much substance. Remarkably, perhaps as many as eleven of the first-year class proceeded to commit suicide.[39]

The extraordinary dominance of the Menninger School in psychiatric training did not last, as university-affiliated departments of psychiatry took advantage of federal largesse and expanded their programs, but the reorientation toward psychodynamic or psychoanalytically inclined professionals most certainly did. By the end of the 1950s, they were one-third of the nation's psychiatrists, and by 1973, they were in the majority. If anything, these raw numbers underestimate their impact, for those working within this tradition occupied the commanding heights of the profession, dominating most academic departments of psychiatry and creaming off the most talented of the profession's new recruits. Their patients were the affluent and the educated, and their incomes higher than those accruing to people in many other branches of the medical profession.[40]

Meanwhile, the less talented recruits were sent off to practice in the mental hospitals and public clinics. Here, for a stigmatized clientele, they provided a second-class variety of psychiatry, often following what Nathan G. Hale, Jr., has described as "incomplete psychiatric residencies or none at all," and perforce relying on brief contacts with patients for whom they prescribed pills and administered shock therapy.[41] Though not reducible to simple statistics, the cultural dominance of analytic approaches was equally clear from best-selling potboilers detailing the miracles wrought within the fifty-minute hour, through popular novels, and on to a variety of Hollywood movies extolling the virtues of the psychiatrists of the couch. Freud may have loathed America, but it was here that his ideas and approach enjoyed their greatest success.

CHAPTER FIFTEEN

Professional Transformations

AMERICAN PSYCHIATRY IN THE YEARS after 1945 became a much more complicated beast than it had been until then. The landscape of academic medicine, and perforce of psychiatry, was irrevocably changed by the enormous expansion of the federal government and its growing involvement in funding scientific and medical research. No one was initially sure what the impact of agencies like the National Science Foundation, the National Institutes of Health, and the National Institute of Mental Health (NIMH) would be, but by the early 1950s, it was clear that America had entered the era of Big Science and Big Medicine and that the previous influence of organizations like the Rockefeller Foundation was waning. Even the enormous endowment of the largest private foundations was dwarfed by the resources that could be mobilized by the modern state, once the requirements of total war had broken the barriers to the expansion of the federal government.

The vast influx of federal dollars, initially to train new practitioners who were now deemed essential, provided a more stable underpinning for academic psychiatry and, with the benefits targeted at veterans, underwrote the rapid increase in the number of psychiatrists. The new practitioners opted, insofar as possible, to distance themselves from the traditional mental hospital and sought to provide outpatient services to a steadily more heterogeneous patient population, many of whom were at once less disturbed and far more affluent than those who had traditionally been institutionalized. Yet institutional psychiatry persisted, stigmatized and all-but-disowned by the glossy psychodynamic psychiatry, whose ambitions were directed elsewhere. The hundreds of thousands of psychotic patients, and the huge sums dispensed by the states to support the empire of asylumdom, ensured as much.

In the divided profession that now emerged, the largely unprecedented market for outpatient services expanded at a remarkable rate. Americans were used to paying for medical care as a commodity, and so the idea of paying for psychotherapy readily made sense. The war had helped promulgate the idea that mental illness and health existed on a continuum, and the psychiatric troubles of soldier-heroes had helped reduce some of the stigma surrounding the wide variety of psychological issues psychiatrists now claimed to treat.

Increased geographical mobility in the booming postwar economy and the baby boom meant that more and more parents had to learn to cope without being able to draw on an extended family network for help. Many turned to the first celebrity pediatrician, Benjamin Spock, for advice. *The Commonsense Book of Baby and Child Care*, first published in 1946, which would go on to sell over fifty million copies, translated Freudian ideas about neurosis into prescriptions for child-rearing, helping to indoctrinate a whole generation of young parents into the psychoanalytic perspective on life. Spock had been one of the first to train at the New York Psychoanalytic Society and Institute, and his devotion to Freudian ideas was lifelong.

His discussion of feeding infants, for example, revolved around the oral needs of the child. Contrary to the harsh and rigid approach to toilet training advocated by behavioral psychologists, he insisted that parents should take a casual approach, never shaming a child about soiling themselves, lest the youngster came "to dread all kinds of dirtiness" and became a fussy, finicky adult, obsessive about everything. Parents should likewise understand that it was quite natural that "boys become romantic toward their mother, girls toward their father," and that their sense of possession will "at times [lead to] wishes that something will happen to" the parent of the opposite sex.[1] These feelings, he assured the parents, help children to grow spiritually and to acquire wholesome feelings toward the opposite sex. But, of course, such sentiments must not be allowed to go too far or persist too long, so the parent must be at once understanding and firm. And so on. Freud was never mentioned, but his perspective was everywhere. Though the book began by urging parents to trust themselves, in becoming the authority figure relied on by millions of them, Dr. Spock helped popularize the notion that all sorts of troubles and distractions could be successfully addressed by professionals.

Hollywood also played an important role in presenting Freud's ideas to a broad public. The film industry displayed an affinity for psychoanalysis from very early on, providing therapists a lucrative market for their wares. German and Viennese analysts dominated the Los Angeles scene from the start, and partly in consequence lay analysts from Europe were allowed to practice, something most American psychoanalysts viewed with dismay. The local clientele was not dissuaded, liking their therapists' foreign accents and personal ties to Freud.

Acting is a deeply problematic profession, and the "talent" at the mercy of the Hollywood studio system frequently sought refuge on the analytic couch, as did more than a few directors. Even the Hollywood moguls, a shallow and venal lot, engaged the services of psychoanalysts. Samuel Goldwyn, convinced that Freudian sexual storylines could prove immensely lucrative and give a semblance of science and respectability to his movies, sailed by ocean liner in 1924 to Europe, determined to make Professor Freud an offer he couldn't refuse. He planned to dangle a check for $100,000 if Freud would only become a screenwriter, only to discover that the great man declined even to grant him an interview. Other moguls sought therapy, either for themselves, for the wives they had betrayed on the casting couch, or for the children they ignored. David Selznick, unexpectedly gripped with depression after the overwhelming commercial success of *Gone with the Wind* in 1939, tried a few sessions himself before giving up and returning to amphetamines and sex. He then sent his wife, Irene, daughter of his great rival, Louis B. Mayer, to the analyst he had tried, May Romm. His wife learned enough to dump him but soon found herself sharing an analyst with one of her husband's many mistresses and her replacement, the actress Jennifer Jones (who married Selznick once she had rid herself of her previous husband).[2]

The success psychoanalysis enjoyed behind the camera was soon replicated in front of it. Moss Hart's *Lady in the Dark,* a musical comedy about the heroine's psychoanalytic treatment, was successfully transferred to the screen in 1941, and in the war years, films with psychoanalytic themes regularly rolled off the production line. In the war's immediate aftermath, Alfred Hitchcock's *Spellbound* marked perhaps the most unembarrassed exploitation of Freud's ideas in the service of popular entertainment. As the titles roll, viewers are informed that the "story deals with psychoanalysis, the method by which modern science treats the emotional problems of the sane. The analyst seeks only to induce the patient to talk about his hidden problems, to open the locked doors of his mind. Once the complexes that have been disturbing the patient are uncovered and interpreted, the illness and confusion disappear . . . and the evils of unreason are driven from the human soul."

Produced by the inimitable David Selznick, who insisted that Hitchcock use May Romm as a consultant, the film exploits the parallels between the work of a detective solving a crime and the role of psychoanalysis in unpacking the hidden layers of the self and past. The solution to the murder mystery depends on the rediscovery of repressed memories by its heroine and her teacher, and a key explication of some of the plot's central puzzles lies in the symbolism of dreams, re-created on the screen (again at Selznick's insistence) by Salvador Dali.

Commercially successful, the film was probably the most overt attempt to educate the masses about the wonders of the talking cure. It was scarcely the last movie to embrace psychoanalytic themes.[3] In 1948, *The Snake Pit,* the highest-grossing film of the year, provided a portrait of a disturbed young woman committed to a dreadfully overcrowded state hospital (a realistic starting point) who is cured by a dedicated psychoanalyst who mysteriously finds the time to use the talking cure, with great success. *Suddenly Last Summer* (1959) employs a similarly sympathetic depiction of the use of analysis to rescue Elizabeth Taylor from a looming lobotomy. Beyond that, viewers of *Rebel without a Cause* (1955) and *Psycho* (1960) were left in little doubt about the sources of pathology and the linkage of adolescent delinquency and adult crime to Freudian family tangles. Hollywood's worship of psychoanalysis continued through *I Never Promised You a Rose Garden* and Robert Redford's 1980 directorial debut, *Ordinary People,* and beyond. Psychoanalytic influence in popular culture was matched among intellectuals. Humanists and social scientists both found Freud's ideas fascinating.

BEYOND A CORE GROUP of officially certified analysts, many who lacked the imprimatur of the established institutes practiced what they termed "psychodynamic psychiatry" on an outpatient basis. In the postwar years, intellectuals across the humanities and social sciences pronounced themselves Freudians, and the first director of the newly established National Institute of Mental Health, Robert Felix, endorsed psychosocial perspectives on mental illness in the most practical of ways: through federal grants underwriting the training of would-be practitioners. Federal research money was also directed to a broad array of projects exploring the psychological and sociological dimensions of mental disorders.[4]

Emboldened by the rapid growth in their numbers, Freudian analysts and their allies took on the old guard, the psychiatrists who still made their living in the state mental hospitals, ministering to the largely impoverished and stigmatized clientele who thronged their wards. William Menninger, whose prestige was at its height after his wartime service, and who found working alongside his older brother Karl in Topeka intolerable, was the chosen candidate of the upstarts. There had been simmering disputes between the brothers all through the 1930s, which Will's absence had temporarily alleviated, but Karl's authority as the older brother was eroded by the successes Will had enjoyed in Washington and the national prominence it had brought him.[5] The jealousy, the petty slights, the barely suppressed hostility each displayed toward the other were perhaps not the best advertisement for the value of psychoanalysis in reconstructing human personalities, but they were intense and threatened the very future of the Menninger Clinic. William Menninger's activism on the national stage, and his subsequent absences fundraising for their joint enterprise, kept some sort of check on what otherwise promised to be a poisonous atmosphere. He was persuaded, with some show of reluctance, to run for president of the American Psychoanalytic Association, a position he took up in 1946.

Menninger's hesitations about taking that job reflected his odd and marginal relationship to classical analysis. His own didactic analysis with Franz Alexander had been brief and unsatisfactory, and the experiences of the next decade had left him even more disenchanted. As he confided to his brother Karl in 1939, "I don't know of any medical group in any place that has as many 'queer birds,' and eccentric individuals in it as the psychoanalytic group. The fact that this has been augmented now by a lot of emigrants only adds to the bizarre nature of this whole group. For that reason, I don't feel any great urgency to identify myself too closely with them."[6] His experiences during the war with psychotherapy had modified but not entirely altered these critical views. And yet the training manual he wrote to systematize the training of army psychiatrists, with the inauspicious title of *War Department Technical Bulletin No, 203*, was full of psychoanalytic language, and its diagnostic categories displayed a heavy Freudian influence.

His ambivalence about assuming the presidency of the American Psychoanalytic Association was not feigned. Indeed, Menninger initially wrote a speech declining the nomination, noting that for years he had had "little contact with the Association's organization and affairs," deterred by its reputation for "discord, disharmony, sectional squabbles and professional disagreements."[7] But when pressed to change his mind, he relented, for reasons that soon became clear.

IT SHOULD OCCASION NO SURPRISE that the analytic community was eager to trade on the greater visibility the war had given to its doctrines. Its new president, however, sought to use the platform his position supplied to push psychoanalysis in a more ecumenical direction. The postwar mental health needs were so great, he contended, and the supply of analysts so inadequate, that the strict boundaries that separated psychoanalysis from mainstream psychiatry ought to be relaxed. Membership ought "to be opened to interested physicians and social scientists and . . . training programs . . . established for psychiatrists in psychoanalytic applications of psychotherapy."[8]

It was a heretical idea the membership promptly vetoed. While Franz Alexander, whose institute in Chicago had already strayed from orthodoxy, endorsed the idea of weakening the barriers between psychoanalysis and other forms of psychotherapy, a parade of the orthodox and their allies—such prominent figures as Ernest Jones, Kurt Eissler, Heinz Hartmann, Phyllis Greenacre, and Rudolph Loewenstein—denounced any attempt to deviate from psychoanalytic truth and Freud's legacy. Rather than endorse Menninger's appeal, on taking his post, "to develop a united front and a spirit of unity towards the enormous jobs to be done that must eclipse personal differences and sectional disagreements," the assembled members voted to make the criteria for membership even more stringent.[9]

Rebuffed on this front, Menninger tried another tack. The American Psychoanalytic Association (APsA) and the American Psychiatric Association (APA) had for some time arranged to coordinate their meetings. Taking advantage of this opportunity to exchange views with like-minded figures in the analytic community, Menninger sought to realize his vision of a socially active, psychodynamically oriented profession by turning to the much larger organization, the APA, and bending it to his vision. Together he and his allies—one of whom, significantly, was the director of the NIMH, Robert Felix—formed a pressure group intended to serve as the catalyst for change in the APA. They called themselves the Group for the Advancement of Psychiatry (GAP).

Membership was to be limited to 150 psychiatrists committed to a thoroughgoing reform and expansion of the professional association, which had hitherto largely confined itself to the publication of an academic journal, the *American Journal of Psychiatry;* and to an annual meeting whose primary purposes often seemed social rather than scientific.[10] GAP sought to change all that, advocating for a sharp increase in dues, the appointment of a full-time medical director, and an outright expansion of the association. Its broader ambition was to serve, in Menninger's words, as a "mobile striking force for American psychiatry."[11]

In its early stages, GAP's single-minded pursuit of its objectives proved remarkably successful. Those committed to Menninger's vision of an eclectic, psychodynamically oriented psychiatry were drawn, in many cases, from the ranks of military psychiatry, along with leading figures from academic psychiatry, which the Rockefeller Foundation had done so much to bring into being. GAP members had prestige and a shared contempt for the backward world of institutional psychiatry, but as yet they were but a minority of the profession. When they secured William Menninger's election as president of the APA in 1947, they only did so because the traditional wing of the profession split its votes between two other candidates, Nolan Lewis of the New York Psychiatric Institute and Arthur Noyes, superintendent of the Norristown State Hospital. In the three-way race, Menninger won with only 41 percent of the total votes, making him simultaneously the president of both the APsA and the APA. GAP members occupied four out of the five positions on the APA Executive Committee and also constituted two-thirds of its council. The association's newly appointed full-time medical director was also a member.

There was ample reason for state hospital psychiatrists, still the majority of the profession, to be alarmed by these developments. Virtually without exception, GAP members were psychiatrists who had no contact with the world of the asylum and who viewed those working there as the most backward members of the profession. Events over the next few years only heightened their concerns, creating a polarization so acute that for a time, in Gerald Grob's words, "the very existence of the APA was called into question."[12] Seeking to reassert their control over the APA, the hospital psychiatrists who sat on the nominating committee put forward a single candidate for the presidency to succeed William Menninger in 1948.

CHARLES BURLINGAME was the superintendent of the Institute of Living in Hartford, Connecticut, an exclusive asylum for the very wealthy. He was closely allied with his state hospital brethren and, like them, an enthusiastic proponent of biological psychiatry and of the various somatic treatments that had emerged in the 1930s, from insulin comas and electroshock to lobotomy. Indeed, he had just opened the first specially designed operating theater to lobotomize those of his "guests" who were suitable candidates for psychosurgery. In a speech subsequently published in the *Journal of the American Medical Association,* he spelled out his allegiance in no uncertain terms, suggesting that "only a stupid person would decry" lobotomy and the shock treatments. As for psychotherapy, it stood alongside the occult and threatened to bring the whole profession into disrepute. The traditionalists quite naturally considered him an ideal candidate to put GAP and its allies in their place. The nonsense GAP was peddling was going to be set to one side, and the medical and biological identity of the profession reaffirmed.[13]

But at the 1948 convention, the institutional psychiatrists found themselves outmaneuvered. Dexter Bullard, the director of the psychoanalytically inclined Chestnut Lodge, considered advocates of lobotomy like Burlingame to be criminals who assaulted and irretrievably damaged the patients in their care. Though not a GAP member himself, he served their purpose by nominating an alternate candidate from the floor, George Stevenson, medical director of the National Committee for Mental Hygiene, a man known for his "progressive" views and his political connections in Washington.[14] From his early work helping to set up child guidance clinics in Minnesota to his years of bureaucratic service in Washington, Stevenson had little direct contact with patients of any sort, and even less with those immured in mental hospitals. His links to the Menningers and their allies had solidified during the war years, and he had played a behind-the-scenes role in drafting the legislation that created the NIMH. In other respects, though, he was an unpromising candidate. As one of his contemporaries noted, in Stevenson's obituary no less, "He was not a colorful character, never smoked or drank and had no small talk or casual conversation . . . and little sense of humor or ability to laugh at the foibles of mankind."[15]

It was thus not Stevenson's personal qualities that prompted his nomination, or his charisma that brought him victory against his gregarious opponent. Voting was confined to the delegates in attendance, and GAP had packed the proceedings with supporters, while many hospital psychiatrists, convinced that the election was a foregone conclusion, had neglected to attend. Their complacency turned out to be a grave mistake. Though institutional psychiatrists constituted the majority of the profession, Stevenson won the presidency by 389 votes to 342.[16]

If that were not bad enough, GAP had set up a number of subcommittees, and several of them began to issue reports that were highly critical of state hospital practices. From September 1947 onward, GAP published a series of bulletins purporting to assess various psychiatric therapies. These were the somatic interventions that were the standard recourse of those practicing in mental hospitals, whose outlook and interventions many members of GAP despised.

The first report GAP issued, in September 1947, was highly critical of electroconvulsive therapy (ECT). Doubt was cast on its therapeutic value, and complaints were voiced about the "overemphasis and unjustified use of electro-shock therapy."[17] A subsequent report on lobotomy was equally critical. Those severing the frontal lobes of their patients were portrayed as irresponsible actors, out of touch with modern psychiatry. Lobotomy, its authors wrote, "represents a mechanistic attitude toward psychiatry which is a throwback to our pre-psychodynamic days."[18] Robert Knight, who had served as chief of staff at the Menninger Clinic before moving to a similar position at Austen Riggs in Massachusetts, brought these criticisms to the lay public via the pages of the *New York Times*, assailing "the indiscriminate use of 'strong-arm' methods of psychotherapy such as electroshock, injections of sodium amytal and lobotomy." A "pernicious" attitude had spread among many of his colleagues that "the patient's illness must be beaten out of him," and the profession had been "prostituted" by those who "know no other methods" and "think only of making the patient give up his complaints and subside."[19]

If their aim was to provoke an angry response from traditional psychiatrists, the members of GAP certainly succeeded. Mental hospital psychiatrists who had initially responded facetiously to the creation of GAP by forming a group of their own dubbed GUP—Group of Unknown Psychiatrists, or Guppies—were now furious. They formed a more serious organization and called themselves the "Preserves," psychiatrists committed to biological psychiatry and somatic treatments. There was open talk of secession, along with mobilization against the analysts and their allies. The structural divisions between academic and office-based psychiatry, and those who had charge of the now half-million patients in mental hospitals, threatened an irretrievable rupture of the whole enterprise. So fierce was the counterreaction that Menninger led a hasty retreat, adding a major proponent of ECT—Lothar Kalinowsky—to GAP's committee on research, and announcing that the report would be revised in a more balanced form.

A compromise candidate was put forward to succeed Stevenson as the APA president, Adolf Meyer's colorless successor at Johns Hopkins, John C. Whitehorn, a man guaranteed to accomplish not much.

GAP'S INITIAL CRUSADES now began to founder. Menninger and his circle had propounded plans for a much broader role for psychiatry in public life, vetting the psychological stability of candidates for public office, applying psychoanalytic insights to the conduct of foreign policy, and reworking "education, industry, recreation" to improve the community's mental health. In many ways, these claims mimicked those made earlier in the century by the proponents of mental hygiene (and Karl Menninger's suggestion in 1940 that it was time for psychiatry to broaden its attentions "beyond a hapless and hopeless few").[20] And like the programs put forward by their predecessors, these suggested social and political interventions mercifully had no substantive content.

They collapsed accordingly in short order. William Menninger kept preaching these doctrines till his death from cancer in 1965, but he cut an increasingly forlorn figure, periodically showing up at the offices of the Rockefeller Foundation and the Commonwealth Fund seeking seed money to implement broad-ranging psychosocial programs at the Menninger Clinic.[21] The Commonwealth Fund had been skeptical of the Menningers' plans even during the 1940s, and though they did not formally reject the proposals they received, neither did they fund them.[22] Internally, the staff was dismayed by the quality of the annual reports forwarded by the Menninger Clinic. One wrote that "even the most conscientious reader is left in a state of bewilderment regarding the accomplishments of the Menninger Foundation."[23] Efforts in 1958 to get support for a marriage counseling service were rebuffed. In 1963, and again in 1964, William Menninger solicited support directly from Malcolm Aldrich, the president of the foundation, and was spurned.[24] For some time, his residual reputation ensured that he was received before his supplications were politely rejected. Eventually, however, exasperated foundation officers begged him not to bother. No money would be forthcoming, so he was asked, in so many words, not to waste their time: "as we told Miss Crim, however, there is no possibility of being of further assistance in the foreseeable future. . . . Under the circumstances I thought it was only fair to let you know the situation so that you would not make a needless call when I know your time in the city is limited."[25]

In the new landscape created by the growing intervention of the federal government in the mental health arena, the Rockefeller Foundation had already embarked on a major reassessment of its commitment to psychiatry. Government programs embraced a very different approach to funding psychiatry (and medicine and science more broadly) than the one the foundation had relied on to dispense its funds. The foundation preferred to pick out prominent and up-and-coming scholars, relying on the instincts and personal judgments of its officers in making its decisions. Its model was now being challenged by something approximating a peer-review system. Within less than a decade, the resources Washington could provide swamped those that could be offered by private philanthropy and entrenched this very different approach to assessing funding priorities.

The long tenure of Raymond Fosdick as the Rockefeller Foundation's president ended in 1948. His replacement, the management theorist Chester Barnard, immediately launched a major review of the foundation's priorities, questioning what its massive investments over the preceding two decades had accomplished. In the case of the Natural Sciences Division led by Warren Weaver, the answer was quite reassuring. Weaver had coined the term "molecular biology," and the grants he had administered had largely created the field.[26] The technologies, laboratories, and scientists Weaver had funded had proved their worth in the just-concluded war, and his own activities in organizing the Applied Mathematics Panel, created to solve mathematical problems related to the wartime military effort, had further cemented his reputation. His investments looked sound and would have massive payoffs in the next decade. Of the eighteen molecular biologists who received a Nobel Prize between 1954 and 1965, fifteen had received funds from the Rockefeller Foundation, and they had received their first support beginning on average nearly two decades earlier.[27] Barnard could feel assured that the Natural Science Division grants had been money well spent.

What about Gregg's program in support of psychiatry? Here, matters were more complicated. The various somatic therapies—malaria therapy, insulin comas, metrazol, ECT, lobotomy, and the like (still mostly seen as having some therapeutic value)—had all originated in Europe and owed nothing to Rockefeller funding. Nor could Gregg and his team point to any other breakthrough that could be traced to their investments. In 1944, Gregg had acquired an able deputy, the neurophysiologist Robert Morison, and four years later, Morison attempted a survey of the state of psychiatry.

What he confronted was the inescapable reality that "a generation of funding [had] yielded painfully little in tangible results."[28] Morison's report did not make for very encouraging reading. Of the approximately $16 million in grants to psychiatry between 1931 and 1946, only about a quarter went to departments that already existed, and little of that money went for research. Nearly a half had been spent on "establishing entirely new or to expanding negligibly small university departments of psychology or psychiatry." What had all this wrought? Morison could point to some progress in the treatment of epilepsy and some "slow but steady progress . . . in the understanding of the elementary functions of nervous tissue. But the total is not distinguished or dramatic." Still, he somewhat unconvincingly concluded, "a sound beginning has been made."[29]

In August 1948, Barnard sent a sharp memorandum to both Gregg and Morison stemming from his reading of the latter's diary and commentary on the state of psychiatry. (Diaries at the Rockefeller Foundation were not private documents but internal records that circulated widely among the officers and were meant to inform policy.) The portrait of the state of psychiatry Morison presented was something he found "terribly disturbing, [though] somehow it wasn't terribly surprising to me. Isn't there a way," he asked, "to blast this situation?"[30]

Morison complained on multiple occasions that the profession's heightened emphasis on psychotherapy had not been accompanied by any effort to test the efficacy of such forms of treatment. Instead of looking for ways to address the issue, psychiatry's leadership, now increasingly consisting of psychoanalysts, seemed to throw up its hands, declare the problem beyond solution, and rely on anecdotal evidence. Barnard was unimpressed. Rather sharply, he confronted his officers: "Doesn't a continued and general refusal to permit or attempt validation of psychotherapeutic methods put everyone concerned, including ourselves, in a position of promoting or carrying on a social racket? How can the charlatans be dealt with if the good men will give no validation but their own individual say-sos?"[31]

A month and a half later, having consulted with Gregg, Morison attempted to answer these pointed questions. His diary entries, he noted, were "to be regarded as a collection of data relevant to the present situation but not necessarily a complete or conclusive description of it." But what followed cannot have made very reassuring reading. Medicine, Morison claimed, had long displayed an almost complete neglect of "the less easily analyzed psychological factors." Combined with "the very rapid increase in scientific knowledge about the organic elements in disease," the upshot was that "the prestige of psychiatry, which had never been very high, declined almost to a disappearing point during the Twenties and Thirties." There had, Morison hastened to add, "been an extraordinary change," "due in part to the interest" of the Rockefeller Foundation. Faced with their professional marginality, "the younger generation of psychiatrists have naturally devoted a large proportion of their energies to gaining acceptance on the part of the rest of the medical profession."[32]

How had they done so? "Since their art was too primitive to be defended on the basis of scientific evidence, psychiatrists have relied largely on rhetorical persuasion in their campaign for recognition. A large part of this persuasiveness has rested upon the revelatory nature of Freudian concepts." This scarcely advanced Morison's defense of the profession very far, as he was immediately forced to concede the limitations of this approach. "It is certainly very difficult to give in any clear and simple way one's reasons for believing that the basic Freudian hypotheses are correct," he conceded. The best he could offer was that "there is no question in my mind . . . that the concept of unconscious motivation has enabled us to understand the meaning of psychiatric symptoms which have hitherto been incomprehensible." But understanding their meaning was not the same as deducing their cause. Morison argued that the acceptance of psychiatry was more tenuous than most practitioners realized. The rest of medicine was still waiting "for evidence of the sort which has validated, for instance, the use of antibiotics. If this is not forthcoming within the next ten to fifteen years, [physicians] may react rather violently, partly out of embarrassment for having extended a welcoming hand to a group which finally failed to produce."[33]

Toward the end of his lengthy assessment of the state of psychiatry, Morison provided direct evidence of the problems he had identified. His focus was a recent report prepared by GAP's research committee on psychotherapy. As he pointed out, this document spent a lot of time talking about "the intrinsic difficulty of doing research in psychotherapy . . . [and] seems more concerned with explaining why it is impossible to do a good job of validation than to find ways of circumventing the difficulties. It would be so much more comfortable if one could only maintain the status quo of acceptance on rhetorical grounds rather than risking the whole reputation of the art by submitting it to scientific study." There was, however, something even more worrisome to which he drew Barnard's attention: "the ease with which the Group for the Advancement of Psychiatry has adopted the committee approach to situations of this sort. There have been several times recently when I have felt that the leaders of American psychiatry are trying to establish the truth on the basis of majority vote. This is, of course, quite contrary to the usual scientific procedure of submitting evidence which can stand on its own merits in a candid world."[34]

Voting would remain a device that organized psychiatry would resort to all the way down to the present. It would be the basis, for example, on which the profession would decide in 1973 that homosexuality was no longer a mental illness; and it has underpinned each successive appearance of the *Diagnostic and Statistical Manual of the American Psychiatric Association* (*DSM*) from the third edition of 1980 all the way down to the fifth edition of 2013.[35]

The skeptical Barnard must have wondered whether the decision to focus the foundation's efforts on remaking psychiatry had been a terrible error. Morison immediately sought some way to avoid such a devastating judgment: "I very much hope that this frank statement of my misgivings about current trends in psychiatry will not give the impression that I feel we have made a mistake in helping these trends to develop," he wrote. "There is absolutely no doubt that something had to be done fifteen years ago to increase the medical interest in psychiatry and to recruit and train personnel. One had to begin somewhere, and it was impossible to start on the basis of tested scientific knowledge." Perhaps conscious that statements like this risked damning the whole program with faint praise, Morison now changed tack. Contrary to the impression his previous remarks surely would have given, he insisted that, looking at the program as a whole,

> The gains so far have really been surprisingly large. For example, it is really of immense importance that the oncoming generation of medical students is being shown that the emotions play an important role in almost all their patients. It is equally significant that there is now a large group of able young men who have been attracted to the field of psychiatry and who may, if properly handled, be able to take the necessary next step. I therefore do not feel that we are supporting a racket when we continue to aid psychiatry in its present, admittedly imperfect state.[36]

It was, in any event, no time for "blasting," but rather for "some less drastic handling," perhaps a shift away from psychiatric teaching toward a greater emphasis on research.

Four years later, in an interoffice memorandum, Morison lamented that "most [psychiatrists] refuse to recognize that the brain may have something to do with the mind." His hopes that the profession's opposition to research on the efficacy of psychotherapy would diminish with time had dimmed. He feared that the "the development of research has lagged badly so that psychiatric practice is still without a scientific foundation."[37] Perhaps the division should move in a different direction. And with that, with a whimper more than a bang, the Rockefeller Foundation essentially exited from its support of psychiatry.

In March 1951, aware that his authority was at an end and that the foundation was moving on, Alan Gregg wrote to Chester Barnard and asked to be relieved of his position as director of the Medical Sciences Division. Barnard proceeded to merge the International Public Health Division and the Division of Medical Sciences into a single entity, and he put Gregg's former deputy, Robert Morison, in charge. A fig leaf was found for public relations purposes to salve Gregg's pride—he was appointed vice chair of the foundation and was charged with writing and speaking about medicine and its broader role in society. He joined various boards and traveled the country, and then in 1956, at the age of sixty-six, foundation policy forced his retirement. A year later, he was dead.

THOUGH GAP HAD BEEN A MAJOR FORCE in psychiatry when Chester Barnard had begun his inquest into what the Rockefeller Foundation's support for psychiatry had accomplished, it was virtually a spent force by the time of Gregg's death. Its demise went unmourned by institutionally based psychiatrists. In the short run, their counteroffensive had headed off their worst fears. Yet in the long run, their fate was sealed. Demographically, the psychiatrists working in the state hospitals were swamped as the numbers of academic psychiatrists and the ranks of those practicing outpatient psychiatry grew. By 1957, only 17 percent of psychiatrists practiced their trade in state hospitals.[38] This rapid shift in the profession's center of gravity occurred in the context of an extraordinary expansion in its absolute size. From fewer than 5,000 APA members in 1948, the association's ranks had risen to more than 27,000 by 1976.

Institutional psychiatrists were badly paid public employees ministering to a largely impoverished and heavily stigmatized patient population, few of whom seemed destined to recover. Their situation contrasted badly with the privileged lives of the professors, and even more so with that of successful psychoanalysts, whose incomes were two to three times those of state hospital psychiatrists and who could inhabit vibrant urban centers and weren't confined to the rural backwaters where asylums were mostly to be found.

Freud's ideas were everywhere in postwar America, reinforced in newspapers and magazines, in popular works of anthropology and sociology, and in novels and movies. New psychiatrists were trained not as they had once been through apprenticeships in state hospitals, but in university departments using textbooks that increasingly emphasized Freudian accounts of the origins of even the most serious forms of mental disorder. Textbooks such as *Practical Clinical Psychiatry,* which had once hewed to a Meyerian line, now embraced psychoanalytic ideas, as did the two-volume *American Handbook of Psychiatry,* which provided passing reference to other approaches but was fundamentally a psychoanalytic text. The second edition of Silvano Arieti's *Interpretation of Schizophrenia* even succeeded in reaching an extra-professional audience, winning the 1975 National Book Award for Science.[39] Though Arieti's suggestion that analysis was the preferred treatment for schizophrenia was a minority view, his claim that pathological families were responsible for creating pathological children became the orthodoxy of the day.

THE ASYLUM MIGHT HAVE BEEN THE SOCIAL SPACE that gave birth to modern psychiatry. Increasingly, however, those who practiced within its walls were the profession's marginal figures—if not quite ostracized, then largely ignored. The impact of the Great Depression on state budgets had been massive, and since mental hospitals constituted the largest, or second-largest, item in most states' expenditures, they fared poorly in the 1930s, a situation compounded by the fact that poverty and want drove increased numbers into these institutions. Then came the war, which deprived the mental hospitals of many of their staff, crimping budgets even further. Desperately short of attendants (to say nothing of doctors and nurses), mental hospitals had filled some of their vacancies by employing conscientious objectors.

Educated middle-class men exposed to the appalling conditions in which nearly half a million fellow citizens were kept reacted with horror to what they observed. After the war, some 2,000 of them collaborated on a best-selling exposé, *Out of Sight, Out of Mind,* a litany of vivid stories reporting, in the words of its editor, a nightmare world of "Inadequacy, Ugliness,

Crowding, Incompetence, Perversion, Frustration, Neglect, Idleness, Callousness, Abuse, Mistreatment, Oppression."[40] A photo essay published the year before in the mass-circulation magazine *Life* had printed some of the pictures smuggled out of hospitals by conscientious objectors. Two of them, taken at Byberry State Hospital in Philadelphia by the Quaker Charles Lord, were particularly shocking. One, of the male incontinent ward, pictured nude men, some slumped on the floor, others picking their way across a concrete floor amid puddles of urine and piles of excrement. A second was of a ward for violent men, known to the attendants as "the death house."[41]

Journalists, some of whom had just returned from Europe, where they had visited Nazi concentration camps, compared the back wards of American mental hospitals to the death camps they had seen.[42] Unlike the Nazis, they noted, the United States had not directly set about killing the mentally ill. Under Hitler, the so-called T-4 program had murdered more than a quarter million mental patients, with the active and enthusiastic participation of many of Germany's leading psychiatrists, eager to relieve the Third Reich of those they contemptuously referred to as "useless eaters." Instead, American institutions neglected and half-starved their patients, relying on nature to take its course.

Perhaps the most widely read of these exposés was a series of newspaper articles on conditions in mental hospitals all across the United States, subsequently published in book form in 1948 as *The Shame of the States*. Its author, the journalist Albert Deutsch, had written the first history of the treatment of the mentally ill in America, a glowing account of how reformers had rescued the mentally disturbed from brutality and neglect in the community, and from confinement and cruelty in jails, to which most of the mentally disturbed had previously been sent. That book was an uplifting story of the passage from ignorance and superstition to enlightenment and science, so congenial to the nation's psychiatrists that they had promptly made its author an honorary member of their association. It might seem paradoxical, then, that its author, of all people, should pen a notably harsh critique of the state of mental hospitals in the postwar era. The fact that the foreword was written by Karl Menninger, and that Deutsch fulsomely praised GAP in his concluding remarks, might suggest that Deutsch had become a critic of institutional psychiatry and come to side with the psychoanalysts. But that conclusion would be badly mistaken.

Deutsch's book provided a series of detailed reports of what he had seen as he had visited a dozen mental hospitals, many of them in liberal northern states that supposedly provided the best public psychiatry. His descriptions were chilling. Everywhere, patients were housed in decayed and overcrowded facilities, bereft of even a semblance of therapy. Beds were crammed together, sometimes stacked two and three high. Food was abominable. Brutality and neglect were rife, with much of the violence inflicted by the attendants: "As I passed through Byberry's wards, I was reminded of the pictures of the Nazi concentration camps at Belsen and Buchenwald. I entered buildings swarming with naked humans herded like cattle and treated with less concern, pervaded by a fetid odor so heavy, so nauseating, that the stench seemed to have almost a physical existence of its own. I saw hundreds of patients living under leaking roofs, surrounded by moldy, decaying walls, and sprawling on rotting floors for want of seats or benches." Byberry, Deutsch noted, had more than 6,100 patients in buildings meant to house no more than 3,400.[43]

Byberry was the first mental hospital Deutsch visited. The rest were as bad or worse. He was oppressed by "the deadly monotony of asylum life, the regimentation, the depersonalization and the dehumanization of the patient, the herding of people with all kinds and degrees of mental sickness on the same wards, the lack of simple decencies, the complete lack of privacy in overcrowded institutions, the contempt for human dignity." Napa State Hospital in California was, he averred, a "grotesque mirage . . . [a] gingercake monstrosity . . . a dangerous firetrap." In Detroit, "Many patients were strapped to their beds by leather thongs. Others sat rigid in chairs to which they had been bound hand and foot. Still others lay tightly wrapped up in 'restraining sheets.' Steel handcuffs restrained the movement of a large number of patients."[44]

Deutsch concluded that "life in a mental hospital, all too often, is a never-ending nightmare."[45] At each of his visits, he had brought along a photographer, and these images provided powerful visual evidence of the horrors his text proclaimed. So here, at first blush, was a powerful indictment of institutional psychiatry. Yet superintendents had thrown open their doors to Deutsch. In some cases, they had personally toured the worst wards with him and his accompanying photographer. So, far from evading the scrutiny he promised to provide, they had positively encouraged the publication of his findings.

They knew their man. For Deutsch sought to use his exposé not to call for the abolition of state hospitals, but to secure huge amounts of new funding to turn them into the therapeutic institutions they purported to be. The conditions he had found were not, he insisted, the fault of the psychiatrists, but ultimately of the politicians and the public at large, both of whom had defaulted on their obligation to provide the amounts of money that humane and effective care necessarily cost. In a concluding chapter on the ideal mental hospital, Deutsch outlined the utopia that increased funding could create: new hospitals for no more than a thousand patients, located in or near urban areas, where easy links could be established with medical schools to bring science to bear on the problems of mental illness. "Maximum application will be made of Freudian psychoanalysis properly modified to meet the needs of psychotic patients."[46] And modern treatments—metrazol shock, ETC, insulin coma treatment, and frontal lobotomies, all in desperately short supply in extant mental hospitals—would be universally available and widely applied.

DEUTSCH AND HIS ALLIES were never able to muster the necessary political support for such a program. Instead, within a decade, as sociologists supported by grants from the new National Institute of Mental Health turned their attention to the state hospitals, a very different set of proposals began to circulate. The sociologist Ivan Belknap suggested in 1956 that mental hospitals "are probably themselves obstacles to the development of an effective program of treatment for the mentally ill" and that "in the long run the abandonment of the state hospitals might be one of the greatest humanitarian reforms and the greatest financial economy ever achieved."[47] It was a judgment echoed by many of his fellow sociologists, most notably the Canadian-American Erving Goffman, whose indictment of the asylum argued that the defects of these establishments were embedded in their very structure and thus could not be removed by any conceivable expenditure or reform. They were total institutions, akin to prisons and concentration camps—a comparison now made not by hyperbolic journalists but by sober social scientists. Far from sheltering and helping to restore the mentally disturbed to sanity, mental hospitals disabled and damaged their clientele.[48]

As the reputation of the mental hospital reached its nadir, institutional psychiatrists faced a bleak future. A diminished presence in their own profession, their claims to expertise and the legitimacy of their calling were under sustained assault. Their far better-remunerated colleagues practiced a very different type of psychiatry on patients who were in almost all respects more desirable, and they also distanced themselves as far as possible from the stigmatized, impoverished, and deeply disturbed psychotics their beleaguered counterparts struggled to cope with. So, far from the increased funding Deutsch and his allies had sought for asylums, states were displaying a growing skepticism about public psychiatry, a shift in sentiment that had begun to manifest itself in the declining number of patients residing in state hospitals on any given day. That decline was slow at first, though it was historically unprecedented since the birth of the asylum in the nineteenth century. But in the 1960s, it would begin to accelerate, and by the Reagan era, mental hospitals and public psychiatry were on the verge of disappearing.

Deinstitutionalization, as the abandonment of traditional mental hospitals came to be called, extended over several decades. The deaths of the asylum and of institutional psychiatry were drawn-out processes. Neither academic psychiatry nor the ever-larger fraction of practicing professionals who made their living from ambulatory patients showed any disposition to criticize, let alone oppose, what was happening. Surely, they concluded, care in the community would prove superior to the horrors of the barracks-asylum. Psychiatrists and their political masters seemed not to notice that both the community and the care were chimeras, as we shall see. So it was that an alternative version of malign neglect became public policy, dressed up in the raiment of reform.

CHAPTER SIXTEEN

A Fragile Hegemony

BY THE MID-1950S, the ideological domination of American psychiatry by Freudian ideas was almost complete. The recruits to the profession who were fortunate enough to secure admission to the burgeoning number of psychoanalytic training institutes and to navigate the complexities of didactic analyses became the professional elite. Though analytic training almost always took place outside university departments of psychiatry (with the Institute at Columbia University the only major exception), the vast majority of academic psychiatrists secured this credential, which rapidly became a sine qua non for professional advancement. Psychoanalysts who chose not to enter academia generally had the most lucrative clientele. Psychiatrists in private practice who could not secure admission to an institute, or failed to complete their training, nonetheless proclaimed themselves practitioners of psychodynamic psychiatry, a watered-down Freudianism that played down the psychosexual elements of Freud's theories and his pessimism but endorsed the view that early experiences had a powerful impact on individual thought and perception. These therapists were satisfied with only one or two therapeutic sessions a week, rather than the five hours required by classical analysis, and often intervened more directly in the therapeutic process than classical analysis allowed.

Even some psychiatrists who found themselves immured in state mental hospitals gave lip service to analytic ideas. Their therapeutic interventions remained heavily somatic but were often given a psychoanalytic gloss, and if psychoanalysis was not embraced in these establishments as a therapy for major mental illnesses, its premises were invoked as the framework for understanding symptoms and their pathogenesis. Perhaps the most striking example of this phenomenon was the effort to invoke the Freudian model of the human personality to explain how lobotomy worked and why it enjoyed the success some psychiatrists claimed it did. The frontal lobes, some alleged, were where the Freudian superego lurked and did its damage, and in severing them, the hospital psychiatrist was freeing the patient from the psychic conflicts that had provoked their misery. A cartoon published in *Life* magazine in 1947 graphically informed its readers of how this surgery of the soul worked: id, ego, and superego were pictured as ruling over different regions of the brain, and the accompanying text summarized how lobotomy solved their conflicts: "[the] surgeon's blade, slicing through the connections between the prefrontal area (the location of the superego) and the rest of the brain, frees his tortured mind from its tyrannical ruler."[1]

Psychoanalysts had a dimensional view of mental illness. Their models of the mind implied that, rather than there being an almost unbridgeable chasm between the mad and the sane, mental illness and health were points along a continuum. The same unconscious drives and conflicts were present in all of us and could be invoked to explain every aspect of human behavior. In the hands of some ambitious analysts, this led to the suggestion that psychoanalysis could be employed to resolve all manner of social problems and political conflicts. Moving from its early focus on therapeutics, the Group for the Advancement of Psychiatry (GAP) formed a Committee on Social Issues, and as early as 1950, this group spoke of "a conscious and deliberate wish to foster those social developments which could promote mental health on a community-wide scale." Psychiatrists, so it claimed, had insights that could resolve "all those problems which have to do with family welfare, child rearing, child and adult education, social and economic factors which influence the community status of individuals and families, inter-group tensions, civil rights and personal liberty."[2] Karl Menninger claimed that psychoanalysis even had a solution to a problem that was preoccupying so many in the 1960s: how to deal with the threat posed by crime and juvenile delinquency. His book *The Crime of Punishment*, a 1966 best-seller, argued that crime was a sickness psychoanalysis could successfully treat, and that punishment was a useless, brutal, and ineffective relic of the past.

Though GAP had by then lost some of the impetus of its early years, its leading figures continued to occupy important positions in American psychiatry. Of the seven presidents of GAP in its first twenty years of existence (1946–1966), five were subsequently elected to the presidency of the American Psychiatric Association. Both the original director of the National Institute of Mental Health (NIMH), Robert Felix, and the two succeeding directors, Stanley Yolles and Bertram Brown, embraced this view of the relevance of psychiatry to a vast range of contemporary issues, and their stewardship of their federal agency reflected this activist bent. A huge range of research was funded between 1949 and 1980, much of it on social issues whose relationship to the central problems of mental health was marginal, at best. Such an ecumenical approach vanished only when the election of Ronald Reagan brought about a dramatic change in the political environment, and an insistence that the NIMH confine its research agenda to the biological sciences. By then, the psychoanalytic dominance of American psychiatry, which had fueled these claims, was on the brink of ending—a development that Reaganite hostility toward expansive governmental programs and social interventionism certainly helped to accelerate.

THE CONFIDENCE IN THE RELEVANCE of psychodynamic psychiatry was not confined to claims about the contributions it could make to solving a vast array of social and political problems. It extended to assertions that psychological factors loomed large in the genesis of a range of illnesses that had traditionally been seen as rooted in the body. The Rockefeller Foundation's limited support for psychoanalysis in the 1930s was fueled in considerable measure by the appeal of Franz Alexander's ideas about psychosomatic illnesses. Alan Gregg and his subordinates were concerned that biological reductionism invited too constrained a view of disease and its treatment. On the other side of the coin, encouraging psychoanalysts to engage with the biological was, they thought, a useful way to temper Freudian excesses and bring psychoanalysis into closer relationship with the medical mainstream.

The founding of the journal *Psychosomatic Medicine* in 1939 was followed three years later by the creation of the American Psychosomatic Society, and in the postwar era, such major problems as heart disease and gastrointestinal disorders were said to have significant psychological components. Asthma, back pain, and allergies were still another group of illnesses whose mysterious etiologies prompted claims for the importance of psychological factors in the genesis of bodily malfunctions. (The German American psychoanalyst Erich Wittkower had popularized the idea of an "allergic personality" in the 1930s.) There was talk of type-A personalities being especially prone to heart attacks, and stress was routinely invoked as a causative factor in the genesis of peptic ulcers and "dyspepsia."[3] "Ulcer types," it was said, were people whose personalities rendered them likely candidates for stomach ulcers. This was one of the first claims Franz Alexander made for psychosomatic medicine (though he was far from the only twentieth-century physician to link emotions and gastric upset).[4] It was a notion that in the early 1980s would fall by the wayside when two Australian physicians, Barry Marshall and Robin Warren, demonstrated that the overwhelming majority of gastric ulcers were caused by a bacterium, *Helicobacter pylori*—a discovery that in 2005 won them a Nobel Prize in medicine.[5]

Asthma was another common and deeply distressing disease with an obscure etiology that psychoanalysts sought to bring within their ambit. Alexander and the American psychoanalyst Helen Flanders Dunbar, who had collaborated to found *Psychosomatic Medicine* before the war, were both convinced that asthma was a classic psychosomatic disorder. Alexander believed that "the asthmatic wheeze was the 'suppressed cry' of a patient suffocated by an over-attentive mother."[6] In her popular treatise *Mind and Body*, published in 1947, Dunbar went even further, asserting that most diseases could be traced back to childhood trauma, often of a sexual sort. Asthma and hay fever were paradigmatic cases:

> There are certain specific emotions which seem to be linked especially to asthma and hay fever. A conflict about longing for mother love and mother care is one of them. There may be a feeling of frustration as a result of too little love or a fear of being smothered by too much. A second emotional conflict characteristic of the allergic is that which results from suppressed libidinal desire, often closely associated with longing for the mother. The steady repetition of this emotional history of "smother love" in the asthmatic is as marked as the contrasting history of hostility and unresolved emotional conflict in the sufferer from hypertension.[7]

If parents were to blame, and hypnosis failed, the remedy for these disorders might be a radical one—parentectomy, or the severing of relationships between parent and child. The 1950s saw the establishment of residential schools for asthmatics to put such doctrines into practice.[8]

Similar etiologies were propounded for diseases that fell more centrally within the psychiatric ambit. Autism as a disease of childhood had been separately identified by the Nazi collaborator Hans Asperger in Vienna and Leo Kanner at Johns Hopkins University in Baltimore. (Later generations have given primary credit to the latter, which is perhaps fortunate now that we know Asperger was complicit in the murder of autistic children.)[9] Kanner's 1943 paper "Autistic Disturbances of Affective Conduct" was a summary of his clinical observations of eleven children who were highly intelligent but displayed "a powerful desire for aloneness," coupled with "an obsessive insistence on persistent sameness."[10] It was a condition then thought to be comparatively rare, though from the mid-1990s onward, the number of children diagnosed with autism would explode.[11]

Kanner borrowed the term "autism" from Eugen Bleuler, famous for being the first to describe and diagnose "schizophrenia." Bleuler had used autism to describe the disconnection of schizophrenics from the outside world. But it was Kanner's paper, and his application of the concept to children, that brought autism to public attention and inspired subsequent generations of researchers.[12] Such children, he argued, exhibited patterns of social withdrawal characterized by restricted social relationships, limited speech, repetitive language and behavior, and obsessions with routine.

Many parents were grateful to Kanner for providing a diagnostic label that helped give some semblance of order to the chaotic world into which their child's social isolation and self-harming behavior had plunged them, and Kanner's formulation encouraged others to attend to and undertake research on the condition. But by the 1950s, that sense of gratitude curdled: Kanner openly entertained the idea that the emotional constipation of the parents, and most especially the mother, explained their children's psychosis. In a 1949 paper on the nosology and psychodynamics of the condition, he claimed that the parents of autistic children had unaffectionate, mechanical relationships with them, prompting their neglected children to "seek comfort in solitude."[13] More vividly, and for a much larger extra-professional audience, in a 1960 interview with *Time* magazine he spoke of cold and distant parents "just happening to defrost long enough to produce a child."[14]

It was a notion he came to repent of and recant by the late 1960s, but for a generation the idea of the "refrigerator mother" inflicted blame and misery on already-traumatized families, largely through the self-promoting efforts of another Austrian refugee, the psychoanalyst and charlatan Bruno Bettelheim, who went one step further and openly compared autistic children to inmates in a concentration camp, with the parents playing the role of sadistic SS guards. "The difference in the plight of prisoners in a concentration camp and the conditions which lead to autism and schizophrenia in children is, of course," he opined in his best-selling *The Empty Fortress*, "that the child has never had a previous chance to develop much of a personality." For anyone in doubt as to parents' culpability, he added, "The precipitating factor in infantile autism is the parents' wish that the child should not exist."[15] Bettelheim suggested that a crucial element in "curing" autism was the same parentectomy other psychoanalysts had recommended for asthmatic children—a severing of all ties between pathological parents and the child they had brought into the world. For decades, Bettelheim put such ideas into practice at the Orthogenic School, an institution with which the University of Chicago saw fit to associate itself, an establishment whose inmates were subjected to mental and physical abuse at the director's hands.[16]

Parents, and especially mothers, were similarly portrayed as the progenitors of perhaps the most serious of all mental disorders, schizophrenia. The analyst Frieda Fromm-Reichmann, a refugee from Nazi Germany, was one of the first Freudians to suggest that psychoanalysis had a place in understanding and treating schizophrenia.[17] As early as 1948 she spoke of "schizophrenogenic" parents, particularly mothers, who displayed a fateful mix of rejection and overprotection that amounted to "malevolence." "The schizophrenic," she suggested, "is painfully distrustful and resentful of other people, due to the severe early warp and rejection he encountered in important people of his infancy and childhood, as a rule, mainly in a schizophrenogenic mother." Before psychoanalysts turned their attention to the disorder, there was mutual incomprehension between psychiatrist and patient. "The thought processes, feelings, communications, and other manifestations of the disturbed schizophrenic seemed nonsensical and without meaning," she continued, but "psychoanalysts know that all manifestations of the human mind are potentially meaningful." She went on to suggest loftily that "it is now recognized that the communications of the schizophrenic are practically always meaningful to him, and potentially intelligible and not infrequently actually understandable to the trained psychoanalyst."[18]

Understandable, perhaps, or at least interpretable within a psychoanalytic paradigm, but not exactly treatable. The best face Fromm-Reichmann could put on the situation was that "the results of the psychotherapeutic efforts with disturbed schizophrenics, so far, are not too discouraging." What did that mean? "Cures have not been to the psychoanalysts' satisfaction as to number or durability"—a result she sought to explain away as "not because of the therapeutic technique used but because of the personal problems of the psychotherapist in his dealings with schizophrenics and because of the personality of the therapist."[19]

Over the following decades, the limitations of talk therapy in the treatment of psychosis turned out to be completely intractable, even in richly endowed private institutions like Chestnut Lodge, where Fromm-Reichmann practiced. A handful of recoveries were offset by the vast majority of failures. The prospect of employing such time-intensive techniques in overcrowded and understaffed state hospitals was, of course, chimerical. Yet the psychoanalytic dominance of psychiatry, and its preeminence in the academy, helped ensure that the idea that the cause of schizophrenia was family pathology became the ruling orthodoxy. The pain and resentment such theories caused the families of schizophrenics remained largely hidden from view so long as psychoanalytic hegemony lasted, not only because of the internalized guilt that having a mentally ill relative created, but because any protests were readily dismissed as emanating from the "malevolent" parent who had fostered the madness in the first place. Once professional doubts began to surface, however, the latent anger and anguish would erupt in a fierce backlash against psychoanalysis, contributing to a rapid embrace among patients' families of the alternative perspective offered by biological psychiatry.

IF PSYCHOANALYSIS WAS ILL-SUITED, as Freud had acknowledged, to the treatment of the hundreds of thousands of patients struggling with acute forms of psychosis who thronged the wards of the state hospitals, it had a much greater appeal to the substantial fraction of psychiatrists who practiced outpatient therapy. As long as sufficient numbers of patients saw analysis as the solution to their neuroses and unhappiness, it promised an attractive career, one that proved appealing to those recruits to medicine who found biomedical reductionism and the lack of sustained contact with their patients distasteful. When cheaper alternatives began to manifest themselves in the therapeutic marketplace, the viability of the profession began to crumble, its ability to defend itself further undermined from within by the schisms that had characterized the Freudian enterprise from its earliest years. Orthodox Freudians at the New York Psychoanalytic Society and Institute poured scorn on the adulterated version of psychoanalysis associated with the Menningers and their followers. There were vicious fights between the groups surrounding such figures as Franz Alexander and Sándor Radó, and between those identifying with Anna Freud and

Melanie Klein. Even within individual psychoanalytic institutes, petty jealousy, tensions, power struggles, and backbiting were the order of the day—scarcely an advertisement for the maturity and self-awareness that being analyzed was supposed to produce.[20]

The gap between promise and performance was troubling and had been evident to some observers very early on. Beyond the internal debates at the Rockefeller Foundation about what its massive funding of psychiatry had actually accomplished, Robert Morison was privately deeply concerned about the propensity of psychoanalysts to rely more on rhetoric than substance in advancing their case.[21] When he succeeded Alan Gregg as head of the foundation's medical division, Morison repeatedly pushed leading analysts to provide systematic evidence of the effectiveness of their interventions, only to be repeatedly spurned. His disenchantment grew, and he soon concluded that "for some time to come it seems likely that university departments of psychology will offer better research possibilities than most departments of psychiatry." Psychology, thanks to a new emphasis on its clinical applications, was emerging as a competitor in the mental health arena. With its focus on cognition and human behavior, it claimed to be more adept at treating the host of anxieties, fears, learning disabilities, and anger-management issues besetting patients, including the traumatized veterans of the war.[22]

THE NATIONAL INSTITUTE OF MENTAL HEALTH, which by the early 1950s had replaced the Rockefeller Foundation as the major source of funding for psychiatric training and research, had reached essentially the same conclusion. The NIMH's first director, Robert Felix, interpreted his mandate broadly, funding not just psychiatrists, but also psychologists and other social scientists.[23] The bulk of the research funding, both in dollars and number of projects supported, went to these other disciplines, including psychology, sociology, and anthropology.

The interest in funding work of this sort in the 1940s and 1950s was boosted by the claims of psychiatric casualties of war, but it also reflected the enormous fiscal and social costs of mental illness among the civilian population, a topic that greatly exercised the governors of individual states.[24] One very important part of the NIMH's intramural research capacity, its Biometry Branch, provided valuable ammunition for those seeking increased funding for basic research by periodically issuing reports estimating extraordinary direct and indirect costs to the economy from the burden of mental illness.[25] Felix could be relied on to testify regularly before his congressional masters about the great progress being made, and imminent breakthroughs about to be realized, provided that the flow of federal dollars was sustained. His sunny optimism was rarely scrutinized, for who could doubt the progressive powers of modern medical science? At times, as the historian of psychiatry Gerald Grob has noted, Felix's colleagues quietly sought to attenuate his enthusiasm. But Felix was evidently more politically sagacious than they, and funding increased at an exponential rate. An initial budget of $9 million in 1949 grew to $14 million in 1955, $50 million by 1959, and $189 million by 1964.[26]

There was, however, no clear road map for spending this cornucopia of research dollars. Under Felix, the NIMH employed a scattershot approach, underwriting research on lobotomy and epilepsy that went primarily to medical investigators, but also giving grants to epidemiologists, to teams of researchers trying to make mental hospitals function as therapeutic institutions, to those examining psychological therapies, and to those proposing basic research with little by way of direct clinical applications. As an in-house history indicates, the decision was motivated in part by a recognition of how weak the current understanding of the etiology of mental illness was, making it "wisest to support the best research in any and all fields related to mental illness."[27]

The very first grant in NIMH's history was awarded to a psychologist, Winthrop Kellogg, in 1949 for a study on the "basic nature of the learning process" and that turned out to be symbolically appropriate, for in subsequent years psychology was the discipline that routinely received most of the research dollars dispensed by the NIMH.[28] In 1964, for example, 62 percent of principal investigators were classified as social and behavioral scientists (overwhelmingly psychologists, who numbered 55 percent of the total number of grantees), and they disposed of 60 percent of that year's grant funding. By way of contrast, psychiatrists were only 12 percent of the principal investigators, and their share of the research budget was a comparatively meager 15 percent.[29]

A WIDE RANGE OF SUBSPECIALTIES shared in the bonanza. Psychobiologists took their share, but so, too, did psychologists working on cognition, perception, personality, and social psychology; on group dynamics, motivation, and development; and on language and behavior, psychotherapeutic interaction, and operant conditioning. In some of these areas, the mental health relevance of the work was clear; in others, attenuated almost to the vanishing point. Central or peripheral, it scarcely seemed to matter to Felix and his subordinates, provided that the grant applications passed muster with peer reviewers. This, surely, was one of psychology's key advantages. For the experimental, laboratory-based, and statistical character of most research in the field, and its conformance with the hypothesis-testing empiricism and mathematical formalism that was seen as the hallmark of "science," made it ideally suited to survive the grant-review process. Over time, as leading cadres became experienced at grantsmanship and incorporated the lessons they had learned into the training of the next generation of psychologists, and as experimental design, data collection, and statistical sophistication advanced, these comparative advantages became self-reinforcing.

The contrast with psychiatry in these same years is illuminating. With positions of power and authority dominated by the psychoanalytically inclined, psychiatry was poorly placed when it came to competing for large research grants. Psychoanalysis, beginning with Freud, developed its theory and technique from individual clinical encounters and the case history in ways that simply did not lend themselves to the experimental, large-scale approaches the NIMH and other government funding agencies quickly came to prefer. Indeed, most scholars working within the analytic tradition were actively hostile toward such modes of knowledge generation, seeing them as deeply flawed and unlikely to address the questions that they argued were central to understanding mental illness. Psychoanalysis suffered also from some self-imposed structural disadvantages. Where academic psychologists were entrenched within the university system, the most prominent and influential psychoanalysts were located in free-standing institutes with no direct connection to academic medicine. Beyond this, the training they provided was oriented toward practice, not research. Leading psychoanalytic clinicians were to be found outside the ambit of the universities—at places like the Menninger Clinic in Kansas, Austen Riggs in the Berkshires, and Chestnut Lodge in Maryland, all establishments with at best tenuous links to the academic world.

In the years of psychoanalytic dominance, the NIMH was generous with funding to train analysts and encourage the expansion of their ranks in academic psychiatry. When it came to research grants, however, much less was forthcoming. Looking back, one psychoanalyst complained, "We couldn't develop grants . . . that satisfied the psychologists and social workers who were running the grant departments at NIMH. I went down to Washington (and) I got nowhere because we couldn't formulate psychoanalytic research in a way that was 'one, two, three, four, five.' Psychoanalysis is not that way."[30]

Between 1948 and 1963, NIMH research grants totaled $156 million. Of this, less than $4 million (about 2 percent) was directed to psychoanalysts or psychoanalytic institutes, and most of this money was, as the historian Nathan Hale, Jr., pointed out, "not directly for research in psychoanalysis; most were for psychoanalysts working in related fields, such as studies of family relationships in schizophrenia, autism, mental hospitals, psychosomatic studies of adults and children, and early infant development." Over a longer time frame, and considering only grants for research on psychotherapy, the marginalization of psychoanalysts was striking: only 7 percent of the $30 million disbursed in this area between 1947 and 1973 went to analysts. By contrast, nearly 50 percent, or $14 million, was awarded to study behavioral therapy, a field dominated by psychologists.[31] In time, psychoanalysts' inability to secure major research funding would help to undermine their standing in medical schools.

THE EXPANSION OF THE UNIVERSITY SYSTEM after the Second World War owed much to the GI Bill, which provided funding for returning soldiers to pay for higher education. But in the long run, it was the ever-increasing research money provided by the federal government that proved a more durable basis for the growth of tertiary education, and transformed its operations. A product of total war and of the Cold War, the big science and big medicine underwritten by federal largesse rapidly transformed research universities into knowledge factories. Institutions and departments were ranked according to the dollars their research entrepreneurs succeeded in capturing from government and industry, and the practitioners making up the various academic guilds found their prestige, their influence, even their salaries, ever more tightly linked to their contributions (or lack thereof) to the pile of treasure to which the seats of modern learning became addicted. For a time, the flow of training dollars hid the danger this represented to psychoanalysis. NIMH-funded training grants increased from $4.25 million in 1948 to $84.6 million in 1965 and reached a peak of $111 million in 1974. But the rapid decline of training grants after that left psychoanalytic psychiatry bereft of the currency that mattered to its academic masters, so when challenges arose from other quarters, it was ill-equipped to defend itself.

Worse still, there was a cuckoo in the psychotherapeutic nest, a creature that owed its very existence to the Second World War. Psychology had emerged after 1945, not just as a highly successful competitor for research dollars, but as an alternative profession offering therapy to the mentally troubled. In the first half of the twentieth century, as a variety of social science disciplines organized and defined themselves within a university context, psychology had been split between the core of the discipline, which consolidated around a vision of an academic, laboratory, and research-based specialization, and a group with a more applied agenda. The progress of what as early as 1907 had been dubbed "clinical psychology" was halting and uncertain for most of this period.[32] It owed what success it had to the development of mental testing, building on the importation of IQ testing, first developed by the French psychologist Alfred Binet in 1905, and its application in such arenas as the identification and control of the "feeble-minded" and the disposition of juvenile delinquents.[33] Attempts to diversify into the treatment of psychological disorders met with fierce resistance from organized psychiatry.[34] Until the war, even as a marginal and stigmatized branch of medicine, psychiatry had far more legitimacy and power than the (heavily female) new discipline of applied psychology, and it was readily able to keep the upstart psychologists in a subordinate position. Reflecting this reality, in 1940 there was still not a single PhD program in clinical psychology.[35] That year, a mere 272 members of the American Psychological Association identified themselves as practicing clinical psychology of any kind, and for the most part this meant mental testing, not administering psychotherapy.[36]

THE SHEER SCOPE OF MENTAL PROBLEMS among the armed forces, and the mismatch between the number of trained psychiatrists and the demand for treatment of psychiatric trauma among the troops, had prompted the recruitment of some psychologists to treatment teams—a move made easier by the fact that the treatments on offer were essentially supportive and psychotherapeutic in nature, and by the existence of military hierarchies that enabled medics to remain in overall charge.[37] William Menninger, desperate for more manpower, welcomed their contributions (though preferring that they concentrate on mental testing rather than psychotherapy).

As Ellen Herman has pointed out, by war's end, there were more than 1,700 psychologists working for the military, many of them providing psychotherapy. "In 1946, a survey of every psychologist and psychologist-in-training who had served in the military showed a striking movement toward clinical work in the war years. Hundreds of them had practiced psychotherapy for the first time and many intended to return to school for further training in this field."[38] Federal funding, originally from the Veterans Administration and then from the newly established NIMH, made that possible on a mass basis. The center of gravity of psychology as a discipline was irrevocably altered.

There was fierce resistance in some departments of psychology to this shift toward clinical psychology. "Applied" work in university settings has traditionally been regarded with suspicion, and "theoretical" work routinely carried the highest prestige.[39] Wedded to this view, and to the laboratory-based, research-oriented model of their discipline, some high-status departments like Harvard, Princeton, and the University of Pennsylvania rejected the very idea of establishing clinical psychology programs.[40] Others, however, were more easily swayed. In short order, that began to change.

Chafing at their subordinate position in the social division of labor, clinical psychologists sought to bolster the legitimacy of their new profession and obtain a greater degree of professional autonomy. Leaders of the movement realized that their legitimacy depended on close ties to university departments and to a curriculum that combined clinical training with demonstrated competence in research methodology. At a 1949 conference held in Boulder, Colorado, and funded by the US Public Health Service's Division of Mental Hygiene, the nuts and bolts of just such a program were hammered out. What became known as the "scientist-practitioner model" was the core of the new approach, which appropriated the mantle of science and combined it with supervised clinical training, emphasizing "the necessity of an academic background in general and experimental psychology as the foundation for training in clinical psychology."[41]

Politically, this was extraordinarily astute. By requiring two years of basic training in academic psychology, the new program encouraged the model's acceptance by existing university departments. The influx of federally funded clinical psychologists brought extraordinary amounts of new funding to the discipline, and the prospect of adding substantial numbers of new research faculty. The "scientist-practitioner" model ritually bowed to the superior knowledge and standing of the researchers, provided a "scientific" basis for the new professionals' practice, and created a means of distinguishing the properly trained from the quack.[42]

As early as 1947, the Veterans Administration was underwriting the training of 200 clinical psychologists. From 1949 onward, the newly established NIMH advanced much larger sums to underwrite graduate and professional training, and while the bulk of the institute's funding was committed to the training of psychiatrists, a substantial sum was diverted to clinical psychology, subsidizing the hiring of additional faculty in psychology departments, and providing stipends to would-be practitioners.[43] The upshot was a dramatic expansion of the field. In 1945, the American Psychological Association had 4,173 members. By 1960, there were more than 18,000—a reflection of the fact that five times as many doctorates in psychology had been awarded in the 1950s as in the preceding decade. By the turn of the century, membership exceeded 80,000, and those numbers continued to rise, reaching a peak of more than 92,000 in 2008. National Science Foundation data suggest that, by 1964, more than two-thirds of American psychologists with a doctorate were working in the mental health field.

Ironically, it was the very dominance of psychoanalytic perspectives in postwar America that had done much to create the social space for psychology to expand its domain beyond the laboratory. The army and the traditional mental hospitals were both hierarchical organizations, and their bureaucratic structure kept psychologists duly subordinate to their medical superiors. William Menninger had told the clinical psychologists who worked for him during the war exactly where they stood: cooperation could proceed only if the psychologists acknowledged and abided by their dependent status. There ought to be no gainsaying that "certain kinds of painstakingly gathered clinical knowledge . . . are prerequisites to carrying on psychotherapy" and these forms of knowledge necessarily needed to be taught by medically trained psychiatrists, those who alone possessed that knowledge in its entirety. Provided that psychologists "can accept the psychiatrist as the quarterback of a team that works together," Menninger made clear, "the bugaboos of status, jurisdiction, equality and subordination become dead issues."[44]

But psychoanalysis, with its office-based practice, removed the possibility of bureaucratic subordination, allowing psychologists to practice independent of psychiatric supervision. The incursion of clinical psychologists into their turf was a development that medically trained analysts resented and fiercely resisted. In 1955 Maxwell Gitelson, who would become president of the American Psychoanalytic Association the following year, wrote that he was "committed to the liquidation of lay therapy in the United States."[45] Yet "liquidation" was quite beyond his powers.

On the contrary, the experimental, laboratory-based, and statistical character of clinical psychology, and its conformance with the hypothesis-testing empiricism that was seen as the hallmark of "science," made it far more capable of developing research programs that satisfied the requirements of peer review. Guided by their academic colleagues, clinically oriented psychologists were soon adept at modeling their grant proposals along these lines. Crucially, this led clinical psychology to develop therapeutic interventions that targeted particular symptom complexes for modification, and that could claim some degree of empirical validation.

On another front, and not coincidentally, the emphasis on a strong research component in the doctoral training of clinical psychologists greatly strengthened their ability to claim for themselves the prestige that accrued to the status of being a "scientist" rather than a mere technician. It also allowed them to boast of their superiority to the average practicing psychiatrist, whose medical education had not included instruction in how to conduct scientific research. As the executive secretary of the American Psychological Association, Dael Wolfle, pointed out in 1949 as the model was being adopted, "The average practicing physician or psychiatrist has neither the research interest nor the research skill that we attempt to develop in the student receiving his Ph.D. in clinical psychology."[46]

THE APPROACH THAT CAME to dominate clinical psychology, and, after the demise of psychoanalysis in the 1980s, to characterize whatever residual dabbling in psychotherapy some psychiatrists made use of, was cognitive-behavioral therapy (CBT), a series of techniques that could be standardized and aimed at narrowly focused treatment of particular symptoms, rather than the more nebulous and hard-to-operationalize global reconstruction of the personality that psychoanalysts promised. This sort of intervention lent itself more readily to quantitative assessment and evaluation, and it was much briefer and more focused (and cheaper) than the notoriously interminable interventions associated with classical psychoanalysis. Though a handful of psychiatrists, most notably Aaron T. Beck, a disillusioned psychoanalyst based at the University of Pennsylvania, contributed in important ways to its development, most of its central exponents and protagonists came from the ranks of the psychologists—figures like Albert Bandura of Stanford University and Albert Ellis, who founded his own training institute in New York.

Many sectors of academic psychology in the years following the Second World War had embraced a behaviorism that had its roots in the work of John B. Watson, a movement whose foremost exponent in the postwar years was B. F. Skinner of Harvard University. A younger generation, however, had become disenchanted with behavioral psychology's singular focus on external actions, and its denial or neglect of human consciousness and the mind. Steering clear of introspection, and thus remaining sharply at odds with psychoanalysis, these psychologists nonetheless gave increasing weight to internal mental states—beliefs, desires, and motives.

The cognitive revolution that swept the field was translated among clinical psychologists into an array of techniques that sought to combine an emphasis on cognition and consciousness with behavioral strategies that could form the basis of new forms of psychotherapy. Thought disorders and maladaptive behaviors that led to repetitive negative thinking were, in their eyes, a primary source of mental disturbance. Focusing directly on symptoms, they attempted to create strategies to modify the cognitive distortions and maladaptive behaviors that they claimed led to emotional distress and self-defeating cognitive patterns. Applied initially to panic disorders and depression, those techniques in time came to be applied to a vast array of psychic troubles: anxiety, eating disorders, phobias, obsessive-compulsive disorders, personality disorders, anger management, even spousal abuse—a host of the kinds of problems clinical psychologists encountered in their growing office practices.[47]

Cognitive-behavioral therapies that presented themselves as having statistical validation would prove an important comparative advantage when psychotherapists sought insurance reimbursement, and these interventions helped clinical psychologists to legitimize their profession. Mental health care in the first four decades of the twentieth century had mostly consisted of treatment in public mental hospitals at state expense. In the postwar era, a new market had emerged for outpatient psychiatry. Those sessions had to be paid for privately, and it helped that Americans were used to treating health care of all sorts as a commodity purchased in the marketplace. Increasingly, many patients could offset some of these costs through employer-provided health insurance. The insurance industry was initially reluctant to pay for the treatment of any kind of mental illness, fearing that the associated costs would prove crippling.[48] Over time, however, more and more middle-class Americans secured at least some degree of insurance coverage for mental health care.[49]

In this new environment, psychoanalysts found themselves facing a distinctly skeptical reception from insurance companies. Their form of psychotherapy required extensive and expensive sessions and could extend over many years, with no obvious end in sight. Analysts were now competing, besides, with a heavily feminized profession in clinical psychology that perforce had to settle for lower financial rewards. The rich might prefer the services of psychoanalysts, but when it came to less affluent patients, Freud's followers were forced to confront the dismaying reality that their rivals provided much shorter treatments, for which they charged considerably less by the hour, and that apparently produced demonstrable results.

THOUGH CBT HAD A VERY DIFFERENT intellectual genealogy, it echoed Adolf Meyer's notion that mental illness was rooted in faulty habits. Unlike Meyer, though, these therapists began to develop techniques for addressing maladaptive patterns of thinking, beliefs, and behaviors. For anxiety disorders, phobias, panic attacks, obsessive-compulsive behaviors, and some forms of depression, the various tools CBT provides—teaching about the nature of fear and anxiety, and offering patients techniques to avoid or alter self-destructive ways of thinking, reduce muscular tension, or desensitize oneself to feared stimuli via controlled exposure—have been shown to be effective, though not universally so. Their usefulness in the treatment of depression is more variable and uncertain, and improvements have not always been found to be lasting.[50]

The distress and disability associated with these conditions is often considerable, so even though the Cochrane review of CBT in generalized anxiety disorder found that fewer than 50 percent of patients showed clinical response to treatment, that degree of improvement is still considerable and helps explain the ability of CBT therapists to attract clients.[51] For some of these patients, CBT is certainly useful and welcome. But just as Meyer's suggestion that psychotic disorders and manic depression could be seen as the product of bad habits was implausible and found few followers, so, too, attempts to extend cognitive-behavioral approaches to encompass these graver forms of mental disturbance have been largely unavailing. As we shall see in Chapter 22, CBT for serious forms of mental illness (schizophrenia, bipolar disorder, and major depression) shows little evidence of broad effectiveness. A further caveat is that Cochrane reviews, the most systematic reviews of the evidence we have, emphasize that the available studies are of low to moderate quality, limiting the confidence one can invest in their findings. CBT can claim that it has some empirical support for its approach, unlike many of its rivals, but its vaunted evidentiary support is less secure than its enthusiasts proclaim.[52]

Analysts insisted that their rivals were playing whack-a-mole, treating the symptoms of mental disorders and not their root causes. Psychodynamic interventions, they claimed, provided a deeper and more lasting transformation of the personality and attacked the roots of psychopathology. For a skeptical insurance industry, these claims were undercut by psychoanalysts' continued inability to provide compelling evidence of the efficacy of their interventions, the same problem that Robert Morison of the Rockefeller Foundation had criticized them for in the late 1940s.

For a quarter century after the end of the Second World War, psychoanalytic ideas enjoyed a remarkable degree of authority in intellectual circles, spreading widely in popular culture. In the 1950s and 1960s, analysts made lucrative livings on both coasts and at a few cities in the interior, their incomes outpacing those of many other medical specialties. They dominated most academic departments in medical schools and attracted the most talented recruits to psychiatry. Many psychoanalysts thought their position was unassailable. But theirs would prove to be a fragile hegemony, one that would be reduced to rubble in the space of little more than a decade.

PART THREE

A PSYCHIATRIC REVOLUTION

CHAPTER SEVENTEEN

The Birth of Psychopharmacology

ONE OF THE CRUCIAL FACTORS undermining psychoanalytic hegemony originated not from academia, or from cognitive-behavioral therapists, but from Europe. Its initial impact was felt most powerfully in the least prestigious region of the psychiatric imperium, and it would ultimately provoke a total transformation of the psychiatric enterprise: its therapeutic practices; the settings in which therapists plied their trade; what counted as mental illness and how its various manifestations were to be defined, delimited, and understood; the training of professionals; and the intellectual orientation of the profession's leading lights. It prompted an influx of research dollars into academic departments that had previously shunned laboratory life and the prestige that bench research might bring. It accompanied, if it did not completely cause, radical changes in the experience of being mentally ill and major shifts in public understanding of the nature of mental illness. It is usually called the psychopharmacological revolution, and that is scarcely an exaggeration. But it was at first a slow-moving revolution, one whose major impacts took decades to materialize.

No one planned the psychopharmacological revolution, and no one understood, in its early years, just how dramatic and far-reaching the effects of the new drug treatments would be. Psychiatrists had often used drugs to help manage their patients. In the early years of the asylum, patients were dosed with cathartics and emetics, among the sheet anchors of an increasingly moribund Hippocratic medicine, whose emphasis on bleeding and purging was soon to disappear. These were eventually supplemented by an array of narcotics: opium and later on morphine, and then, from the 1860s onward, by newer drugs like the sedatives chloral hydrate and bromide.

There was no pretense that these interventions were anything other than management tools, designed to quiet disturbed patients or even render them comatose. During the 1920s, barbiturates were employed as "sleep therapy" and, following the discovery of insulin to treat diabetes, its ability to induce hypoglycemia and comas had likewise been exploited, as we have seen, with extravagant claims for its efficacy belied by the remorseless rise in institutionalized patients. But deep-sleep and insulin therapy, with their associated comas and the risk that these might prove irreversible, bore little relationship to the standard medical practices of the time. By contrast, the drug now introduced to treat psychosis seemed precisely modeled on conventional medical practice. Chlorpromazine, known as Largactil (or "mighty drug") in Europe, and trademarked as Thorazine in the United States, was the product of research conducted by a French pharmaceutical house, Rhône-Poulenc, and its psychiatric applications were purely serendipitous. Yet within eighteen months of its coming to market in the United States in 1954, two million patients were prescribed Thorazine to treat their mental illness.[1]

The modern pharmaceutical industry developed in the second half of the nineteenth century, gradually disentangling itself from the excesses of the so-called patent medicine sector, which advertised its secret remedies to a gullible public. "Ethical" drug manufacturers advertised only to the medical profession, and by the end of the nineteenth century, they increasingly allied themselves with the science of chemistry. German chemists were particularly adept at creating new compounds, many of them originally derived from work on new dyes created for the textile industry. These synthetic dyes proved extraordinarily lucrative for the companies that discovered and marketed them. Some of these substances, many derived from coal tar, were accidentally found to have therapeutic applications in medicine, and the companies that had started out making dyes for clothing eventually became major actors in the pharmaceutical industry, as some remain to this day. (Aspirin and the chloral hydrate that was widely used as a sedative in mental hospitals were among the many therapeutic agents discovered largely serendipitously. Bayer and Sandoz [now Novartis] are two important examples of companies that moved from dyestuffs to manufacturing drugs.)[2] One of the compounds German scientists synthesized, phenothiazine, had no obvious use at the time but would, in the late 1940s, become of interest to the burgeoning pharmaceutical industry.[3]

As late as the 1930s, nearly half of all medicines were compounded at small local pharmacies, and most were sold over the counter, without the need for a doctor's prescription. The pharmaceutical industry had begun to secure a larger presence in the marketplace as the decade wore on, but it was the impact of the war that completely transformed its prospects and importance, most especially, though not exclusively, through the discovery of penicillin's therapeutic applications in the early 1940s and the solving of the problem of its manufacture on a commercial scale. A scramble was on to locate other antibiotics and to find other magic bullets with commercial possibilities. A period of rapid expansion ensued on both sides of the Atlantic, as pharmaceutical houses began to invest on an ever-larger scale in research, development, and marketing.

When Rhône-Poulenc began to search for possible therapeutic uses for chemical derivatives of phenothiazine, it had no intention of developing a drug to treat psychiatric illnesses. No such market existed, and few dreamed of creating one. Instead, having previously had some commercial success with an antihistamine, Phenergan, the company decided to explore whether it could find other drugs of this sort. There were suggestions that this class of drugs might prove useful in the treatment of Parkinson's disease, as a sedative, or as an anti-emetic, and that one or more of them could alleviate the problem of circulatory shock during and after surgery, in part by acting as a catalyst or potentiator for general anesthetics. In December 1950, Rhône-Poulenc's chemists synthesized a new derivative of phenothiazine. Chlorpromazine, so-called because of the addition of a chlorine atom to the original compound—internally referred to as 4560 RP—was now a drug in search of a disease it could be used to treat, and a broad spectrum of clinical applications was scrutinized for therapeutic and commercial possibilities.

As that search began, Rhône-Poulenc was anxious to find an American corporation willing to buy up rights to its new compound. In the twenty-first century, the pharmaceutical industry is dominated by huge multinational conglomerates, but that consolidation was a long way off. A European corporation seeking to sell its wares in the United States faced substantial structural obstacles, not the least of which was the pronounced skepticism that American physicians exhibited toward European science and medicine. Rhône-Poulenc's efforts to offset its development costs by selling rights to the American market encountered severe difficulties. The first two corporations it approached rejected its overtures, for this new drug had no obvious market niche they could exploit. But in early April 1952,

Smith, Kline & French (SK&F), based in Philadelphia, a company that had been attempting to secure reciprocal deals with Rhône-Poulenc, agreed to buy the North American rights to chlorpromazine. It was a relatively small gamble—the company initially spent a mere $350,000 on research and development costs—but it was a wager that paid off in spades.[4]

Both Rhône-Poulenc and SK&F were initially quite uncertain as to chlorpromazine's prospects. There were a number of promising leads. By 1953, trials had been conducted to see whether it could be used as an anti-emetic, as a treatment for skin irritations, as a general sedative, and as an aid in surgery. None of these applications produced sufficiently clear-cut results, though an SK&F internal memorandum dated April 8, 1953, indicated that "nausea and vomiting are still felt to be the most appropriate indications on which to conduct rapid clinical testing to try to get marketing clearance by the F.D.A."[5] That did indeed prove to be a useful market for chlorpromazine. Pediatric journals in the late 1950s are replete with advertisements for its value in treating childhood vomiting—a disturbing phenomenon, given what we now know about the drug's side effects. But within less than a year, SK&F had rapidly changed tactics, seizing on serendipitous findings in both France and Canada to home in on a very different set of applications for the drug, ones that would massively increase its profits.

SK&F had originally risen to prominence by bringing amphetamines to market in the late 1930s, promoting them as a nasal decongestant, an appetite suppressant, even as an antidote for depression. Such claims rested on extraordinarily flimsy evidence—five published articles reporting on a total of fewer than 150 patients, conducted under conditions that rendered them useless as scientific tests of efficacy.[6] Yet despite mounting reports of serious side effects from 1939 onward, ranging from addiction to psychosis and death, the resultant profits had led to the company's initial rise to prominence, aided by the use of amphetamines by air force pilots in the Second World War to heighten vigilance under combat conditions. Only many decades later would amphetamine abuse and methamphetamine addiction lead to serious regulatory action.[7] Chlorpromazine, or Thorazine, to give it its American trade name, came to market in a rather similar fashion.

Few rules constrained the conduct of pharmacological research in the middle decades of the twentieth century, and the standards for assessing the value of new drugs were remarkably lax. Even the trial of streptomycin as a treatment for tuberculosis, often cited as the origin of the later insistence on double-blind, randomized, and placebo-controlled assessments of new therapies, was neither blind nor placebo-controlled. Those exploring the therapeutic possibilities of new drugs had extraordinary latitude in how they proceeded and how they reported their results. Drug companies seeking possible applications for new compounds made them widely available to physicians, often with little serious attention to how they were used. The testing of chlorpromazine conformed to this pattern, and it was precisely this cavalier approach to clinical testing that led to the discovery of its psychiatric applications.

Henri Laborit was a young French naval surgeon with few discernible qualifications to conduct academic research and even less interest in abiding by scientific norms. In the casual way of the times, he was given some chlorpromazine to try on his patients. His interest in reducing surgical stress dovetailed with Rhône-Poulenc's interest in seeing whether the compound might be useful as an anesthetic potentiator. (He would be sent to the United States in November 1953 in an attempt to persuade skeptical American surgeons of the usefulness of chlorpromazine in surgery; the trip turned into a disaster when most of the dogs he was using to demonstrate his technique at Harvard died of heart failure during the operations.) Laborit's findings were inconclusive, because he mixed 4560 RP with two other drugs to create what he called an artificial hibernation in surgical patients given the mixture. His suggestions that 4560 RP might be useful for an extraordinary range of conditions—treating burns, reducing stress in those suffering from infectious diseases, heart attacks, obesity, and reducing the doses of anesthetic needed in surgery, especially in obstetrics—were seen as idiosyncratic and of little use. But along the way, Laborit had noticed that patients given 4560 RP became uninterested in what was going on around them. He indicated that one of his colleagues thought that "4560 may produce a veritable medicinal lobotomy."[8]

Laborit had urged his psychiatrist colleagues at the Val-de-Grâce hospital in Paris to try 4560 RP on their psychiatric patients. A few did so, usually in combination with sleep or shock therapy, though their reports attracted little attention. But a fellow surgeon spoke of Laborit's findings to his brother-in-law Pierre Deniker, who was a psychiatrist at the Centre Hospitalier Sainte-Anne, the largest psychiatric hospital in Paris, with some 4,000 patients. Deniker persuaded his superior, Jean Delay, that the two of them should use 4560 RP on a broad spectrum of their patients. After a few desultory experiments with artificial hibernation, they began using 4560 RP by itself on a sustained basis.

Jean Delay occupied an important position in French psychiatry. He was a professor of psychiatry at the Faculty of Medicine in Paris, and Chief of Medicine at Ste. Anne's. He and Deniker produced a series of papers that, somewhat cautiously, touted the merits of the new medication, especially in calming patients displaying violent agitation. Patients given the drug, Deniker noted, "look as though they have been turned to stone. . . . [T]hey are usually indifferent to themselves and their environment, they are stuporous or prostrate"—a description that resembles Laborit's observation that chlorpromazine produced a "chemical lobotomy."[9] Delay and Deniker also noted some improvement among the half-dozen schizophrenics to whom they had given the drug but added that "the relatively small number of treated cases and the brevity of the remission do not permit any estimate of the possible usefulness of the method in this most severe affliction."[10]

Delay and Deniker painted an optimistic picture of the changes administering chlorpromazine had brought to the atmosphere on their disturbed wards, emphasizing that it had allowed them to eliminate the old means of restraint. Despite their efforts, though, they found that "their work with CPZ [chlorpromazine] in 1952–1953 aroused little interest among their colleagues in Paris" and the drug was slow to take hold in France.[11] Rhône-Poulenc, however, was not oblivious to Delay and Deniker's findings, and it tried to get its sales staff to use reprints of their work to induce other psychiatrists to follow their lead. Oddly, that effort bore little fruit in France but had a major impact in North America, in a curiously roundabout fashion.

Quebec was a deeply divided Canadian province, whose Anglophone elite despised the French-speaking Catholic majority. Yet it was a psychiatrist at the Verdun Protestant Hospital in Montreal, Heinz Lehmann, who took a meeting with the Rhône-Poulenc salesman and subsequently read Delay and Deniker's publications, by his account while taking a bath.[12] A refugee from Nazi Germany who had arrived in Canada in 1937, Lehmann had married a French Canadian and was fluent in three languages. What he read prompted him to obtain samples of chlorpromazine to experiment with, and those experiments did much to encourage SK&F to seek FDA approval for Thorazine as an antipsychotic drug, the first of its kind.

Like Laborit, Lehmann had no training as a researcher—indeed, he had no permanent license to practice medicine in Canada. Only his staff position at a provincial mental hospital allowed him to treat patients. His situation was hardly unique. As the least attractive and most poorly paid part of the profession, institutional psychiatry had a difficult time recruiting physicians to fill vacant posts, and many state hospitals recruited psychiatrists from abroad with training that did not meet American standards. Working for the state allowed these physicians to evade the normal requirements for licensing, but kept them trapped in the state hospital system.

Lehmann's position gave him the freedom to treat patients as he saw fit. He had previously enthusiastically embraced ECT and insulin coma therapy, and he had used barbiturates and paraldehyde to control the agitated and violent. These were, of course, treatments widely used by his professional colleagues, but Lehmann had also shown himself keen to conduct therapeutic experiments on the patients in the Verdun hospital. He had tried subjecting some of them to the inhalation of pure nitrous oxide, causing seizures and loss of consciousness. He then substituted carbon dioxide, which produced extreme fear among those subjected to the experience, but to no avail. He injected sulphur oil to create high fevers, an experiment that proved extremely painful and once again useless. And he injected turpentine into the abdominal muscles of other patients, deliberately creating abscesses in the forlorn hope that an induced infection would somehow drive out the demons of madness.[13] Lehmann was, in other words, content to try any novel treatment that struck his fancy and was in a position to do so. Now he had another weapon at his disposal, for Rhône-Poulenc eagerly provided him with a substantial quantity of chlorpromazine.

Nearly two decades later, in an interview with Judith Swazey, Lehmann boasted about how easy it had been to conduct the study: "There were plenty of patients available to us, because none of them were on any other particular therapy. . . . [W]e didn't need, or get, the patients' consent before we started giving them the drug." Their experiment began in late 1953. Having no clue how long to administer the drug, Lehmann and his resident, T. E. Hanrahan, settled on four weeks, the period they generally allowed for a course of shock therapy. As for dosage, where patients failed to respond to the initial 150 milligrams per day, the experimenters simply upped the dosage till the patient became drowsy, even when that meant giving them "many times the dosage reported in the literature." Worrying side effects surfaced—three cases of jaundice raising concerns about impaired liver function, and "some Parkinsonian symptoms. . . . [W]e got the scare of our lives because we had no way of knowing if the symptoms would abate."[14]

Undeterred, Lehmann and Hanrahan continued the experiment, and wrote up their findings for the *Archives of Neurology and Psychiatry.*

One of the most striking things about Lehmann and Hanrahan's work, besides the fact that the trial on a haphazard collection of seventy-one patients ranging in age from eighteen to eighty-two had employed no criteria of improvement besides the clinical opinion of the two men conducting it, and had no controls, is what they claimed were the major consequences of administering chlorpromazine. Lehmann's reading of Delay and Deniker's papers prompted him to see whether the new drug could do a better job "of controlling acute and chronic states of psychomotor excitement—disruptive and uncontrollable behavior on the wards that made the lives of the staff and other patients a misery." Better, that is, than the existing treatments, principally repeated doses of electroconvulsive therapy (ECT) to produce confusion and somnolence, or deep sleep therapy, with either barbiturates or insulin employed to put the disruptive patients into a coma. Sleep therapies, Lehmann conceded, required enormous medical and nursing resources and were associated with side effects that "are serious and manifold in nature . . . [including] pneumonia, hyperpyrexia, and cardiovascular collapse." ECT was simple to administer, but "immediate relapses are frequent"; it has "a disorganizing effect on the higher brain functions [and can produce] confusion, restlessness, and aggression"—not to mention "the unpleasant side-effect of amnesia."[15]

Chlorpromazine, by contrast, produced "the most gratifying results," often within twenty-four hours:

> Patients receiving the drug become lethargic. Manic patients often will not object to bed rest, and patients who present management problems become tractable. Assaultive and interfering behavior ceases almost entirely. The patients under treatment display a lack of spontaneous interest in the environment, yet are easily accessible and respond relevantly to questions even if awakened from sleep. As Delay and Deniker described them, they tend to remain silent and immobile when left alone and to respond to questions in a slow monotone. . . . Although a patient under the influence of chlorpromazine at first glance presents the aspect of a heavily drugged person, one is surprised at the absence of clouding of consciousness.[16]

Archives of Neurology and Psychiatry sat on Lehmann's paper for several months and only moved to publish it in 1954 when he threatened to send it elsewhere. In the interim, apprised of the findings, Rhône-Poulenc and SK&F quickly reproduced the paper and circulated it to other psychiatrists practicing in mental hospitals. As the companies immediately recognized, a drug that could ease the management of hundreds of thousands of inpatients had considerable commercial prospects.

As other studies of chlorpromazine accumulated, the initial assessment of the potential market would prove to be an underestimate of major proportions. As late as December 1953, SK&F had tested the drug on only 104 psychiatric patients (all in uncontrolled trials), whereas its potential value as an anti-emetic, for use in treating motion sickness or the side effects of anesthesia or chemotherapy, had been tested on over a thousand patients. Finding psychiatrists willing to test the drug was a source of enormous frustration within the company. It reflected the skepticism of most, who doubted that pharmaceutical remedies would be useful for disorders that the bulk of the profession now believed to have a psychodynamic origin. In Heinz Lehmann's words, "No-one in his right mind in psychiatry was working with drugs. You used shock or various psychotherapies."[17]

It is a measure of how lax the FDA's procedures were in those years that by March 26, 1954, SK&F had secured federal approval to market the substance it now called Thorazine for use as an anti-emetic and antinausea drug and also as a drug to treat psychiatric illnesses.[18] Fortunately for SK&F, in the years before the Thalidomide tragedy, the standard they had to meet was not a demonstration of efficacy, but rather one of safety, and apart from some internal worries about the potential for liver toxicity, nothing had yet surfaced to raise questions for the FDA staff. The major side effects would only come to light some years after Thorazine had come to market.

Within a year of its introduction, Thorazine had increased SK&F's total sales volume by more than a third. A major portion of the corporation's subsequent growth, from net sales of $53 million in 1953 to $196 million in the two years after Thorazine came to market, and $347 million in 1970, was directly attributable to this enormously profitable product. The overwhelming bulk of these sales came not from its use as an anti-emetic, but from its psychiatric applications.[19] That explosive growth was no accident. SK&F's senior management grasped very early on the financial bonanza that this novel marketplace represented, and they moved aggressively to create and dominate it, though the first advertisement for Thorazine's psychiatric applications did not appear until July 1955. Significantly, it touted the medicine's value in suppressing the outbursts of violent and disruptive patients, as a replacement for earlier management techniques. "Thorazine," the company boasted, "reduces the need for electroshock therapy."[20]

Company strategies often remain opaque to outsiders. In this case, however, the architects of the program were keen to enlighten the public. They shared their account of how they proceeded with the author of a celebratory study of the drugs revolution, Judith Swazey. The product manager for Thorazine, Charles Bolling, and the head of the hospital sales department, Frazier Cheston, spoke at length about the approach they adopted.[21] At that time, SK&F's total sales force numbered 300 people. When the decision was made to promote the use of Thorazine for psychiatric illnesses, they told Swazey, "We found that this group was not adequate to deal with the introduction of Thorazine into mental hospitals, in terms of geographic distribution, the special funds needed for this type of work, or the specialized knowledge and communications network that would be needed. A professional representative, we discovered, could make no inroads at a single hospital. We had to work with the whole system, from the legislature in a state through the entire public mental hospital system and civic groups."[22] To accomplish this massive task, the best fifty salesmen were selected to form a task force whose full-time job was to create the market and sell the drug.

From the outset, SK&F had essentially written off working with "office practice" psychiatrists. They were generally separable into two groups: the electroshock people, and the analysts and related psychotherapists, both with a great commitment to their years of experience and basic training philosophy and some resistance to the use of drugs. Mental hospitals, which in the early 1950s contained more than half a million patients, were seen as a much more promising market, and the early reports of Thorazine's ability to calm the agitated and the violent, and to make patient management infinitely easier, had an obvious attraction to beleaguered hospital psychiatrists. But sunk in "a heavy atmosphere of custodial care" and lacking any money to spend on drugs, the hospital authorities had little ability to respond to SK&F's marketing efforts: "[We] saw there would be a need to educate supervisors, business managers, and staff people about the potential therapeutic value of Thorazine and the vast administrative savings it offered in reduced damage to plant and reduced inpatient population. We also saw that we would have to go to work with, and similarly educate, state legislators of the need for higher drug budgets for the state hospitals."[23]

The task force was kept in place for six years. The propaganda efforts were sophisticated and sustained. Films were made to demonstrate Thorazine's dramatic effects. A speakers' training bureau was set up to coach hospital administrators on how to approach their political masters and persuade them to provide the necessary funds for drug purchases. Statistics were created to show how "factors such as less staff turnover, and less cost for such things as broken windows and damaged furniture" were correlated with various dosage levels of Thorazine. And, to encourage mass prescription of the drug, states were offered quantity discounts.

Still, the SK&F executives commented, there was resistance in many quarters:

> We found that some individual legislators, and in certain cases entire state legislatures, were so apathetic to the idea of funding intensive-treatment programs that the Task Force sometimes had to use "drastic" procedures. In one state, for example, through the efforts of SK&F and other interested groups, a special legislative session took place at one of the state mental hospitals, with the governor's and the legislative leaders' blessings. The entire session was filmed by the "Today" show, and, in that state, it was the breakthrough that eventually committed the legislature to funding an intensive-treatment program for the state hospital system.

It was, Bolling and Cheston insisted, "not lobbying per se" but rather "a true educative effort."[24]

SK&F's "education" of politicians and state hospital psychiatrists played a major role in the psychopharmaceutical revolution, and, as the first mover, the company enjoyed a substantial initial advantage in the marketplace. Not content to watch from the sidelines, other pharmaceutical companies rapidly sought to enter the fray. Just as chlorpromazine was coming to market, a potential rival appeared on the scene: reserpine, a drug synthesized from *Rauwolfia serpentina*, a plant known as Indian snakeroot, or Devil Pepper. It was the subject of experiments run by an ambitious young psychiatrist, Nathan Kline, at Rockland State Hospital in suburban New York.

Traditional Indian medicine had long used concoctions of *Rauwolfia* to treat a wide variety of ailments, including hypertension and insanity. Another New Jersey drug company, CIBA, had derived reserpine from the original plant material and begun exploring whether it could be used to treat hypertension. Hearing of this, Kline obtained a supply to experiment with on his psychiatric patients. His report of promising results appeared in 1954, just six weeks after Thorazine came to market, and for a time some psychiatrists tried both drugs.[25] But reserpine's effects were slower to materialize, and, unlike Thorazine, it had no massive marketing campaign behind it. Moreover, its use was associated with dangerous drops in blood pressure, and in the early stages of treatment it produced restlessness and agitation, in contrast to the initially calming action of Thorazine. In short order, reserpine faded from view.

It turned out that phenothiazine, the drug from which Thorazine had been developed, was readily susceptible to other manipulations in the laboratory. Soon a number of copycat drugs that were claimed to possess similar antipsychotic properties were on the market. Hospital psychiatrists reported enthusiastically that "the atmosphere of disturbed wards has been completely revolutionized. Patients now remain clothed; they are quiet; they do not annoy each other; they conform to the conventions, take an interest in their personal appearance and in the appearance of the ward."[26] Testimony of this sort reinforced the sense that a dramatic change had occurred. The impression was strengthened by further papers suggesting that besides their success as major tranquilizers, the drugs were also effecting positive changes among schizophrenic patients, previously among the most refractory and troublesome patients. Major drug companies—SK&F, Rhône-Poulenc, CIBA, Sandoz, Geigy—helped to create this mounting optimism, funding a series of national and international conferences designed to highlight the new drugs and to encourage the creation of the new discipline of psychopharmacology.[27]

For several years, the vast majority of the scholarly literature on the phenothiazines, while highly favorable, was deeply flawed. Tests of the drugs' efficacy continued to be conducted on the basis of nonblind or single-blind designs (in which only the patient was nominally unaware of whether he or she was receiving either the active substance or the placebo). Reported sample sizes were extremely small, sometimes just a single patient, and sometimes the number of patients under treatment was not even recorded. All sorts of potentially relevant confounding factors—duration of illness, previous and concomitant treatments, differential treatment of treated and untreated patients—were ignored, and assessments of whether improvement had taken place were frequently purely subjective. Most reports also covered only brief periods, with long-term follow-up studies notable by their absence.[28] As studies began to be undertaken in a more systematic fashion, and double-blind assessments became the norm, early estimates of the drugs' efficacy began to decline.

The initial enthusiasm was not entirely misplaced. Thorazine and its competitors did indeed calm severely disturbed patients. They decreased the hallucinations and delusions that, along with distortions of perception and peculiar thoughts, are the most conspicuous manifestations of psychosis. The diminution of violence and destructive acting out that mental hospitals had long struggled to contain were not inventions of the professional mind. Those forms of symptomatic relief were real and dramatic, and their importance should not be minimized. Revolutionary in their way, they continue to underpin psychiatry's faith in psychopharmacology to this day.

But, as gradually became apparent, the symptomatic relief was far from universal. Many patients failed to respond to the drugs, and even among those who did, many paid a heavy price in side effects—an issue to which I shall return. Worse still, the phenothiazines turned out to be largely ineffective in treating the less dramatic, but in many ways more devastating deficits that are characteristic of schizophrenia—those that psychiatrists term "negative" symptoms. The blunted affect, the poverty of speech, the absence of spontaneity and initiative, the failure to connect with others, the anhedonia, or apparent inability to experience or feel pleasure: cumulatively, these have catastrophic effects on people's quality of life and their ability to function independently. And, as with the cognitive deficiencies that mark the disorder, the new drugs left these deficits largely untouched. These were uncomfortable realities that psychiatrists preferred to ignore, emphasizing instead, to themselves as well as to outsiders, the gains that the antipsychotics brought in their train.

As usual, science journalists provided the public with enthusiastic accounts of the latest scientific breakthroughs. At least as important in persuading people that mental troubles could be dissolved by taking a pill was the marketing, in 1955, of Miltown and Equanil (both the identical chemical meprobamate). If Thorazine was a drug for the serious forms of mental illness, these "minor tranquilizers," as they were soon called, were chemical balm for the ordinary troubles of daily life. Within a year, one in twenty Americans had tried them, and by 1957, they accounted for a third of all the prescriptions written in the United States. The following year, 500 tons of them were manufactured, and in its first decade on the market, the Carter-Wallace drug company sold 14 billion tablets of Miltown, filling 500 million prescriptions for the drug.[29]

Milton Berle, the first major star of the new television era, started calling himself "Uncle Miltown" and heavily promoted the drug. Later joined by benzodiazepines (the first ones were trademarked as Valium and Librium), it accustomed a whole generation to the idea of medicating away their troubles.[30] Though these were drugs that fostered dependence, and brought a host of other troubles in their wake (memorably captured for a mass audience by the Rolling Stones' song "Mother's Little Helper"), they were immensely profitable and played a vital role in ushering in the psychopharmacological revolution. Librium and Valium's sales exceeded those of Equanil and Miltown by the mid-1960s. By 1971, these two drugs alone accounted for $200 million of Hoffmann-Laroche's $280 million in US sales. They were soon joined by copycat benzodiazepines (the most successful of these was Upjohn's Xanax, launched in 1981), as competing pharmaceutical houses rushed to share in the bonanza. By 1977, 8,000 tons of these drugs were consumed in the United States alone.

The massive commercial success of the tranquilizers was no accident. Arthur Sackler, the doyen of the Sackler family who would subsequently become enormously wealthy from marketing Oxycontin, one of the major sources of America's opioid addiction crisis, played a vital role in marketing both drugs. Alongside his career as a pharmaceutical executive, Sackler had pioneered ways of marketing pharmaceuticals directly to physicians. He correctly perceived that by clever marketing in medical journals and via promotional literature distributed directly to doctors' offices, he could have a massive effect on drug sales. The advertisements looked like serious clinical information provided by one doctor to another, and they were enormously effective as a result. Alongside these direct forms of advertising, he began publishing a biweekly newspaper, the *Medical Tribune*, which eventually reached some 600,000 physicians. Here was another way of persuading physicians to adopt certain products, in stories that were ostensibly noncommercial. The advertising campaigns for Librium and Valium were perhaps his greatest marketing triumph and brought him a fortune.[31]

Psychoanalysts had reacted with relative equanimity to the advent of Thorazine and its competing drugs. Few analysts, after all, ventured to treat the psychotic. But many of those who did comment on drug treatment were dismissive, calling the drugs a placebo, a Band-Aid, or, worse, actively harmful. "Drug therapy," these analysts argued, "impaired the patient's progress by increasing his magical reliance upon medical treatment, fostering his dependency upon the physician and blunting his capacity for insight."[32] Others adopted a less confrontational position, claiming the drugs were a useful adjunct, diminishing florid symptomatology so that the real work of analysis could proceed.

As long as the phenothiazines were seen as tranquilizers, drugs that calmed patients and reduced violent and disruptive behaviors, such ideas enjoyed some purchase. Increasingly, however, as the enthusiasm for the drugs grew, some psychiatrists began to conceive of them in more ambitious terms: they were no longer "major tranquilizers" but rather "antipsychotics," disease-specific treatments for schizophrenia and the major psychoses.[33] It was a subtle but important shift of perspective, one that was both welcomed and encouraged by the pharmaceutical industry. It implied that rather than simply suppressing or masking mental distress, the new drugs were somehow acting on the disease process itself. To the extent that this belief spread, biology rather than psychodynamics became not only the place one needed to look to in order to treat mental illness but also the sole and singular cause of mental illness.

Provided that heretical notion remained confined to what the psychoanalysts regarded as the second-class psychiatrists who manned the back wards of the nation's mental hospitals, they could complacently assure themselves that their hegemony would persist largely untouched. But the drug treatments insidiously captured the public imagination, with a large assist from the "educational" efforts of the pharmaceutical industry. In 1964, the *Archives of General Psychiatry* published a collaborative study funded by the National Institute of Mental Health, which claimed to demonstrate that Thorazine and two rival phenothiazines had disease-specific effects when given to schizophrenic patients. Its conclusions that "the phenothiazines should be considered to be 'anti-schizophrenic' in the broad sense" and that "it is questionable whether the term 'tranquilizer' should be retained" was prophetic of a coming seismic shift in American psychiatry.[34] Within two decades, biology and Big Pharma would essentially complete their capture of the psychiatric profession.

EVEN BEFORE RHÔNE-POULENC had been alerted to the potential psychiatric value of the phenothiazines, another compound had been serendipitously discovered to have a therapeutic impact on another major form of mental illness, mania. In 1949, an obscure Australian physician, John Cade, who ran a small mental hospital for chronic patients in a Melbourne suburb and had, as he later confessed, "no research training, primitive techniques and negligible equipment," announced in a weekly medical journal largely ignored outside his own country that he had discovered a sovereign remedy for this debilitating and often fatal disorder: lithium.[35]

The metal lithium had been discovered in 1817, and its salts had become a common element in many patent medicines later in the nineteenth century, especially for the treatment of gout and a variety of "nervous" complaints. Silas Weir Mitchell and William Alexander Hammond, both prominent late nineteenth-century neurologists, had used lithium salts to treat their patients.[36] But lithium salts are toxic. In excess, they produce tremors, nausea, seizures, damage to the kidneys, and, in cases of serious overdose, death. Even at therapeutic levels, they can cause kidney and thyroid damage. Accumulating evidence of these problems eventually caused lithium to be banned from sale in the United States in 1949.

Cade guessed—and it was simply an unfounded guess—that mania might be caused by some toxic substance produced in the body and excreted in the urine. He attempted to prove this hypothesis by injecting the urine of manic patients into guinea pigs. The guinea pigs promptly died. Undeterred, Cade sought to break down the various components of urine to see if one of them would prove his speculation. Uric acid, however, had to be made soluble to make it injectable, and Cade chose lithium salts to accomplish the task.[37] On a whim, he decided to inject some guinea pigs with lithium carbonate alone, and the excitable animals at once became tranquil. As we've seen, the obstacles to moving immediately to trials on mental patients were nonexistent, and Cade claimed that once he did so, trying it out on an array of schizophrenic, depressed, and manic patients, the effects on one subset of these verged on the miraculous: manic patients calmed down dramatically.[38]

The medical world shrugged. Cade's claims were ignored for several years, before a Danish psychiatrist, Mogens Schou, happened upon Cade's paper. Many of Schou's relatives had symptoms of manic depression (later relabeled bipolar disorder), so he was motivated to try the treatment, subjecting it to one of the first controlled trials in psychiatry. He proclaimed the trial a success and lithium a valuable prophylactic, diminishing the risk of manic episodes. That alone did not suffice to create a market for the drug, in part because, as a natural substance, it was unpatentable, giving no drug company an incentive to market it. Perhaps, had it not been for an explosive and long-running dispute with another prominent pioneer of controlled trials in psychiatry—Michael Shepherd of London's Institute of Psychiatry—the drug might have fallen by the wayside. But the dispute was lasting and vituperative. It attracted the attention of others, and further controlled trials showed that lithium indeed was somewhat useful in the treatment of mania.[39]

It was not until 1970, however, that the FDA licensed its use in the United States, and it took even longer before it began to be prescribed on a routine basis to patients with bipolar disorder. It has significant side effects, including increased urination, tremors, diarrhea and vomiting, and, if the therapeutic dose is exceeded (and the therapeutic / toxic ratio is narrow), hypothyroidism and kidney damage. Consequently, it requires careful monitoring of levels in the blood.

The discovery of the therapeutic effects of lithium was pure happenstance. Its ability to reduce the chances of relapse or suicide is obviously important, and it can head off much misery for those patients for whom it works. Just as Thorazine and its analogues unquestionably acted in many patients to relieve some of the more disturbing features of schizophrenia (those that psychiatrists label, oddly, its positive symptoms, most notably delusions and hallucinations), so, too, lithium has positive effects. It diminishes the rate of suicide and reduces the chances of recurrence of what is generally a relapsing disorder. But it remains a partial and imperfect treatment, and episodes of mania and depression often recur even among those taking the drug.[40] It is curious, though, given its usefulness, that a recent study of drug-prescribing patterns shows that prescriptions for lithium and other mood stabilizers have fallen dramatically in the twenty-first century. Prescribed in 62.3 percent of outpatient visits from 1997 to 2000, they were prescribed in only 26.4 percent of cases from 2013 to 2016.[41]

Two decades ago, David Healy, a psychiatrist who is also one of the leading historians of the psychopharmacological revolution, made a bold prediction. He noted that "there was then and still is now no theoretical basis for lithium's use, no rationale that could then or can now be used to sell it. As a result, the use of lithium will almost certainly end when Schou dies. Another agent, probably of lesser efficacy, will displace it by virtue of a marketing strategy that depends on offering a 'biological rationale.'"[42] Mogens Schou died on September 29, 2005. Healy's comments increasingly seem prophetic.

CHAPTER EIGHTEEN

Community Care

THE MENTAL HOSPITALS where the drugs revolution got its start represented a vast investment of social and intellectual capital. From the 1830s onward, the notion that prompt institutionalization represented the best or only response to serious forms of mental disturbance had become the ruling orthodoxy. On this foundation, states had been induced to spend large amounts of treasure to build and maintain ever-larger asylum systems to which the "insane" were consigned. The psychiatric profession at once emerged from the creation of the mental hospital and, for more than a century, ran and justified its existence. By 1950, more than half a million inmates filled the wards of state hospitals, and if any promise of cures seemed evanescent, the public had by now become thoroughly used to the idea that the proper response to psychosis was institutionalization.

Throughout its history, the asylum had had its critics. Some patients from the very outset proclaimed that the psychiatric emperor had no clothes, and that their confinement was a form of imprisonment.[1] But they were easily dismissed, their credibility undermined by their status as mental patients. Muckraking journalists from Nellie Bly (Elizabeth Seaman), who in 1887 tricked her way into the Women's Asylum on Blackwell Island in New York, to Albert Deutsch, whose exposés in *The Shame of the States* brought public attention to the deficiencies of these establishments just after the Second World War, exposed the scandals that lurked behind their facades, but the populations housed in them grew relentlessly year on year—until one day they didn't.[2]

The decline in the hospital census was slow at first, but by the second half of the 1960s, it was rapidly accelerating. Between 1955, when the number of residents reached an all-time high of 558,922, and 1960, the inpatient census declined by just over 4 percent to 535,540. The next five years saw a slightly faster rate of decline, but the dramatic decrease took place between

1965 and 1980, when numbers fell from 475,202 to 132,164. By then, the remnants of the old institutions led, in the words of the Harvard psychiatrist Richard Mollica, a "lingering existence as demoralized and impoverished facilities." Mollica warned of "a threatened disintegration of the public system."[3] Two decades later, the traditional mental hospital had virtually disappeared from the scene. So, too, had institutional psychiatry. All that had once seemed so solid melted into air.

In what must rank as an astonishing act of clairvoyance, the nineteenth-century British physician Andrew Wynter once speculated on what would become of the empire of asylumdom, at a time when it was rapidly expanding: "As we see wing after wing spreading, and story after story ascending, in every asylum throughout the country, we are reminded of the overgrown monastic system, which entangled so many interests and seemed so powerful that it could defy all change, but for that very reason toppled and fell by its own weight, never to be renewed. Asylum life may not come to so sudden an end but the longer its present and unnatural and oppressive system is maintained, the greater will be the revolution when it at last arrives."[4]

As the decline of the asylum accelerated, the public was assured that deinstitutionalization constituted a grand reform. Patients who had languished on the back wards of mental hospitals were being returned to the community, thanks to the advent of modern drug therapy. The horrors of the madhouse were rapidly becoming a thing of the past. Community care would prove at once more humane and more effective. Mental hospitals were portrayed as places where social skills atrophied, the opportunity to make autonomous choices evaporated, and patients lost the capacity to cope with everyday life. They damaged rather than rehabilitated those they confined. A more tolerant and welcoming society would spare the mentally ill these dehumanizing effects of involuntary confinement and also provide an environment where, with skilled psychiatric assistance, they could live relatively normal lives. Jack Ewalt, chair of the department of psychiatry at Harvard, and his colleague Gerald Caplan were particularly active in spreading the doctrines of community care.

In 1964, Robert Felix, a former president of the American Psychiatric Association and the director of the National Institute for Mental Health from its foundation in 1949 until 1964, articulated the consensus of those who now called themselves "community psychiatrists": "Many forms and degrees of mental illness can be prevented or ameliorated more effectively through community oriented preventative, diagnostic, treatment, and rehabilitation services than through care in the traditional—and traditionally isolated—state mental hospital. . . . [I]t will be possible to reduce substantially, within a decade or so, the numbers of patients who receive only custodial care—or no care at all—when they could be helped by the application of one or more of the modern methods of dealing with emotional disturbances and mental illness."[5] Embracing community care offered the prospect, Felix explained, "of ending forever the neglect and isolation which has been the lot of the mentally ill, both in and out of hospital, since the dawn of time."[6]

It was a fairy tale. The decanting of mental patients into the community (and later the tightening of commitment statutes to make admission to the mental hospital increasingly difficult) took place with virtually no advance planning or provision for the housing or other needs of those with disabling mental illnesses. "Community care," it transpired, was a shell game with no peas. In place of forcible confinement in publicly run asylums, the chronically mentally ill were abandoned to their fate.

ON ONE ACCOUNT, the demise of the state hospital and the rise of treatment in the community was the product of a simple technological fix. The advent of the new drugs for schizophrenia (and soon for depression as well) transformed the prospects for treating mental illness and made possible both the deinstitutionalization of patients already in the mental hospitals and the management of newer cases of mental illness on an outpatient basis. This was an account swiftly embraced by many psychiatrists, and it is still accepted in many quarters today. However, it ill accords with a careful analysis of these dramatic developments.

It is obvious that at the national level the first declines in state-hospital populations took place at the very moment when Thorazine (and its competitors) entered the marketplace. But temporal coincidence tells us little about causation. As it happens, a variety of data undermines the assertion that the introduction of psychotropic drugs caused the emptying of mental hospitals. In at least seventeen states, inpatient censuses had begun to decline between 1946 and 1954, a period when antipsychotics had not yet reached the marketplace. That pattern of extreme differences among states in the rate of discharge of mental patients persisted for at least a decade after the introduction of psychotropic drugs, with at least ten states experiencing increases in their inpatient populations during this period.[7] This suggests that there was no immediate or necessary connection between the introduction of psychotropic drugs and the decline in mental hospital populations. Two contemporaneous studies that sought to understand how the two phenomena interacted reinforce that conclusion. The first is a series of papers by Henry Brill and Robert Patton that examined the decline in New York State hospital populations in this period, often cited as "proving" that drugs caused the decline. In reality, the papers demonstrate nothing of the sort. The authors note the temporal coincidence and then state that "we know of no other major change in operating conditions which took place between 1954–55 and 1955–56" that could have caused hospital populations to fall. But the other data they report undermine even this tautological statement, for they note that only a distinct minority of patients were given the drugs in this period, and there were wide variations in the proportions of patients receiving drug therapy (33.9 percent of women, for example, versus only 20.9 percent of men—yet again suggesting a greater willingness to experiment on women). And when they examined discharge statistics, they acknowledged that "no quantitative correlation could be shown to exist between the percentage of patients receiving drug therapy in a given hospital or a given category [of mental illness] and the amount of improvement in releases."[8]

On the opposite coast, in California, state mental hospitals differed widely in how quickly they adopted the new drug treatment, and they kept records of which patients did and did not receive phenothiazines. L. J. Epstein and his colleagues were able to examine these detailed records, and, focusing on white male schizophrenics, they found that the relationship between drug treatment and release was a negative one. That is, patients given the drugs consistently spent longer in hospital than those who were not so treated. Both within and between hospitals, "the drug treated patients tended to have longer periods of hospitalization." Hospitals that were the most enthusiastic adopters of drug treatment had lower discharge rates than those that were more cautious about the new drug therapy, and the evidence as a whole showed no relationship, they concluded, between drug treatment and "the more rapid release rate that has been observed in recent years."[9]

A series of retrospective studies that subsequently explored this issue all came to essentially the same conclusion.[10] In damping down the florid symptomatology of psychosis, drugs visibly affected the atmosphere inside mental hospitals, persuading psychiatrists that management in the community might prove possible. But the primary driver of deinstitutionalization lay elsewhere, in the loss of political support for mental hospitals and in changes in social policies brought about as limited postwar moves toward a welfare state changed the environment in which state policy makers operated.

IN THE UNITED STATES, the mentally ill had always been the responsibility of the individual states. The costs of housing and providing even minimal levels of treatment for this population were a huge burden on state budgets, and threatened to become an even more serious problem in the postwar period, as the old infrastructure crumbled and the unionization of state employees forced up costs. As long as mental patients were housed in state facilities, the federal government disclaimed all responsibility for them. But as the more astute and entrepreneurial states like New York and California began to realize (and as more backward states like Alabama more slowly began to appreciate, sometimes prodded by lawsuits), changes in the safety net provided by Congress dramatically altered this picture, providing an opportunity to transfer many of these costs off the state and local budgets and on to the federal government. In a famous case brought to federal court in the state of Alabama, *Wyatt v. Stickney*, members of the newly emerging mental health bar argued that patients involuntarily confined in a mental hospital had a right to treatment, which Alabama's massively overcrowded state hospitals failed to provide. The judge in the case, Frank Johnson, endorsed this view and insisted that the state had to provide the minimum level of staffing laid down by the American Psychiatric Association. Governor George Wallace avoided a massive increase in staffing costs by simply discharging most of the inpatients, thus providing the ratio of staff to patients that the court demanded for those left behind in the institutions.

There were no improvements in drug therapy or dramatic breakthroughs in psychopharmacology in the period between 1965 and 1980, when the dramatic decrease in state-hospital populations took place. Indeed, well-placed observers have argued that there have been no major advances with respect to antipsychotic drugs all the way down to the present, a contention that will be discussed later in this book. There were, however, major innovations in social policy, and these are reflected very closely in the pattern that deinstitutionalization took in this crucial decade and a half.

As part of his Great Society program, and in the teeth of opposition from the American Medical Association, Lyndon Johnson secured the passage of Medicare and Medicaid, signing the bills into law on July 31, 1965. Primarily directed at the elderly, and secondarily at the poor, these benefits were not paid to any one resident in a state institution. As state bureaucracies came to realize, albeit not all at once, these funds could replace state expenditures on the mentally ill, provided patients were discharged from the state hospitals and placed elsewhere.

Looking at the period between 1965 and the early 1970s, what stands out is the huge fraction of discharged patients that came from the ranks of those over the age of sixty-five. Some states, like Wisconsin, passed legislation directly prohibiting the admission of those over sixty-five into state hospitals. Others simply adopted administrative policies that produced the same effect. Between 1955 and 1975, California cut the number of inpatients over the age of sixty-five by 94.6 percent. In 1969, Illinois appointed a new director of its department of mental health, a fiscal and management specialist. By September of that year the state mental health code had been revised to exclude most of the elderly, and plans were rapidly drawn up to discharge 7,000 to 10,000 elderly patients over an eighteen-month period. Massive discharges followed. The sequence was soon repeated in New York State, as part of a general shift of responsibility from the mental health system to the welfare system.[11]

Many patients were transferred from state-run and state-financed mental hospitals to private, profit-making nursing homes. Between 1963 and 1969 alone, the numbers of elderly patients with mental disorders living in nursing homes increased by nearly 200,000, from 187,675 to 367,586.[12] By 1972, with some younger patients added to the mix, the mentally disturbed population housed in nursing and board-and-care homes had risen to 640,000, and two years later, it rose to 899,500.[13]

Younger patients only began to be discharged or diverted away from mental hospitals in large numbers following the addition of Supplemental Security Income (SSI) to the Social Security program. SSI provided a guaranteed income to those unable to work by reason of age or disability, and unlike Social Security, those payments did not depend on prior work history. Some ex-patients, less disturbed than others and victims of an earlier tendency to overhospitalize, were able to assimilate reasonably successfully back into society, but they were a distinct minority. Others initially returned to their families of origin. Isolated and without much support or advice, these families faced an onerous task. The afflictions and the misery they endured were largely hidden, not least because of their understandable desire to shrink from the stigma that so closely clung to mental illness. They quickly discovered that public authorities had failed to provide aftercare facilities and refused to countenance rehospitalization. In the long run, most families found the burdens all but intolerable, and these ex-patients and those who in an earlier era would have been hospitalized found themselves sharing the fate of the much larger group of those who had no family or whose relatives from the outset had refused to assume responsibility for them.

One consequence of this situation was the creation of a new group of institutions, reminiscent of the handful of private, profit-making madhouses that had emerged in eighteenth-century England to cope with the mentally disturbed, albeit on a far larger scale.[14] Ownership of these twentieth-century equivalents of the madhouse—the nursing home, the board-and-care home, the welfare hotel—was open to all who possessed or could borrow the necessary capital and could negotiate or bypass certain licensing requirements. In an ironic twist, some of these "community" facilities were opened and operated by those who had previously worked on the back wards of the old state hospitals. Like their eighteenth-century forebears, these facilities largely operated free of state supervision or oversight.[15]

For elderly patients, the transfers often meant premature death.[16] For many ex-patients, young and old alike, one set of institutions had been substituted for another, and the logic of the marketplace all but ensured neglect, for the less the operators of these facilities spent on the inmates, the larger their profits. State and federal payments to this burgeoning entrepreneurial class were in any event scarcely munificent, and at best it sufficed to purchase the most basic forms of subsistence care. As one contemporary study documented, "A typical day for a mentally ill person in a nursing home was sleeping, eating, watching television, smoking cigarettes, sitting in groups in the largest room, or looking out the window; there was no evidence of an organized plan to meet their needs."[17]

Quite soon, the massive numbers discharged from or refused admission to the state hospitals exceeded the capacity of the nursing home system to absorb more bodies. Many of the chronically mentally ill thus found themselves in a variety of other, still less salubrious settings—group houses, foster-care homes, halfway houses, room-and-board facilities, and "welfare" hotels. Others began living on the streets, and the sidewalk psychotic became an increasingly familiar feature of the American urban scene.

For a time, as this new geography of madness established itself, it was possible to leave these lost souls to decay, physically and otherwise, in the most blighted portions of cities, essentially invisible to the better off. The first decades of deinstitutionalization thus entailed, according to one study of the move from the asylum to the street, "the growing ghettoization of the returning ex-patients, along with other dependent groups in the population; the growing succession of inner-city land use to institutions providing services to the dependent and needy . . . [and] the forced immobility of the chronically disabled within deteriorated urban neighborhoods."[18] Zoning laws were invoked to exclude such "undesirables" from places frequented by the "respectable classes," though usually the costs of housing in these areas, and absence of social services, were sufficient to keep them out. The alleged advantages of community treatment were expected to materialize in areas with high crime rates, abandoned buildings and substandard housing, and a pervasive social anomie.

Such developments did not occur without implicit and sometimes explicit state sponsorship and encouragement. Remarkably, they persisted even when scandals erupted. In New York State, the corrupt links between the board-and-care industry and the political establishment eventually surfaced thanks to a *New York Times* exposé in 1975, forcing a full-scale inquiry and subsequent prosecutions.[19] Pennsylvania, with remarkable foresight, repealed its provisions for inspecting boarding homes in 1967, the same year that it began "a massive deinstitutionalization program aimed at moving patients out of mental hospitals into community programs."[20] Hawaii faced a major shortage of beds in licensed boarding homes when it adopted a policy of accelerated discharge of mental hospital patients. The problem was solved, with unusual bureaucratic flexibility, through a proliferation of unlicensed facilities actively promoted by the state mental health department.[21] Nebraska at first shied away from such a laissez-faire approach, deciding that some form of state oversight was called for. Accordingly, in a splendidly original variation on the ancient practice of treating the mad like cattle, the state placed the licensing of board-and-care homes in the hands of the state department of agriculture. Subsequent citizen complaints about the resultant conditions led to second thoughts about the desirability of taking official notice of the board-and-care operators' practices, so the state withdrew the licenses, but not the patients, from "an estimated 320 of these homes, leaving them without state supervision or regulation."[22] Missouri simply noted the existence of some "755 unlicensed facilities in state housing more than 10,000 patients" and continued to dispense the funds on which their operators depended.[23] Other states, like Maryland and Oregon, opted for perhaps the safest course of all—no follow-up on those they released and hence a blissful official ignorance about their subsequent fate.[24] So it was that states all across the country abandoned the publicly funded asylums for a new form of privatized neglect.

A curious political alliance supported the dissolution of the state hospitals. On the left, a fierce opposition to the incarceration of the mentally ill in places that resembled prisons or warehouses for the unwanted, a profound skepticism of psychiatry's claims to expertise, and a conviction that institutionalization dehumanized and damaged those caught up in its coils fueled a desire to abolish these Goffmanian total institutions.[25] On the right, a libertarianism that hated the public provision of services of any sort, coupled with the promise of fiscal savings, made closure of state hospitals equally irresistible.[26] And once asylumdom had essentially vanished from the scene, there was little chance there would ever be substantial support for its revival.[27]

ORGANIZED PSYCHIATRY WATCHED the dissolution of the vast establishments that had given birth to the profession in virtual silence. A few isolated voices raised an alarm about what the changes meant in practice, but these were few and far between. Thomas Reynolds, for example, who headed one subdivision of the vast federal St. Elizabeths Hospital in Washington, DC, warned that

> if we examine the actual quality of the lives of these people we now return so easily to the community, in their homes and foster homes and halfway houses, taking their Thorazine tablets or their Prolixin injections, we will discover that the great majority, while outwardly sane and tractable, are living utterly barren and blasted lives. . . . We have created a kind of slow spiritual euthanasia with chemical agents, whose primary function is get the patients away from us so that by not seeing the poverty of their lives, we may cease feeling any responsibility for the matter.[28]

His warnings went unappreciated and ignored, and the situation soon became ever more dire.[29]

THE THOUSANDS OF PATIENTS who had once thronged the wards of the state hospitals had always been a visible reminder of psychiatry's near-impotence when it came to managing severe and chronic forms of mental illness. The seriously psychotic had become a professional liability. Few professionals wanted to practice among such an impoverished, clinically hopeless clientele. Hence the alacrity with which the profession moved to distance itself from the socially contaminating effects of an overly close association with these wards of the state. Psychiatric involvement with the chronically disabled could in the future be limited to the periodic prescription of antipsychotic medications, often dispensed by others. In place of trying to cope with the problems of the seriously disturbed, socially deprived, often physically decrepit people who had previously languished on the back wards of state hospitals, psychiatry found it far preferable to minister to the needs of socially functioning patients (or at least those covered by insurance policies) who found themselves struggling with anxiety or depression or eating disorders; to children exhibiting moderate to severe behavioral problems; to the abused and abusive; and those trapped in the coils of substance abuse.

The welfare reforms (a euphemism for the retrenchment of the social safety net) that began in the Reagan years and have been a feature of the neoliberal consensus that has persisted ever since, under Republican and Democratic administrations alike, have in many ways worsened the problems confronting the psychotic. Programs like Aid to Families with Dependent Children were systematically dismantled, amid claims that they fostered dependence and did more harm than good. President Clinton's welfare "reforms" limited lifetime benefits to five years, and in place of earmarked funds for particular programs, block grants to the states became the norm, weakening the social safety net even further. Where the market rules, the outlook is bleak for those unable or ill-equipped to fend for themselves.

Casualties of the hostility to social provision included the federal transfer programs that once underwrote and encouraged the neglect and destruction of state hospitals. These programs have become more restrictive and limited where they have not been eliminated altogether. State governments have lacked the resources and political will to ameliorate the situation to any significant extent, a problem exacerbated by the stigma attached to mental illness and the low priority accorded to programs for the chronically dependent.[30] Funds that once paid for the miserable routines of the board-and-care homes have dried up. The number of people confined in mental hospitals in 2000 (54,836) amounted to less than a tenth of those in 1955, though in this period the population of the United States had increased by more than two-thirds. These numbers mask a drastic shift in the makeup of the institutionalized population, an ever-increasing proportion of which is composed of those with criminal justice histories (often committed by the courts based on incompetence to stand trial, or sex offenders whose sentences have expired but who are held in mental health facilities because they are deemed a continuing threat to the community).[31] One can add to these a small group of patients too disabled to be discharged and also others who cycle through the system for brief periods before being dismissed, often back on to the streets from whence they came.

Though the mentally ill are scarcely the only source of the epidemic of homelessness that now characterizes urban America, they are a major component of the problem. Informed estimates suggest that they constitute as much as a quarter or a third of the homeless.[32] In place of incarcerating inconvenient and often intolerable people who only incidentally violate the law in mental hospitals, twenty-first-century America has adopted a policy of repressively tolerating those who would once have been consigned to these establishments. For such lost souls, cycling between the streets and shelters, with periodic trips to jails when their behavior becomes too disturbing and threatening, has become a routine part of their existence. Tolerance of a sort (one that ignores the problems of the mentally disturbed and the burden they impose on the urban social order) alternates with repression, and then more tolerance, better described as neglect. On occasion, a brief trip to a psychiatric ward is thrown into the mix, where medications are adjusted in an attempt to damp down florid symptomatology. But discharge is rapid, followed by a return to the streets.

The largest providers of psychiatric services, if that is the correct descriptor, have become carceral institutions such as the Los Angeles County Jail, Rikers Island in New York, and the Cook County Jail in Chicago.[33] Citing figures supplied by the Los Angeles County Sheriff's department, Joel Braslow and his colleagues indicate that, on an average day, more than 5,000 prisoners with serious mental illness are housed in the Los Angeles County Jail, making it the world's largest psychiatric institution.[34] African Americans constitute "9.6% of the county's population, yet they constitute 31% of LA County jail prisoners, and 43.7% of those diagnosed with 'serious mental illness' requiring special jail housing."[35] Those statistics hint at the persistence of the differential treatment of mentally ill patients by race that has a history that extends all the way back to the asylum years, and that has undoubtedly been exacerbated by the collapse of public provision for the social consequences of serious mental illness.

Racial disparities have been shockingly understudied when it comes to contemporary mental health issues, though there is evidence that these figures from Los Angeles are part of a larger pattern: Blacks are disproportionately affected by this warehousing of the mentally ill in jails, part of the larger discrimination that pervades the criminal justice system.[36] One would like to know whether drugs were and are disproportionately prescribed by race. But mental hospitals stopped collecting racial data in the mid-1950s, just as the psychopharmaceutical revolution got under way. In the words of one scholar who has focused on mental illness in urban environments, "The inaccessibility of patient records [for privacy reasons] places limits on the archival evidence. Unfortunately, any racial differential in the use of chlorpromazine and reserpine must remain speculative."[37] The dearth of serious research on the subject is itself a telling commentary on the neglect of the social dimensions of serious mental illness that has characterized the last four decades.

Black Americans disproportionately lack social and financial capital—indeed, the differences in wealth are extreme.[38] Their access to decent housing and health care is sharply restricted.[39] All too many of them live in deprived neighborhoods and have little control over their life circumstances.[40] Black men are incarcerated at seven times the rate of white men: one in fifteen over the age of eighteen is in jail or prison.[41] And both adults and children are systematically exposed to higher levels of trauma: emotional or physical neglect and abuse; parental separation or death; witnessing or suffering from gun violence; not to mention violence at the hands of the police.[42] Recent research has suggested that the concatenation of these problems has profound implications for the mental health of Black Americans.[43]

In the early stages of deinstitutionalization, a policy of neglect was largely invisible to the better-off among us. Unwanted and unloved, discharged patients lived an appalling existence in the blighted regions of our cities, where they were left to decompose in areas the well-to-do studiously avoided. But their numbers have grown. It is no longer possible to contain them in the skid rows and ghettoes of our urban centers. On the contrary, their presence is now all too visible to the more fortunate members of our society, for whom they represent a recurrent source of nuisance, alarm, and danger. Though the situation has provoked some political backlash, as yet it has not prompted any major new policy initiatives or even plausible proposals about how to respond to the problems we confront. Deinstitutionalization continues to define public policy toward serious forms of mental illness, notwithstanding the cumulative evidence of its disastrous impact on the lives of many of those supposed to benefit from it.[44]

Under the circumstances, it should come as no surprise to learn that those afflicted with serious mental illness have a life expectancy of between fifteen and twenty-five years less than the rest of us.[45] "Community care" has turned out to be an Orwellian euphemism masking a nightmare. That is a disturbing commentary, not just on the failures of American psychiatry but on the politics and priorities of twenty-first-century America.

CHAPTER NINETEEN

Diagnosing Mental Illness

THE COLLAPSE OF INSTITUTIONAL PSYCHIATRY ought to have proved a boon to the bulk of the profession that now practiced in outpatient settings. In reality, however, the period between 1960 and 1980 saw rising challenges to psychiatry and to the Freudian elite who had dominated the profession since 1945. Some of these attacks were overt and obviously threatening. Other doubts were initially purely intraprofessional concerns debated by a minority of psychiatrists. These drew little broader notice at the time but later exploded into public view in an especially damaging fashion.

As early as 1961, the renegade psychiatrist Thomas Szasz, a professor of psychiatry at the State University of New York at Syracuse, claimed that mental illness was nothing more than a myth, an imaginary entity conjured up by his fellow professionals that had no biological reality. People had "problems in living," but these were not illnesses. Medicalizing them was simply a way of surreptitiously exercising social control of a particularly insidious sort over troublesome people, depriving them of their rights and freedoms in the name of "helping" them. Beyond this, psychiatrists were pathologizing more and more varieties of human behavior, masking their role as enforcers of conventional morality by asserting they were acting in the name of science. Claiming to occupy the other end of the political spectrum from the right-wing Szasz, the Scottish psychiatrist R. D. Laing drew a great deal of attention a few years later for his heretical claim that schizophrenia was some sort of super-sanity, a voyage of discovery that should be indulged and encouraged. It was society, not the mental patient, who was sick.[1]

Sociologists chimed in to assert that mental illness was all a matter of labels, not individual pathology. It was a societal reaction to transient departures from social norms that stabilized rule-breaking behavior.

Psychiatrists cemented the process by applying their scientific-sounding labels, often on the basis of the briefest and most casual of encounters.[2]

The contention that psychiatrists for decades had damaged and deformed those they purported to treat began to morph into a more general skepticism about the profession's claims to expertise.[3] Such contentions were taken up by members of the newly emerging mental health bar, public-interest lawyers who at the end of the 1960s began suing psychiatry on multiple fronts. The most prominent of these attorneys, Bruce Ennis of the American Civil Liberties Union, soon co-authored a long law-review article dismissing psychiatrists' claims to be experts in the diagnosis and treatment of mental illness as scientifically indefensible and verging on the fraudulent. Psychiatrists who weighed in on questions of sanity fared no better, he charged, than a trained monkey flipping a coin.[4]

In the heyday of psychoanalysis, Hollywood had produced a string of movies extolling the virtues of the new psychiatry and trumpeting Freud's insights into the human psyche. Miloš Forman's *One Flew over the Cuckoo's Nest,* adapted from Ken Kesey's novel and released in 1975, almost single-handedly changed all that. Still widely watched nearly a half century after it was first shown, the film constituted a sustained assault on psychiatry's competence, beginning with the failure to recognize that the roustabout Randle P. McMurphy was merely feigning insanity. As the film proceeded, a dark portrait of psychiatry emerged: it was a vicious, repressive, antitherapeutic enterprise whose "group therapy" was a form of sadism and whose physical interventions, from drugs to electroshock to lobotomy, were simply weapons employed to cow dissent and discipline the unruly.[5]

Critiques of psychiatry's competence found support in internal concerns that some psychiatrists had begun to articulate about the unreliability of psychiatric diagnoses. During the 1960s, a series of sober academic studies conducted by leading lights in the profession had repeatedly documented the problem. As early as 1959, the prominent American psychiatrist Benjamin Pasamanick and his associates had drawn attention to the fact that "commonly promulgated definitions of mental . . . illness are still so vague that they are frequently meaningless in practice. . . . [It is] an even stronger indictment of the present state of psychiatry, that equally competent clinicians as often as not are unable to agree on the specific diagnosis of psychiatric impairment. . . . Any number of studies have indicated that psychiatric diagnosis is at present so unreliable as to merit very serious question when classifying, treating and studying patient behavior and outcome."

Their own study of the issues demonstrated that the clinical and theoretical commitments of the treating psychiatrist were of greater importance than the symptoms presented by the patients. In particular, "the greater the commitment to an analytic orientation, the less the inclination toward diagnosing patients as schizophrenic."[6]

The University of Pennsylvania psychiatrist Aaron T. Beck's subsequent review of systematic studies of reliability, undertaken some three years later, essentially confirmed these findings. Setting aside organic cases such as delirium or dementia, where inter-rater reliability might reach 80 percent or more, in functional cases of mental disorder, psychiatrists at best agreed with one another just over 50 percent of the time, and often agreement was far less than this. Diagnosis was, he acknowledged, vital for research, treatment, and teaching, and yet the highest agreement on specific diagnoses that he found in these studies was only 42 percent. Perhaps, he suggested, though this situation was a major problem for research and epidemiological work, it mattered less in clinical settings, since most psychiatrists were "seldom bound by the actual diagnosis . . . [and may] simply regard the clinical diagnosis as an additional bit of information (unreliable as it may be) which may support the therapeutic decisions made on the basis of other factors."[7]

Psychoanalysts disdained diagnostic labels, treating them essentially as an irrelevance. Indeed, Karl Menninger, whose best-selling books had made him perhaps the most famous psychodynamic psychiatrist of the postwar era, argued in 1963 that diagnostic labels should be abandoned altogether.[8] Not only was it a charade, since the labels had no discrete meaning, but affixing a diagnosis actively harmed patients, he contended, turning them into objects, not people, and stigmatizing them, while providing nothing useful for the clinician. Small wonder, then, as the acerbic psychopharmacologist Donald Klein put it, that "for the psychoanalysts, to be interested in descriptive diagnosis was to be superficial and a little bit stupid."[9]

AMERICAN PSYCHIATRY HAD PRODUCED a *Diagnostic and Statistical Manual* in 1952, and a second edition was produced in 1968, both testaments to the postwar dominance of psychoanalysis.[10] They were slight documents, barely over a hundred pages long. The second edition cost $3.50, which was more than most psychiatrists thought it was worth. Though the broad distinction made between neuroses and psychoses was uncontroversial, the manuals' content was otherwise lightly regarded and little consulted. For those who saw the treatment process as involving an inquiry into the precise psychodynamics of the individual case, diagnostic labels were irrelevant, artificial creations that added nothing of substance to the understanding of a patient's problems or the treatment process. That the primary forms of psychosis, schizophrenia and manic-depressive psychosis, derived from the work of Freud's bête noir, Emil Kraepelin, probably further alienated the analysts from the whole process.

One of the most striking demonstrations of psychiatry's inability to agree on diagnoses appeared in 1972, when a systematic cross-national study of the diagnostic process in Britain and the United States was published by Oxford University Press. John Cooper and his colleagues presented results that laid bare just how uncertain the status of psychiatric diagnoses was, how variable and subject to the whims of local culture. They looked at the two most serious forms of mental breakdown, schizophrenia and manic-depressive illness, and measured the cross-national differences in the ways these conditions were diagnosed.

Scientific knowledge is meant to be universal and to travel easily across national boundaries. No one would expect large variations in the diagnosis of, say, tuberculosis or pneumonia. It proved to be quite otherwise in the psychiatric realm. Schizophrenia, it turned out, was diagnosed far more frequently in the United States than in Britain. Contrariwise, the diagnosis of manic-depressive illness was embraced far more frequently by British psychiatrists. The contradictions were massive. New York psychiatrists diagnosed nearly 62 percent of their patients as schizophrenic, while in London, only 34 percent received this diagnosis. And while less than 5 percent of the New York patients were diagnosed with depressive psychoses, the corresponding figure in London was 24 percent. Detailed reexamination of the patients suggested that these diagnoses were not rooted in differences in their symptoms but were a by-product of the preferences and prejudices of each group of psychiatrists. Yet these differences had real-world consequences, producing major differences in the treatments the patients received.[11]

All these studies, and more, were couched in dry academic prose. They were of concern to a subset of psychiatrists who worried about their implications, but laymen did not read the psychiatric journals or monographs written for a handful of specialists and priced accordingly. Given the dismissive attitude most psychiatrists adopted toward diagnosis, few at the time expected these critiques to cause a major upheaval in the psychiatric enterprise. And then, quite suddenly, another publication attacking the profession's diagnostic competence turned the psychiatric world upside down.

On January 19, 1973, *Science* (alongside *Nature* the most influential general science journal in the world) published an article that instantly captured major media attention. In itself, that is not unusual, for science journalists often use *Science* as a source of copy, but what *was* somewhat unusual was that the paper in question was authored by a social scientist, not someone from the biological or natural sciences. This particular paper has enjoyed an unusually long half-life. Nearly a half century after its appearance in print, it continues to attract hundreds of citations a year and to be a staple of undergraduate textbooks in both psychology and sociology.[12] More remarkably still, one can plausibly argue that its findings had an extraordinary real-world impact, playing a major role in transforming a common subspecialty in medicine in ways that continue to resonate all the way down to the present.

The paper was by David Rosenhan, a Stanford professor of psychology and law, who had had a fitful academic career during the 1960s—a string of temporary teaching appointments along the East Coast, before landing a more promising job in 1968 at Swarthmore College. Two years later, he was invited to visit Stanford for a year, which led to the offer of a permanent post there.[13] At Stanford, he co-authored a textbook on abnormal psychology with Martin Seligman—one that went through four editions between 1984 and 2000—and set himself up as an expert advising on jury selection. But his paper in *Science* was his one significant contribution to the social psychological literature, albeit one that made him famous for decades. He never revisited the topic in any academic journal or published anything of comparable impact for the rest of his career.

"On Being Sane in Insane Places" purported to report the results of an experiment involving eight subjects, one of whom was Rosenhan himself. (Rosenhan reported that there had been a ninth participant, but he had been dismissed from the study for violating the experimental protocol.) These volunteers, who had been screened to eliminate anyone with mental health issues, were instructed to show up at a variety of mental hospitals claiming that they were hearing voices and to seek admission. Those were the sole symptoms they were to report, and they were strictly enjoined to behave normally postadmission and to inform their doctors that they were no longer symptomatic. Together, the volunteers had approached a total of twelve mental hospitals (some participants engaged in the charade more than once). The hospitals, Rosenhan reported, were spread across five states and represented a wide spectrum of facilities, from isolated, run-down rural state hospitals to public facilities that were modern and relatively well staffed, as well as a single private mental hospital linked to an academic department of psychiatry. Uniformly, whatever institution they approached, the subjects of Rosenhan's experiment were admitted as inpatients and then spent anywhere from seven to fifty-two days in the hospital before they were discharged (an average of nineteen days). The private mental hospital diagnosed its lone patient as manic-depressive, a relatively favorable diagnosis. By contrast, all of those admitted to public facilities were given the label of schizophrenia, and Rosenhan reported that, upon discharge, they were noted to be "schizophrenics in remission."

On the basis of these results, he claimed that "we cannot distinguish the sane from the insane in psychiatric hospitals." As he began to give presentations of his findings prior to publication, staff at a local teaching and research hospital "doubted that such an error could occur in their hospital." Rosenhan's response was to inform them that he would send along pseudopatients over the following three-month period and see whether they could detect the imposters. It was a trap. "Forty-one patients were alleged, with high confidence, to be pseudo patients by at least one member of staff. Twenty-three were considered suspect by at least one psychiatrist." Rosenhan rather gleefully reported that he had sent not a single pseudo-patient.[14]

A furious correspondence ensued. Psychiatrists from all over the country lined up to criticize the study and reject its findings. A number of correspondents were incensed at Rosenhan's use of the terms "sanity" and "insanity," objecting that these were legal, not medical terms. In reality, psychiatrists made use of "insanity" as a medical term well into the twentieth century, and terminological disputes were in any event irrelevant to the question at hand: the damage Rosenhan's findings had created for the public's view of the profession. What these complaints missed was that by adopting these vernacular terms, Rosenhan (doubtless deliberately) had invited greater lay attention to his findings. Rosenhan's work had appeared in the most prestigious of places, presumably after strict peer review, so who could doubt its integrity? The implications of his findings were profound. If psychiatry could be so easily duped and would assign the most devastating of diagnoses—schizophrenia—on the basis of such superficial grounds, it was surely an emperor with no clothes.

None of the earlier critiques of psychiatry's problems with diagnosis had attracted any attention outside the profession, and most within it had treated the problem as trivial. "On Being Sane in Insane Places" altered the landscape at once, and quite fundamentally. Rosenhan's findings attracted massive media attention all across the country. At least seventy newspapers, both regional and national, gave prominent attention to his study. Television and radio shows interviewed Rosenhan. A major commercial publisher offered him a lucrative contract for a book based on his research, an offer Rosenhan accepted with alacrity. Harvard even sent out feelers about a possible appointment to its faculty. Rosenhan's exposure of psychiatry's flaws caused a sensation. No wonder so many practitioners rushed to register their objections in the pages of *Science,* which, quite extraordinarily, devoted nine pages of a subsequent issue to their howls of protest and to Rosenhan's response.[15] That in itself was a measure of how powerfully this exposé resonated outside the cloistered world of academia.

Thanks to some astonishing historical detective work by a New York journalist, Susannah Cahalan, we now know something remarkable. David Rosenhan, as she meticulously shows, perpetrated one of the most egregious and successful academic frauds of the twentieth century.[16] It is highly probable that several of the pseudo-patients were simply figments of Rosenhan's imagination. In any event, Cahalan provides extensive documentary evidence of falsified and distorted data, and gross departures in the conduct of the study from what the published findings claimed had happened. Rosenhan was himself a pseudo-patient in his study, and Cahalan quotes from his own medical records to document how fraudulent his published account was of what transpired when he sought admission to Haverford State Hospital in Pennsylvania. Far from confining himself to reporting three discrete aural hallucinations and otherwise behaving perfectly normally, Rosenhan (who had identified himself as David Lurie) gave ample evidence of deep intellectual and emotional disturbance. Besides grimacing and twitching during his intake interview, and the dull, halting speech pattern he exhibited, he indicated that the radio was broadcasting to him and that he could "hear" other people's thoughts. It an attempt to quiet the voices, he had taken to wearing a copper pot over his ears. He was depressed and frightened and had been unable to work for months. Outpatient treatment with drugs had failed to improve matters. Visibly "tense and anxious," he thought he was worthless and had contemplated "suicide as everyone would be better off if he was not around."[17] These are an infinitely more serious and extensive set of pathological symptoms than the ones he recounted in his *Science* article.

Had Rosenhan told the truth about his presentation at the hospital, no one would have been surprised that a psychiatrist would have decided to admit him or to diagnose the patient as schizophrenic. In his *Science* paper, Rosenhan further claimed that, once admitted, the pseudo-patients (himself included) immediately stopped displaying symptoms and behaved normally. Again, the surviving medical records show that in his case this is quite false. In the days after his admission, two other psychiatrists examined him at some length. Both documented the depths of the pathology Lurie was complaining of.

Cahalan's book, *The Great Pretender,* provides many more examples of assertions Rosenhan made that turn out to be pure fiction. Her exposure of this far-reaching scientific fraud is a remarkable accomplishment, all the more so because of how hard it was to discover the truth decades after the fact. But Rosenhan managed to take his secret to his grave, and long before the truth emerged, his study had served as the catalyst for a revolution in the orientation and practice of psychiatry, one whose effects have dominated our approach to mental illness for almost four decades now.

THE SERIOUSNESS OF THE CRISIS the profession faced as soon as Rosenhan's paper was published was immediately recognized by psychiatry's elite. Within weeks of the article's appearance, on February 1, 1973, the board of trustees of the American Psychiatric Association (APA) called an emergency meeting in Atlanta. How could they respond to "the rampant criticism" that enveloped the profession, not least to the perception (or, rather, the reality) that its practitioners could not reliably make diagnoses of the mental illnesses they claimed to be expert at treating?[18] Over three days they debated how to proceed and, after prolonged discussion, came to a decision: the association would set up a task force charged with evaluating and reworking the *Diagnostic and Statistical Manual* (*DSM*).

Before that task force could be established, however, another controversy arose. For decades, psychiatry had held that homosexuality was a form of mental illness—a claim with deep roots in Freudian doctrine that reinforced strongly held prejudices in the public at large. Now, prompted by the civil rights revolution, gay activists, including closeted gay psychiatrists, revolted and demanded that the profession reverse its previous position. After much internal debate and discussion, the APA resolved the issue by means of a postal ballot—an approach that solved a wrenching political issue but that invited public ridicule and provoked further commentary about the reliability and scientific standing of psychiatric diagnoses.[19] The ballots showed that 5,854 psychiatrists voted to remove homosexuality from the *DSM* and 3,810 to retain it.

The Columbia psychiatrist Robert Spitzer had played a large role in brokering the "solution" that put the controversy over the status of homosexuality to bed. Soon thereafter, after some behind-the-scenes lobbying, a grateful association appointed him to head the task force charged with revising the *DSM*. Most psychoanalysts continued to regard the whole project as silly and unworthy of their time, which allowed Spitzer great leeway in determining the working group's makeup. The one psychoanalyst in its midst, finding himself marginalized and ignored, soon ceased attending its sessions.

The transformational impact that *DSM III* would have was not clear at all when Spitzer obtained his appointment. Most of the profession's elite disdained what they saw as the dull, intellectually uninteresting task of constructing a new nosology for the field. They had, as they saw it, far more interesting intellectual puzzles to pursue. Spitzer demonstrated an extraordinary far-sightedness and great political skill in putting together the membership of the task force, guiding its members toward consensus, and then persuading a skeptical profession to adopt its work product.[20]

In the absence of the impetus provided by Rosenhan's study, one wonders how eager psychiatry would have been to revise its diagnostic procedures. We do not live in that counterfactual universe, however. The revision of the *DSM did* take place, and the publication of the third edition, 494 pages long compared with the 134 pages of its predecessor, transformed American psychiatry irrevocably. It accelerated the decline of psychoanalysis and secured its replacement by a biologically oriented psychiatry that claimed that the "diseases" the manual identified and listed were akin to those that mainstream medicine diagnosed and treated. The *DSM III* provided an almost mechanical approach to the diagnostic process, one that, at least in theory, sharply raised the odds that psychiatrists in Topeka or Walla Walla, or San Francisco or New York, would attach the same label when confronted by the same patient.

DURING THE 1960S AND 1970S, there was a single major exception to the psychoanalytic domination of academic departments of psychiatry—

Washington University in St. Louis. The academics there remained heavily committed to the notion that mental illness was in fact physical, a pathology of the body like any other illness. For them, psychoanalysts were either charlatans or medical men who had badly lost their way, imagining that a medical disorder could be cured by talk therapy when they should have been searching for biomedical treatments that acted on the body.[21] Though other university departments contained the occasional somatic psychiatrist (seen by colleagues as distinctly odd), no other department clung so stubbornly to the idea that mental illness was a brain disease. At Washington University, this was an article of faith, and Robert Spitzer, who had once flirted with psychoanalysis himself, found in its faculty, and in some of the psychiatrists it had trained, the core members of the *DSM* Task Force he assembled.

The anomalous commitment of Washington University's department of psychiatry to an approach that rejected psychoanalysis was an accident—or, rather, a reflection of the rules the university imposed on the faculty of its medical school. Those who took regular academic positions there had to give up any prospect of earning private clinical income. They were to live solely on their university salaries. This was an arrangement—the so-called strict full-time system—that the Rockefeller Foundation had urged on medical schools as it underwrote the reform of medical education in the 1910s and 1920s. Unsurprisingly, it had proved unpopular among many would-be faculty and had largely been abandoned by most medical schools. Washington University was an exception, and the psychoanalysts of St. Louis were having none of it. If the university sought to deny them the rich rewards of private practice, why, then, the university could do without their services. So it did.

If psychoanalysts thought diagnostic labels a waste of time, the psychiatrists at Washington University were committed to them, provided they could be refined and made more coherent. That process, by carefully identifying distinct psychiatric disorders, would facilitate the return to biology that they were certain was the key to developing an effective response to major mental illnesses. In pursuit of this goal, they had collectively sought to develop ways of distinguishing among mental disorders, looking for an approach that might facilitate research rather than worrying about its usefulness in clinical interventions. John Feighner, then a resident, later acknowledged that his mentors had concluded that "it seemed imperative . . . that we refine our diagnostic criteria to assist us in selecting specific treatments for specific patients and to improve communication between research centers."[22] His senior colleagues sought to define such criteria for a variety of psychiatric disorders and collectively produced a paper published in the *Archives of General Psychiatry.* By departmental convention, their chief resident was assigned first authorship, so the distinctions the paper laid out were known thereafter as "the Feighner criteria."[23] It became, according to Hannah Decker, "the most cited paper ever published in a psychiatric journal."[24]

When Robert Spitzer was charged with rewriting the *DSM,* it was to this group of outsiders that he turned to compose his committee, a decision that was left up to him because the Freudians thought the whole exercise a waste of time. He later commented that, in putting together the Task Force, he had "selected a group of psychiatrists and consultant psychologists committed primarily to diagnostic research and not clinical practice. With its intellectual roots in St. Louis instead of Vienna, and with its intellectual inspiration derived from Kraepelin, not Freud, the task force was viewed from the outset as unsympathetic to the interests of those whose theory and practice derived from the psychoanalytic tradition."[25] It was an accurate perception. From the beginning, the plan was to eliminate what its members regarded as the fanciful Freudian etiologies that had been embedded in the two earlier editions of the *DSM* and to strip out all references to neuroses and other psychoanalytic language.

SPITZER SHARED THE ST. LOUIS GROUP'S COMMITMENT to biology and to reconnecting psychiatry to the medical mainstream, though his ambitions were greater than theirs. Rather than just creating labels that might be useful in psychiatric research, he was committed to writing a manual that would guide clinical practice. It was an ambition that at times threatened to cause rifts with many of his Task Force, but Spitzer proved a skilled and effective political operator, and eventually he managed to secure broad acceptance within the working group for his plans.

Crucially, he understood, as the St. Louis group did not, that if he were to persuade the members of the American Psychiatric Association (most of whom were clinicians) to endorse the new *DSM,* he had to produce a document that found some place for the whole range of problems that brought patients to the psychiatric waiting room. Samuel Guze, the dominant figure in the St. Louis group, urged Spitzer to produce a severely truncated manual, one that included only a relative handful of well-validated conditions. Spitzer dismissed his suggestion out of hand. "If we do what you are proposing, which makes sense to us scientifically," he countered, "we will give the insurance companies an excuse not to pay us."[26] Instead, he made sure that "if any group of clinicians had a diagnosis that they thought was very important, with a few exceptions, we would include it. That's the only way to make it acceptable to everyone."[27] This was the "logic" that ensured the relentless growth in the number of psychiatric "illnesses" that would become a feature of each successive edition of the *DSM*.

One particularly striking example of how this expansion of diagnostic categories came about is the inclusion of post-traumatic stress disorder (PTSD) in the new manual. The diagnosis was in one very important respect an anomalous category in the new *DSM*, for as part of their attempt to break with the Freudian overtones of the two previous diagnostic manuals, Spitzer's group jettisoned the purported psychodynamic origins of various disorders. Claims about the etiology of schizophrenia or depression were dismissed as just so much unscientific speculation. *DSM III* was to remain resolutely agnostic about the causes of the disorders it included. The new disorder that was PTSD, however, was explicitly tied to a particular source: trauma and its effects on the psyche. Like the undead, or so its proponents argued, memories of past horrors refused to remain buried and were so disturbing, intrusive, and disruptive that long after the event they overwhelmed an individual's capacity to cope.

The pressure to include this new disorder in the spectrum of psychiatric illnesses initially arose from the ranks of disaffected veterans of the Vietnam War. The military brass had claimed that, unlike the earlier wars of the twentieth century, embedding psychiatrists in the combat zones had ensured that "psychiatric casualties need never again become a major cause of attrition in the United States military in a combat zone."[28] It was a stance that embittered opponents of that official narrative fiercely rejected. Joined by two sympathetic psychoanalysts, Robert Jay Lifton and Chaim Shatan, and then by others, Vietnam Veterans against the War argued that battlefield traumas had left them with lasting psychic scars that constituted still another consequence of an evil and immoral war. Theirs was a demand for official recognition and recompense for the serious psychological damage that lingered in their ranks and persisted years after the trauma that had brought it about.[29]

Shatan's 1972 op-ed in the *New York Times* on "Post Vietnam Syndrome" was an early statement of their aims, and when Spitzer and his task force began to revamp psychiatry's diagnostic system, they became an obvious target for those seeking recognition of this novel disorder.[30] Initially, Shatan and his allies met with resistance. It was not just the evident political overtones of their arguments that provoked pushback, but the fact that their proposed diagnosis was sharply at odds with the whole approach the Task Force aimed to put in place.[31] But unlike the rigid St. Louis group, Spitzer was a pragmatist, willing to satisfy any sizable constituency demanding inclusion of its preferred diagnosis if by doing so he ensured the success of his overall project. The diagnosis was legitimized, as the sociologist Wilbur Scott concluded, "because a core of psychiatrists and veterans worked consciously and deliberately for years to put it [in the manual]. They ultimately succeeded because they were better organized, more politically active, and enjoyed more lucky breaks than the opposition."[32]

Faced with a sustained and highly organized pressure group, Spitzer ultimately gave way, but with a major proviso: the new diagnostic category he agreed to include in *DSM III* was not post-Vietnam syndrome, but a much broader and less specific stress-related disorder, post-traumatic stress disorder. A whole variety of traumas, not just those stemming from military conflict, were now recognized as possible triggers of lasting forms of mental disorder—sexual violence and assault prominent among them.

In time, the adoption of the PTSD diagnosis would open a Pandora's box. Some enterprising psychiatrists and psychologists would soon uncover a whole host of alleged victims of trauma, those who remembered not too much, but too little. In their hands, patients whose early sexual traumas were so powerful that they had repressed all memory of them now learned to recall them in vivid detail. It was the return of the repressed with a vengeance. And vengeance it unleashed, with increasingly elaborate "recovered memories" wreaking havoc on the lives of many who stood accused of horrific crimes against their children (or even other people's children, as in the case of the McMartin preschool scandal that rocked Southern California in the 1980s).[33] For a decade and more, moral panics like these spread, ruining lives and reputations. Criminal trials and civil suits abounded, even as a growing volume of research on how memory works undermined the core tenets of the recovered-memory advocates.[34] And then, in the late 1990s, almost as swiftly as the recovered-memory movement had gained public attention and credibility, it collapsed. As much as anything, the sociologist Allan Horwitz argues, it vanished when the major proponents of the syndrome lost a series of countersuits and were forced to pay staggering damages.[35]

If the recovered-memory movement has now largely faded from view, the same cannot be said of trauma-related diagnoses. In *DSM III*, the stressor was conceptualized as being a major and life-threatening event that "would evoke significant symptoms of distress in almost everyone."[36] Psychiatrists and clinical psychologists argued that in those developing PTSD, trauma produced involuntary, recurrent, and intrusive memories, hypervigilance, and persistent negative emotions. These in turn were often associated with reckless or self-destructive behavior.

In later editions of the *DSM*, however, the boundaries became more elastic. By 1994, the precipitating trauma did not need to be something so awful as combat exposure, seeing one's parent or child being shot, being raped, or suffering other forms of sexual trauma. The emotional impact of hate speech, sexual harassment, witnessing a fight, indirectly learning of the death of a family member, even watching a disaster unfolding on mass media: all these came to be seen as sufficiently traumatizing to trigger PTSD.[37] The upshot, unsurprisingly, has been the creation of "a largely autonomous profession that studies and treats trauma," accompanied by a massive explosion of research on the subject, and the entrenchment of trauma counselors "in schools, hospitals, corporations, the military, the judicial system, and disaster relief organizations."[38] Post-Vietnam syndrome had expanded beyond all recognition.

RESPONDING TO THE IMPERATIVE to make sure that psychiatrists faced with the same case would attach the same diagnostic label, Spitzer and his team early on set aside any concern with validity—that is, whether their labels corresponded to divisions found in nature. As they recognized, they could not demonstrate convincing chains of causation for any major form of mental disorder, nor were there any biological tests that could be used for diagnostic purposes. Perforce, they had to rely on symptoms to distinguish among the disturbances that confronted them, a situation akin to that of eighteenth-century physicians trying to create an orderly classification from the confusing mass of clinical material that confronted them. If psychiatrists were to be brought to agreement, the criteria for diagnosing mental illness had to be consistent and straightforward. The Task Force's emphasis thus fell on creating lists of symptoms that allegedly characterized different species of mental disorders; using that list then created a "tick the boxes" approach to the problem of diagnosis. That way, at least in theory, the embarrassing disagreements about diagnosis exposed by Aaron Beck and John Cooper and his associates (let alone the nightmare of Rosenhan's pseudo-patients) would become a thing of the past.

The members of the Task Force presented themselves as data-oriented. In reality, theirs was a thoroughly political exercise. Spitzer asserted that they were "committed to the rigorous application of the principles of testability and scientific verification." But in fact, as Robert Morison of the Rockefeller Foundation had complained about an earlier generation of psychoanalytic leaders, matters were resolved by taking votes and manipulating verbiage to gain consensus. As the historian of psychiatry Hannah Decker's reconstruction of the process shows, decisions repeatedly relied on political horse-trading and settling on what "felt right"—which often meant what felt right to Robert Spitzer. So it was, for example, that when deciding how many of a laundry list of symptoms made someone eligible for the label "schizophrenic"—an enormously consequential decision—Spitzer's group settled on six out of a list of ten possibilities. Left unsaid was that this meant that two people allegedly afflicted with this condition might share only two of this long list of symptoms and yet be given the same diagnosis. As to how many psychiatric illnesses to accept, and which ones, these were again questions that aroused much debate, with the answers the subject of politicking and votes by the members of the Task Force.

The Columbia psychiatrist Donald Klein, who was one of the most influential members of the group, did not bother to hide how the sausages were made:

> We had very little in the way of data, so we were forced to rely on clinical consensus, which, admittedly, is a very poor way to do things. But it was better than anything else we had. We thrashed it out, basically. We had a three-hour argument. There would be about twelve people sitting down at the table, usually there was a chairperson and there was somebody taking notes. And at the end of each meeting there would be a distribution of events. And at the next meeting some would agree with the inclusion, and the others would continue arguing. If people were still divided, the matter would be eventually decided by a vote. . . . [T]hat is how it went.[39]

Psychoanalysts, who had initially ignored what they regarded as a tedious and anti-intellectual exercise, gradually began to express some alarm about what Spitzer and his team were up to. There were complaints about the "linguistic and conceptual sterility" that marked early drafts of the revised manual. "*DSM III,*" it was alleged, "gets rid of the castles of neurosis and replaces it with a diagnostic Levittown."[40] In words that dripped with contempt, another psychoanalyst compared the depth and sophistication of Freudian perspectives with the jejune ideas Spitzer and his group appeared to be wedded to: "It is unreasonable . . . to treat equally the carefully reproduced work of thousands of psychoanalysts and psychodynamic clinicians and the relatively recent learning theorists or esoteric fantasies about the etiology of psychopathology."[41]

As events would show, Spitzer's group, far from treating these two elements as equal, regarded the work of those thousands of analysts as unworthy of serious consideration. In April 1977, Otto Kernberg, a member of the executive council of the American Psychoanalytic Association, prophetically warned his colleagues that what they were inclined to dismiss as "a joke" was, on the contrary, "a straitjacket and a powerful weapon in the hands of people whose ideas are very clear, very publicly known, and the guns are pointed at us."[42] His warnings were largely ignored, though Spitzer, sensing the need to head off psychoanalytic opposition if he could, handpicked two analysts, John Frosch and his nephew William Frosch, to add to the Task Force. Within a year, facing ridicule and hostility from the original members and having no effect on the group deliberations, John Frosch gave up and resigned.

NOT UNTIL A FEW MONTHS before the American Psychiatric Association was about to vote on whether to accept the radically new *DSM* did analysts finally attempt to mobilize to protect their interests. They decided to launch a symbolic fight to rescue the term "neurosis" from the oblivion into which Spitzer proposed to cast it, and they insisted that the concept be included in the *DSM,* along with an explanation of the underlying psychic conflicts that psychoanalysis held were responsible for its existence. Without such changes, they threatened to mobilize votes and secure the rejection of the Task Force's work. For a brief period, Spitzer worried that the psychoanalysts might succeed and that his room for maneuver was sharply limited. The St. Louis–based members of his Task Force were in no mood to allow him to compromise. They had fought to create a document that deleted all references to psychoanalytic ideas and were not prepared to see them reemerge at the last moment. Caught between these conflicting forces, the politically skillful Spitzer eventually found a diplomatic solution. After certain entries, a parenthesis would appear: "Anxiety Disorder (or Anxiety Neurosis)," for example. On May 12, 1979, by voice vote, with those modifications, the *DSM III* was approved as the official stance of the American Psychiatric Association. It would appear in print the following year and become an unexpected best-seller, something every mental health practitioner felt compelled to own, and a major contributor over the years to the coffers of the association.[43]

IT QUICKLY BECAME APPARENT that the analysts who had warned of the dangers the new manual would present for their branch of the profession had underestimated how deadly the new *DSM* would prove. The dominance the Freudians had exercised over American psychiatry withered so rapidly over the next few years that one might almost argue that the publication of *DSM III* marked the pronouncement of the last rites over what turned out to be a largely moribund enterprise. Psychoanalysts and their sympathizers were rapidly defenestrated from the elite positions they had for a quarter century occupied in the profession. Academic departments appointed biologically oriented psychiatrists as their chiefs, or imported neuroscientists into these posts. Spitzer's ostensibly theory-neutral classification system in fact underwrote a rapid shift to a psychiatry that embraced a biologically reductionist model of mental disorder, one that had no truck with psychodynamic or psychoanalytic approaches and instead embraced psychopharmacology as the way forward.

There had been signs, even before the publication of *DSM III*, that the position of psychoanalysis in American psychiatry was under threat. In the mid-1960s, 50 percent of the psychiatric residents at UCLA were training in psychoanalysis. A decade later, only 27 percent sought psychoanalytic training. Over the following decade, the number of medical students opting to specialize in psychiatry again fell substantially.[44] At the Menninger Clinic, once heavily committed to psychoanalytic treatment, the numbers receiving psychotherapy had fallen from 62 percent in 1945 to 23 percent in 1965, a pattern also evident at other private psychiatric hospitals that had once employed it as a first-line treatment.[45]

On another revealing front, besides the National Institute of Mental Health (NIMH), a main source of research support for psychiatry was the Foundations Fund for Research in Psychiatry, established in 1953 by a rich patient grateful for the psychoanalytic treatment of his depression by

Lawrence Kubie. Initially its grants were, not surprisingly, given to many prominent analysts. But between 1962 and 1973, the foundation's priorities changed drastically. By the early 1970s, most of its money was flowing toward research on somatic treatments, and only 9 of its 194 awards went to psychoanalysts. Between 1973 and 1978, that number dwindled to zero.[46] Meanwhile, even analysts were losing faith in their ability to treat schizophrenics, and skepticism was increasingly being voiced about the outcomes of psychoanalytic treatment of neuroses.[47] Making matters worse, university officials resented the fact that, with few exceptions, psychoanalytic institutes were organized and controlled by private practitioners and existed wholly outside the university's orbit and control.[48] That resentment, and the failure of analysts to secure research money, weakened whatever support might have been forthcoming from medical school administrators. Once the mainstream of psychiatry moved away from psychoanalysis toward a biological psychiatry that began to attract serious research support and was based in university facilities, the psychoanalytic elite found their previous dominance rapidly undermined.

The collapse of psychoanalytic supremacy was spectacularly swift. By the mid-1980s, psychoanalytic institutes were facing an extraordinary dearth of medically trained recruits to their programs. It was a situation that sharply curtailed the incomes even of the leading members of the institutes, who had always been able to rely on fees from neophytes seeking admission to the guild. Internal conflicts flared.[49] In 1988, after seventy-five years of fiercely resisting the idea of training nonmedically qualified recruits, the institutes began to admit them. They were compelled to do so, to be sure, by an antitrust lawsuit launched by clinical psychologists in 1985, but that decision did provide another source of apprentices.[50]

Department chairs in American medical schools are extremely hard to budge, for they exercise great power over their faculty and routinely use their patronage to secure their positions against critics. But by 1990, just over ten years after the publication of *DSM III,* only three of the top ten departments of psychiatry were still headed by trained psychoanalysts or members of psychoanalytic organizations.[51] That same year, when the American Psychoanalytic Association surveyed its members, they were seeing on average two patients a week for analysis, scarcely the basis for a secure living. Two decades on, the *Journal of the American Psychoanalytic Association* reported that there had been a 50 percent decline in the number of applicants for training since 1980 and "an even more precipitous decline in applications from psychiatrists."[52] Partly as a consequence, the profession was aging rapidly. In 2012, the International Psychoanalytic Association announced that 70 percent of the membership of its component societies were between fifty and seventy years of age; 50 percent were older than sixty, and as many as 20 percent of training analysts were over the age of seventy. Five years later, only 15 percent of its members were under the age of fifty.

IF THE *DSM III* SEEMED to miraculously create a reincarnated medical psychiatry, one of the midwives of the rebirth was the insurance industry. Private health insurers had increasingly begun to provide some degree of coverage of mental health issues during the 1960s. Federal employees in the Washington, DC, area had enjoyed particularly generous coverage, including relatively extensive coverage of psychotherapy. That proved extremely costly. The insurance companies had no easy way to limit the length of treatments and were confronted with paying for therapies whose efficacy was unsupported by more than anecdotal evidence directed at ill-defined, amorphous pathologies about which there appeared to be little consensus.[53] By contrast, the new manual claimed to identify distinct diseases that could then be linked to discrete treatments.

To the extent that psychotherapy continued to be employed, insurance companies strongly preferred the cognitive-behavioral therapies that sought the rapid alleviation of symptoms to open-ended psychoanalysis—a preference reinforced by the fact that cognitive-behavioral therapy could be offered by people who were not physicians and thus could be paid much less.[54] Hence the growing influence of clinical psychologists and psychiatric social workers, and hence the declining interest of the psychiatric profession in psychotherapy. In a world where the imperatives of managed care were taking hold, insurance companies proffered such low rates for treatments of this sort that fewer and fewer medically qualified personnel continued to offer such services, unless they had access to a clientele willing and able to pay privately for them. Increasingly, therefore, psychiatrists concentrated on forms of treatment for which their monopoly power was legally enforceable—and that could fit the strict requirements of a managed care regime: running through checklists, assigning a diagnostic label, and prescribing the relevant psychotropic medication or medications.

Given the complex nature of psychiatric illnesses, this silencing of patients' voices and lack of sustained attention to their mental states was a major loss, as one of the major architects of the new *DSM* later acknowledged.

Rather than seeing the categories of the new manual as the best approximations available: "*DSM* came to be given total authority in training programs and health care delivery systems. Since the publication of *DSM III* in 1980, there has been a steady decline in the teaching of careful clinical evaluation that is targeted to the individual person's problems and social context and that is enriched by a good general knowledge of psychopathology. Students are taught to memorize the *DSM* rather than to learn complexities." It was an approach, Nancy Andreasen ruefully concluded, that "had a dehumanizing impact on psychiatry."[55]

THE SECOND MIDWIFE OF PSYCHIATRY'S REBIRTH in radically changed form was the pharmaceutical industry. For these corporations, the existence of stable diagnostic categories could play a vital role in the testing of new drugs that needed FDA approval. One of the most crucial and consequential legacies of *DSM III* was the creation of ever-closer links between psychiatry and pharmaceutical corporations, and the money that flowed from that connection until recently greatly improved psychiatry's standing among medical school deans.

Psychoanalysts had greatly broadened the range of conditions they claimed to treat successfully, but their refusal to make diagnostic distinctions a priority, not to mention their resistance to any attempts to demonstrate statistically the usefulness of their interventions, had sharply curtailed the ability of the pharmaceutical industry to design the necessary trials and produce evidence that their innovative treatment worked. The new diagnostic manual might have been designed for the purpose. It incorporated categorical distinctions about different types of mental illness, and the number of such supposedly different syndromes multiplied rapidly. If a subset of patients appeared to respond to a medication under trial, soon enough a new label was attached to these patients, and a new psychiatric disease was born. Rather than diseases calling forth remedies, remedies began calling forth new "diseases."

A final source of validation for the new *DSM* came from two federal agencies. The NIMH embraced the new system, seeing in its scientific-seeming diagnostic system a way to fend off political attacks on the social orientation it had adopted in previous decades.[56] Equally critical, however, was the FDA's embrace of the *DSM*'s assumption that mental illnesses had the same form as physical illnesses, a decision that ensured that drug companies would test and advertise their products as treatments for specific diseases.[57] Those endorsements ensured the triumph of the approach Spitzer had championed for decades to come.

IN 1987, seven years after *DSM III* was published, a revised edition, again under Spitzer's leadership, was published (though it was called *DSM III R* and not a fourth edition). All remaining references to analytic ideas were purged. With no discernible resistance, the fig leaf Spitzer had offered in 1980, the parenthetical gesture that saw "(or neurosis)" added to some of the disorders, vanished. Seven years after that, a new edition officially labeled *DSM IV* appeared, edited this time by Spitzer's protégé, Allen Frances. That edition was superseded again in 2000 by what was officially called *DSM IV TR* (Text Revision). All adhered to the same logic that had inspired *DSM III*. They relied on symptoms to divide and subdivide the world of mental pathology. If that led to much overlap, as patients qualified for more than one disorder, either they could be allocated the most serious disease they qualified for, or the whole embarrassment could be solved by calling them victims of "co-morbidity." The Freudians had regarded symptoms as just the visible sign of underlying disorders that required treatment, and they argued that to treat symptoms alone was to play a game of whack-a-mole. In all the successive editions of the *DSM*, from the third edition onward, symptoms became the very markers of disease, the key to deciding what ailed the patient.

Editions of the manual grew ever larger and gave birth to an ever-longer laundry list of types of mental disorder. *DSMs I* and *II* had been modest little documents of little interest to the profession at large, spiral-bound pamphlets of 132 and 119 pages, respectively. The number of possible diagnoses grew from 128 in the first edition to 193 in the second, assuming anyone paid much attention to the labels they provided. *DSM III* appeared between hard covers and ran to 494 pages. It listed 228 separate diagnoses, and now every psychiatrist and clinical psychologist was forced to employ its categories if he or she wished to be reimbursed by insurance companies. *DSM III R* grew to 567 pages and 253 "diseases," while *DSM IV* was 943 pages long and encompassed 383 officially recognized disorders. *DSM II* had earned a modest $1.27 million for the American Psychiatric Association during its twelve-year run. By contrast, *DSM III* brought in $9.33 million; *DSM III R*, $16.65 million, and *DSM IV*, an astonishing $120 million.[58]

Psychiatrists could now match particular medicines to particular diseases, as physicians do with other forms of illness. Regularizing the diagnosis of mental illness allowed linkages to develop to standardized modes of treatment. This meant that uncertainty could be replaced by predictability, and, at least in principle, finite limits to insurance coverage could be set. Large-scale clinical trials of psychotropic drugs became possible for the first time, and instead of being marginalized in medical schools because of their inability to generate research dollars, psychiatrists became their deans' darlings. Antipsychotic medications and (within a few years) antidepressant pills became huge sources of profit for Big Pharma, regularly among the most lucrative products it produced. While these helped some people lead a less tortured existence, they at best provided a measure of symptomatic relief, albeit at the risk of incurring significant side effects. For the pharmaceutical industry, that outcome had its advantages. Chronic diseases are chronically profitable, for those suffering from them seldom succeed in dispensing with their medications.

So it was that psychiatry, reembracing its medical identity, recommitted itself to the study of the brain as the key to understanding the mysteries of mental disorders. Where the superego once ruled, the usurpers brandishing molecular biology, genetics, and neuroscience now exercised their power. During the Reagan presidency, funding for psychosocial research on mental illness was almost eliminated and funding for Social Security payments to the mentally disabled was cut. The NIMH research budget, however, grew 84 percent, to $484 million annually, with the bulk of that money now directed at neuroscientific work on the most serious forms of mental disorder. This reorientation grew even more pronounced during the 1990s, years that President George H. W. Bush announced were to be the decade of the brain.[59]

Where drug-company dollars flowed, federal dollars followed. In 2013, Barack Obama launched his own BRAIN initiative (Brain Research through Advancing Innovative Neurotechnologies), aimed at developing new ways to understand brain function and how to treat, cure, and even prevent mental illness. Long before then, in the words of Steven Sharfstein, the incumbent president of the American Psychiatric Association, the discipline had moved from a bio-psycho-social model of mental disorder to a bio-bio-bio model.[60] More accurately, it had effectively abandoned the psychosocial approach to understanding and treating mental illness in favor of a near-exclusive focus on biology. Or, as the Harvard psychiatrist Leon Eisenberg put it, the profession has traded "the one-sidedness of the 'brainless' psychiatry of the past for that of a 'mindless' psychiatry of the future."[61]

CHAPTER TWENTY

The Complexities of Psychopharmacology

THE DRUGS REVOLUTION THAT BEGAN to transform psychiatry and the experience of patients in the 1950s was an accidental revolution. Few foresaw its arrival, and the discovery of a new approach to the treatment of schizophrenia and then of manic-depressive psychosis, depression, and a host of less devastating psychiatric syndromes was the product of serendipity, not rational design. Chance observations of the impact of chlorpromazine on psychiatric patients led to its introduction into the psychiatric arena. The equally fortuitous finding that tubercular patients treated with iproniazid or isoniazid became much more cheerful, even when faced with a grim diagnosis, led to the introduction of a novel therapeutics for depression. In the late 1950s, these so-called MAOIs (monoamine oxidase inhibitors) were joined by a new class of antidepressants, the tricyclics.[1] Both of these types of drugs were seen as treatments for what was thought to be a relatively rare but devastating form of mental illness, melancholia or endogenous depression. The malignant character of this disorder—with its overwhelming feelings of sadness and guilt, social isolation, periodic psychotic features marked by hallucinations and delusions, characteristic anhedonia, and increased suicidality—distinguished it from other forms of depression and anxiety that were far more common but were, at the time, viewed as best treated with psychotherapy. Since melancholia was seen as a comparatively rare disorder, drug companies were slow to bring these drugs to market and were ambivalent about doing so. It would be more than two decades before a new view of depression emerged. Then its prevalence began to skyrocket, until it came to be seen as the common cold of psychiatry.

Taken collectively, the antipsychotics, the first generation of antidepressants, and the minor tranquilizers like Miltown marked a radical shift in society's response to mental disorders, minor and profound. No one knew why these new pills worked. That pharmaceutical interventions modified mental symptoms over time led many to reembrace earlier notions that mental illnesses were rooted in the body, but it was wholly unclear why the drugs had the effects they did, or what the hypothesized biological origins of mental disturbance might be.

These puzzles would become central research questions for academic psychiatry. Both federal-grant monies and drug-company funds underwrote this new agenda, and the hiring of a new generation of research psychiatrists who focused on the brain, or on the supposed genetic roots of mental illness, created a new field of academic psychiatry that bore little resemblance to its midcentury predecessor. A large gap—a veritable chasm—separated the world of these academic researchers from the great majority of the profession, who were still facing the clinical complexities of managing mental illness. The clinicians embraced the pharmaceutical interventions that the academic-industrial alliance produced, and they borrowed from the hypotheses of the neuroscientists to lend an aura of science to their day-to-day practice. Managed care in any event left them little choice. Psychotherapy was far too time-consuming and reimbursement rates from the insurance companies far too small to make that an option, save for those whose practices permitted them to collect fees directly from their patients. Ten- to fifteen-minute consultations to prescribe and check medications soon became psychiatrists' standard modes of practice.[2]

When Thorazine was introduced to the marketplace in 1954, it was thought that the cells of the brain communicated through electrical impulses.[3] By the 1960s, a rival hypothesis had essentially replaced that electrical model. Neurotransmitters, chemicals that had previously been seen as important only in the peripheral nervous system, were now cast as the principal underpinnings of activity in the brain. The gaps between the 100 billion or so neurons making up the human brain were, it became increasingly clear, bridged by chemical messengers that excited or inhibited other neurons, binding to receptors that absorbed the signal they provided. Perhaps the new psychiatric drugs produced their effects by acting on neurotransmission, altering either the production of the transmitters or the number and sensitivity of the neurons' receptors. It was a hypothesis that emphasized the potential connections between psychopharmacology and basic research on the functioning of the brain, leading to the recruitment of increasing numbers of neuroscientists into academic departments of psychiatry.[4]

Though there had been a long history of research on the structure of the brain and nervous system, dating back at least as far as the work of Thomas Willis—the man who coined the word "neurologie" in the seventeenth century—the term "neuroscience" was first used at the Massachusetts Institute of Technology in 1962.[5] Later that decade, self-identified neuroscientists created their first professional organizations—the British Neuroscience Association in 1968 and its American counterpart, the Society for Neuroscience, in 1969. Explosive growth of the field followed. The first meeting of the Society for Neuroscience in Washington, DC, in 1971 had 1,400 attendees. Its fiftieth meeting, in 2020, was expected to attract more than 30,000 attendees before COVID restrictions forced its cancellation. Departments of psychiatry have played no small part in the expansion of the field, and the presence and prominence of neuroscientists in academic psychiatry have contributed massively to the biological turn in psychiatry over this period.

An important indirect stimulus to the neuroscientific turn was provided by the work of the Swedish neuroscientist Arvid Carlsson. In the late 1950s, Carlsson discovered that dopamine functioned as a neurotransmitter and could control movement—a discovery that won him a Nobel Prize in 2000. He suggested that defects of dopamine might cause the symptoms of Parkinson's disease, and though that suggestion was initially greeted by skepticism, postmortem examinations of patients dying from that disease later revealed a significant depletion of dopamine in their midbrains. In short order, others experimented with levodopa (L-DOPA), a precursor of dopamine that crosses the blood-brain barrier, as a treatment for Parkinson's disease, and by 1970, it had become the standard therapy for the disorder. It remains so today, despite its serious side effects, and notwithstanding the fact that, while initially effective, it eventually fails to control patients' symptoms, as the destruction of the nerve terminals that produce dopamine in the brain reaches critical levels.[6]

If a shortage of dopamine was behind the tremors of Parkinson's disease, and those tremors could be alleviated by L-DOPA, might whatever therapeutic benefits Thorazine and other psychoactive substances produce also take place via some hitherto undiscovered transformations of the brain's neurotransmitters? It was a reasonable hypothesis, and that dopamine might be involved was suggested by some of the side effects the phenothiazines were known to produce, which closely resembled the motor disorders characteristic of Parkinson's disease. Many early enthusiasts for

Thorazine had welcomed the appearance of these symptoms, and, thinking that they were critical to the drug's therapeutic effects, had deliberately increased patients' dosage until they appeared.[7] It was a connection that strengthened as research continued to show that the phenothiazines acted to block dopamine receptors in the brain.[8]

That connection, which in light of decades of confirmatory research seems indisputable, soon prompted a more daring hypothesis: if Parkinson's disease was the product of a deficiency of dopamine, perhaps schizophrenia was caused by too much dopamine or an overstimulation of dopamine receptors in the brain.[9] For those seeking to replace psychoanalytic speculations with biological accounts of mental disorder, this was a theory with great attractions. It promised to root the most serious forms of mental disorder in the body, simultaneously to include psychiatry in one of the most exciting emerging scientific fields, and to link it to the laboratory world of chemistry and pharmacology. Not coincidentally, it was a theory that drew a warm welcome from the pharmaceutical industry, for whose products it provided powerful marketing copy.

Attempts to validate the theory, however, have produced a litany of disappointments.[10] If there was excessive dopamine in the brains of schizophrenics, then their cerebrospinal fluid should show evidence of this excess. It does not. Hormones produced by the pituitary gland are responsive to dopamine levels in the brain, but once again the great bulk of the studies that have addressed this question have been unable to demonstrate differences between schizophrenics and control subjects on that front. Nor have postmortem studies of the brains of schizophrenics and controls shown evidence of elevated levels of dopamine or its metabolites in the schizophrenic patients.

In an attempt to rescue the theory, some suggested that it was not excess dopamine that produced schizophrenia, but the proliferation of receptors, which heightened the effects of a given level of dopamine. That more receptors have been found in the brains of schizophrenics lent some credence to the theory, except that other research shows that the blockade of dopamine, which is a principal effect of treatment with many antipsychotic drugs, precisely stimulates the production of such additional receptors. In other words, rather than being evidence of something that causes psychosis, the proliferation of these receptors is more plausibly seen as an iatrogenic process—the product of drug treatment, not evidence for its necessity. Then there was the awkward finding, as Joanna Moncrieff put it in her study of the marketing of antipsychotics, "already apparent in the 1970s that some drugs, such as Clozapine and thioridazine (Mellaril), which had relatively weak dopamine-blocking properties, were as effective as other antipsychotics."[11] Finally, there have been more recent attempts to suggest that variants in genes involved in the dopamine system influence the susceptibility to schizophrenia, but the large majority of studies of this sort have produced negative results, and in the studies that buck this trend, the gene effects are acknowledged to be very small.

Yet despite a paucity of research providing support for the dopamine theory (or theories) of schizophrenia, it remains almost an article of faith among many psychiatrists.[12] Like those who once clung to a heliocentric view of the universe, instead of abandoning the dopamine hypothesis, psychiatrists have repeatedly modified it in an ad hoc fashion to prevent its being disconfirmed.[13]

The dopamine theory of schizophrenia is not the only chemical theory proffered to explain a serious mental illness. Depression, too, came to be explained in similar terms. Indeed, whereas the suggestion that schizophrenia was caused in some fashion by dopamine was primarily embraced by the psychiatric profession, the claim that depression was caused by an imbalance of the chemical soup in the brain was adopted far more broadly, and this idea continues to be regularly echoed in the media and believed by the public at large.

The hypothesis that depression is caused by a deficiency of serotonin (another of the first neurotransmitters to be identified) was mooted as an explanation for why an earlier generation of antidepressants, the MAOIs, might alleviate its symptoms. This thesis was put forward with renewed vigor from 1988 onward, when a new class of antidepressant drugs came to market. The rapid spread and acceptance of Prozac and such copycat drugs as Zoloft and Paxil owed much to the pharmaceutical industry's ability to promote the idea that depression was a purely physical ailment. It was, so their marketing insisted, a disorder that could be eliminated by making adjustments to the biochemistry of the brain.[14] The drug companies proceeded to bombard physicians and then the public at large with advertisements promoting these claims. In the words of one of the early direct-to-consumer pamphlets produced to market Prozac, "Prozac doesn't artificially alter your mood and it is not addictive. It can only make you feel more like yourself by treating the imbalance that causes depression."[15]

Patients and their families proved a particularly alluring target, and advertising directed toward them helped sales to soar. The FDA had begun allowing direct-to-consumer advertising in the mid-1980s, and then significantly loosened the rules surrounding such advertising in 1997.[16] Drug companies swiftly took advantage of the new regime, and the so-called SSRI antidepressants (selective serotonin reuptake inhibitors) were sold to the public as newly discovered wonder drugs that attacked the chemical basis of unhappiness, giving the brain back its serotonin and banishing the blues. Depression, so the advertising copy made it seem, could be banished, and patients made "better than well," by ingesting these pills. Peter Kramer's *Listening to Prozac,* published in 1993, and the first to advance this claim, was a publishing sensation, soon matched by Elizabeth Wurtzel's memoir, *Prozac Nation: Young and Depressed in America.* Both did much to spread the gospel of the SSRIs.[17]

Pfizer was somewhat restrained in its Zoloft advertisements, though most would-be consumers probably overlooked its carefully hedged conditional clause. "Scientists believe," the marketing copy read, "that [depression] could be linked with an imbalance of a chemical in the brain called serotonin." The manufacturer of its rival, Paxil, showed no such restraint. Consumers were assured that "with continued treatment, Paxil can help restore the balance of serotonin."[18] Such bald assertions were permitted by a complaisant FDA, though, as scientists have since made clear, "there is no such thing as a scientifically established correct 'balance' of serotonin."[19] We possess no way of measuring serotonin levels in people's brains, let alone of determining what "normal" levels of that neurotransmitter might be. As with the dopamine hypothesis, we must perforce rely on indirect evidence, and that indirect evidence fails to support the drug companies' claims.[20]

Once again, the claimed connection between serotonin and depression was "rescued" by a series of ad hoc modifications attempting to explain why, for example, serotonin levels rise within a day or two of ingesting the drug, while changes in mood take weeks to materialize; or why measures of serotonin metabolites in the cerebrospinal fluid of depressed patients are all over the map, half of them within "normal" limits, a quarter higher than normal, and a quarter lower—or why SSRIs perform no better in clinical trials than the older tricyclic drugs or other pills such as Wellbutrin, that rely on a completely different mode of action.[21] (Wellbutrin is a norepinephrine-dopamine reuptake inhibitor, with no effects on serotonin, unlike SSRIs like Zoloft.)

Soon enough, having convinced the FDA that SSRIs could be used to treat social anxiety disorder, obsessive-compulsive disorder, and premenstrual dysphoric disorder, the Zoloft and Paxil websites aimed at consumers claimed that these "diseases" too were the product of serotonin deficiency. Kramer's *Listening to Prozac* asserted that we had entered a new era of "cosmetic psychopharmacology." Adding more serotonin to our brains would increase our self-confidence, our happiness, our creativity, our energy levels, our success in life. "Shy, Forgetful? Anxious? Fearful? Obsessed?" *Newsweek* asked its readers in a cover story on the wonders of Prozac, only to promise the secrets of "how science will let you change your personality with a pill."[22] Vice President Al Gore's wife, Tipper, recounting her own experience with depression, provided just one prominent example of how quickly the chemical theory of depression had spread: "It was definitely a clinical depression, one that I was going to have to have help to overcome. What I learned is that your brain needs a certain amount of serotonin and when you run out of that, it's like running out of gas."[23]

How one ubiquitous neurotransmitter acted as life's panacea, or alternatively could cause such a variety of psychiatric problems, was left wholly unexplained by both the pharmaceutical industry and by the journalists who uncritically served as its echo chambers. The claims about serotonin depended on reasoning backward from observations about the efficacy of the drugs, and knowledge of some of their effects on the brain. This backward reasoning came from the pharmaceutical industry, not psychiatry, and though it was an effective marketing ploy, it is a deeply unsatisfactory argument, as we can readily see by some counterexamples. Many people are shy and withdrawn in social situations. Alcohol often lowers their inhibitions (sometimes with disastrous consequences) and has obvious physiological effects, as well as temporarily alleviating social awkwardness. Yet no one would sensibly argue that shy and introverted people are suffering from a deficiency of alcohol in their brains. Similarly, aspirin relieves headaches, but not because the brain is suffering from a deficiency of aspirin. The mechanism by which drugs work often does not neatly map on to the roots of the underlying pathology.

THE FACT THAT THORAZINE and its rivals produced symptoms that resembled Parkinson's disease was swiftly acknowledged by psychopharmacologists, as we have seen. The emergence of these uncontrollable tremors was in many quarters seen as inescapably linked to the drugs' efficacy. It was a price patients would have to pay to be relieved of their delusions and hallucinations. The new drugs were also found to blunt emotions, decrease initiative, and curtail movement. Fritz Freyhan, a pioneering psychopharmacologist, acknowledged that psychiatrists actively sought "transitions from hypermotility to hypomotility, which, in a certain proportion of patients, progressed to the more pronounced degrees of Parkinsonian rigidity"—and argued that "clinical evidence . . . indicated that the therapeutic function of chlorpromazine and reserpine could not be separated [from these effects]."[24] The Parisian psychiatrist Pierre Deniker, who played a major role in the introduction of phenothiazines into psychiatry, was blunter. He spoke of patients who "look as though they have been turned to stone" and argued that clinicians must "resolutely and systematically aim to produce neurological syndromes to get better results than can be obtained when neuroleptic drugs are given at less effective doses."[25]

Those "neurological syndromes," it turned out, were far more widespread and serious than these distressing echoes of Parkinson's disease. A 1964 National Institute of Mental Health (NIMH) study of the safety and efficacy of the new antipsychotic drugs pronounced, on the basis of a six-week trial, that their unwanted side effects were "generally mild or infrequent."[26] That was a gravely mistaken conclusion. Reports of dizziness and drowsiness might be seen as relatively minor problems. Weight gain might be (mistakenly) dismissed. But many patients given the drugs became pathologically restless and unable to keep still, pacing up and down, exhibiting symptoms of extreme anxiety, often extending to panic and even violence and thoughts of suicide. Akathisia, as this syndrome was dubbed, sometimes persisted for months after the drugs were discontinued. More serious still was a condition known as tardive dyskinesia. As its name indicates, this syndrome emerged only after some time passed and sometimes was masked as long as the patient remained on antipsychotics. But in cases of long-term treatment, it afflicts between 20 and 60 percent of patients to varying degrees, and it is often irreversible.

Tardive dyskinesia is a profoundly disturbing and stigmatizing affliction. It involves facial tics, grimacing, grunting, protrusion of the tongue, smacking of the lips, rapid jerking and spasmodic movements, or sometimes slow writhing of the limbs, torso, and fingers. Naive observers often regard these as signs of mental illness. And it appears to be worsened by the prescription of drugs to control the symptoms of parkinsonism that accompany the use of antipsychotics. Data in one careful study suggested that 26 percent of older patients taking the drugs developed the disorder within a year of beginning therapy, and another 60 percent did so within three years, with 23 percent being diagnosed with "severe" symptoms of a disorder for which, even now, few effective treatments exist.[27]

Remarkably, during the first two decades of antipsychotic prescription, these serious problems were ignored or minimized by most of the psychiatric profession. Four years after the NIMH collaborative study had dismissed the side effects of phenothiazines as trivial and rare, Nathan Kline, referred to by some as "the father of psychopharmacology" and once a serious candidate for a Nobel Prize, asserted that these movement disorders were common in schizophrenia and reiterated that tardive dyskinesia was "not of great clinical significance."[28] It was a judgment echoed five years later by the long-term editor of *Archives of Neurology and Psychiatry*, Daniel X. Freedman. Dismissing the importance of tardive dyskinesia, he insisted that prevalence rates were low—3 percent to 6 percent—and the affliction was the "unavoidable price to be paid for the benefits of prolonged neuroleptic therapy."[29]

The drugs' ability to damp down the florid symptomatology of schizophrenia outweighed any concern over their side effects. Not until the Maryland psychiatrist George Crane published a paper on the subject in the pages of *Science* in 1973 did that complacency begin to evaporate. (Two of his earlier papers in 1967 and 1968, in which he had sought to warn his colleagues of an ongoing iatrogenic disaster, had been largely ignored.)[30] By 1980, the profession had moved from "curiosity and mild concern to panic" about what seemed to be an epidemic.[31] The American Psychiatric Association's task force on the problem accepted that the incidence of tardive dyskinesia was at least 20 percent in adults, and 40 percent or more in the elderly, and that in as many as two-thirds of such patients, the symptoms persisted when treatment was discontinued.[32] Accordingly, it recommended careful monitoring of patients taking antipsychotics, and minimizing the doses received—recommendations that seem to have been ignored by most practitioners, judging by the growing volume of prescriptions for antipsychotics during the 1980s and the doses that patients received.[33] Even clinicians operating in a university setting were shown to be singularly poor at diagnosing tardive dyskinesia and other so-called extrapyramidal side effects, including uncontrollable muscular contractions, uncontrolled restlessness, and parkinsonism.[34]

IT WAS IN THIS CONTEXT that the Swiss pharmaceutical company Sandoz sought to bring a very old antipsychotic drug with a different mode of action back to the marketplace. Clozapine had been synthesized in 1956 by the Swiss company Wander, one of a number of compounds that drug companies developed in this period in an effort to compete with chlorpromazine. It was brought to market in Europe in the early 1970s, after Wander had been taken over by Sandoz. The delay was partly because of concerns about Clozapine's propensity to cause abnormally low blood pressure and seizures. Ironically, there had also been hesitation about introducing the drug because it had many fewer extrapyramidal side effects, which were then considered to be an essential precondition for efficacy. (Extrapyramidal effects are physical symptoms, including tremors, slurred speech, uncontrollable restlessness [akathisia], uncontrollable muscular contractions [dystonia], and marked slowing of the thought processes, all of which are often accompanied by considerable anxiety and distress.) Before Sandoz could release Clozapine in the United States, evidence accumulated that it could also cause an often-fatal condition called agranulocytosis—a loss of white blood cells—in a significant fraction of the patients under treatment. In 1975, it was therefore voluntarily withdrawn from sale.

By the end of the 1980s, precisely because Clozapine was less likely to produce tardive dyskinesia and other extrapyramidal side effects (probably because it had relatively weak effects on dopamine and dopamine receptors in the brain), Sandoz changed course and requested permission from the FDA to introduce its drug to the American market. To bolster its case, it proffered evidence that Clozapine appeared to have positive effects in cases of schizophrenia that had not responded to other antipsychotics.[35] With the assurance that this medication would only be used when other drugs had failed, and with careful weekly monitoring of patients' blood to head off cases of agranulocytosis, the FDA gave approval for Clozapine to come to market, and it began to be prescribed in 1990.[36]

Over the ensuing decade, Clozapine was joined by a succession of other new compounds that the pharmaceutical industry cleverly sold as "atypical" or "second generation" antipsychotics.[37] In clinical trials, these medications were matched against Haloperidol, a first-generation antipsychotic, and the data appeared to show they were less likely to provoke tardive dyskinesia and other side effects. By now, the first-generation drugs were off-patent, while these newer medications obviously were not and were thus hugely more profitable. Psychiatrists, insurance companies, and the federal government, seduced by claims that the new pills rescued patients from the debilitating side effects of the earlier drugs, and by assertions that they were clinically superior in the bargain, migrated rapidly to the new medications, and drug-company profits soared. Abilify, Risperdal, Seroquel, Zyprexa, and others became the new coin of the realm.

Almost simultaneously, SSRIs were transforming the market for antidepressants, replacing the MAOIs and the tricyclics that had hitherto been used to treat depression. That market was in any event poised to explode for an altogether different set of reasons.

IN COMPILING THE THIRD EDITION of the *Diagnostic and Statistical Manual of Mental Disorders* (*DSM III*), Robert Spitzer and his colleagues had been determined to erase all traces of psychoanalysis from what became psychiatry's operating manual. In the process, they erased the prior distinction between endogenous and neurotic depression, substituting a new "illness" they dubbed major depressive disorder (MDD). In so doing, they created a new landscape, where depression acquired a far greater salience for psychiatrists and patients alike. The generalized anxiety and unhappiness, the tension and worries of those previously seen as suffering from neurotic depression, were subsumed under the same diagnostic umbrella as what had previously been described as melancholia or endogenous depression, a far rarer and more disabling disorder. To become eligible for the new diagnosis of MDD, symptoms had to last only two weeks, accompanied by an unhappy or dysphoric mood.

The checklist of symptoms that justified the diagnosis made no distinction between what most outside observers would regard as major disturbances (recurrent thoughts of suicide or death; overwhelming feelings of guilt and worthlessness; a profound hopelessness that was unresponsive to external changes; retarded movement, speech, and thought; and a more or less complete inability to think or make decisions about one's life) and such things as insomnia or excessive sleep; loss of appetite and / or weight; decreased libido; a general feeling of fatigue or loss of energy; agitation or its opposite; and a general sense of lethargy. One could now be diagnosed with major depression after two weeks of unhappiness, unaccompanied by any of the major forms of disturbance. As Allan Horwitz notes, "The ease with which people could fulfill the five-symptom two-week criteria made MDD the target of the wildly popular SSRI antidepressant drugs that entered the market in the late 1980s."[38]

Though unintended, the consequence of creating such a heterogeneous diagnosis was entirely predictable: the rapid rise in the prevalence of depression and its elevation to the status of one of the most common psychiatric diagnoses. By 2017, NIMH reported that 17.3 million adults in America had experienced a major depressive episode in the preceding year, along with a further 3.2 million adolescents between the ages of twelve and seventeen. Years after the fact, one of the *DSM* Task Force's most prominent and influential members, Donald Klein, lamented "the plague of affective disorders that have descended on us"—a complaint he had raised at the time, only to be dismissed by Spitzer.[39]

An enormous market beckoned. As Edward Shorter notes, "The incidence of major depression in the United States more than doubled in the 1990s, rising from 3.3 [percent of the adult population] in 1991–92 to 7.1 per cent in 2001–2."[40] Drug companies moved rapidly to respond to the opportunity. Drugs under patent were obviously more desirable to them, because of the enormously increased profits they generated. SSRIs thus became the focus of an intense and highly focused marketing campaign. As with the second-generation antipsychotics, the move to prescribing the newer drugs was sold to psychiatrists as a way to avoid the well-documented side effects that dogged the earlier medications. MAOIs quite commonly produce dry mouth, dizziness, insomnia, or drowsiness and are also associated with weight gain, low blood pressure, and reduced sexual desire. But much more serious are the interactions with certain foods and drinks, including beer, aged cheese, cured meats, and fermented foods, which can cause extremely high blood pressure, as well as the risks that, if taken alongside certain other drugs and herbal supplements, MAOIs might provoke life-threatening complications.[41] Tricyclics, the second group of antidepressants to be introduced in the 1950s, could cause disorientation or confusion, increased or irregular heartbeat, and a heightened propensity for seizures, as well as such problems as weight gain, blurred vision, dry mouth, constipation, and low blood pressure on standing.[42] SSRIs, the pharmaceutical industry claimed, avoided most of these problems and yet restored brain chemistry to its "natural" state.[43]

Because atypical antipsychotics and SSRIs enjoyed patent protection, drug companies could charge enormous sums for these products, provided psychiatrists could be persuaded of their merits. That partially explains the huge growth of expenditures for these medications, though it is not the whole story. Looking first at the market for antipsychotics, the shift from the first-generation to the second-generation drugs was swift. Between 1996 and 2005, the percentage of the US population taking first-generation medications fell from 0.6 percent to 0.15 percent. Simultaneously, the percentage taking second-generation pills increased from 0.15 to 1.06 percent, or nearly sevenfold.[44] By 2004, sales of atypical antipsychotics in the United States had reached $8.8 billion, and by 2008, that had risen again to $14.6 billion, making them the best-selling drugs in the country by therapeutic class. Five years later, annual sales of just Abilify, then the best-selling atypical antipsychotic, had reached $7 billion.

The market for antidepressants saw a similar pattern of explosive growth, paralleling ever-increasing estimates of the prevalence of depression. At the turn of the century, 7.7 percent of Americans over the age of twelve were taking antidepressants. By 2013, this had risen to 12.7 percent, an increase of almost two-thirds. By the latter date, nearly one in five people over the age of sixty were taking them. Two-thirds of those taking these drugs had done so for two years or more, and a quarter of them had been taking the pills for a decade or more.[45] NIMH data for 2017 again reported a marked difference by gender, with an incidence of major depression of 8.7 percent among women eighteen and older, and 5.5 percent among men.

Such numbers suggest how deeply the everyday practice of American psychiatry had become intertwined with psychopharmacology. But they also reflect the fact that much of the increased prescription of these powerful chemicals came from primary-care physicians, who encountered patients with a range of presenting problems and increasingly responded with doses of antidepressants or even antipsychotics.[46] A substantial part of the growing market for psychiatric drugs of all sorts had been prompted by a further broadening of the kinds of disorders for which these pills were prescribed. Antipsychotics were sold as remedies for the increasingly popular and protean diagnosis of bipolar disorder, which was extended, as we shall see, to the very young. Abilify and other atypical antipsychotics were then touted as adjunctive therapy for patients suffering from depression, to be added to the doses of antidepressants that patients were already on if their symptoms failed to improve.

THE MARKET FOR ANTIDEPRESSANTS ballooned as all sorts of everyday anxieties, emotional upsets, and phobias became targets for the drug companies. Zyprexa, for example, was marketed as the solution to patients displaying "complicated mood symptoms," such as "anxiety, irritability, disruptive sleep, and mood swings"—relatively minor issues that extended far beyond the criteria for diagnosing schizophrenia and bipolar disorder, the only "illnesses" for which FDA approval had been granted. Quite correctly, Eli Lilly saw these mild problems as a powerful way "to expand our market," part of its drive to make Zyprexa "the world's number one neuroscience pharmaceutical in history."[47]

On occasion, these domains expanded after the pharmaceutical industry obtained FDA approval for broadened applications for their products. Very often, though, the expansion came through the subtle encouragement of "off-label" applications for drugs already on the market—applications for which no approval had been sought and for which no systematic evidence of efficacy and safety had been offered or obtained.[48] Once the FDA has approved the prescribing of a drug for one set of applications, there is little to preclude doctors from prescribing it for other purposes. They simply have to be persuaded to do so. At one end of the age spectrum, difficult and demented elderly patients were an almost irresistible target for interventions of this sort. At the other, a diagnosis of childhood bipolar disorder was manufactured, soaring in popularity.[49]

Poor children from families on Medicaid are especially likely to receive a psychiatric diagnosis and to be placed on psychotropic medications at a very young age. A cohort study of over 35,000 newborns born to poor families on Medicaid found that by the age of eight, 19.7 percent of the children had received a psychiatric diagnosis. and 10.2 percent had received a psychotropic medication.[50] Between 2003 and 2011, the numbers diagnosed with attention-deficit / hyperactivity disorder (ADHD) rose by 41 percent, to 11 percent of all children. A recent study found that the United States accounted for more than 92 percent of the worldwide expenditures for ADHD-treatment drugs. Nearly one in five American high school boys and one in eleven American high school girls had been diagnosed with ADHD, by far the highest incidence in the world.[51]

Drug companies themselves were legally barred from promoting these off-label uses (though this prohibition was frequently circumvented when the financial temptations proved overwhelming).[52] As a rule they preferred to rely on prominent academic psychiatrists, so-called thought or opinion leaders, to promote new uses for their products beyond those the FDA had already approved. Such figures can prove invaluable to the marketing departments of Big Pharma. Apparently disinterested and independent, but in fact deeply indebted to the companies who fund their research, pay them consulting fees, and advance their careers, these academics can transform the range of conditions the drugs they promote are used for, all the while allowing their corporate sponsors to disclaim responsibility for the recommendations they make.

Joseph Biederman, chief of the Clinical and Research Programs in Pediatric Psychopharmacology and Adult ADHD at the Massachusetts General Hospital and professor of psychiatry at Harvard Medical School, and his associates at Harvard almost single-handedly transformed bipolar disorder from a condition seldom found among young children to a disorder of epidemic proportions.[53] In 1994, the fourth edition of the *DSM* revised the definition of the syndrome. The most important difference was a move from seeing bipolar disorder as a single monolithic entity with only one set of diagnostic criteria to a division into two, with each having its own separate set of diagnostic criteria. A decade later, the number of young patients with this diagnosis had multiplied forty-fold.[54] Biederman and his team pioneered the off-label prescribing of powerful antipsychotic and antidepressant drugs for these children (even preschool-aged children), though for many of these chemicals there had not been even a modicum of testing for safety and efficacy in this population, and for others, such as paroxetine (Paxil), the tests had shown that the side effects of the drugs outweighed their benefits.[55] The Harvard academics received millions of dollars in drug-company fees and funding, while concealing their conflict of interest from the university.[56]

Biederman was equally aggressive in promoting the use of stimulants like Adderall, Concerta, and Ritalin for children diagnosed with ADHD—another disorder whose prevalence has skyrocketed. Data from the Centers for Disease Control and Prevention show that 15 percent of high school children are now diagnosed with the condition, and 3.5 million children now take these medications. According to the *New York Times*, Biederman is

> well known for embracing stimulants and dismissing detractors. Findings from Dr. Biederman's dozens of studies of the disorder and specific brands of stimulants have filled the posters and pamphlets of pharmaceutical companies that financed the work. Those findings typically delivered three messages: The disorder was underdiagnosed; stimulants were safe and effective; and unmedicated A.D.H.D. led to significant risks for academic

> failure, drug dependence, car accidents and brushes with the law. . . . Drug companies used the research of Dr. Biederman and others to create compelling messages for doctors. "Adderall XR Improves Academic Performance," an ad in a psychiatry journal declared in 2003, leveraging two Biederman studies funded by Shire [the drug's manufacturer]. A Concerta ad barely mentioned A.D.H.D., but said the medication would "allow your patients to experience life's successes every day."[57]

Uncontroverted evidence shows that Biederman engaged in a prolonged campaign to get Johnson & Johnson (J&J) to fund a research center at the Massachusetts General Hospital, promising in documents revealed in a court filing that one of its goals would be to "move forward the commercial goals of J&J." An internal email from one of the company's executives explained that "the rationale for this center is to generate and disseminate data supporting the use of risperidone" for the treatment of ADHD and bipolar disorder in children and adolescents.[58] The pitch was successful. In 2002 alone, J&J provided $700,000 to the center. Biederman delivered. Ahead of conducting a proposed trial in 2004 of the use of Risperdal to treat ADHD, he promised that the research "will support the safety and effectiveness of risperidone in this age group." Equally important from J&J's standpoint, he assured the company that the trial "will clarify the competitive advantages of risperidone vs. other neuroleptics."[59] He was clairvoyant. A year later, he produced a paper comparing risperidone to Zyprexa, a rival neuroleptic manufactured by Eli Lilly. The paper concluded that risperidone, but not Zyprexa, improved children's depressive symptoms.[60]

In the words of a Columbia child psychiatrist, David Shaffer, the author of a classic textbook on child developmental psychology, "Biederman was a crook. He borrowed a disease and applied it in a chaotic fashion. He came up with ridiculous data that none of us believed"—but that was swallowed by many in the media and by desperate parents. "It brought child psychiatry into disrepute and was a terrible burden on the families of the children who got that label."[61]

When these revelations surfaced, they prompted an editorial in the *New York Times* decrying Biederman's "appalling conflicts of interest. . . . [I]t is hard to know whether he has been speaking as an independent expert or a paid shill for the drug industry."[62] *Nature Neuroscience* echoed these concerns, speaking of an "ethical crisis" that was "particularly dangerous in child psychiatry as the potential consequences of treating the developing mind with powerful drugs are both less well understood and potentially more severe than in adults."[63] Biederman's response was that Harvard was currently reviewing the claims of conflicts of interest "and fairness dictates withholding judgment until that process has been completed."[64] Three years later it was. Biederman and his two associates were found to have violated Harvard's policies. As punishment, they received what amounted to a slap on the wrist: they were required to refrain from paid-industry sponsored activities for a year and undergo ethics training: they were also told they might suffer a delay of consideration for promotion and advancement—a meaningless sanction for Biederman since he was already a full professor.[65]

Demented and disruptive elderly patients, particularly those confined in nursing homes, were likewise potentially an extremely lucrative market, but the pharmaceutical companies had a difficult time securing FDA approval for the use of antipsychotics and antidepressants in such cases. Those attempting to cope with these patients, however, were an easy mark for drug salesmen, who could tout the drugs' presumed ability to calm and pacify their aggressive and agitated charges. A 1992 study found that 25 percent of nursing home residents were being prescribed antipsychotic drugs.[66]

Eli Lilly had initially hoped to obtain FDA approval for the use of Zyprexa for dementia, but by 2003, unable to muster the necessary data, it abandoned the pursuit—but not its efforts to exploit the commercial opportunity this largely captive population represented.[67] Lilly was scarcely alone. Other major drug houses were equally keen to access this market. The advent of second-generation antipsychotics intensified these efforts, since this group of drugs was alleged to mitigate the risks of adverse effects when compared with the original antipsychotics. Unfortunately, as trials were conducted, they revealed a heightened mortality rate for those on the drugs, with a pattern of accelerated cognitive decline among those being medicated. Nor did the pills have any demonstrable advantage over placebos, except when used on a short-term basis to sedate angry and aggressive patients.[68]

FOR THE YOUNG AND OLD, and everyone in between, treatment at the hands of most psychiatrists now revolves around the prescription of a variety of drugs. A remnant of the profession still employs some forms of psychotherapy, and if they can attract a sufficiently affluent patient population, these psychiatrists will employ the kinds of psychosocial interventions that fifty years ago were the profession's bread and butter. Mostly, though, those interventions have become the province of clinical psychologists and social workers, who accept lower pay and must strive to attract clients in an environment where the predominant message is that mental troubles are brain diseases for which drugs are the logical form of treatment.

There is, it turns out, a not-inconsiderable market for their services. Parents worried about the prescription of drugs with uncertain long-term effects on their children often turn to those employing a variety of psychotherapeutic techniques to treat their offspring. In many cases, except in the most serious forms of disturbance, the cognitive-behavioral therapy and related techniques clinical psychologists provide prove sufficiently successful in resolving the behavioral problems that bring families to the consulting room. Adults who obtain little relief from drug treatments, or find the side effects of psychopharmacology intolerable, provide still another source of clients. So a parallel market for psychological counseling persists and flourishes, and many families profess themselves satisfied.

In the first decades of the twenty-first century, psychiatrists, whether academics or clinicians, have found their fate closely bound up with psychopharmacology and the drug industry. Over the past half century, they have tied their practice to a diagnostic system that has implied a biologically reductionist approach to mental illness. It is a diagnostic system that proved highly useful to the pharmaceutical industry, cementing the notion that the various forms of mental illness were discrete diseases, each potentially treatable with its own class of chemicals. And for a time, it helped resolve an earlier crisis of legitimacy, when it seemed that psychiatrists had trouble reaching basic agreement about what was wrong with a particular patient—or whether anything was amiss at all.

Those seeming certainties are now crumbling. As we shall see, the activities of the pharmaceutical industry and their products have been coming under sustained critical and legal scrutiny, and the limitations of psychopharmacology have become ever-more manifest. Having underwritten psychiatric practice and academic psychiatry for several decades, Big Pharma is increasingly distancing itself from the search for improved chemical remedies for mental illness. Simultaneously, attempts to rescue and improve the *DSM* are foundering. Once more, psychiatry is in crisis.

CHAPTER TWENTY-ONE

Genetics, Neuroscience, and Mental Illness

MOST AMERICAN PSYCHIATRY remains firmly committed to a biologically reductionist view of mental disorder. Yet the hunt for the physical roots of mental disturbance has not led to the decisive breakthroughs its enthusiasts were convinced it would. The result is a profession facing an existential crisis, one that in my view cannot be resolved by doubling down on its wager that mental disturbance is nothing but brain disease.

The bet on the body is not a novel strategy. As we have seen, late nineteenth-century alienists were convinced that mental illness represented a form of degeneration, of biological evolution run in reverse. Defective heredity was what accounted for the explosion of insanity that accompanied the birth of the asylum, and simultaneously explained (and explained away) the profession's inability to produce much in the way of cures. In subsequent decades, the rediscovery of Gregor Mendel's work transformed vague notions of inheritance into a new science of genetics. Particularly in Germany, in the Munich laboratory established by Emil Kraepelin and run by Ernst Rüdin, a series of family and twin studies purported to give scientific substance to the idea that mental illness was inherited. It was a seductive idea that attracted substantial financial support from the Rockefeller Foundation throughout the 1930s, notwithstanding the awkward fact that Rüdin was an ardent Nazi and a key architect of Hitler's program to slaughter the mentally ill. After the war, except among a handful of psychiatrists, the association of genetic theories of mental illness with mass murder led to the marginalization of such work—a development that also reflected the growing dominance of psychoanalysis.[1]

Eric Kandel, a Nobel Prize–winning psychiatrist, recalls that in 1965, during his psychiatric and psychoanalytic training at the Massachusetts

Medical Center, the major teaching hospital of the Harvard Medical School, he and several of his fellow residents "tried to recruit a psychiatrist in the Boston area to speak about the genetic basis of mental illness. We could find *no one;* not a single psychiatrist in all of Boston was concerned with or even had thought seriously about that issue."[2]

Genetic research began to reemerge as part of the general revival of biological psychiatry in the 1970s and 1980s. Along with neuroscience, it appeared to be the most promising route to understanding major mental illness. The earlier twin and family studies, conducted by eugenicists who knew in advance what conclusions they wished to reach, had been riddled with methodological errors that cast serious doubt on their scientific credibility.[3] Now new studies were conducted, supposedly avoiding these errors.[4]

The centuries-old folk belief that madness ran in families seemed to be confirmed by these new studies. People whose relatives had been diagnosed with serious mental illness were shown to have a much higher risk of having other relatives enter treatment for such disorders. Renewed efforts were made to examine monozygotic twins and compare their susceptibility to schizophrenia or manic-depressive illness and major depression to the incidence among dizygotic twins. Among genetically identical twins, concordance for schizophrenia (i.e., both twins being diagnosed with the disorder) was about 50 percent. Among dizygotic twins, the risk that both would develop schizophrenia was much lower, but, even here, these twins appeared to be about twice as likely to develop schizophrenia as ordinary siblings.[5] Using mostly Scandinavian data, which provided access to the health records of entire populations, some of these studies extended the analysis to adopted children, showing that their risk of developing schizophrenia closely resembled that of their biological families, not their environmental families.

Skeptics might note that the well below 100 percent concordance among monozygotic twins demonstrated that biology was not destiny and that families seem to "inherit" many things that are clearly not prompted by their biology (such things were religious beliefs, television viewing habits, and political preferences, for example). This should remind us that asserting "heritability" is not the same as proving that disorders are biologically determined. Family studies failed to provide any insight into the molecular genetic mechanisms that might lie behind the purely correlational data they produced.[6] Still, the conviction that genetics might provide a compelling window into the biology of mental disorders grew among many research psychiatrists in the 1990s. The elucidation of the genetic origins of Huntington's chorea in 1993 encouraged the belief that the genetics of schizophrenia, depression, and other major varieties of psychiatric disorder would soon become clear.[7]

The National Institute of Mental Health (NIMH) and other funding agencies began to funnel money into this line of research. The development of the polymerase chain reaction (PCR) technique in 1985, and its commercial licensing in 1989, created a promising new tool for genetic analysis, by permitting the easy creation of millions or billions of copies of short segments of a target piece of DNA. The race to decode the complete human genome was entering its final stages—its successful completion was announced in April 2003. The genetic basis of several severe forms of mental illness was something that previously could only be inferred from family, twin, and adoption studies. Now, optimists were captivated by the idea that the newly deciphered human genome might at last reveal the genetic roots of their patients' troubles. This paralleled the conviction of some early twentieth-century alienists that the discovery of the syphilitic origins of General Paralysis of the Insane would soon lead scientists to uncover the infectious origins of other forms of insanity.

The neuroscientists who made up a large fraction of academic psychiatry were meanwhile pursuing their own pathway to heightened understanding of the biological basis of mental disorder. Alongside the massive expansion of work on neurotransmitters and the operations of the brain, research projects facilitated by connections to psychopharmacology and advances in imaging technology were spurring attempts to map brain activity and brain structures in novel ways. If schizophrenia, mania, and major depression were disorders of the brain—now an article of faith in academic psychiatry—here was another set of medical technologies that might soon uncover the etiology of the major forms of mental disturbance, dispelling the fog of confusion that had long enveloped the field. There were reports of shrinkage and structural abnormalities in the brains of schizophrenics, and when the technique of functional magnetic resonance imaging (fMRI) arrived in the early 1990s, it was proclaimed a way to decipher brain activity in real time, giving hope that it would display the differences between mad and sane brains in living technicolor. (fMRI imaging measures brain activity by detecting changes in cerebral blood flow, using these data as an indirect measure of the activation of neurons.)

CONFIDENT THAT THESE ADVANCES in neuroscience and genetics were on the brink of revealing the biological bases of a host of mental disorders, leading figures in psychiatry began to press for a new revision of the *Diagnostic and Statistical Manual of Mental Disorders* (*DSM*). This time, they claimed, the purely symptomatic approach to diagnosis that had formed the basis for the three editions since 1980 would be supplanted by a completely different approach. The process was launched at a conference jointly sponsored by the American Psychiatric Association (APA) and the NIMH in 1999. The director of the NIMH, Steven Hyman, was determined to "introduce neuroscience into diagnosis," and this was an approach the researchers launching the process endorsed.[8] This suggested a sharp break with "the heuristic and anachronistic" approaches that had found their way into previous editions of the *DSM,* for "as new findings from neuroscience, imaging, genetics and studies of clinical course and treatment response emerge, the definitions and boundaries of disorders will change."[9] The categories that formed the basis for previous versions of the manual had been based on symptoms, with all the emphasis placed on ensuring that psychiatrists reached the same diagnosis when confronted by the same patient. Now, validity would take center stage. Rather than using categories that might or might not identify diseases that existed in nature, the emerging understanding of the biological basis of mental illness would drive the process and transform the ways mental illness was categorized. Henceforth, the would-be architects of *DSM 5* asserted, the way forward was "to recognize the most prominent syndromes that are actually present in nature."[10] Their version of the *DSM,* they promised, would seek to provide "scientific hypotheses rather than inerrant Biblical scripture"—a not-so-subtle jab at the pretensions of their predecessors.[11]

Their confidence that they could reconstruct the whole basis of psychiatric diagnosis, and do so by 2012, proved misplaced. The conviction that the combination of PCR testing and the decoding of the human genome would rapidly produce revelations about the genetic basis of the whole spectrum of psychiatric disorders—uncovering the reasons mental illnesses seemed to run in families and revealing just how schizophrenia, bipolar disorder, and major depression reflected someone's genetic code—simply collapsed.[12] Science couldn't deliver the evidence it had promised. The massive resources that had been poured into uncovering the genetics of mental disorder produced more frustration than enlightenment. Or, rather, the enlightenment genetic researchers produced threatened to undermine the standing and legitimacy of the "diseases" that psychiatrists had believed in for more than a century—concepts that had woven themselves deeply into the ways the lay public had been taught to view mental illness. To acknowledge that the distinctions between schizophrenia and bipolar disorder were spurious, that those constructions might have to be abandoned—but with what to put in their places?—threatened the very foundations of the psychiatric profession's claims to expertise. If such fundamental building blocks of the psychiatric universe crumbled on close inspection, what was left?

IN THE EARLY DAYS OF THE RENEWED EMPHASIS on genetics, there were periodic claims to have discovered the Holy Grail, a genetic basis for schizophrenia. But such assertions have repeatedly failed the test of replication.[13] The University of North Carolina geneticist Patrick Sullivan used genetically realistic simulated data to show how easily positive associations could be derived from the kinds of statistical techniques these scholars relied on "merely by the play of chance." As often as 96.8 percent of the time, when what seemed to be compelling or intriguing linkages surfaced, they turned out to be false positives.[14] Genetic researchers had identified a series of candidate genes that seemed likely contenders to explain the genetic roots of schizophrenia and major depression, and they produced papers that purported to document such linkages. Those findings have failed to survive, products of precisely the statistical artifacts to which Sullivan had pointed.[15]

As research proceeded, it became impossible to keep believing that there was a single gene for schizophrenia or for any other form of major psychiatric disorder.[16] The field as a whole, as Kenneth Kendler put it, had to absorb some "painful lessons" and acknowledge that "despite our wishing it were so, individual gene variants of large effect appear to have a small to non-existent role in the etiology of major psychiatric disorders."[17]

The search for a genetic basis for mental illness came to be recognized as extremely complex and elusive. Researchers more and more relied on an approach that scoured the entire sequence of the human genome, looking for correlations between particular stretches of genetic code and any susceptibility to schizophrenia, bipolar disorder, or major depressive disorder. So-called genome-wide association studies (GWAS) look as broadly as possible for anomalies associated with these disorders. Such studies make no prior assumptions about the etiology of the disorder under study and treat every genomic variant equally, examining hundreds of thousands of sites looking for possible linkages. It's a brute-force technique applied to a whole spectrum of psychiatric disorders, and by now researchers have accumulated data covering tens of thousands of patients. Whereas family studies inevitably conflated genetic and environmental influences, and scientists had difficulty separating the two, GWAS labored under no such handicap.

As studies of this sort accumulated, hopes that genetic research would uncover clear pictures of an underlying biology of mental disorders quickly faded. It was not just that the maximal claim—that schizophrenia, for example, was a Mendelian disorder—was quickly shown to be false, but even an alternative hypothesis, that "a substantial proportion of the [hypothesized] genetic signal could have been concentrated in a few large-effect genes," was soon rejected.[18] In the vast majority of cases, what materialized was a muddle. Crunching the data has shown that hundreds of genetic variants may (or may not) contribute to the diagnosis of a particular case. Each of these potential variations is individually of small effect and may be present without giving rise to the disease. Many carriers of these genetic variations fail to exhibit signs of mental disorder, suggesting that these are polygenic risk factors for mental illness, not differences that inevitably or even probably lead to schizophrenia or affective disorders. Besides, even when aggregated together, the genetic variants uncovered by GWAS account for a tiny percentage of the variance. When the International Schizophrenia Working Group published its findings in 2014, it had looked at 36,989 patients and 113,075 controls, and it asserted that it had found 108 loci associated with schizophrenia. Taken together, however, these 108 sites accounted for a total of 4 percent of the variance in the diagnosis of schizophrenia.[19] Subsequent collaborative work has expanded the number of genetic loci to as many as 270, and by incorporating this expanded array, these loci account for around 7.7 percent of the observed variance.[20]

Each of these variations contributes to risk, but only in a very minor way, and in total, this huge array of factors accounts for a small fraction of the risk for schizophrenia. A person may harbor any number of these genetic variations without ever developing mental illness. Conversely, "despite its high heritability, the majority of individuals with schizophrenia do not have a first-degree relative with schizophrenia."[21] The same pattern holds with other mental disorders. For example, a collaborative study of major depressive disorder, "the largest yet conducted," fared poorly, finding nothing that reached genomewide significance and acknowledging that the work was "unable to identify robust and replicable findings."[22]

Genes are not fate, and the thousands of alleles that contribute a small additional risk of illness do not operate "in a simple deterministic manner."[23] Developmental and environmental factors play a crucial role in whether the "nudge" of these alleles manifests itself in mental disorder.

A recent systematic review of the GWAS data on schizophrenia examined the various genetic polymorphisms that had previously been identified as likely to be associated with the disease. With minor exceptions, the authors found that "schizophrenia candidate genes are no more associated with schizophrenia than random sets of control genes," such as those for height or type 2 diabetes. Their conclusion was stark and unambiguous: "Our results suggest that the statistical rationale for further prioritization of these genes is weak. . . . [O]ur results do not support the original hypotheses involving the most-studied candidate genes, and thus provide no reason to believe that rare variants in these genes, or the trans-elements that regulate them, will be particularly relevant to schizophrenia."[24]

The pattern was identical when the research on major depression was scrutinized: "As a set, depression candidate genes were no more associated with depression phenotypes than noncandidate genes." That meant, Richard Border and his colleagues concluded, "that the large number of associations reported in the depression candidate gene literature are likely to be false positives."[25]

Despite extensive funding and vast teams of scientists cooperating to generate reams of data, in the words of two leading psychiatrists, Rudolf Uher and Michael Rutter, "Molecular genetic studies of psychiatric disorders have done a lot to find very little. In fact, in the era of genome-wide association studies, psychiatric disorders have distinguished themselves from most types of physical illness by the absence of strong genetic associations."[26] Steven Hyman, who did so much as director of NIMH to fund these studies and is now the director of the Stanley Center for Psychiatric Research at the Broad Institute and a professor of stem cell and regenerative biology at Harvard, has spoken of how the early confidence that "we might identify causal mutations" has dissolved into uncertainty. Instead, he suggested that it was incumbent to acknowledge that "the genetic, epigenetic, and other environmental risks of psychopathology are etiologically complex and heterogeneous." It was now clear that "the many risk genes that might contribute to schizophrenia are not transmitted together across generations." These findings required stem-cell researchers to take account of "a plethora of other disconcerting observations."[27]

NOT ONLY DID GENETICS fail to deliver the promised results but such findings as it did produce threatened to destabilize the psychiatric universe from another direction. For rather than revealing one set of vulnerabilities for schizophrenia and others for manic-depressive disorder or major depressive disorder, most such risks as it identified seemed common to a whole range of mental disorders.[28] "The high degree of genetic correlation . . . among attention deficit hyperactivity disorder (ADHD), bipolar disorder, major depressive disorder (MDD), and schizophrenia," the international Brainstorm Consortium reported, provides important "evidence that current clinical boundaries do not reflect distinct underlying pathogenic processes. . . . This suggests a deeply interconnected nature for psychiatric disorders."[29]

Biology, it seems, predisposed some to a heightened liability to mental disorder, but most of that liability is not disorder-specific. Modern geneticists would recoil from the language the late nineteenth-century alienist Henry Maudsley used to summarize his beliefs about the inheritance of madness, but their own chastened assessment of the relations between genetics and mental disorder are, *mutatis mutandis*, not so very distant from his: "It is not that the child necessarily inherits the particular disease of the parent . . . but it does inherit . . . a constitution in which there is a certain inherent aptitude to some kind of morbid degeneration . . . an organic infirmity which shall be determined in its special morbid manifestations according to the external conditions of life."[30]

Or perhaps, and this was the more disturbing implication, the distinctions psychiatrists derived from symptoms lacked validity and required rethinking.[31] The logic, such as it was, of the various *DSM* editions was to divide mental illness into a series of discrete and specific "illnesses," each of which was supposed to constitute a more or less homogeneous entity. That had been one of the major attractions of Robert Spitzer's system for the pharmaceutical industry, because such constructed "facts" (periodically added to as necessary) provided the underlying conditions it needed to convince the FDA of the value of its products.[32] Further entrenching the *DSM* hegemony, its distinctions led the FDA to embrace the notion that psychiatric disorders were strictly comparable to other illnesses. By requiring drug companies to test and advertise their products as treatments for specific

DSM diagnoses, the agency inadvertently contributed to the reification of its categories.[33]

To the extent that one can derive useful conclusions from recent work on the genetics of mental illness, two seem to stand out: what we have learned strongly supports the adoption of a dimensional rather than categorical view of psychiatric disturbances; and its findings demonstrate that psychotic patients are not genetically different in kind from the rest of us. The latter point is a sharp corrective—indeed, a reversal—of the conclusion many drew from the research of early twentieth-century psychiatric genetics, that the mentally ill were biological degenerates. Those studies from a century ago heightened the stigma attached to mental illness. Perhaps the recognition that the sane are not fundamentally different from those suffering from serious forms of mental disturbance will have the opposite effect, though I can't say I'm sanguine this will be the case.

Some optimists have suggested that one way to connect genetic findings to clinical applications would be to derive polygenic risk scores from the existing data and use these findings to guide interventions. Aggregating all the small genetic risks that recent research has uncovered can quite cheaply be used to provide a measure of heightened risk for future mental illness, just as it can be used to assess risks of developing (say) type 2 diabetes. The 10 percent of people with the highest such score for potentially developing schizophrenia, for example, appear to face three times the risk that confronts the general population. The authors suggest that because there is some data that suggest that recreational drug use exacerbates the risk of future psychosis, at-risk individuals could then be counseled to avoid such exposure.[34]

Unfortunately, however well intentioned such suggestions might be, major problems would follow any such move. Epidemiological statistics suggest that one out of every one hundred people will be diagnosed as schizophrenic over the life course. Granting for the sake of argument the claim that psychiatric genetics shows that among a select subpopulation the risk is three times as large, for every one hundred people who are informed of their heightened susceptibility, only three will ultimately be diagnosed as schizophrenic. It scarcely requires much imagination to understand the anxiety and other problems such information would create or to grasp the heightened possibilities of stigma and exclusion such people would then face.[35]

Perhaps further advances in genetics may provide guidance about treatment choices if such knowledge helps predict responsiveness to medications and susceptibility to their side effects, to guide the treatment of future patients. One can also put an optimistic gloss on the way genetic findings have forced a reassessment of existing diagnoses in psychiatry. The Brainstorm Consortium makes precisely these suggestions. The most recent genetic findings, it argues, "are promising steps toward reducing diagnostic heterogeneity and eventually improving the diagnostics and treatment of psychiatric disorders."[36] Once more, these are promissory notes, speculations that may or may not prove well founded in years to come. Their discounted current value is very small, especially given the dishonored promissory notes that litter psychiatry's past.

NEUROSCIENCE HAS PROVED similarly unhelpful to date. Much excitement originally arose from a series of findings comparing the brains of schizophrenic patients and controls. These demonstrated greater volumes of brain shrinkage in the patients and were initially hailed as "proof" that schizophrenia was a brain disorder. Setting aside concerns about the comparability of the healthy controls, whose brains may have been affected by the fact that they are "typically living more active lives, interact with more people, are wealthier and have more control over their environment"—all factors likely to positively affect brain structure and functioning—a yet more powerful confounding factor had been overlooked in the excitement: the psychiatric patients had all been taking large quantities of antipsychotic drugs over the years before brain volume was being measured.[37] And as a whole series of subsequent studies, both animal and human, would show, there are clear links between the ingestion of these drugs and transformations in the structures of the brain.[38] Rather than being the product of their disease, changes in these patients' brains were probably an artifact of their treatment. Robin Murray, a leading British researcher, puts it more strongly: "It is clear that high-dose antipsychotics contribute, not to the subtle brain changes present at the onset of schizophrenia, but to the subsequent progressive changes thereafter." Unfortunately, Murray adds, "some psychiatrists refuse to accept the evidence and cling to the nihilistic view that there is an intrinsically progressive schizophrenic process, a view greatly to the detriment of their patients."[39]

Among the neuroscience community, much excitement attended the arrival of functional magnetic resonance imaging (MRI) technology. Such scans, which, as I've noted, measure blood flow in the brain as an indirect proxy of its neuronal activity (albeit with a lag), have proved of some value in basic research, but their clinical utility in psychiatry to date has been essentially nonexistent.[40] It will surprise no one—not even the immaterialist philosopher George Berkeley—to learn that when I move, speak, think, or experience an emotion, these are correlated with changes in my brain. But such correlations tell us nothing about the causal processes involved. Nor do we possess any means to translate "heightened activity" in different regions of the brain to the contents of people's thoughts. Weak inferences based on statistical averages of gross changes in brain function in large groups of subjects provide little guidance when it comes to the individual subject, and they are a poor substitute for tying together behavioral symptoms and brain defects in a robust causal fashion. Such esoterica served only to widen the chasm separating academic researchers from the clinicians dealing directly with the problems posed by mental illness.

In a review of the possible clinical utility of brain imaging for major psychiatric disorders, Jonathan Savitz and his colleagues reached an unsurprising conclusion: "There are currently no brain imaging biomarkers that are clinically useful for establishing diagnosis or predicting treatment outcomes in mood disorders."[41] This, along with the fading hopes of breakthroughs from genetics and imaging, forced the *DSM 5* Task Force to abandon its grand plans and to settle for one more round of revisions to the elephantine device that was the *DSM*, relying once more on symptoms to guide diagnosis. Hyman, who had done much to underwrite the shift to biology, neuroscience, and genetics, had by 2007 concluded that "it is probably premature to bring neurobiology into the formal classification of mental disorders."[42]

In 1886, the American alienist Pliny Earle lamented that "in the present state of our knowledge, no classification of insanity can be erected on a pathological basis, for the simple reason that, with but slight exceptions, the pathology of the disease is unknown. . . . Hence . . . we are forced to fall back upon the symptomatology of the disease."[43] Nearly a century and a half later, nothing, it seems, had substantially changed.

THE PROBLEMS ASSOCIATED WITH THE CREATION of the new manual were far from over. On the contrary, they multiplied in the years that followed, spilling over into the public arena in ways that threatened to discredit the whole enterprise. When leaks from Task Force deliberations suggested that its members were contemplating reining in the boundaries of autism—a diagnosis whose prevalence had exploded in the preceding twenty years—a buzz saw of protest ensued.[44] Parents, whose access to social and educational services for their children were dependent on the diagnosis, reacted with fury. The Task Force beat a hasty retreat. It nominally abolished the previous diagnosis of Asperger's syndrome, but, in reality, it simply relocated it into a new category called autism spectrum disorder. That made sense given that the Task Force concluded that the existing dividing lines were hard to justify and difficult to draw, but patients' families were clearly concerned that insurance companies and government entities might stop paying for services for those at the milder end of the spectrum. To head off further protests, the Task Force issued a "clarification": "Individuals with a well-established DSM-IV diagnosis of autistic disorder, Asperger's disorder, or pervasive developmental disorder not otherwise specified should be given the diagnosis of autism spectrum disorder."[45]

Far more serious was a campaign of denigration directed at the whole *DSM* enterprise that came not from outsiders, but from the very heart of the profession. It was led by the two previous directors of the enterprise, Spitzer and Allen Frances. Their sustained assault delayed the appearance of the new manual by a year, to May 2013, and inflicted major damage on its reputation. Their objections had two main foci: the secrecy with which they claimed that the revisions were being undertaken by people with deep ties to the pharmaceutical industry; and the threat that the new manual would drastically expand the range of emotions and behaviors that were seen as forms of psychiatric pathology, transforming the normal vicissitudes of everyday life into "illnesses" requiring psychiatric intervention and treatment.

Critics pointed out that the ties between those constructing the diagnostic manual and the pharmaceutical industry had become even tighter. Fifty-seven percent of the *DSM IV* Task Force had financial ties to industry; for *DSM 5* that percentage rose to 80 percent. In the subsection dealing with mood disorders, the largest market for psychiatric services, that number had reached 100 percent. There was ample reason to worry about the conscious and unconscious biases this situation might introduce. To cite one example, meta-analyses of the reported outcomes of drug trials show a powerful relationship between the identity of the trial's sponsor and the published results of what purport to be double-blind trials. Trials sponsored by drug companies consistently produced more favorable outcomes than those with independent sponsorship, a pattern also seen when one drug company pitted its product against those of its competitors.[46]

Spitzer was particularly exercised by the secrecy surrounding the development of the new manual. He had discovered that the members of the various working groups responsible for the revisions had been forced to sign nondisclosure agreements, ensuring that the sorts of political debates that lay behind the final text would remain hidden from view and from outside scrutiny.[47] In my view, these complaints were something of a red herring. The deliberations of Spitzer's Task Forces had likewise taken place behind closed doors, where horse-trading and arm-twisting were the norm.[48] As a good politician, Spitzer did periodically discuss the progress of the manual with different audiences, but the same can be said of those working on *DSM 5*, where the availability of the internet also allowed work groups to post draft documents and to solicit outside comments, which were often extensive. But while the objection about secrecy may have been misplaced or overblown, there can be no doubting its effectiveness as a public relations tactic.[49]

Frances's concern, shared by others, was the vast expansion of the psychiatric imperium that would result if some leaked proposals about changes to the manual were eventually implemented.[50] He warned:

> DSM-5 will turn temper tantrums into a mental disorder. . . . Normal grief will become Major Depressive Disorder. . . . The everyday forgetting characteristic of old age will now be misdiagnosed . . . creating a huge false positive population of people. . . . Excessive eating 12 times in 3 months is no longer just a manifestation of gluttony and the easy availability of really great tasting food. DSM-5 has instead turned it into a psychiatric illness. . . . DSM-5 has created a slippery slope by introducing the concept of Behavioral Addictions that eventually can spread to make a mental disorder of everything we like to do a lot. . . . Many millions of people with normal grief, gluttony, distractibility, worries, reactions to stress, the temper tantrums of childhood, the forgetting of old age, and "behavioral addictions" will soon be mislabeled as psychiatrically sick.[51]

With Spitzer increasingly sidelined by his advancing Parkinson's disease, it was Frances who took the lead in amplifying their critique. In online blogs, repeated critiques in the profession's monthly newsletter, the *Psychiatric*

Times, and op-eds and media interviews, he relentlessly promulgated his criticisms.[52]

Joining forces with the president of the APA, Alan Schatzberg of Stanford, the leaders of the *DSM 5* Task Force struck back with an ad hominem attack. Spitzer's and Frances's criticisms, they insisted, should simply be discounted. *DSM 5* was just as scientific as its predecessors. Objections stemmed from disreputable motives that neither critic had disclosed. "Both Dr. Frances and Dr. Spitzer have more than a personal 'pride of authorship' interest in preserving the *DSM-IV* and its related case book and study products. Both continue to receive royalties on *DSM-IV* associated products."[53]

In a further attempt to stifle criticism, the *DSM 5* leadership took steps to ensure that "all the people at the top of the previous *DSMs* were completely excluded from their deliberations."[54] It was a strategy that proved wholly counterproductive. The media seized on such a public demonstration of splits within the profession—"a dispute that puts the Hatfield-McCoy feud to shame" as a blog in the *Psychiatric Times* put it—and so, far from squashing the controversy, the attack on Spitzer and Frances only amplified it.[55] Critics took particular delight in pointing out that Schatzberg had accused Frances of being corruptly influenced by $10,000 a year in royalties, while he himself had just been forced to relinquish his position as principal investigator on an NIMH grant over a conflict of interest running into the millions of dollars.[56]

Behind the scenes, the process of assembling the new *DSM* was dissolving into chaos. Spitzer and Frances had exercised an iron control over the manuals they oversaw, ensuring at least a continuity of approach. Their successors, David Kupfer and Darrel Regier, had adopted a hands-off approach to the various work groups, which responded to their uncertain mandate in often-conflicting fashion. As one participant complained, "I get aggravated with Kupfer and Regier sometimes, where I want to say, 'For God sakes, you have to tell us how many dimensions we can have.' I mean these are things where you really need someone to make the decision about what the parameters are so that you can work. These guys are just way too open and flexible for us."[57] Some opted to loosen the criteria for particular diagnoses (most notably the group working on major depressive disorders), while others took the opposite tack.

As Allan Horwitz has recorded in *DSM: A History of Psychiatry's Bible,* an invaluable guide to the construction, contents, and problems surrounding the various editions of the *DSM,* the result was alarm at the highest levels of the APA. Concerned with the lack of progress on the manual, and the disorganization and the controversy surrounding its work, the board of trustees of the APA appointed an oversight committee to oversee the work groups, and then, a year later, "a Scientific Review Committee that was independent of the DSM revision structure to review all the proposed changes to the manual and make recommendations directly to the APA President and Board of Trustees."[58]

While Kupfer and Regier had by now abandoned the idea that they could completely transform the foundations of the *DSM*, they still clung to the hope that they could substitute a dimensional for a categorical approach to the definition of the various mental illnesses. That is, they wanted to incorporate the idea that these disorders, rather than being sharply distinguished from mental health, fell along a range, varying with respect to intensity, severity, and duration. Here, too, with the partial exception of autism spectrum disorder, where there was a broad agreement that sufferers were arrayed on a continuum, their efforts to introduce major changes to the *DSM* failed. Many individual working groups simply ignored their directives. Then, at the annual meeting of the APA in 2012, the clinicians who formed the vast majority of the profession openly rebelled. For them, the dimensional approach threatened to be unwieldy or even impossible to administer in practice. There was alarm among those who feared that insurance companies would deny reimbursement for cases that fell on the milder end of the spectrum. They recognized, as the researchers who formed virtually the entire membership of the Task Force clearly had not, that "diagnostic categories make mental disorders seem more *real* to the public, physicians in other medical specialties, insurance companies, and federal regulators."[59] Accordingly, when the assembly of the APA convened in May 2012, it unanimously voted to consign the talk of dimensions to an appendix of the manual, safely insulated from the sections clinicians used in their daily practice.[60]

DSM 5 (as it now called itself—the Roman numerals replaced by Arabic numbers to permit easy interim additions and revisions, like the modifications of software) was scheduled to be unveiled at the APA annual meeting in May 2013, a year late. It not only fell far short of the ambitions its progenitors had originally set for it, but also endured years of controversy and even ridicule.

The promise to create professional consensus on diagnosis came under threat. Spitzer had used a statistical measure of agreement among clinicians (kappa) that takes into account the possibility that concordance could occur simply by chance.[61] He used kappa to measure interrater agreement in field trials of his new manual to document the heightened reliability the *DSM III* produced. When the *DSM 5* Task Force conducted its own field trials, it used the same statistic. When they reported their findings, Frances immediately cited them as further evidence of the defects of their work. Comparing the two sets of data, he pointed out that kappa fell from 0.81 in the *DSM III* trials for schizophrenia to only 0.46 in the new trials, and for major affective disorders the results were worse: kappa here declined from 0.80 to 0.25.[62]

At first blush, that seems a devastating critique. Defenders of the *DSM 5*, however, rightly pointed out that there were crucial differences in the methods used in 1980 and 2012. Spitzer used two interviewers who had been highly trained in the use of the new manual, and he had them examine the same patient at the same time. It was an artificial test of reliability of the *DSM* categories that (one is tempted to say by design) increased the chances of interrater agreement. By contrast, the *DSM 5* field trials employed briefly trained interviewers who performed separate evaluations at different points in time (between four hours and two days apart), an approach almost guaranteed to produce less agreement, albeit one that much more closely resembled the results to be expected when the manual was deployed in the real world.[63]

That difference in methodology makes direct comparisons between the results of the two misleading.[64] Those involved in the *DSM 5* field trials asserted that their aim was "to provide the greatest possible assurance that those with a particular disorder will have it correctly identified (sensitivity) and those without it will not have it mistakenly identified (specificity)."[65] Yet with the exception of major neurocognitive disorder and post-traumatic stress disorder, few trials showed even a "fair" degree of agreement among clinicians. Even those in charge of the 2012 trials acknowledged that the degree of agreement they found for major depression was "questionable," and for schizophrenia, the results were marginally better but scarcely cause for celebration. With the quest for validity abandoned, the new manual failed even to deliver intraprofessional agreement.[66]

That was not the worst of it. The doyens of NIMH, who had helped launch the revolution *DSM 5* was supposed to bring, and who served, alongside the drug companies, as the joint paymasters of academic psychiatry, delivered a damning and all-too-public verdict on what the profession had wrought. They had refused to help underwrite the reworking of the manual, as they had in the case of prior revisions, pointing out that the APA was making millions of dollars from selling it and thus did not deserve public subsidy. Hyman had co-convened the conference that had launched the revision process. Now the director of Harvard's Stanley Institute for Psychiatric Research, Hyman did not mince words in an interview with the *New York Times:* "[*DSM 5* is] totally wrong in a way [its authors] couldn't have imagined. . . . What they produced was an absolute scientific nightmare. People who get one diagnosis get five diagnoses, but they don't have five diseases—they have one underlying condition."[67]

His successor at NIMH, Thomas Insel, was even more savagely dismissive. On April 29, 2013, a week before *DSM 5* was officially published, he complained publicly that "the final product involves mostly modest alterations of the previous edition." That was not intended as a compliment. "In the rest of medicine," he suggested, "this would be equivalent to creating diagnostic systems based on the nature of chest pain or the quality of fever. . . . [S]ymptom-based diagnosis, once common in other areas of medicine, has been largely replaced in the past half century as we have understood that symptoms alone rarely indicate the best course of treatment." *DSM 5* set itself up as psychiatry's Bible, he reflected, but "biology never read that book" and "patients with mental disorders deserve better."[68]

It was simply astonishing, he averred, that psychiatrists should practice in this fashion. Most of his psychiatric colleagues "actually believe [that the diseases they diagnose using the *DSM*] are real. But there's no reality. These are just constructs. There is no reality to schizophrenia or depression."[69] Henceforth, Insel announced, NIMH would alter its approach to studying mental illness, since "we cannot succeed if we use DSM categories as the 'gold standard.' . . . That is why NIMH will be re-orienting its research away from DSM categories."[70] In particular, he suggested, "we might have to stop using terms like depression and schizophrenia, because they are getting in our way, confusing things."[71] One sees what motivated such a statement (and it must have been greeted with glee by the Scientologists), but the phrasing was distinctly unfortunate. The labels may need to go (with who knows what consequences for psychiatry's reputation), yet the distress and pathology those traditional labels seek to capture will not disappear with them.

Insel's comments incited a furor, with both medical and scientific journals and the mass media hastening to report his skepticism.[72] He had hoped to use the controversy to advance his own pet project, something he called research domain criteria (RDoC), an attempt to install a research framework based on biology, and more particularly on a nebulous notion that related mental illness to a mysterious something called "brain circuits." But RDoC was not ready for prime time. No other entity engaged in psychiatric research endorsed his hobbyhorse, and it has faced fierce criticism since Insel stepped down as NIMH director in 2015.[73] Meanwhile, the political problem Insel and Hyman had created urgently needed a solution, or at least a public relations response. The latter duly arrived: Insel huddled with Jeffrey Lieberman, the president of the APA, and on May 14, 2013, they issued a joint press release.

Closing ranks, they argued that *DSM 5* and the International Classification of Diseases "remain the contemporary consensus standard for how mental disorders are diagnosed and treated. . . . Patients, families, and insurers can be confident that effective treatments are available and the *DSM* is the key resource for delivering the best available care." To justify their earlier statements and position, they then clarified, "What may be realistically feasible today for practitioners is no longer sufficient for researchers." Hence, the need for RDoC. That, coupled with a promise to continue "to work together . . . improving outcomes for people with some of the most disabling disorders in all of medicine," represented the best the two could do by way of damage control, passing over in silence perhaps the most damaging of Insel's remarks, the assertion that conditions such as schizophrenia and depression were no more than artificial constructs that ought to be discarded.[74]

CHAPTER TWENTY-TWO

The Crisis of Contemporary Psychiatry

CONDITIONS SUCH AS SCHIZOPHRENIA and bipolar disorder are engraved on our collective consciousness. For a century and more, psychiatry has insisted on their reality, and the misery and suffering these terms have encapsulated are real and often terrifying. The prospect of abandoning them would surely shake the foundations of psychiatry. Yet prominent voices within the profession are beginning to express doubts about what most would surely assume to be "real" diseases. Robin Murray, former dean of the London Institute of Psychiatry, a Fellow of the Royal Society, and one of the most widely cited researchers in the field, recently confessed, "I expect to see the end of the concept of schizophrenia soon. Already the evidence that it is a discrete entity rather than just the severe end of psychosis has been fatally undermined. Furthermore, the syndrome is already beginning to break down. . . . Presumably this process will accelerate, and the term schizophrenia will be confined to history, like 'dropsy.'"[1] The symptoms and suffering will not disappear, but if they do not correspond to a distinctive disease, then trying to discover the cause or causes of that nonexistent disease will necessarily be a fruitless task.

The problem that flows from this, of course, is that we can scarcely hope to find "the cause" of something if that something simply does not exist.[2] Kenneth Kendler, who has performed important studies on the genetics of schizophrenia, reluctantly concluded that "individual psychiatric disorder[s] are clinical-historical constructs not pathophysiological entities." This was not easy to accept. "The historical traditions on which we rely," he bravely acknowledged, "give no guarantee about biological coherence—that underlying the clinical syndrome is a single definable pathophysiology."[3]

There is a paradox here, because today, few would argue that syndromes such as schizophrenia and depression are single, homogeneous diseases.

And yet, when it comes to clinical research, including clinical trials, both are still almost always treated as such. For example, studies continue to be published on the genetics of both of these syndromes despite the fact that there will never be a robust genetics of either condition, as the nature and severity of specific symptoms are too heterogeneous across individuals to have any consistent genetic correlations.[4]

Perhaps this helps explain the dismal return the National Institute of Mental Health (NIMH) has enjoyed on all the money it has poured into research on the neuroscience and genetics of mental disorder. In September 2015, no longer directing the institute and dispensing billions of dollars for research, Thomas Insel insouciantly summed up what all these dollars had purchased. "I spent 13 years at NIMH really pushing on the neuroscience and genetics of mental disorders, and when I look back on that I realize that while I think I succeeded in getting lots of really cool papers published by cool scientists at fairly large cost—I think $20 billion—I don't think we moved the needle in reducing suicide, reducing hospitalizations, [or] improving recovery for the tens of millions of people who have mental illness."[5]

IF PSYCHIATRY'S CLAIMS to diagnostic competence seem increasingly threadbare, what of the weapons it possesses to treat the various forms of mental distress? Here the picture is mixed. It would be a serious mistake not to acknowledge that for some patients, antipsychotics and antidepressants provide real relief from some of the symptoms that cause so much distress and suffering. Additionally, as I shall discuss shortly, a not-inconsiderable number of patients improve when treated with cognitive-behavioral therapy (CBT) or interpersonal therapy, though usually these patients are not those with the most serious and debilitating symptoms.

Yet it must also be pointed out that, for many patients, the therapeutic interventions the profession relies on have only limited efficacy. Accumulating evidence published during the last quarter century indicates that large fractions of those diagnosed with schizophrenia, bipolar disorders, and depression are not helped by the available medications, and it is difficult to know in advance who will respond positively to drug treatments. For many patients, a further major problem is that such relief as the pills provide must be set against the serious, debilitating, and sometimes life-threatening side effects that often accompany the ingestion of these medications. In this connection, academic psychiatry's close ties to the pharmaceutical industry and clinicians' overreliance on drugs as their primary treatment modality have created another profound set of concerns for the future of the profession.

THE FIRST DECADE OF THE NEW MILLENNIUM brought a sea of legal troubles for the pharmaceutical industry. Not all the problems revolved around psychopharmacology. The scandal surrounding the painkiller Vioxx is one particularly prominent example: having suppressed data that showed that its use was associated with a long-lasting heightened risk of heart attack and stroke, its manufacturer, Merck, was forced to pull the drug from the market and to pay damages totaling nearly $5 billion.[6] Then there is the ongoing litigation surrounding the opioid crisis, and the alleged role played by the Sackler family and their corporation, Purdue Pharmaceuticals, in creating and profiting to the tune of billions of dollars from the epidemic of opioid addiction.[7] Johnson & Johnson has also been held legally liable for their activities in this area and has paid hundreds of millions of dollars in fines to date, with more to come. But many of the most highly publicized and financially costly cases involved drugs prescribed for mental illness. GlaxoSmithKline pleaded guilty to criminal charges of consumer fraud in marketing its antidepressant Paxil, paying a fine of $3 billion in 2012 (a little more than a year's worth of sales).[8] The following year, Johnson & Johnson was fined $2.2 billion for concealing the risks of weight gain, diabetes, and stroke associated with its antipsychotic drug Risperdal.[9] The year after that, Pfizer paid $430,000 in damages to settle a single lawsuit over its illegal promotion of its antiseizure drug Neurontin for psychiatric disorders, and further legal complaints raised the damages it paid to nearly $1 billion.[10] Bristol Myers Squibb settled similar claims in 2007 for illegal marketing of its drug Abilify (an atypical antipsychotic) for more than $500 million.

Though these settlements could be and were seen as simply a cost of doing business (the fines constituted only a fraction of the billions these drugs had contributed to the companies' bottom line), they did inflict a good deal of reputational damage. Worse, they attracted political attention from Chuck Grassley, chair of the United States Senate's Finance Committee, who commenced an inquiry into psychiatry's links to the drugs industry in 2007. And through the legal process of discovery, lawsuits brought to light some of the industry's behind-the-scenes practices.

It was Senator Grassley's investigations that unearthed Joseph Biederman's financial ties to the pharmaceutical industry, as well as Alan Schatzberg's concealed financial stake in the very corporation whose product he was "disinterestedly" researching with NIMH money.[11] Charles Nemeroff, chair of the department of psychiatry at Emory University, was likewise found to have failed to report well over $1 million in income from consulting fees from drug makers, and he had routinely violated federal rules designed to avoid conflict of interests in conducting funded research.[12]

Nemeroff, it turned out, was a serial offender. On at least two previous occasions, his university had been notified of problems with his outside consulting arrangements. In 2004, an internal review found that he had committed "multiple 'serious' and 'significant' violations of regulations intended to protect patients" and had failed to disclose conflicts of interest in trials of drugs from Merck, Eli Lilly, and Johnson & Johnson.[13] The report was buried. In 2006, new problems arose. Nemeroff used his position as editor of a prominent journal to tout the merits of a controversial medical device, neglecting to disclose that he had financial ties to the corporation that made it, Cyberonics. When a university dean accused him of producing "a piece of paid marketing," Nemeroff blamed the episode on "a clerical error."[14] Earlier, when concerns had arisen about Nemeroff's membership on a dozen corporate advisory boards, he had responded by reminding the university of the benefits it derived from the arrangement: "Surely you remember that Smith-Kline Beecham Pharmaceuticals donated an endowed chair to the department and that there is some reasonable likelihood that Janssen Pharmaceuticals will do so as well. In addition, Wyeth-Ayerst Pharmaceuticals has funded a Research Career Development Award program in the department, and I have asked both AstraZeneca Pharmaceuticals and Bristol-Meyers [*sic*] Squibb to do the same."[15]

The cases of Biederman, Schatzberg, and Nemeroff were scarcely unique. In acting as they did, and then denying wrongdoing, they appear to have expected to suffer few consequences for their actions. And at least at first, their expectation that they could violate conflict-of-interest regulations and suffer few consequences seemed well founded. Stanford initially declared that the university saw nothing wrong with Schatzberg's behavior. Belatedly and only temporarily it removed him as principal investigator on the grant that had provoked Grassley's ire, but two years would pass before Schatzberg relinquished the chairmanship of his department. Harvard resisted taking action against Biederman and his associates. An internal inquiry found that, despite a university regulation requiring researchers to accept no more than $10,000 a year from companies whose products were being assessed, Biederman and his colleagues had received a total of $4.2 million. Harvard announced it would take "appropriate remedial actions." Biederman and his colleagues were banned from "industry-sponsored activities" for twelve months. In addition, the university imposed some minor administrative penalties. None of these sanctions had any substantial effect on Lieberman, and to this day he remains a tenured full professor.[16]

Nemeroff's fate is still more revealing of university priorities. Grassley's staff had discovered that, while overseeing a federal grant designed to evaluate a GlaxoSmithKline drug, Nemeroff had received more than $960,000 from the company, while reporting only $35,000 of that money to the university. Once again, Emory seemed disposed to ignore the infraction, until it became apparent that its access to federal-grant money might be curtailed or eliminated. Only then was Nemeroff stripped of his department chairmanship—a highly unusual action in the world of academic medicine. It was announced that he would not be allowed to apply for federal grants and had been forbidden to accept more drug-company largesse.

A belated comeuppance it would seem. Subsequent events, however, showed that it was anything but. In short order, Nemeroff was offered and accepted the position of chair of the department of psychiatry at the University of Miami. In the negotiations, Miami promised to waive all the restrictions he had faced at Emory.[17] Sometimes, we seem to have entered Alice's looking-glass world. When Donna Shalala's administration at the University of Miami was negotiating with Nemeroff over his move from Emory, there was serious talk of him heading an institute to promote ethics in academia and industry. Nine years later, the University of Texas came calling, and Nemeroff moved to another department chairmanship, higher up in the academic pecking order, and became simultaneously the director of its Institute for Early Life Adversity. He was once again supported by grants from the National Institutes of Health and the pharmaceutical industry, and he continues to play an outsized role in the American Psychiatric Association.

In some respects, those who express surprise at this sequence of events are either naive or, like Captain Renault in *Casablanca,* pretending to be "shocked, shocked to find that gambling is going on here." Research universities depend for their existence on the constant flow of research dollars into the institution. The levels of funding from other sources are grossly inadequate to sustain their operations. The Nemeroffs, the Biedermans, and the Schatzbergs are experts at securing millions of dollars to fund their research, and those grants come with 50 or 60 percent overhead charges that accrue to the university budgets. No wonder America's research universities have learned to turn a blind eye to ethical failings if the money on offer is sufficiently tempting. As these scandals proliferate, they threaten to inflict major damage on the legitimacy of psychiatry and on the medical-industrial complex more generally. The whole value of university research and the endorsement of particular therapeutic interventions by leading academics comes from the fact that they are thought to reflect disinterested "pure" science.

THE DEEPER SCANDAL OF PSYCHIATRY'S incestuous relationship with the pharmaceutical industry lies elsewhere, however. We live in an era that purports to be governed by something called evidence-based medicine. In principle, this is a development we should applaud. The centuries when physicians relied on "clinical judgment" to assess the worth of therapeutics gave us bloodletting and purges, vomits and blisters. In the twentieth century, psychiatrists relied on clinical experience to justify a long series of destructive interventions, from surgical evisceration and insulin comas to lobotomies. Empirical evidence that refuted entrenched collective wisdom played a vital role in consigning these interventions to the scrap heap, though in many cases only after a sustained battle with practitioners wedded to what "clinical experience" had taught them.

In the second half of the twentieth century, the major mechanism for tempering therapeutic enthusiasms became the double-blind controlled trial, with patients randomly assigned to receive the active treatment under study or either a placebo or an existing approved therapy.[18] Particularly in the aftermath of the thalidomide tragedy in 1962, the FDA made evidence of safety and efficacy based on such trials the linchpin of its process for approving new therapies.[19] To bring a new drug to market, the FDA requires companies to produce evidence from two trials that show that it is significantly more effective than either the placebo or an existing treatment.

That is the theory, at least. In practice, other pressures often subvert the process. One of the least controversial parts of the psychiatric universe is its attempts to do something for patients with dementia or Alzheimer's. That lack of controversy is not the product of the profession's success in treating or alleviating these conditions. On the contrary, these are probably the most treatment-resistant of all the disorders psychiatry confronts. Even efforts to palliate the enormous problems patients and their families face over years, even decades sometimes, have proved unavailing. The burden for individuals and society as a whole is massive and forecast to grow exponentially as the population ages. An estimated six million people now suffer from Alzheimer's, and by 2050, that number is expected to approach thirteen million.

Few doubt that the roots of Alzheimer's are biological, and there is general recognition of just how recalcitrant it has proven to be. Over the course of more than a century, it has defeated our best efforts to unravel its mysteries or solve the practical and therapeutic challenges it presents. Psychiatry has escaped blame for these failures precisely because everyone recognizes the enormity of the challenge Alzheimer's and dementia represent. Yet that does not diminish the desperation either of those slipping into mental darkness or of their nearest and dearest as they struggle with the ensuing devastation.

If the travails of those suffering from other major psychiatric disorders have licensed and encouraged the employment of desperate measures to cure them, those same pressures are increasingly felt with respect to a condition even more resistant to successful interventions. Doing nothing does not seem an option. Hence a marketplace for peddlers of "alternative medicine" and other quack remedies: diet, herbs, acupuncture, light therapy—the list is long and the results are dismal. In more conventional quarters there are enormous pressures to find something, anything, that promises progress. For the pharmaceutical industry, Alzheimer's represents a potential financial bonanza, if only its chemists can discover a pill that works.

Biogen claims that it has. On June 6, 2021, the FDA approved the first drug that purports to treat Alzheimer's. It did so while admitting that the data from the two clinical trials that had been run left "residual uncertainties regarding clinical benefit." Quite explicitly it made an exception to the usual requirements for approving a new medication because of the utter absence of any alternatives. The decision to grant approval for the drug came under a rarely used procedure, the so-called Accelerated Approval pathway, designed to provide a way to license drugs that seem "potentially valuable" for a disease for which alternative treatments don't exist. As the FDA put it, they acted because "there is an expectation of clinical benefit despite some residual uncertainty regarding that benefit."[20]

Aducanumab (or to give it its trade name, Aduhelm) was nearly abandoned when two large clinical trials proved unpromising. One showed slight benefits; the other, no benefit at all. An intravenous drug, Aduhelm targets the clumps of amyloid-beta proteins that led Alois Alzheimer to identify the disease, and brain scans reveal that it does indeed shrink the clumps in question. What remains unclear is whether it actually slows the disease or improves patients' quality of life.

The FDA's own advisory panel concluded that the evidence of its therapeutic effectiveness was far too weak to license its release, a conclusion strengthened by the risks of the drug's side effects. (Patients who receive high doses in the trials experienced episodes of bleeding or swelling in the brain.) The FDA's decision to overrule its own expert panel is in consequence highly controversial. Three members of the advisory group resigned in protest, and the controversy shows no sign of dying down. Adding to the concerns about whether the drug is efficacious, its enormous cost (a list price of $56,000 a year, plus the great expense of regular scans looking for brain hemorrhages), likely to be borne for the most part by the taxpayer via Medicare, has heightened opposition in many quarters to the FDA's decision. On the other hand, desperate families and patients seem eager to try the experimental treatment, lacking anything else that works.[21] Meanwhile, Biogen's stock rose in the aftermath of the announcement by over 38 percent, adding over $16 billion in market value.[22] Those with a sense of history will share the alarm of the scientists who recommended against approval. The FDA will have to hope its decision doesn't backfire as the experimental drug enters clinical practice.

EVEN WHEN THE FDA STICKS to its usual practice and demands two controlled trials that provide empirical evidence of a new drug's effectiveness, its procedures have often produced evidence-biased medicine rather than evidence-based medicine. Over time, most clinical trials have come to be funded and owned by the pharmaceutical houses. All sorts of unfortunate consequences have flowed from this situation. The size of patient groups recruited for trials has expanded greatly. Trials are typically run across multiple centers and very often across national boundaries. In the ordinary course of business, the data produced are owned by the company funding the research, and access to the full data set is carefully controlled and protected. Large numbers of participants make it easier to secure statistical significance for a finding, which may be far different from discovering a change that has clinical significance. (The statement that drug X is "significantly" better than a placebo means only that it produces some greater degree of improvement than can be explained by chance, not that it actually makes a useful difference to patient outcomes.) The data the company owns can be, and are, cherry-picked to find favorable results that form the basis for the studies the FDA requires. Unfavorable outcomes, including both trials that fail to show positive findings and records of serious side effects, are suppressed.[23] That pattern of hiding damaging information even extends to fatal side effects. These data have generally surfaced only through lawsuits, and the associated process of discovery, which has forced the disclosure of unpublished studies and internal company deliberations that reveal the depths of the deceit. Unfortunately, recent Supreme Court decisions have imposed severe restrictions on future use of this weapon.[24]

A particularly revealing study is Glen Spielmans and Peter Parry's "From Evidence-Based Medicine to Marketing-Based Medicine," which looked at what could be learned by mining industry documents. One of the documents they reproduce is a Pfizer communication to its sales force about how to market sertraline (Zoloft), which is at once blunt and chilling. Headed "Data Ownership and Transfer," it asserts that "Pfizer-sponsored studies belong to Pfizer, not to any individual." As for the "science" they contain, the "purpose of data is to support, directly or indirectly, marketing of our product. . . . Therefore commercial marketing / medical need to be involved in all data dissemination efforts." Among those efforts, it explicitly references the way "publications can be used to support off-label data dissemination." These are, needless to say, standard industry practices. An Eli Lilly document lays out its plans to "mine existing data to generate and publish findings that support the reasons to believe the brand promise." The drug referenced here is Zyprexa, an antipsychotic.[25]

When studies are written up for publication, professional ghostwriters are routinely employed to present findings in the most favorable possible light. Major academics are recruited to add their names as "authors"—a process designed to give an appearance of legitimacy to the papers and to secure publication in the most prestigious medical journals.[26] Studies can, of course, be designed in ways that appear neutral but that are in reality biased in favor of the drug the company wishes to promote. When selectively reported, with the suppression of negative data and trials, what is meant to provide "evidence-based medicine" has allowed the marketing of useless or actively harmful interventions.

Here are two examples, but they are substantial and revelatory. One concerns the creation of spurious evidence of the superiority of so-called second-generation antipsychotics over their predecessors. The other examines the data used to justify the prescription of antidepressant drugs.

Industry-sponsored trials of second-generation antipsychotics typically used haloperidol, a first-generation antipsychotic, as the comparison drug, administering it generally in high doses. Haloperidol is known to have a high likelihood of inducing movement disorders and other distressing side effects, making it likely that the newer drug will seem to have a superior side-effect profile and receive FDA approval.[27] Similarly, when selective serotonin reuptake inhibitor (SSRI) antidepressants are compared to placebos, their side effects are likely to reveal to both the doctor and patient who is receiving the active treatment and who is not, vitiating the presuming blinding of the study and rendering its results suspect to some indeterminate degree.[28]

Setting these concerns to one side—though they are weighty concerns that speak poorly to the psychiatric enterprise—what can we make of the evidence we do possess about the drug treatments psychiatrists have at their disposal? Second-generation or atypical antipsychotics were heavily marketed as improvements over the drugs that were introduced in the 1950s, when the psychopharmacological revolution began. In the first decade of the new millennium, however, major doubts began to surface about such claims.

A paper by John Geddes and his colleagues published in 2000 reviewed a range of studies and concluded that the evidence for the superiority of atypical antipsychotics was largely mythical.[29] But it was two subsequent studies, both government funded, which inflicted the greatest damage on the presumption that atypical antipsychotics represented a superior approach to the treatment of schizophrenia. The Clinical Antipsychotic Trials of Intervention Effectiveness (CATIE), funded by the NIMH, was designed to compare four atypical antipsychotics (with annual sales of over $10 billion) with a single older (and far cheaper) first-generation antipsychotic. The project was led by Jeffrey Lieberman, chair of the department of psychiatry at Columbia University. The highly anticipated study attracted major attention when it appeared in 2005 in the *New England Journal of Medicine.*

Throughout the preceding decade and a half, the pharmaceutical houses had claimed that "the introduction of second-generation or 'atypical' antipsychotics promised enhanced efficacy and safety" and on that basis, such drugs had acquired "a 90 percent market share in the United States, resulting in burgeoning costs."[30] The CATIE study, however, showed unambiguously that the atypical antipsychotics offered few benefits over the older drug with which they were being compared. Subsequent experience has shown that among most atypicals, the incidence of tardive dyskinesia has dropped by 50 or 60 percent, a welcome development, though the problem remains a serious one, and "many clinicians may have developed a false sense of security when prescribing these medications."[31] On the other hand, Lieberman and his colleagues found that the second-generation medications carried the same risk of neurological symptoms such as stiffness or tremors and had also brought other potentially life-threatening complications in their wake.[32] Their use was associated with major weight gain and metabolic disorders. Those taking olanzapine averaged a weight gain of thirty-six pounds if they remained in treatment for the full eighteen months of the study and displayed signs of developing metabolic syndrome. These medications also led to an increased incidence of diabetes and a greater susceptibility to heart disease. A subsequent careful meta-analysis of the prevalence of antipsychotic adverse effects confirmed that "the AE [adverse effect] profiles of the newer antipsychotics are as worrying as the older equivalents for the patient's long-term physical health." The researchers' review of the literature also prompted a sense of alarm about a broader issue: "Despite the frequent use—both on-license and 'off-label'—of antipsychotics, the scientific study of their AEs has been neglected."[33]

As for efficacy, Lieberman and his colleagues found that "antipsychotic drugs . . . have substantial limitations in their effectiveness in patients with chronic schizophrenia." Nor were the newer compounds an improvement in this respect: with the partial exception of olanzapine, "there were no significant differences in effectiveness between perphenazine and the other second-generation drugs." The condition of a not-insignificant proportion of those under treatment took a turn for the worse: depending on which of the drugs they were taking, between 11 and 20 percent of the study participants had to be "hospitalized for an exacerbation of schizophrenia."[34]

The CATIE study reported another major finding that constituted a sobering commentary on the limits of contemporary psychopharmacology. Between two-thirds and more than 80 percent of those taking part in the study had dropped out, primarily for two reasons: the drug they were taking was not working, or the side effects of the medication were intolerable. Averaged across all five drugs, the dropout rate was 74 percent, ranging from 64 percent of those assigned to olanzapine (Zyprexa) to 82 percent of those given quetiapine (Seroquel). The authors of the CATIE report tried to normalize this rather extraordinary set of numbers by observing that these dropout rates "are generally consistent with those previously observed."[35] And in this they were correct.[36] But such numbers constitute a stark reminder that many chronic schizophrenics find the therapy they are offered useless, or intolerable, or both.

Contemporaneously with the CATIE study, the British National Health Service was funding research comparing the effects of first- and second-generation antipsychotics. Its authors, aware of the growing preference among clinicians for the second-generation drugs, predicted that patients on them would enjoy an improved quality of life at a one-year interval, and they used a generally accepted and wide-ranging instrument to assess patients' status. To their expressed surprise, their research decisively rejected their working hypothesis. If anything, patients on the first-generation medications fared better. "We emphasize," the authors concluded, "that we do not present a null result: the hypothesis that SGAs [second-generation antipsychotics] are superior was clearly rejected."[37]

TO PSYCHIATRISTS, the discovery of chlorpromazine and related compounds in the 1950s seemed to promise "both therapeutic progress and significant probes of brain function. Looking back, the picture is painfully different. The efficacy of psychotherapeutic drugs plateaued by 1955."[38] The older drugs, like their modern-day descendants, clearly altered the course of psychotic disorders, but it turns out that the improvements they provide are far more modest than the early enthusiasm for them suggested. Assessment of those effects has gradually become more systematic, often now using scales that attempt to assess the patient's quality of life, or to measure improvements with respect to the so-called positive and negative symptoms of schizophrenia—using an instrument known as PANSS (Positive and Negative Syndrome Scale).[39] "Positive" symptoms include such things as hallucinations, delusions, and racing thoughts, while "negative" symptoms include apathy, blunting of affect, disorganized thoughts, poverty of speech and thought, and poor or nonexistent social functioning. These rating scales include many unrelated items, and it is easy for scores to improve without the patient experiencing any functional improvement. PANSS, for example, contains a number of measures of tension, excitement, and anxiety—all likely to be present when someone is in an acute psychotic state. The sedating effects of antipsychotics are likely to produce a calmer patient, who will then be counted as improved, and yet there has been no change in the crucial psychotic symptoms, and the disorder remains untouched.

The largest effects are seen in patients experiencing their first psychotic episode. Unfortunately, though, "about 80% of patients relapse within five years."[40] Looking at results as a whole, antipsychotics unambiguously produce improvements in PANSS scores when compared to placebos. But most of this improvement is so limited as to verge on the clinically insignificant—what the authors of the largest systematic review of outcomes to date call a "minimal" response. On average, 51 percent of those given antipsychotics obtain this degree of improvement versus 30 percent of the placebo group. When it comes to a reduction in symptoms large enough to be termed a "good response" (not remission), the statistics are far less reassuring: 23 percent of the patients receiving antipsychotics obtain this result versus 14 percent of those in the placebo group.[41] Schizophrenia, or the array of mental disturbances gathered under that label, clearly remains a malignant and devastating problem.

IF MANY WOULD NOW ARGUE that what we call schizophrenia is a label that lumps together a variety of heterogeneous phenomena, this seems even more obviously the case of what psychiatrists call the affective disorders. The criteria in the fifth edition of the *Diagnostic and Statistical Manual of Mental Disorders* (*DSM 5*) for a diagnosis of major depressive disorder require the presence of five or more of nine symptoms in a patient for two weeks or more. But one of the criteria is either insomnia or sleeping too much, so there are in effect ten criteria, and two patients can receive the identical diagnosis while sharing not a single symptom. Perhaps that is one reason why the antidepressants prescribed to deal with these disorders appear to have such limited efficacy. Yet 12 percent of the American population over the age of twelve currently uses antidepressant medications, and the percentage taking them for two years or more rose from 3 percent in 1999 to 7 percent by 2010.[42]

One might expect that drugs prescribed and used on such a massive scale must have obvious positive effects, and indeed it is routinely the case that studies comparing antidepressants to placebos in controlled trials show the superiority of the drugs, as measured by conventional rating scales.[43] As with antipsychotics, many patients find taking these pills a blessing. That said, in the words of two Columbia University psychiatrists and a professor of psychological and brain sciences at Dartmouth College, "Even with maximal treatment, many patients will not experience sustained remission of their depression. The cumulative percentage of patients achieving remission after four sequential antidepressant trials in the National Institute of Mental Health (NIMH) sponsored Sequenced Treatment Alternatives to Relieve Depression (STAR*D) study was only 51%."[44]

Antidepressants consistently outperform placebos in published clinical trials, and many patients swear by them, but the findings that these drugs work come with some serious caveats. All drug trials are now supposed to be prospectively registered with the FDA, a system that evolved to try to counter manipulation of the clinical trials system by the pharmaceutical industry. That has allowed us to see that a large fraction of these trials—more than 30 percent—are never published. The unpublished trials almost all produced negative assessments of the drugs under study and then never saw the light of day. It has also been possible to compare the data submitted to the FDA with those that appear in published papers, and that comparison, too, raises red flags: "According to the published literature, it appeared that 94% of the trials conducted were positive. By contrast, the FDA analysis showed that 51% were positive." The problems this situation creates for the assessment of antidepressants are obvious. "Selective publication of clinical trials, and the outcomes within those trials, lead to unrealistic estimates of drug effectiveness," as researchers have pointed out in an article in the *New England Journal of Medicine,* while "the true magnitude of each drug's superiority to placebo was less than a diligent literature review would indicate."[45] Less is not nothing, but this is a disturbing finding.

The literature reviews that have been conducted add to one's disquiet. They routinely show that treatment with antidepressants is statistically superior to the results obtained by administering a placebo. But the observed differences are not large—on average an additional 10 or 15 percent improvement over what placebo treatment produces.

Many instances of what are now labeled as major depressive disorder—50 percent or more—seem to resolve spontaneously over time.[46] Perhaps that helps explain the substantial placebo responses these clinical trials regularly report. Two recent reviews of these published results survey the degree of improvement patients registered on the widely used Hamilton Rating Scale for Depression, independently finding that those who were given placebos registered an 8.3-point improvement in their symptoms. Those given antidepressants fared better, and their symptoms improved by a further 1.8 points. However, clinically speaking, an average improvement of this sort verges on the trivial.[47] That situation is masked, of course, from most clinicians and patients, since what they observe is the amount of improvement that follows after the prescription of the pills: the larger placebo effect; and the additional increment the drugs produce. The natural tendency under these circumstances is to embrace the post hoc ergo propter hoc fallacy and to assume that the entire improvement can be attributed to the medication.

One of the most curious phenomena when one examines the research on antidepressants is that the difference between placebo and active treatment has diminished sharply over time. Ordinarily, one would expect the gap between drug and placebo to widen. Instead, a systematic review in 2012 of major studies demonstrated that "antidepressant-placebo differences have decreased alarmingly over the past three decades." A difference of 6 points on the Hamilton scale in 1982 had fallen to less than 3 points by 2008. The authors noted that their review had only been able to examine published studies. That was a significant problem, because "negative trials (meaning lower antidepressant-placebo differences) are more likely to remain unpublished." "We can only surmise," they concluded, "that including these unpublished trials would further reduce the mean antidepressant-placebo difference across all trials."[48]

Arif Khan and his colleagues, who conducted this study, suggest that some of this decline may reflect a difference in the kinds of patients being studied: cases of serious endogenous depression in the early studies of antidepressants versus a more heterogeneous population created by the more expansive definitions of major depression embodied in *DSM III* and its successors. They also point to other methodological differences that have arisen as the FDA has modified its criteria for clinical trials. Still, as Khan acknowledges in a more recent paper, looking at the data

> about the relative potency of antidepressants compared to placebo, . . . it became evident that the magnitude of symptom reduction was about 40% with antidepressants and about 30% with placebo. . . . [W]here the investigators and their staff were blinded to the design and execution of the trials . . . the symptom reduction with each treatment was of smaller magnitude and the differences among the various treatments and controls were

> also smaller. . . . [W]hen the level of blinding was high and it was difficult for the investigators, their staff and the depressed patients to guess treatment assignment, the differences between these treatments, controls and placebo became quite small.

Though interpreting all these data is "difficult and confusing," their best estimate of "the effect size of current antidepressant trials that include patients with major depressive episode is approximately 0.30 (modest).[49]

What is troubling about this situation is that antidepressants, like antipsychotics, are not benign drugs. They are associated with a whole series of very troublesome side effects that ought to weigh heavily in any cost-benefit analysis of their usefulness. Among adults, these include sexual dysfunction, weight gain, nausea, apathy, and sleep disturbance, which for many prove intolerable, and these problems may not disappear after the drugs are discontinued.[50] For children, adolescents, and young adults, the heightened risks of suicidality (serious thoughts of taking one's own life, suicide plans, and suicide attempts) and of violence prompted the FDA to issue a black-box warning about the issue in 2004—the most serious step short of withdrawing a medication from the marketplace. Published papers attesting to the safety of prescribing antidepressants to children, as the discovery process associated with lawsuits against GlaxoSmithKline showed, had been based on manipulated data that hid the issue of suicide and that transformed data that showed the drugs did not work into claims that they were safe and effective, a dramatic example of how drug-company ownership and suppression of data can produce evidence-biased medicine. The papers were ghostwritten and had concealed the fact that taking Paxil was associated with a three times greater risk of suicidal ideation.[51]

There is a further problem that needs to be taken into account. For a not-inconsiderable fraction of those prescribed antidepressants, discontinuing the drugs proves problematic. While for some the resulting symptoms are mild and dissipate over a period of a few weeks, for others they are serious and persistent, representing a major iatrogenic problem. In effect, these patients may find themselves trapped into an endless cycle of antidepressant usage.[52]

GIVEN ITS HIGHLY NEGATIVE reputation in the court of public opinion and the decreasing teaching about and use of the procedure by most psychiatrists in the 1970s, most observers at the time suspected that electroconvulsive therapy (ECT) would soon suffer the same fate as virtually all the somatic treatments introduced between 1920 and 1940: it would essentially disappear. This has not happened. While it remains unclear just how extensively it is employed in contemporary psychiatric practice, ECT has enjoyed something of a revival, particularly in the treatment of severe treatment-resistant depression and, to a lesser degree, mania. To some extent, that revival occurred as growing attention began to be paid to "the dangers . . . and other toxic effects of the presently available antidepressants" and "the long-term negative effects of neuroleptics."[53] Gradually, though, better-designed studies with controls and blind assessments of mental status convinced many psychiatrists that ECT acted more rapidly than drugs and was at least as efficacious as other available therapies.[54]

There were patients and psychiatrists who still swore at ECT and insisted it had brain-disabling effects and should be banned.[55] But from the 1980s onward, a larger number of patient memoirs from those who had suffered from extreme depression sang praises about the treatment and credited ECT for rescuing them from a life of utter misery.[56] Prominent entertainers like Dick Cavett and Carrie Fisher joined forces with prize-winning writers like Simon Winchester and Sherwin Nuland. All spoke of how it had helped lift their bouts of depression. Granted, there were others who lamented how badly ECT had affected their memory, but these patients too were often willing to tout the treatment.[57] Meanwhile, a series of somewhat better-designed studies appeared to provide stronger grounds for supporters of ECT to claim it worked in serious depression and did so more quickly and more effectively than antidepressant medications.[58]

Unfortunately, for many patients, the symptomatic improvement ECT produces does not last. As Ana Jelovac and her colleagues report, "Relapse rates following ECT are disappointingly high and appear to have increased over time. In patients treated with continuation pharmacotherapy . . . nearly 40% of ECT responders can be expected to relapse in the first six month and roughly 50% by the end of the first year." Without ongoing treatment, either with drugs or more ECT, "relapse rates were even higher, approaching 80% at six months."[59]

As for side effects, modified ECT has essentially eliminated the problem of fractures in the course of treatment. For memory and cognition, the picture is much cloudier. Lothar Kalinowsky's chief protégé and the major advocate for ECT in the contemporary era, the New York psychiatrist Max Fink, bluntly acknowledges that the patients in the 1940s and 1950s had reason to complain about the damage ECT had done to them: "The prevalent belief that electroshock impairs memory is based on the early experiences of patients who were treated without anesthesia or oxygen [the vast majority of patients through the mid-1950s]. Such treatments were associated with severe and persistent memory losses." But he insists that those problems are a thing of the past. In *Electroshock: Restoring the Mind,* first published in 1999 and reissued in several revised editions since, he points to improvements in the dosage and administration of the therapy. An unbridled enthusiast for ECT, he insists "there is no longer any validity to the fear that electroshock will erase memory or make the patient unable to recall her life's important events or recognize family members or return to work."[60]

Not everyone is as convinced that the problems with memory have disappeared. Harold Sackeim, who for decades has insisted on ECT's usefulness, accepts that problems with side effects remain, even if fractures are no longer an issue. "Virtually all patients," he claims, "experience some degree of persistent and, likely, permanent retrograde amnesia." Those complaints could not be dismissed as the product of hysteria: "There is no dearth of patients who have received ECT who believe the treatment was valuable, often life-saving, who are not litigious, who return to productive activities, and yet report that a large segment of their life is lost."[61]

Some ECT practitioners seek to minimize the loss of memory through the use of unilateral ECT, which is designed to affect one rather than both hemispheres of the brain. That may have helped skeptics to embrace the procedure. Such modifications were by no means universal, however, and many continued to use the traditional bilateral approach, despite accumulating evidence of greater long-term cognitive deficits in the aftermath of treatment. These deficits showed a direct relationship to the number of treatments given and the techniques used. A New York study by an enthusiast for ECT concluded that "some forms of ECT have persistent long-term effects on cognitive performance" and acknowledged that the study's "findings do not indicate that the treatments with more benign outcomes are free of long-term effects."[62]

ECT has thus had a slight renaissance, but it is hardly free of controversy. The history of its abuse in mental hospitals and the extraordinarily negative image it acquired in popular culture have proved hard to overcome. In the last quarter century, it has once more obtained a place in the psychiatric armamentarium—mysterious as the basis for its apparent therapeutic usefulness remains. How secure that place will prove to be remains uncertain. Though the American Psychiatric Association issued carefully couched endorsements of the procedure, ECT is regarded in many quarters with suspicion.[63] One practical measure of its ambiguous standing is its tepid reception among those running the NIMH, the paymaster of American academic psychiatry. As of the end of the twentieth century, Edward Shorter and David Healy report, the NIMH had "supported 171 drug trials for depression, 21 trials for acupuncture, and only 4 for ECT."[64] And the consensus conference called by the NIMH to examine ECT's place in modern psychiatry reflected that lack of enthusiasm in the very wording of its endorsement of the procedure: "ECT is demonstrably effective for a narrow range of severe psychiatric disorders in a limited number of diagnostic categories: delusional and severe endogenous depression, and manic and certain schizophrenic syndromes."[65]

RECOGNIZING THAT THE PSYCHOPHARMACOLOGICAL revolution has been seriously oversold should not prompt us to despair or to dismiss drug treatment out of hand. There are substantial numbers of patients for whom drugs provide considerable relief from their suffering and permit them to resume some semblance of a normal life. Importantly, the side effects this group of patients experiences are mild or moderate, and on balance the treatment, for them, is clearly preferable to the disease.

One needs to remember that all medications carry risks. That is as true of the pills the rest of medicine prescribes as it is of the drugs psychiatrists employ to treat mental illness. In medicine, as elsewhere in life, there is no free lunch. Awareness of side effects and the need to balance potential benefits and harms is crucial in all of medicine, and psychiatry is no exception. Still, one could argue that, given the side-effect profiles of many psychiatric drugs and their inability to cure, caution is especially called for here.

For those patients who respond well to psychiatric drugs and who experience few negative effects, the choice to take them is clear. Symptomatic relief is often all that general medicine can offer: think, for example of treatments for diabetes, for arthritis, for autoimmune diseases, or for AIDS. Many of those symptomatic treatments work far better than their psychiatric equivalents, to be sure, so one should not push the parallel too far. Insulin for type 1 diabetes and triple therapy for AIDS can transform and extend lives in ways psychiatric medications simply cannot. Still, in cases where psychiatric medications successfully diminish the pain and suffering of depression or anxiety, or the hallucinations and delusions that torment schizophrenics, we should acknowledge that and be grateful for the limited but important degree of progress the psychopharmacological revolution has brought in its wake. Lithium treatment to treat mania and reduce the chances of recurrent episodes of bipolar disorder is another advance, though curiously American psychiatrists seem to be using it less and less.[66] And we have reasonably good evidence that SSRIs can be used to alleviate the symptoms of obsessive-compulsive disorder (as does CBT).[67]

Unfortunately, however, many of those who come to psychiatry seeking relief from their suffering fall into one of two other camps: there are those for whom whatever degree of symptomatic relief the drugs provide is offset by the side effects produced by the medications—so troubling and unbearable that these patients very often refuse further treatment or become noncompliant and fail to take their prescriptions—and another very large group for whom the drugs simply don't work at all. Members of the last group either drop out of treatment and remain a source of distress and disturbance for themselves and those around them, or they suffer the problems psychiatric medications bring in their wake, without obtaining any corresponding benefit.

In some sense, speaking of first- and second-generation antipsychotics is quite misleading. Both categories are highly heterogeneous. The drugs combined under these labels are pharmacologically and clinically distinct. Efficacy, side effects, and costs all vary substantially. The degree to which patients tolerate different drugs, and the balance between benefits and side effects, can be substantial and in conditions that are so distressing and disabling, as Stefan Leucht, of the Techische Universität at Munich and one of the leading practitioners of meta-analysis, argues, "even a small benefit could be important."[68]

At present, though, psychiatrists have no way of knowing which response any given patient will have when a particular pill is prescribed. Taking antipsychotics or antidepressants is thus a game of craps, and when the gamble turns up snake eyes, patients are usually either switched to another drug or have other drugs added to the cocktail of chemicals that are prescribed for them—so-called polypharmacy.[69] But the odds of either approach working diminish rapidly. Worse still, the price of polypharmacy—treating patients as neurochemistry experiments—is a greatly increased chance of suffering major side effects, with little likelihood of a positive outcome. Yet again, those who suffer are offered desperate remedies and confronted with desperately poor outcome statistics. The sobering reality is that we are very far from possessing psychiatric penicillin, and we should not be seduced into thinking, as Jeffrey Lieberman put it in *Shrinks*, that "the modern psychiatrist now possesses the tools to lead any person out of a world of mental chaos into a place of clarity, care, and recovery."[70] Sadly, we don't.

ACKNOWLEDGING THE LIMITATIONS of psychopharmacology, clinical psychologists and some psychiatrists have attempted to use either CBT or interpersonal therapy, which were initially developed as treatments for the milder forms of mental distress seen in office practice, as an additional, adjunctive therapy for psychosis and graver sorts of mental illness.[71] The hope was that these forms of psychotherapy could lead to better outcomes than drugs alone. Small, unblinded trials produced initially encouraging results, prompting groups like NICE (the National Institute for Health and Care Excellence)—the organization that decides which interventions the British National Health Service should provide—to endorse its use. Unfortunately, larger and better-designed studies, particularly those where the ratings of degrees of improvement are made blindly, have suggested that we should be much less sanguine about CBT's value in treating psychosis. Indeed, some have gone further, dubbing the "vigorous advocacy of this form of treatment . . . puzzling" in light of accumulating evidence of its weak or nonexistent effects on the positive and negative symptoms of schizophrenia and on the propensity to relapse.[72]

Reviewing the results of CBT in schizophrenia, one well-designed meta-analysis of the published research found that it produces only a small improvement in patient functioning, which was not sustained at follow-up. As for its ability to have beneficial effects on patients' subjective feelings of distress and their overall quality of life, here its contributions were weak at best. Indeed, not a single CBT trial has ever reported a rise in the quality of life among patients diagnosed with schizophrenia.[73] A Cochrane review, widely seen as authoritative, likewise found CBT had "no advantage" in treating general psychiatric symptoms, delusions, and other so-called positive symptoms of schizophrenia. Nor did CBT have positive effects on the likelihood of relapse or rehospitalization.[74]

The evidence supporting the use of CBT in the treatment of bipolar disorder is similarly underwhelming, if slightly more positive. In the short run, it appears to reduce the risk of relapse and the severity of mania (though not of the depressive episodes). Longer term, however, these moderately positive outcomes disappear, and CBT does not appear to have persistent positive effects.[75] While CBT appears to be more effective in major depression, the positive effects that have been recorded are unfortunately quite small.[76] There is, it is true, considerable heterogeneity in the reported results, possibly because of the incoherence of the category of major depressive disorder, which lumps together a congeries of conditions with only an artificial label in common.

In most respects, the overall record of CBT in the treatment of psychosis is far from encouraging. In one very important respect, however, CBT does offer some promise. Whatever the drawbacks and failures of psychoanalysis, its attempts to grapple with psychosis did force its practitioners to listen to patients and to try to make sense of their condition. The rise of the *DSM* and of psychopharmacology reduced the psychiatric encounter to brief consultations that revolved around the prescription of medications and management of side effects. That essentially ended any serious attempt to listen to a patient's concerns. In contrast, by its very nature, CBT, like other forms of psychotherapy such as interpersonal therapy and psychodynamic psychotherapy, does require actually listening to patients and giving legitimate weight and attention to their thoughts, feelings, and experiences. That, assuredly, is something we ought to welcome, even while recognizing that serious forms of mental disorder remain, for many, resistant to our best efforts to treat them.

Epilogue
Does Psychiatry Have a Future?

PSYCHIATRY IS NO LONGER A PROFESSION confined as securely as its patients in warehouses of the unwanted. Asylums no longer haunt our imaginations, though their demise occurred essentially behind the backs of the profession, more the consequence of changes in social structures and shifts in public policy than of professional initiative. The abandonment of the mentally ill to euphemistically named "board-and-care" homes, to the street, and to jail has been paralleled by the collapse of the once-dominant fraction of the profession that practiced public psychiatry. In the contemporary neoliberal environment, the stigmatized souls who suffer from major mental disturbances find few friends in the halls of government, and, perhaps unfairly, are seen as a standing reproach to a profession whose core constituency they once were.

Still in some ways a barely acknowledged stepchild of the larger medical enterprise, psychiatry has acquired a substantial presence in academic medicine and now dispenses remedies that, at first blush, look more like the therapies employed by mainstream medicine than they once did: no more insulin comas, surgical eviscerations, or destruction of the frontal lobes; and no more talk of sex and the id either. Talk therapy has been largely ceded to therapists who practice without medical credentials, and biology and pills have become the dominant ways in which psychiatrists explain and respond to mental disorders.

But the wager on biology is a bet whose payoff has been far more limited than its architects promised. Though the advances in basic science have been noteworthy, their bearing on the fate of the mentally ill has been slight. The neo-Kraepelinian revolution of the 1980s, named after the German psychiatrist who established the importance of descriptive psychiatry and authored the original distinction between dementia praecox (schizophrenia) and manic-depressive psychosis (bipolar disorder), was premised on granting a new prominence to questions of diagnosis, a categorical view of mental disorders that portrayed them as distinct and separable illnesses, sharply distinguishable from normality—and, in theory, from one another. Though serviceable for a time, resolving the public embarrassment of a profession unable to agree on how to label the problems it confronted, it privileged reliability over validity. By relying on a checklist of symptoms, it contained the seeds of its own destruction. "Illnesses" proliferated, inviting ridicule. Unexpectedly, as the research on biology and genetics proceeded, the distinctions between disorders crumbled, and the notion that such artificial constructs could identify distinct diseases became ever-more implausible.

The world of the *Diagnostic and Statistical Manual* (*DSM*) survives, but barely. For patients and their families, being able to put a name to their troubles provides a measure of reassurance, and for doctors it provides some guidance about how to proceed. Yet this unwieldy edifice has been gnawed away from within and assaulted from without. It survives only because there is nothing to put in its place—and by dint of its crucial role in securing reimbursement from insurers. The *DSM* provides the comfort of a diagnosis to which doctor and patient can cling—a reassurance not to be minimized—and it serves as the rationale for drug treatments and for the research programs that support the academic wing of the profession. But in the words of a prominent psychiatrist who played a vital role in constructing its third and fourth editions, "DSM diagnoses have given researchers a common nomenclature—but probably the wrong one. . . . DSM diagnoses are not useful for research because of their lack of validity."[1]

Genetics and neuroscience have flourished within the confines of universities, but their therapeutic payoff has been minimal. Vast sums have been invested in studies using neuroimaging to try to intervene more successfully in the treatment of mental disorders. By some estimates, more than 20,000 papers have been written and published. To what end? In the words of a major figure in the field, Raymond Dolan, "it is sobering to acknowledge that functional neuroimaging, in particular modalities such as functional magnetic resonance imaging (fMRI) and magnetoencephalography/electroencephalography (MEG/EEG) plays no role in clinical psychiatric decision making, nor has it defined a neurobiological basis for any psychiatric condition or symptom dimension."[2] This situation may change as further work is done, but it is equally possible that those sponsoring these programs may tire of funding investigations that show few signs of producing practical advances.

Psychopharmacology has provided treatments that help some, but only some, patients to mitigate their symptoms without disabling side effects. This is an advance not to be sneered at. But it is limited progress nonetheless, and that progress now seems to have stalled. In the words of the former head of neuroscience at Eli Lilly and Amgen, "Psychopharmacology is in crisis. The data are in, and it is clear that a massive experiment has failed: despite decades of research and billions of dollars invested, not a single mechanistically novel drug has reached the psychiatric market in more than 30 years."[3]

Psychological interventions like cognitive-behavioral therapy and interpersonal therapy have provided relief for many patients suffering from anxiety disorders, eating disorders, post-traumatic stress disorder, and some forms of depression, but these therapies can do little for those with the most serious forms of disturbance. That, too, is progress as grateful patients attest, although many of those so treated discover that the relief doesn't always last. One wishes it were otherwise, but it is foolish to exaggerate how far our understanding and our capacities to intervene have advanced.

Meanwhile, the pharmaceutical industry, having extracted vast profits from the pills it has provided to treat a whole array of mental illnesses, seems to be abandoning the search for novel remedies and treatments.[4] In the last ten years, GlaxoSmithKline has all but closed its psychiatric laboratories, AstraZeneca has essentially dropped internal research on psychopharmacology, and Pfizer has dramatically reduced its spending in the psychiatric arena. Perhaps these companies have been put off by the reputational damage some less-than-salubrious activities in this arena have brought. Or perhaps, more likely, not seeing any obvious path forward and finding therapies for other forms of pathology a more likely source of future profits, they decided to move on.[5]

What of the future? Professions derive their legitimacy from their claims to possess unique and valuable specialized knowledge, combined with a capacity for action—cognitive and practical talents that render the problems they address susceptible to expert intervention.[6] Within the medical universe, it is occasionally possible to obtain elevated status largely on a foundation of diagnostic skills and prognostic precision, even in the absence of effective means of intervening in the course of a disease. Neurology, for much of its history, occupied just such a niche. For its practitioners, diagnostic refinement went hand in hand with therapeutic impotence, and yet neurology enjoyed a relatively high status in the medical profession and among the public at large. More generally, however, medics are expected to combine an ability to identify the troubles they are called on to address with effective treatments. Like most of the rest of medicine, psychiatrists must present themselves as experts both at diagnosing the patients who seek their assistance and at alleviating their distress.

On both fronts, psychiatry is in trouble. Its diagnoses are an increasingly frail reed upon which to rest its claims to expertise and its recipes for intervention in the most serious forms of mental illness are at best Band-Aids. Band-Aids are better than nothing, and the remedies we have are certainly better than those of earlier generations. But too many people are taking medications with little or no effect and suffering from often serious side effects. Defenders of psychiatry (and of clinical psychology) rightly point to evidence of greater success in dealing with some of the milder and more widespread forms of mental distress, yet the problems posed by the gravest illnesses are real and pressing, and the origins of psychosis and depression remain almost as obscure as ever.

IN THE YEARS AFTER THE SECOND WORLD WAR, the leaders of American psychiatry disdained biology and the brain and looked to the psyche —both to untangle the roots of mental disorder and to treat it. The hegemony of the "mind twist" crowd ended abruptly at the end of the 1970s. Since then, "brain spot" psychiatry has enjoyed an almost unchallenged supremacy, instructing us to see mental illness as brain disease.[7]

I think monism of both sorts is deeply misguided. As I noted in the opening pages of this book, I would be astonished if biology made no contribution to the genesis of many serious forms of mental disorder, unsatisfactory and frustrating as two centuries of effort to unravel that mystery have proved. Equally, though, I am convinced that madness cannot successfully be divorced from the cultural, social, and psychological matrix in which human beings exist. To deny that social factors play a major role in the genesis and course of mental illness is to blind oneself to an enormous volume of evidence, epidemiological and otherwise, that teaches us that the environment powerfully matters.

To an extent unprecedented in any other part of the animal kingdom, humans' brains continue to develop postnatally in ways heavily conditioned by the environment. Culture and society, on both a grand and a microscopic scale, interact powerfully with our lifestyle choices and biology, and in all sorts of complex ways the physical structure and functioning of our brains are shaped by psychosocial and other sensory inputs. Human plasticity extends far beyond childhood. As we now know, the very shape of the brain and the neural connections that develop within it—the biology that constitutes the physical-substrate underpinnings of our emotional and cognitive existence—are profoundly influenced by social and psychological stimulation, most crucially of all by the familial environment within which human beings grow. Thinking, feeling, and remembering are in a multitude of ways the product of complex networks and interconnections that form in the maturing brain. Developmental and environmental factors are crucial elements in determining whether someone becomes mentally ill and what forms their disturbance might take.

To think of the brain as an asocial organ is thus profoundly mistaken.[8] So, too, is the crude parallel notion that mental illness is just brain disease. The obverse is surely also true: to dismiss any role for biological factors is to don a different set of blinkers. Leon Eisenberg put it well: "Psychiatry is all biological and all social. There is no mental function without brain and social context. To ask how much of mind is biological and how much social is as meaningless as to ask how much of the area of a rectangle is due to its width and how much to its height."[9]

Chemicals, even far better and more precisely administered than the ones we presently possess, will never provide a wholly satisfactory answer to the riddles of mental illness or the challenges of responding to the disruptions that follow in its train. Developmental and environmental factors play a crucial role in the genesis and the character of mental disturbance, and addressing these effectively requires a different, multifaceted approach. In the meantime, what remedies we have treat symptoms rather than cure, and they are wildly uncertain in their effects in any individual case. Some researchers believe advances in genetics may improve this situation, providing a means of predicting which medications will provide some degree of relief for a particular patient while avoiding a nasty constellation of side effects. There are some flickering signs of progress on identifying patients who are liable to suffer the worst side effects, but establishing who may respond best to particular drugs is still a distant hope.

In any event, biological advances can take us only so far. For fortunate souls, antipsychotics and antidepressants do provide some measure of solution to the devastating tragedy that otherwise envelops them. Palliative measures are assuredly better than nothing, so long as they do not bring new medically induced pathologies that outweigh the problems they began with. With time, too, just as they did before the psychopharmacological revolution and generally just as mysteriously, some mental illnesses remit. Depression lifts, mania subsides, even a diagnosis of schizophrenia is not always as unyielding a proclamation of one's fate as is oftentimes assumed. But for far too many patients, such positive outcomes remain out of reach. Treatments are ineffectual or else bring in their wake a host of new and threatening pathologies. What then?

If we are to confront the challenges that mental disorders present to all of us, we shall have to take account of social and political realities. As we have seen, the decisions to confine the mentally ill to the madhouse and, more recently, to decant them into unwelcoming "communities" have drastically affected what it means to be mentally ill. Institutionalization and deinstitutionalization were both driven by powerful social and political imperatives. For practitioners and their patients, these are the larger matrices within which they live and by which they are confined.

We need a very different approach. Neuroscientists and geneticists have used their command of academic departments of psychiatry to attract large amounts of funding for their research, and they have rewarded their sponsors with some useful basic science. The clinical utility of their work, however, has verged upon a nullity, and their dominance has created a remarkable gap between the worlds of academic and clinical psychiatry. The phenomenological and social dimensions of mental illness have all but disappeared as questions worthy of serious and sustained attention.[10] That is an imbalance that has had profoundly negative effects on psychiatry and, more important, on the prospects of advancing the clinical care of patients.

On the positive side, there are signs that some psychiatrists have begun to recognize the problems that confront us. Many of the criticisms of contemporary psychiatry in the closing chapters of this book rest, after all, on the findings of research being done in the field. Eventually, one must hope, the weight of evidence that the profession has gone astray will become too hard to ignore. Psychiatry has never been a monolith, and though there have been periods dominated by either the mind or brain, there have always been those who have resisted the dominant paradigm or who have found both kinds of reductionism difficult to swallow.

There is, however, a deeper difficulty that we confront as a society as we seek more effective responses to the problems of serious mental illness. The closing of asylums coincided with the abandonment of any serious public effort to ameliorate the sufferings of those gravely disabled by mental disturbances of all kinds. In a society that valorizes the market as a universal solvent, and that attributes failure to the shortcomings of the individual, those with serious mental illness face a harsh future. The malign neglect that has for nearly three-quarters of a century constituted public policy in this arena was not instituted at the behest of psychiatry, though with a few exceptions the demise of public psychiatry drew little protest from within the ranks of the profession.

Dealing effectively with serious mental illness requires a major commitment to housing, supporting, and sheltering people who are incapable, for the most part, of providing for themselves. It necessitates serious engagement with research about the best ways to provide these things. Families, for perfectly understandable reasons, often find these burdens impossible to bear, and in other cases, patients flee their families. In either case, the alternatives are grim.

We ought to recoil from arrangements that condemn helpless and suffering human beings to the street and the jail, and stop pretending that chemistry is the sole and singular way forward. Those afflicted with serious forms of mental illness have been cast into the wilderness—a brutal and often fatal outcome for many with few resources of their own. These are people who lack the capacity to function in an environment in which they are seen as little more than a drain on the public purse. Chronically dependent on the not-so-tender mercies of a shrinking welfare state, they are doubly stigmatized: for their illness; and because they show few signs of reform or recovery.

I divide my time in the United States between San Diego and San Francisco. In both cities, the problem of the homeless mentally ill has festered for decades now. It is, of course, a nationwide problem and in recent years it has created a significant backlash, as public order and the very things people value about urban life are significantly degraded. Hence the spectacle of new mayors in New York, Los Angeles, Seattle, and Portland, Oregon, announcing their intent to tackle the multiple problems this situation creates. If not driven by compassion for the mentally ill, the increasingly harsh reaction of the electorate seems to be creating pressure for solutions. Whether effective policies will emerge remains to be seen. However, I recently visited the Netherlands and several Scandinavian countries, places with much more extensive social safety nets and much less income inequality than in the United States. What was remarkable was how little—almost none—that one saw in the way of begging and homelessness, and how much more vibrant city life was in these countries. Perhaps there is a lesson there for America, though whether it will be heeded remains to be seen.

The idea that we bear a collective moral responsibility to provide for the unfortunate—indeed, that one of the marks of a civilized society is its determination to provide certain minimum standards of living for all its citizens—has never enjoyed widespread support in the United States. Most Americans have embraced an ideology far more comforting to the privileged, but one that also resonates among the masses: the myth of the benevolent "Invisible Hand" of the marketplace and its corollary, an unabashed moral individualism. There is little place (and less sympathy) within such a worldview for those who are excluded from the race for material well-being by chronic disabilities and handicaps—whether physical or mental, or those who suffer the more diffuse but cumulatively devastating penalties accruing to those belonging to racial minorities or living in dire poverty.

The punitive sentiments directed against those who must feed from the public trough extend only too readily to people suffering from the most severe forms of psychiatric misery. Those who seek to protect a long-term mental patient from the opprobrium visited on the welfare recipient may do so by arguing that the patient is both dependent and sick. But I fear this approach has only a limited chance of success. After all, despite two centuries of propaganda, the public still resists the straightforward equation of mental and physical illness. Moreover, the long-term mental patient, in many instances, will not get better and often fails to collaborate with his or her therapist to seek recovery. Such blatant violations of the norms of self-care make it unlikely that people with severe mental illness will be extended the courtesies accorded to the conventionally sick. Those incapacitated by psychiatric disability all too often find themselves the targets of policy makers and pundits who would abolish social programs because they consider any social dependency immoral. That has to change.

If our goal is a revival of a psychiatry that attends to the psychological, physical, and social dimensions of mental disorder, one must recognize just how difficult that transformation is likely to prove. For the professional elite in the academy, the pressures to secure grant money and build scientific careers make sustained interest in the social and environmental dimensions of mental disorder difficult and unattractive. Clinicians in private practice have few incentives to embrace this group of patients or to attend to these issues. But a far larger difficulty is the unpropitious nature of the larger political environment. Where is the political will to break with the conventional wisdom about mental illness and its treatment? And where are the massive resources that would be required were we to take seriously the parlous situation of those whose lives have been devastated by serious mental illness, or address the multiplicity of their needs? Psychiatry faces a difficult way forward. So do we.

Madness remains a mystery that stubbornly refuses to bend itself to the rule of reason. Yet, despite the obstacles, it is a riddle we must continue to strive to solve. Major mental illnesses constitute some of the most profound forms of human suffering. They are, as the historian Michael MacDonald once put it, at once the most solitary of afflictions and the most social of maladies.[11]

The overwhelming social dislocations associated with the epidemic of COVID-19 have emphasized the foolishness of neglecting the connections between the social environment and mental illness. Two groups at opposite ends of the age spectrum have had their lives disrupted in particularly powerful ways. Old people, often living isolated and lonely lives, have been cut off even more extensively than most from ordinary forms of social interaction. Meanwhile, the young—both those in the critical years of schooling and social development and those seeking to launch independent lives and careers—have experienced massive losses that are likely to have lifelong effects. It would be astonishing if the social and economic devastation wrought by the pandemic did not contribute to more mental distress and breakdowns among those most exposed to its ravages. Preliminary data suggest that these consequences are already surfacing.[12] The fear, isolation, job losses, and financial insecurity that so many have experienced have been reflected in a wide variety of survey data that show huge increases in reported levels of anxiety and depression.[13] And while we know that "most people are resilient and will not succumb to psychopathology," in others the psychosocial effects of these stresses can be expected to be profound and long-lasting even after the pandemic subsides, predisposing them to heightened rates of mental disturbance.[15] How we respond to this as a society is enormously important. The costs of failing to do so adequately will obviously fall heavily on those whose mental equilibrium is under threat. But those costs are also inescapable and devastating for patients' families and loved ones and, beyond the domestic circle, for society at large. I have stressed throughout this book the immense suffering and pain that serious mental illness brings in its train. It is long past time for us to devote adequate resources to alleviating the resulting torment and distress.

NOTES

Preface

1. In 2019, Black maternal mortality was two and a half times White maternal mortality (D. L. Hoyert, "Maternal Mortality Rates in the United States, 2019," Health E-Stats, April 2021, National Center for Health Statistics, Centers for Disease Control [CDC]); 2021 CDC data show that the life expectancy of Black males is six years less than that for White males. E. Arias, B. Tejada-Vera, and F. Ahmad, "Provisional Life Expectancy Estimates for January through June, 2020," Vital Statistics Rapid Release, Report No. 10, US Department of Health and Human Services, February 20, 2021, 3.

2. Office of the US Surgeon General, *Mental Health: Culture, Race and Ethnicity: A Supplement to Mental Health: Report to the Surgeon General* (Rockville, MD: US Department of Health and Human Services, 2001).

3. American Psychiatric Association, "APA's Apology to Black, Indigenous and People of Color for Its Support of Structural Racism in Psychiatry," January 18, 2021, https://www.psychiatry.org/newsroom/apa-apology-for-its-support-of-structural-racism-in-psychiatry.

4. A. E. Kass, R. P. Kolko, and D. E. Wilfley, "Psychological Treatments for Eating Disorders," *Current Opinion in Psychiatry* 26 (2013): 549–555; C. A. Fisher, S. Skocic, A. Rutherford, and S. E. Hetrick, "Family Therapy Approaches for Anorexia Nervosa," *Cochrane Database of Systematic Reviews* (May 1, 2019), https://www.cochranelibrary.com/cdsr/doi/10.1002/14651858.CD004780.pub4/full; A. Zeeck, B. Herpertz-Dahlmann, H.-C. Friederich, et al., "Psychotherapeutic Treatment for Anorexia Nervosa: A Systematic Review and Network Meta-Analysis," *Frontiers of Psychiatry* 9 (2018), art. 158. The conclusion of these reviews is that the relevant research is of poor quality and that biased findings are common. G. E. Hunt, N. Siegfried, K. Morley, D. Brooke-Sumner, and M. Cleary, "Psychosocial Interventions for People with Both Severe Mental Illness and Substance Abuse," *Cochrane Database of Systematic Reviews* (December 12, 2019), https://www.cochranelibrary.com/cdsr/doi/10.1002/14651858.CD001088.pub4/full, concludes that "there is currently no high-quality evidence to support any one psychosocial treatment over standard care for important outcomes such as remaining in treatment, reduction in substance

use or improving mental or global state in people with serious mental illnesses and substance misuse." See also J. Sin, D. Spain, M. Furuta, T. Murrells, and I. Norman, "Psychological Interventions for Post-Traumatic Stress Disorder (PTSD) in People with Severe Mental Illness," *Cochrane Database of Systematic Reviews* (January 4, 2017), https://www.cochranelibrary.com/cdsr/doi/10.1002/14651858.CD011464.pub2/full.

5. A. Garakani, J. W. Murrough, R. C. Freire, et al., "Pharmacotherapy of Anxiety Disorders: Current and Emerging Treatments," *Frontiers of Psychiatry* 11 (December 23, 2020), https://doi.org/10.3389/fpsyt.2020.595584.

6. P. Cuijpers, S. Quero, C. Dowrick, and B. Arroll, "Psychological Treatment of Depression in Primary Care: Recent Developments," *Current Psychiatry Reports* 21, 129 (November 23, 2019): 1–10, 7; H. A. Whiteford, M. G. Harris, G. McKeon, et al., "Estimating Remission from Untreated Major Depression: A Systematic Review and Meta-Analysis," *Psychological Medicine* 43 (2013): 1569–1585.

7. For some important reviews of these issues, see the following. J. Barth, T. Munder, H. Gerger, et al., "Comparative Efficacy of Seven Psychotherapeutic Interventions in Patients with Depression: A Network Meta-Analysis," *PLoS Medicine* 10, no. 5 (2013), https://doi.org/10.1371/journal.pmed.1001454; Jonathan Shedler, "The Efficacy of Psychodynamic Psychotherapy," *American Psychologist* 65 (2010): 99–109, which claims that over the long term, this approach produces more lasting results than a shorter course of cognitive-behavioral therapy (CBT); A. C. James, T. Reardon, A. Soler, G. James, and C. Cresswell, "Cognitive Behavioural Therapy for Anxiety Disorders in Children and Adolescents," *Cochrane Database of Systematic Reviews* (November 16, 2020), https://www.cochranelibrary.com/cdsr/doi/10.1002/14651858.CD013162.pub2/full; V. Hunot, R. Churchill, V. Teixeira, and M. Silva de Lima, "Psychological Therapies for Generalised Anxiety Disorder," *Cochrane Database of Systematic Reviews* (January 24, 2007), https://www.cochranelibrary.com/cdsr/doi/10.1002/14651858.CD001848.pub4/full; P. Cuijpers, A. van Straten, G. Andersson, and P. van Oppen, "Psychotherapy for Depression in Adults: A Meta-Analysis of Comparative Outcome Studies," *Journal of Consulting and Clinical Psychology* 76 (2008): 909–922.

8. The architects of the three previous editions of psychiatry's diagnostic manual became some of the fiercest critics of its latest incarnation, and the director of the National Institute of Mental Health and his immediate predecessor both dismissed it as a scientific travesty. For a review of these events, see Gary Greenberg, *The Book of Woe: The DSM and the Unmaking of Psychiatry* (New York: Blue Rider Press, 2013), and the discussion in Chapter 21.

9. Judd Marmor, *Psychiatrists and Their Patients: A National Survey of Private Office Practice* (Washington, DC: Joint Information Service of the American Psychiatric Association and the National Association for Mental Health, 1975), 24, 69–70. Office-based psychiatrists who had not undergone a training analysis and thus did not offer classical psychoanalysis preferred to refer to themselves as practicing "psychodynamic" psychiatry. Some of them offered the sort of "commonsense" psychotherapy that the Swiss American psychiatrist Adolf Meyer—the dominant figure in American psychiatry in the interwar years—had recommended, attempting to train patients to adopt better habits. Others simply offered a watered-down version of classical psychoanalysis, involving less frequent sessions and less emphasis on Freud's insistence on the psychosexual origins of psychopathology.

10. Ramin Kojitabai and Mark Olfson, "National Trends in Psychotherapy by Office-Based Psychiatrists," *Archives of General Psychiatry* 65 (2008): 962–970.

11. Mark Olfson and Steven Marcus, "National Trends in Outpatient Psychotherapy," *American Journal of Psychiatry* 167 (2010): 1456–1463, 1456.
12. Gardiner Harris, "Talk Doesn't Pay, So Psychiatry Turns Instead to Drug Therapy," *New York Times,* March 5, 2011.
13. Jeffrey Lieberman, T. Scott Stroup, Joseph P. McEvoy, et al., "Effectiveness of Anti-psychotic Drugs in Chronic Schizophrenia," *New England Journal of Medicine* 353 (2005): 1209–1223. A better title for their findings might have been "The Ineffectiveness of Anti-psychotic Drugs in Most Patients with Chronic Schizophrenia."
14. Thomas Scheff, *Being Mentally Ill: A Sociological Theory* (Chicago: Aldine, 1966).
15. Thomas Szasz, *The Myth of Mental Illness* (New York: Harper and Row, 1961).

CHAPTER 1 Mausoleums of the Mad

1. The Scottish alienist William Alexander Francis Browne promulgated equally utopian views in his published lectures: W. A. F. Browne, *What Asylums Were, Are, and Ought to Be* (Edinburgh: Black, 1837).
2. The best biography of Dix is David Gollaher, *Voice for the Mad: The Life of Dorothea Dix* (New York: Free Press, 1995).
3. The term "psychiatry," coined in 1808 by the German physician Johann Christian Reil and taken from the Greek word for soul (*psykhē*), gradually became the preferred term in Germany, but it was only adopted in America and Britain in the late nineteenth century.
4. Pliny Earle's *The Curability of Insanity* (Philadelphia: Lippincott, 1887) ought to have been called *The Incurability of Insanity,* such was its pessimistic message.
5. For a discussion of historical trends in mental hospital populations, see Atlee Stropu and Ronald Manderscheid, "The Development of the State Mental Hospital System in the United States: 1840–1980," *Journal of the Washington Academy of Sciences* 78 (1988): 59–68.
6. Gerald Grob, *Mental Illness and American Society, 1875–1940* (Princeton, NJ: Princeton University Press, 1983), 180–187.
7. Frederick Mott, *Syphilis of the Nervous System* (Oxford: Oxford University Press, 1914).
8. H. Noguchi and J. W. Moore, "A Demonstration of Treponema Pallidum in the Brains of Cases of General Paralysis," *Journal of Experimental Medicine* 17 (1913): 232–238.
9. Morton Kramer, "Long-Range Studies of Mental Hospital Patients: An Important Area for Research in Chronic Disease," *Milbank Memorial Fund Quarterly* 31 (1953): 253–264, 255.
10. William G. Rothstein, *American Physicians in the Nineteenth Century: From Sect to Science* (Baltimore, MD: Johns Hopkins University Press, 1983), 108–114.
11. Useful discussions include Bonnie Blustein, "New York Neurologists and the Specialization of American Medicine," *Bulletin of the History of Medicine* 53 (1979): 170–183; Bonnie Ellen Blustein, *Preserve Your Love for Science: Life of William A. Hammond, New York Neurologist* (Cambridge: Cambridge University Press, 1991), esp. ch. 4; and Barbara Sicherman, "The Uses of a Diagnosis: Doctors, Patients and Neurasthenia," *Journal of the History of Medicine and Allied Sciences* 32 (1977): 33–54.
12. Blustein, *Preserve Your Love for Science;* Charles Rosenberg, "The Place of George M. Beard in Nineteenth Century Psychiatry," *Bulletin of the History of Medicine* 36 (1962): 245–259.

13. Bonnie Blustein, "'A Hollow Square of Psychological Science': American Neurologists and Psychiatrists in Conflict," in *Madhouses, Mad-Doctors and Madmen: The Social History of Psychiatry in the Victorian Era,* ed. Andrew Scull, 241–270 (Philadelphia: University of Pennsylvania Press, 1981).
14. E. C. Spitzka, "Reform in the Scientific Study of Psychiatry," *Journal of Nervous and Mental Disease* 5 (1878): 201–229, 209, 202.
15. W. A. Hammond, "The Non-asylum Treatment of the Insane," *Transactions of the Medical Society of New York* (1879): 280–297, 282.
16. Eugene Grissom, "True and False Experts," *American Journal of Insanity* 35 (1878–1879): 1–36. Hammond promptly sued Grissom for libel. Grissom's response was conspicuous for not responding to the substance of Hammond's critique.
17. For a careful analysis of the case, see Blustein, *Preserve Your Love for Science,* 86–93.
18. W. A. Hammond, "The Construction, Organization and Equipment of Hospitals for the Insane," *Neurological Contributions* 1 (1879): 1–22, 4–6, 13; Spitzka, "Reform in the Scientific Study of Psychiatry," 224.
19. As Spitzka conceded in his textbook, "in the present state of our knowledge, it is impossible to frame a definition of insanity which, while it meets the practical everyday requirements, is constructed on *scientific* principles." E. C. Spitzka, *Insanity: Its Classification, Diagnosis and Treatment* (New York: Bermingham, 1883), 18. See also W. A. Hammond, *A Treatise on Diseases of the Nervous System* (New York: Appleton, 1871), 266.
20. Silas Weir Mitchell, "Address before the Fiftieth Annual Meeting of the American Medico-Psychological Association, Held in Philadelphia, May 16, 1894," *Journal of Nervous and Mental Disease* 21 (1894): 413–437, 413, 415.
21. Weir Mitchell, "Address before the Fiftieth Annual Meeting," 414, 415, 418.
22. Weir Mitchell, "Address before the Fiftieth Annual Meeting," 420, 422, 424.
23. Weir Mitchell, "Address before the Fiftieth Annual Meeting," 426, 427.
24. Weir Mitchell, "Address before the Fiftieth Annual Meeting," 430.
25. Adolf Meyer, "Reminiscences and Prospects at the Opening of the New York Psychiatric Institute and Hospital," *Psychiatric Quarterly* 4 (1930): 25–34, 28.
26. Walter Channing, "Some Remarks on the Address Delivered to the American Medico-Psychological Association by S. Weir Mitchell, M.D., May 16, 1894," *American Journal of Insanity* 51 (1894): 171–181, 171, 172.
27. Channing, "Some Remarks on the Address," 176.
28. Channing, "Some Remarks on the Address," 176, 180.
29. Council of State Governments, *The Mental Health Programs of the Forty-Eight States: A Report to the Governors' Conference* (Chicago: Council of State Governments, 1950).
30. Charles Wagner, in *Annual Report of the New York State Commission in Lunacy* 12, 29–30, quoted in Gerald M. Grob, "Rediscovering Asylums: The Unhistorical History of the Mental Hospital," in *The Therapeutic Revolution,* ed. Morris J. Vogel and Charles E. Rosenberg, 135–158 (Philadelphia: University of Pennsylvania Press, 1979), 146.
31. Wise to Blumer, January 13, 1894, Blumer Papers, box 28, Isaac Ray Historical Library, Butler Hospital, Rhode Island, quoted in Ian Dowbiggin, *Keeping America Sane* (Ithaca, NY: Cornell University Press, 1997), 48.
32. Blumer later left his post at Utica State Hospital for a position as superintendent at the private Butler Hospital in Rhode Island, where he was spared an interfering state bureaucracy and massive overcrowding, and he could minister to a more educated clientele.

33. Grob, *Mental Illness and American Society*, 268.
34. Drs. Dodds, Strahan, and Greenlees, "Assistant Medical Officers in Asylums: Their Status in the Specialty," *Journal of Mental Science* 36 (1890): 43–50.
35. As one eminent alienist acknowledged, the assistant physician "sees the best years of his life slipping away from him without any advancement of his interests or improvement of his prospects." Charles Mercier, *Lunatic Asylums: Their Organization and Management* (London: Griffin, 1894), 246.

CHAPTER 2 Disposing of Degenerates

1. Freud reverted back to this earlier view of the connections between madness and civilization, holding that it was the conflict between the demands of civilized morality and the unreconstructed id that prompted neurosis. See Sigmund Freud, "Die 'kulturelle' Sexualmoral und die modern Nervosität," *Sexual-Probleme* 4 (1908): 107–129.
2. A point emphasized in Charles Rosenberg's seminal essay, "The Bitter Fruit: Heredity, Disease, and Social Thought," reprinted in Charles E. Rosenberg, *No Other Gods: On Science and American Social Thought*, rev. and expanded ed., 25–53 (Baltimore, MD: Johns Hopkins University Press, 1997).
3. It was Galton who coined the term "eugenics" in 1883. For an insightful discussion of this aspect of his work, see Theodore Porter, *Genetics in the Madhouse: The Unknown History of Human Heredity* (Princeton, NJ: Princeton University Press, 2018), 221–236.
4. Charles Darwin, *The Descent of Man, and Selection in Relation to Sex*, vol. 1 (London: John Murray, 1871), 168.
5. S. A. K. Strahan, "The Propagation of Insanity and Allied Neuroses," *Journal of Mental Science* 36 (1890): 325–338.
6. E. C. Spitzka, *Insanity: Its Classification, Diagnosis and Treatment*, 2nd ed. (New York: E. B. Treat, 1883), 369.
7. Cesare Lombroso, *Criminal Man*, trans. Mary Gibson and Nicole Hahn Rafter (Durham, NC: Duke University Press, 2006); original, *L'Uomo delinquente* (Milan: Hoepli, 1876).
8. Translation of B. A. Morel, *Traité des dégénérescences physiques, intellectuelles, et morales de l'espèce humaine* (Paris: Baillière, 1857), 5; and Morel at a November 26, 1860, meeting of the Societé Médico-Psychologique, translated and quoted in Ian Dowbiggin, "Degeneration and Hereditarianism in French Mental Medicine 1840–90: Psychiatric Theory as Ideological Adaptation," in *The Anatomy of Madness*, ed. W. F. Bynum, Roy Porter, and M. Shepherd (London: Tavistock, 1985), 1: 192–193.
9. Author's translation of Morel, *Traité des dégénérescences*, 46.
10. Translation of *Annales medico-psychologiques* 12 (1868), 288, in Dowbiggin, "Degeneration and Hereditarianism," 193.
11. Henry Maudsley, *The Physiology and Pathology of Mind* (London: Macmillan, 1867), 83; Henry Maudsley, *The Pathology of Mind* (London: Macmillan, 1879), 88.
12. Henry Maudsley, *Body and Mind* (London: Macmillan, 1870), 61, 63.
13. Henry Maudsley, "Insanity and Its Treatment," *Journal of Mental Science* 17 (1871): 311–324, 323–324, 324.
14. Maudsley, *Physiology and Pathology of Mind*, 204–205. For similar sentiments, see Charles Mercier, *A Textbook of Insanity*, 2nd ed. (London: Allen and Unwin, 1914), 17.
15. Daniel Hack Tuke, *Insanity in Ancient and Modern Life* (London: Macmillan, 1878), 90–95, 152.

16. Strahan, "The Propagation of Insanity," 331, 332, 334.
17. Richard Greene, "The Care and Cure of the Insane," *Universal Review* (August 1889): 493–508, 503.
18. For a biography of one leading American neurologist who made a fortune from tonics and extracts, see Bonnie Ellen Blustein, *Preserve Your Love for Science: Life of William A. Hammond, American Neurologist* (Cambridge: Cambridge University Press, 1991).
19. George Beard, who invented the diagnosis of neurasthenia, was an early enthusiast. See his *The Medical Use of Electricity, with Special Reference to General Electrization as a Tonic in Neuralgia, Rheumatism, Dyspepsia, Chorea, Paralysis, and Other Affections Associated with General Debility, with Illustrative Cases* (New York: William Wood, 1867). Beard's title provides a useful précis of the sorts of nervous complaints that brought patients to the neurologist's waiting room.
20. For useful discussion of medical electricity, see I. R. Morus, "Marketing the Machine: The Construction of Electrotherapeutics as a Viable Medicine in Early Victorian England," *Medical History* 36 (1992): 34–52; and Sander Gilman, "Electrotherapy: Then and Now," *History of Psychiatry* 19 (2008): 339–357. As Gilman points out, in his early years as a neurologist Freud tried using electrotherapy on some of his patients, with little success, before turning to hypnosis and eventually to psychoanalysis.
21. As one London physician put it, "In a great many cases a rapid cure will be effected by this faradic treatment, the patient being encouraged to speak after the current has been turned off three or four times." William Harris, *Electrical Treatment* (London: Cassell, 1908), 214. Treatments like this, administered beyond the limits that paying patients would tolerate, were, as we shall see, adapted and used in the First World War to treat shell shock.
22. Silas Weir Mitchell, *Wear and Tear: Or, Hints for the Overworked* (Philadelphia: Lippincott, 1871); S. Mitchell, *Fat and Blood: An Essay on the Treatment of Certain Forms of Neurasthenia and Hysteria* (Philadelphia: Lippincott, 1877).
23. Charlotte Perkins Stetson [Gilman], "The Yellow Wallpaper," *New England Magazine* 11 (1892); Virginia Woolf, *Mrs. Dalloway* (London: Hogarth, 1933). Weir Mitchell's patient, the social reformer Jane Addams, was another woman who unsuccessfully underwent the rest cure.
24. For Whitman's positive response to his time out West, see Walt Whitman, *Specimen Days* (Glasgow: Wilson and McCormick, 1883). For a feminist view, see Suzanne Poirier, "The Weir Mitchell Rest Cure: Doctor and Patients," *Women's Studies* 10 (1983): 15–40. See also Ellen Bassuk, "The Rest Cure: Repetition or Resolution of Victorian Women's Conflicts," *Poetics Today* 6 (1985): 245–257; and Michael Blackie, "Reading the Rest Cure," *Arizona Quarterly* 60 (2004): 57–85.
25. Massachusetts State Board of Charities, *8th Annual Report* (Boston: Wright and Potter, 1872), xli.
26. Hartford Retreat, *44th and 45th Annual Reports* (Hartford, CT: Case, Lockwood, and Brainard, 1870), 16. The cost of the improvements totaled $150,000, but it was an investment the asylum rapidly recouped. From 1871 to 1873 alone, profits were over $50,000. Lawrence Goodheart, *Mad Yankees: The Hartford Retreat for the Insane and Nineteenth-Century Psychiatry* (Amherst: University of Massachusetts Press, 2003), 162–163.
27. Edward Brush to Blumer, March 12, 1893, quoted in Ian Dowbiggin, *Keeping America Sane* (Ithaca, NY: Cornell University Press, 1997), 46.

28. G. Alder Blumer, "Presidential Address," *American Journal of Insanity* 60 (1903): 1–18, 16, 14.

29. Butler Hospital, *Annual Report,* 1904, 15–16, quoted in Dowbiggin, *Keeping America Sane,* 87.

30. Butler Hospital, *Annual Report,* 1912, 41–45, quoted in Dowbiggin, *Keeping America Sane,* 87.

31. Blumer to William Foster, August 8, 1916; Blumer to L. Vernon Briggs, March 4, 1920, quoted in Dowbiggin, *Keeping America Sane,* 93.

32. Blumer made these remarks in a discussion at the annual meeting of his professional colleagues. See "Discussion," *American Journal of Insanity* 65 (1908): 35.

33. Ronald Numbers, *Prophetess of Health: A Study of Ellen G. White* (New York: Harper and Row, 1976).

34. Howard Markel, *The Kelloggs: The Battling Brothers of Battle Creek* (New York: Pantheon, 2017), 139; "Kellogg, John Harvey," in *The Oxford Encyclopedia of Food and Drink in America,* ed. Andrew Smith, 2nd ed. (New York: Oxford University Press, 2012), 536.

35. The breakfast-cereal business was the brainchild of the younger Kellogg, Will Keith, who had previously operated in the shadow of his sibling. John Harvey Kellogg held his brother in low esteem and initially rejected Will's plans to enter the cereal business, though he later benefited financially from its success. Markel, *The Kelloggs,* provides the best overview of the sibling rivalry and operations of the sanitarium and cereal business.

36. Alan Gregg diary, October 8, 1934 entry, in Alan Gregg diary, 1934, RG 12 (Rockefeller Foundation Officers' Diaries), Rockefeller Foundation Records Collection, Rockefeller Archive Center, Sleepy Hollow, NY.

37. "Obituary: Walter Channing M.D.," *Boston Medical and Surgical Journal* 185 (December 15, 1921): 731–732.

38. Strahan, "The Propagation of Insanity," 334.

39. Dowbiggin, *Keeping America Sane,* 82–83. As Dowbiggin notes, Blumer had studied under Clouston in 1877, corresponded with him regularly over a long period, and remained a fervent admirer. On another aspect of Maudsley's influence on America, see Nicole Hahn Rafter, *Creating Born Criminals* (Urbana: University of Illinois Press, 1997), 81–82.

40. Blumer to Goodwin Brown, November 19, 1897, quoted in Dowbiggin, *Keeping America Sane,* 83.

41. G. Alder Blumer, "Marriage in Its Relation to Morbid Heredity," lecture delivered to the Providence Friday Night Club, September 20, 1900, quoted in Dowbiggin, *Keeping America Sane,* 84.

42. W. Duncan McKim, *Heredity and Human Progress* (New York: Putnam, 1900), iii, 188, 1, 129, 3, 7–8, 193.

43. "Review of W. Duncan McKim, *Heredity and Human Progress,*" *The Nation,* November 1, 1900, 849–850, 850.

44. Madison Grant, *The Passing of the Great Race* (New York: Scribner, 1918), 49.

45. Quoted in the *Washington Post,* November 18, 1915.

46. Davenport to Madison Grant, April 7, 1925, quoted in Charles Rosenberg, *No Other Gods: On Science and American Social Thought* (Baltimore, MD: Johns Hopkins University Press, 1976), 95–96. Davenport became the director of the Cold Spring Harbor Laboratory in 1904 and founded the Eugenics Record Office there in 1910. He maintained connections and corresponded with many Nazis from the 1930s until his death in 1944.

47. Michael Burleigh, "Psychiatry, German Society and the Nazi 'Euthanasia' Programme," *Social History of Medicine* 7 (1994): 213–228; Robert N. Proctor, *Racial Hygiene: Medicine under the Nazis* (Cambridge, MA: Harvard University Press, 1988); R. J. Lifton, *The Nazi Doctors: Medical Killing and the Psychology of Genocide* (New York: Basic Books, 1986).

48. Philip R. Reilly, *The Surgical Solution: A History of Involuntary Sterilization in the United States* (Baltimore, MD: Johns Hopkins University Press, 1991).

49. Francis A. Walker, "Restriction of Immigration," *Atlantic Magazine,* June 1896.

50. Quoted in Gerald Grob, "Class, Ethnicity, and Race in American Mental Hospitals, 1830–75," *Journal of the History of Medicine and Allied Sciences* 28 (1973): 207–229, 218.

51. Massachusetts, where "No Popery" sentiments were particularly rife, was especially exercised by the problem of Irish lunatics. See the discussion in Gerald Grob, *The State and the Mentally Ill: A History of Worcester State Hospital,1830–1920* (Chapel Hill: University of North Carolina Press, 1966), 136–166.

52. On race and asylums in the American South, and race and American psychiatry more generally, see Elodie Edwards-Grossi, *Bad Brains: La psychiatrie et la lute des Noirs américains pour la justice raciale* (Rennes, France: Presses Universitaires de Rennes, 2021). For an important study of segregated asylums in post–Civil War Virginia, see Wendy Conaver, *The Peculiar Institution and the Making of Modern Psychiatry, 1840–1880* (Chapel Hill: University of North Carolina Press, 2019).

53. Grob, "Class, Ethnicity, and Race in American Mental Hospitals," 228–229.

54. Martin Summers, *Madness in the City of Magnificent Intentions* (Oxford: Oxford University Press, 2019), 165–167.

55. Elodie Edwards-Grossi, "A Patient Labor: Le travail des patients noirs et les pratiques de résistance dans les asiles psychiatriques du sud, 1870–1940," *Revue Française d'Études Americaines* 3 (2019): 200–214; Summers, *Madness in the City,* 180–185.

56. T. O. Powell, "A Sketch of Psychiatry in the Southern States," Presidential Address, *American Journal of Insanity* 54 (1897): 21–36, 29. Powell was the superintendent of the Milledgeville State Asylum in Georgia.

57. J. F. Miller, "The Effects of Emancipation upon the Mental and Physical Health of the Negro of the South," *North Carolina Medical Journal* (1896), available at https://docsouth.unc.edu/nc/miller/miller.html. Miller was the superintendent of a segregated asylum.

58. Slave states like Virginia and Maryland attempted to stop the freeing of mentally ill slaves by passing laws forbidding it. See Summers, *Madness in the City,* 20–22.

59. See the discussion of the profession's deliberations on segregation in Summers, *Madness in the City,* 29–34.

60. Mary O'Malley, "Psychoses in the Colored Race: A Study in Comparative Psychiatry," *American Journal of Insanity* 71 (1914): 309–337, 310. For the nineteenth-century origins of these ideas, see Elodie Grossi, "Truth in Numbers? Emancipation, Race, and Federal Census Statistics in the Debates over Black Mental Health in the United States, 1840–1900," *Endeavour* 45 (2021), https://doi.org/10.1016/j.endeavour.2021.100766.

61. W. M. Bevis, "Psychological Traits of the Southern Negro with Observations as to Some of His Psychoses," *American Journal of Psychiatry* 78 (1921): 69–78, 70.

62. Andrew Scull and Diane Favreau, "'A Chance to Cut Is a Chance to Cure': Sexual Surgery for Psychosis in Three Nineteenth-Century Societies," *Research in Law, Deviance and Social Control* 8 (1987): 3–39.

63. See the discussion of these events in Andrew Scull, *The Insanity of Place/The Place of Insanity* (London: Routledge, 2006), ch. 11.
64. Carroll Smith-Rosenberg and Charles Rosenberg, "The Female Animal: Medical and Biological Views of Woman and Her Role in Nineteenth Century America," *Journal of American History* 60 (1973): 332–356.
65. F. Hoyt Pilcher, superintendent of the Kansas State Home for the Feeble Minded, performed 58 castrations and 150 sterilizations in total, beginning in 1894, before being removed from his position. See Michael Wehmeyer, "Eugenics and Sterilization in the Heartland," *Mental Retardation* 41 (2003): 57–60; and Mark Largent, *Breeding Contempt: The History of Coerced Sterilization in the United States* (New Brunswick, NJ: Rutgers University Press, 2007), 22.
66. Philip R. Reilly, "Involuntary Sterilization in the United States: A Surgical Solution," *Quarterly Review of Biology* 62 (1987): 153–170.
67. Alexandra Stern, "Sterilized in the Name of Public Health: Race, Immigration, and Reproductive Control in Modern California," *American Journal of Public Health* 95 (2005): 1128–1138, 1129.
68. Paul Popenoe, "Eugenic Sterilization in California," *Canadian Medical Association Journal* 18 (1928): 467–468; Joel Braslow, "In the Name of Therapeutics: The Practice of Sterilization in a California State Hospital," *Journal of the History of Medicine and Allied Sciences* 51 (1996): 29–51; Wendy Kline, *Building a Better Race: Gender, Sexuality, and Eugenics from the Turn of the Century to the Baby Boom* (Berkeley: University of California Press, 2005).
69. Margaret H. Smyth, "Psychiatric History and Development in California," *American Journal of Psychiatry* 94 (1938): 1234, quoted in Joel Braslow, *Mental Ills and Bodily Cures* (Berkeley: University of California Press, 1997), 58–59.
70. Caroline Wilhelm, a Red Cross social worker, to Albert Priddy, superintendent of the Virginia State Colony for Epileptics, May 1924, quoted in Harry Brunius, *Better for All the World: The Secret History of Forced Sterilization and America's Quest for Racial Purity* (New York: Knopf, 2006), 53.
71. Oliver Wendell Holmes, writing for the majority in *Buck v. Bell*, 274 US 200, 1927, 207.
72. I have drawn extensively here on the account presented in Brunius, *Better for All the World*, 23–77.

CHAPTER 3 Psychobiology

1. I follow here the argument made in Joel Braslow, *Mental Ills and Bodily Cures* (Berkeley: University of California Press, 1997), ch. 3.
2. Mary Baker Eddy, *Science and Health with Key to the Scriptures* (Boston: A. V. Stewart, 1912), xi.
3. See the discussion in Eric Caplan, *Mind Games: American Culture and the Birth of Psychotherapy* (Berkeley: University of California Press, 1998), ch. 6.
4. Charles A. Dana, "Discussion of 'Rest Treatment in Relation to Psychotherapy' by S. Weir Mitchell," *Transactions of the American Neurological Association* 34 (1908): 217. See also Edward Wyllys Taylor, "The Attitude of the Medical Profession toward the Psychotherapeutic Movement," *Journal of Nervous and Mental Disease* 35 (1908): 401–415.
5. Bernard Sachs, "Commentary," *Transactions of the American Neurological Association* 34 (1908): 218.

6. Bernard Sachs, "Commentary on 'The Attitude of the Medical Profession toward the Psychotherapeutic Movement," *Journal of Nervous and Mental Disease* 35 (1908): 405, comments that were echoed by Francis Dercum, J. K. Mitchell, and others.

7. See Andrew Scull, *Madhouses, Mad-Doctors and Madmen* (Philadelphia: University of Pennsylvania Press, 1981), 21–23; Caplan, *Mind Games*, 89–147; Nathan G. Hale, Jr., *Freud and the Americans: The Beginnings of Psychoanalysis in the United States, 1876–1917* (New York: Oxford University Press, 1971), 71–97, 116–150.

8. William F. Bynum, *Science and the Practice of Medicine in the Nineteenth Century* (Cambridge: Cambridge University Press, 1990).

9. Eric Engstrom, *Clinical Psychiatry in Imperial Germany* (Ithaca, NY: Cornell University Press, 2003).

10. For the emerging consensus, see Emil Kraepelin, *Psychiatrie: Ein Lehrbuch*, 7th ed. (Leipzig: Barth, 1904), 2: 376. But for a late dissent, see Max Nonne, *Syphilis und Nervensystem* (Berlin: Karge, 1902).

11. See Ludwig Fleck, *The Genesis and Development of a Scientific Fact* (Chicago: University of Chicago Press, 1979).

12. H. Noguchi and J. W. Moore, "A Demonstration of Treponema Pallidum in the Brain in Cases of General Paralysis," *Journal of Experimental Medicine* 17 (1913): 232–239. See Susan Lederer, *Subjected to Science: Human Experimentation in America before the Second World War* (Baltimore, MD: Johns Hopkins University Press, 1997).

13. Ira Van Gieson, "Notes and Comment," *American Journal of Insanity* 54 (1897–1898): 618–641, 618.

14. Richard Noll, *American Madness: The Rise and Fall of Dementia Praecox* (Cambridge, MA: Harvard University Press, 2011), provides the most thorough survey of the popularity of these speculations about an infectious origin of psychoses at the turn of the century.

15. See, for example, Adolf Meyer, "A Review of the Signs of Degeneration and of Methods of Registration," *American Journal of Insanity* 52 (1895): 344–363.

16. See Meyer to J. C. Sheller (a Zwinglian pastor and friend of Meyer's father), March 1893, 3, quoted in Noll, *American Madness*, 38.

17. See the discussion of these events in Noll, *American Madness*, ch. 2.

18. Adolf Meyer, "Aims and Plans of the Pathological Institute for the New York State Hospitals," in *The Collected Papers of Adolf Meyer*, ed. Eunice E. Winters, 4 vols. (Baltimore, MD: Johns Hopkins University Press, 1951), 2: 93.

19. Eunice E. Winters, "Adolf Meyer's Two and a Half Years at Kankakee," *Bulletin of the History of Medicine* 40 (1966): 441–458.

20. Theodore Lidz, "Adolf Meyer and the Development of American Psychiatry," *American Journal of Psychiatry* 123 (1966): 320–333, 323; Ruth Leys, "Meyer, Watson, and the Dangers of Behaviorism," *Journal of the History of the Behavioral Sciences* 20 (1984): 128–149.

21. Adolf Meyer, "Twenty-Fourth Anniversary of the Henry Phipps Psychiatric Clinic," in Winters, *The Collected Papers of Adolf Meyer*, 2: 228. See also Adolf Meyer and Edward B. Titchener, *Defining American Psychology: The Correspondence between Adolf Meyer and Edward Bradford Titchener*, ed. Ruth Leys and Rand B. Evans (Baltimore, MD: Johns Hopkins University Press, 1990), 40, 61, 83–89.

22. William James, *Pragmatism* (1907; New York, Meridian, 1958), 26.

23. Gerald Grob, *The State and the Mentally Ill: A History of the Worcester State Hospital in Massachusetts, 1830–1920* (Chapel Hill: University of North Carolina Press, 1966),

268–316, provides a detailed examination of the circumstances surrounding Meyer's appointment and work at Worcester, emphasizing that the Massachusetts State Board of Lunacy and Charity was responding in part to Weir Mitchell's stinging criticisms when it created his post.

24. Meyer to Blumer, October 23, 1894, Adolf Meyer Collection, Series I (Correspondence), Unit I / 355 (G. Alder Blumer), folder 1 (1894–1900), Alan Mason Chesney Medical Archives, Johns Hopkins Medical Institutions (hereafter Adolf Meyer Correspondence, Chesney Archives).

25. Noll, *American Madness*, 57.

26. Kraepelin's inaugural lecture at Dorpat, quoted in Eric J. Engstrom and Kenneth Kendler, "Emil Kraepelin: Icon and Reality," *American Journal of Psychiatry* 172 (2015): 1190–1196, 1191.

27. In an essay on "patterns of mental disorder" (Die Erscheinungsformen des Irreseins), Kraepelin acknowledged how difficult it was to separate dementia praecox and manic-depressive psychosis in clinical practice, but he insisted that "the two disease processes themselves are distinct." Emil Kraepelin, "The Manifestations of Insanity" (1920), trans. D. Beer, *History of Psychiatry* 3, no. 12, pt. 4 (1992): 509–529, 527.

28. This point is emphasized both by Engstrom and Kendler, "Emil Kraepelin"; and by Noll, *American Madness*.

29. Meyer to Hoch, October 15, 1896, Unit I / 1725 (August Hoch), folder 2 (1896–1898), Adolf Meyer Correspondence, Chesney Archives. Hoch (1868–1919) was then an assistant physician at the McLean Asylum in Boston, after a stint working under William Osler at Johns Hopkins Medical School. He had taken a two-year tour of European psychiatric facilities, including Kraepelin's clinic, between 1893 and 1895, with the blessing of McLean's superintendent.

30. Meyer to Hoch, October 15, 1896.

31. Grob, *The State and the Mentally Ill*, 287, 297–298.

32. For Van Gieson's vision of the Pathological Institute's future, see his "Remarks on the Scope and Organization of the Pathological Institute of the New York State Hospital," *State Hospitals Bulletin* 1 (1896): 255–274, 407–488.

33. Ira Van Gieson, "Correlation of Sciences in the Investigation of Nervous and Mental Diseases," *Archives of Neurology and Psychopathology* 1 (1898): 25–262, 48, quoted in Theodore M. Porter, *Genetics in the Madhouse: The Unknown History of Human Heredity* (Princeton, NJ: Princeton University Press, 2020), 262.

34. Van Gieson, "Correlation of Sciences," 64.

35. Gerald Grob, *Mental Illness and American Society 1875–1940* (Princeton, NJ: Princeton University Press, 1983), 128. He was not the last to flirt with this notion, as we shall see.

36. Brush to Alder Blumer, February 1, 1900, quoted in Ian Dowbiggin, *Keeping America Sane* (Ithaca, NY: Cornell University Press, 1997), 51n88.

37. Gerald Grob provides a valuable overview of Van Gieson's tenure and professional demise in Grob, *Mental Illness and American Society*, 127–129. I have supplemented this with Ian Dowbiggin's excellent *Keeping America Sane*, 51–58, which displays a thorough understanding of the relevant professional politics and infighting. David Rothman, *Conscience and Convenience: The Asylum and Its Alternatives in Progressive America* (Boston: Little, Brown, 1980), 127–131, 293–294, provides a waspish assessment of the fate of the renamed Psychiatric Institute from 1930 onward, by which time it had abandoned its Wards Island location for more desirable premises at Columbia Presbyterian Hospital, where it

confined its clinical attentions to a carefully selected affluent, young, and more mildly disturbed patient population and largely severed its ties to the state hospital system.

38. See the useful discussions in Rothman, *Conscience and Convenience*, 128–131; and Ruth Leys, "Adolf Meyer: A Biographical Note," in Meyer and Titchener, *Defining American Psychology*, 43–46.

39. For details of Meyer's involvement, see Armond Fields, *Katherine Dexter McCormick: Pioneer for Women's Rights* (Westport, CT: Praeger, 2003).

40. Smith Ely Jelliffe to Harry Stack Sullivan, June 1, 1937, Correspondence 1889–1940, box 19, Jelliffe Papers, Library of Congress, Washington, DC.

41. Franklin Ebaugh, "The Crisis in Psychiatric Education," *Journal of the American Medical Association* 99 (1932): 703–707.

42. Adolf Meyer, "Teaching Psychobiology," in *The Commonsense Psychiatry of Dr. Adolf Meyer*, ed. Alfred Lief, 433–448 (New York: McGraw Hill, 1948), 436.

43. Phyllis Greenacre, *Emotional Growth* (New York: International University Press, 1971), xxii.

44. Susan Lamb, *Pathologist of the Mind: Adolf Meyer and the Origins of American Psychiatry* (Baltimore, MD: Johns Hopkins University Press, 2014), 150.

45. Frank G. Ebaugh, "Adolf Meyer's Contribution to Psychiatric Education," *Bulletin of the Johns Hopkins Hospital* 89, supp. (1951): 71.

46. For a discussion, see Michael Gelder, "Adolf Meyer and His Influence on British Psychiatry," in *150 Years of British Psychiatry, 1841–1991*, ed. German Berrios and Hugh Freeman, 419–435 (London: Gaskell, 1991).

47. Jelliffe to Sullivan, June 1, 1937.

48. Adolf Meyer, "The Role of Mental Factors in Psychiatry" (1908), in Winter, *The Collected Papers of Adolf Meyer*, 2: 586.

49. William James, *Talks to Teachers on Psychology* (a course of lectures first delivered in the early 1890s), quoted in Caplan, *Mind Games*, 99.

50. William James, "Habit," *Popular Science Monthly* 30 (1886–1887): 437, quoting from William B. Carpenter, *Principles of Mental Physiology* (New York: Appleton, 1875).

CHAPTER 4 Freud Visits America

1. Quoted in Saul Rosenzweig, *Freud, Jung, and Hall the Kingmaker: The Historic Expedition to America (1909)* (Seattle: Hogrefe and Huber, 1992), 58, 55.

2. Quoted in Rosenzweig, *Freud, Jung, and Hall the Kingmaker*, 79.

3. Freud to Hall, December 29, 1908, reprinted in Rosenzweig, *Freud, Jung, and Hall the Kingmaker*, 342.

4. Freud to Ferenczi, January 10, 1909, in Ernest Jones, *The Life and Work of Sigmund Freud*, vol. 2: *The Years of Maturity* (New York: Basic Books, 1955), 54.

5. Jung to Freud, January 7, 1909, reprinted in *The Freud/Jung Letters*, ed. William McGuire (Princeton, NJ: Princeton University Press, 1974), 193–194. The French psychologist Pierre Janet was Freud's great contemporary rival in the field of psychotherapeutics. It was he who coined the terms "dissociation" and "subconscious," and he had given a total of fifteen lectures on his theories at Harvard in 1905 and 1906, the visit to which Jung was referring. Thirty years later, Harvard awarded Janet an honorary doctorate.

6. Meyer's 1907 report and his April 29, 1909, letter on the case to Stanley's mother are reprinted in Richard Noll, "Styles of Psychiatric Practice, 1906–1925: Clinical Evaluations of the Same Patient by James Jackson Putnam, Adolph Meyer, August Hoch, Emil

Kraepelin, and Smith Ely Jelliffe," *History of Psychiatry* 10 (1999): 145–189, 155–161, quotes on 155–156, 159.

7. For details on Meyer's machinations, see Armond Fields, *Katharine Dexter McCormick: Pioneer for Women's Rights* (Westport, CT: Praeger, 2003).

8. For transcripts of the examinations and reports of some of the other psychiatrists who examined Stanley McCormick over the years, see Noll, "Styles of Psychiatric Practice." Kraepelin's conclusions about the patient's prognosis, quoted here, appear on 171–172.

9. Noll, "Styles of Psychiatric Practice." Kemp's involvement, under the supervision of his former boss, William Alanson White, superintendent of St. Elizabeths Hospital in Washington, DC, is recorded in William A. White, personal correspondence, NA identifier 2645713, NARA RG 418 (St. Elizabeths Hospital), National Archives, Washington, DC. White was paid a monthly retainer of $500 (approximately $7,500 today), plus a fee of $5,000 for each visit to Riven Rock.

10. Deirdre Bair, *Jung: A Biography* (Boston: Little, Brown, 2003), indicates that Jung's private opinion was very different. Medill, he concluded, was being driven mad by his mother, Kate McCormick, a "real power devil," someone "whose possessiveness had crushed her son's spirit and defeated all his efforts at independence." Medill's diagnosis as a chronic alcoholic was designed to free him from his job at the *Chicago Tribune* and, not coincidentally, from the clutches of his mother. All quotes in this note are from 709n63.

11. At one point, Medill McCormick informed his wife that Jung had recommended he acquire a string of mistresses as part of his therapy, a form of extracurricular activity Jung could recommend from extensive personal experience.

12. Rosenzweig, *Freud, Jung, and Hall the Kingmaker,* 280n23.

13. Hall to Freud, February 16, 1909; Freud to Hall, February 28, 1909, reprinted in Rosenzweig, *Freud, Jung, and Hall the Kingmaker,* 343, 344.

14. There is persuasive evidence that Hall did not initially consider Freud to be the most important figure in the field he sought to bring to Worcester. When he first wrote to invite Freud, he simultaneously invited his own mentor, the German psychologist Wilhelm Wundt. Wundt he tempted with a $750 honorarium, to no avail, whereas Freud was originally offered $400. Only after receiving Wundt's rejection did Hall increase his offer to Freud.

15. Boas had begun his American academic career at Clark in 1888 but had resigned four years later in protest at what he thought were Hall's attempts to interfere with his academic freedom. By now, he headed a department at Columbia and was regarded as the leading anthropologist in North America.

16. On Meyer's complicated relationship with Freud, see Ruth Leys, "Meyer's Dealings with Jones: A Chapter in the History of the American Response to Psychoanalysis," *Journal of the History of the Behavioral Sciences* 17 (1981): 445–465; and Nathan Hale, *The Rise and Crisis of Psychoanalysis in the United States* (Oxford: Oxford University Press, 1995), 168–172.

17. Quoted in Jones, *Life and Work of Sigmund Freud,* 2: 80.

18. Freud to Jung, January 17, 1909, *The Freud/Jung Letters,* 196.

19. Freud to Zweig, March 5, 1939, *The Letters of Sigmund Freud and Arnold Zweig,* ed. Ernst L. Freud (New York: Harcourt Brace, 1970), 170. See also Ernest Jones, *Free Associations: Memories of a Psychoanalyst* (London: Hogarth Press, 1959), 191.

20. Freud to Jones, September 25, 1924, in *The Complete Correspondence of Sigmund Freud and Ernest Jones, 1908–1939,* ed. R. Andrew Paskauskas (Cambridge, MA: Belknap Press of Harvard University Press, 1993), 552.

21. Sigmund Freud to Anna Freud, December 6, 1920, quoted in Peter Gay, *Freud: A Life for Our Time* (New York: Norton, 2006), 570.
22. James to Theodore Flournoy, September 28, 1909, in *The Letters of William James*, ed. Henry James (Boston: Atlantic Monthly Press, 1920), 2: 328.
23. The pseudonym Anna O. was created by moving her initials (B.P.) back one place in the alphabet.
24. Albrecht Hirschmüller, *Physiologie und Psychoanalyse im Leben und Werk Josef Breuers* (Bern: Hans Huber, 1978); Dora Edinger, *Bertha Pappenheim* (Highland Park, IL: Congregation Solel, 1968), 15.
25. For an overview, see Ward Cromer and Paula Anderson, "Freud's Visit to America: Newspaper Coverage," *Journal of the History of the Behavioral Sciences* 6 (1970): 549–557. The coverage in the *Boston Evening Transcript* was particularly positive, possibly because the reporter it sent, Adelbert Albrecht, spoke German and was apparently already familiar with Freud's work.
26. The lectures are reprinted and retranslated in full in Rosenzweig, *Freud, Jung, and Hall the Kingmaker*, 397–498, and I quote from this text.
27. Josef Breuer and Sigmund Freud, *The Standard Edition of the Complete Psychological Works of Sigmund Freud*, Vol. 2., *Studies on Hysteria* (189 studies on hysteria on line5) (London: Hogarth Press), 160.
28. See Eric Caplan, "Embracing Psychotherapy: The Emmanuel Movement and the American Medical Profession," in *Mind Games: American Culture and the Birth of Psychotherapy* (Berkeley: University of California Press, 1998), ch. 6.
29. Interview with Adelbert Albrecht, *Boston Evening Transcript*, September 11, 1909.
30. Freud, quoted in Bair, *Jung: A Biography*, 267–268.
31. George Makari, *Revolution in Mind: The Creation of Psychoanalysis* (London: Duckworth, 2008), 249.
32. James Joyce, who was briefly a recipient of Edith Rockefeller McCormick's largesse while in Zürich, commented scathingly about "a sanatorium where a certain Dr Jung (the Swiss Tweedledum who is not to be confused with the Viennese Tweedledee, Dr Freud) amuses himself at the expense (in every sense of the word) of ladies and gentlemen who are troubled with bees in their bonnets." When Edith abruptly cut off his allowance without explanation, Joyce took his revenge in his characteristic way, reinventing her as Mrs. Mervyn Talboys, a "whip-cracking, sadistic dominatrix," in *Ulysses*, Episode 15: Circe. Compare Bair, *Jung: A Biography*, 302–303.
33. I have drawn the materials in the preceding paragraphs from the standard, certainly the lengthiest, tome on its subject, Bair, *Jung: A Biography*, 267–273.
34. I have relied here on the extended discussion of Edith Rockefeller McCormick in Ron Chernow's biography of John D. Rockefeller, *Titan: The Life of John D. Rockefeller, Sr.* (New York: Random House, 1998); and Bair, *Jung: A Biography*.
35. Gay, *Freud*, 541–543.

CHAPTER 5 The Germ of Madness

1. Russell Maulitz, "'Physician versus Bacteriologist': The Ideology of Science in Clinical Medicine," in *The Therapeutic Revolution: Essays in the Social History of American Medicine*, ed. C. E. Rosenberg and M. Vogel, 91–108 (Philadelphia: University of Pennsylvania Press, 1979).
2. "When we recall the known role which infections and intoxications play in the production of delirium, confusion, and stupor, it is not going too far to infer that de-

mentia praecox is probably due to a toxin—a toxin which at first calls forth by its action upon the cortical neurons hallucinations and their dependent delusions, and later on, in given cases, brings their destruction." F. X. Dercum, "The Heboid-Paranoid Group (Dementia Praecox)," *American Journal of Insanity* 63 (1906): 541–559, quoted in Richard Noll, *American Madness: The Rise and Fall of Dementia Praecox* (Cambridge, MA: Harvard University Press, 2011), 131.

3. Farrar stayed at Trenton from 1913 until 1916 and would go on to occupy the powerful post of editor of the *American Journal of Psychiatry* from 1931 to 1965, as well as becoming medical director of the Toronto Psychiatric Hospital.

4. Independently, a Chicago physician, Bayard Holmes, whose son had succumbed to dementia praecox, also pursued surgical remedies for the disorder. His first patient was his son, who died during the operation. Holmes persisted and published his claims and theories in a long series of papers. Richard Noll, who brought his work to modern notice, acknowledges that he was simply ignored by his fellow professionals and had no major influence on American psychiatry. See Noll, *American Madness*, ch. 7.

5. Frank Billings, *Focal Infection: The Lane Medical Lectures* (New York: Appleton, 1916). Billings served as president of the American Medical Association in 1902.

6. Billings, *Focal Infection*, 131, 128–129. The enormous vogue for tonsillectomies that began in the 1910s to a large degree reflected the broad credence doctors gave to the idea of focal sepsis. On Billings, see Edwin Hirsch, *Frank Billings: The Architect of Medical Education* (Chicago: University of Chicago Press, 1966).

7. William Frederick Braasch, *Early Days in the Mayo Clinic* (Springfield, IL: Charles Thomas, 1969), 33, 40–50, 68–69, 74–75; Cotton to Meyer, April 8, 1918, Adolf Meyer Collection, Series I (Correspondence), Unit I/767 (Henry A. Cotton), folder 14 (1918), Alan Mason Chesney Medical Archives, Johns Hopkins Medical Institutions (hereafter Adolf Meyer Correspondence, Chesney Archives).

8. Henry Cotton, *The Defective, Delinquent, and Insane* (Princeton, NJ: Princeton University Press, 1923), 42.

9. Henry Cotton, "The Relation of Chronic Sepsis to the So-Called Functional Mental Disorders," *Journal of Mental Science* 69 (1923): 434–465, 437.

10. Cotton, "The Relation of Chronic Sepsis."

11. Cotton, *The Defective, Delinquent, and Insane*, 66.

12. *Annual Report of the New Jersey State Hospital* (Trenton, NJ, published by the state, 1921), 23, available at New Jersey State Library Digital Collections, https://dspace.njstatelib.org/xmlui/handle/10929/46204; Henry A. Cotton, "The Etiology and Treatment of the So-Called Functional Psychoses," *American Journal of Psychiatry* 79 (1922): 157–210, 182–183.

13. Mikhail Rotov, "The History of Trenton State Hospital, 1848–1976," unpublished paper, n.d., 46, in possession of author.

14. Hubert Work, speech at the opening of the new psychopathic wards at Trenton State Hospital, quoted in Burdette G. Lewis, "Winning the Fight against Mental Disease," *Review of Reviews* 65 (April 1922): 11; "Association and Hospital Notes and News," *American Journal of Insanity* 79 (1922): 114–115.

15. Royal Copeland, "Your Health," *Trenton Evening Times*, September 23, 1925; Royal Copeland, "Your Health," *Trenton Evening Times*, October 21, 1925; "Physicians Defend Trenton Asylum; Copeland and Other Witnesses Say It Leads the World in Caring for the Insane," *New York Times*, September 24, 1925. Copeland was equally enthusiastic three years earlier.

16. See Kellogg to Cotton, March 26, 1926, Trenton State Hospital Archives.

17. See the complaints from Cotton's psychiatric colleagues that they were being besieged by families "demanding that we shall adopt these theories and follow the methods pursued at Trenton." Unsigned, "Discussion—Functional Psychoses," following Cotton, "Etiology and Treatment," and other papers, *American Journal of Psychiatry* 79 (1922): 195–210, 199, 204.
18. Quoted in Albert Shaw, "Physical Treatment for Mental Disorders: A Summary of Expert Comments on Dr. Cotton's Work at Trenton," *Review of Reviews* 66 (December 1922): 625–636, 634.
19. Quoted in Shaw, "Physical Treatment for Mental Disorders," 634. *Review of Reviews* was the quintessential magazine of Progressive Era reformers who sought to introduce modern "science" into a political system polluted by patronage and machine politics.
20. *Annual Report of the New Jersey State Hospital* (Trenton, NJ, published by the state, 1920), 29, https://dspace.njstatelib.org/xmlui/handle/10929/46203, 18–19, 21.
21. Henry Cotton and J. W. Draper, "What Is Being Done for the Insane by Means of Surgery: Analysis of One Hundred and Twenty-Five Laparotomies," *Transactions of the Section on Gastroenterology and Proctology of the American Medical Association* 71 (1920): 143–157, 144.
22. Thomas Quinn Beesley, "When the Brain Is Sick," *New York Times*, June 18, 1922. Cotton's lectures were published as Cotton, *The Defective, Delinquent, and Insane*.
23. Cotton, "The Relation of Chronic Sepsis," 438, 458–459.
24. Cotton, "The Relation of Chronic Sepsis"; Unsigned, "Chronic Sepsis and Mental Disease," Occasional Note, *Journal of Mental Science* 69 (1923): 502–504.
25. "Medico-Psychological Association of Great Britain and Ireland," Notes and News, *Journal of Mental Science* 69 (1923): 528–580, 555–557, 559, 555.
26. "Medico-Psychological Association of Great Britain and Ireland," 553–560.
27. "Medico-Psychological Association of Great Britain and Ireland," 569–570.
28. Adolf Meyer to George Kirby, January 15, 1919, Unit I / 2110 (George H. Kirby), folder 9 (1918–1919); Cotton to Meyer, January 18, 1919, and Meyer to Cotton, January 21, 1919, Unit I / 767 (Henry A. Cotton), folder 16 (1918–1920); all in Adolf Meyer Correspondence, Chesney Archives.
29. Joseph Schumpeter, *Ten Great Economists from Marx to Keynes* (New York: Oxford University Press, 1951), 223.
30. Cotton, *The Defective, Delinquent, and Insane*, 157. Margaret Fisher is identified here as "Case 24."
31. Cotton, *The Defective, Delinquent, and Insane*, 157.
32. Bloomingdale clinical records, as summarized in Cotton, *The Defective, Delinquent, and Insane*, 158.
33. Cotton, *The Defective, Delinquent, and Insane*, 158–159.
34. Cotton, *The Defective, Delinquent, and Insane*, 159.
35. I. Fisher to M. H. Fisher, August 16, 1919, Irving Fisher Papers, MS 212, Manuscripts and Archives Repository, Yale University (hereafter Fisher Papers, Yale University).
36. I. Fisher to M. H. Fisher, August 18, 1919, Fisher Papers, Yale University.
37. Cotton, *The Defective, Delinquent, and Insane*, 159.
38. I. Fisher to M. H. Fisher, October 7, 1919, Fisher Papers, Yale University.
39. Cotton, *The Defective, Delinquent, and Insane*, 159.
40. Robert Loring Allen, *Irving Fisher: A Biography* (New York: Wiley, 1993).
41. Henry Cotton, "The Relation of Oral Infections to Mental Diseases," *Journal of Dental Research* 1 (1919): 269–313, 273.

42. Quoted in Rotov, "A History of Trenton State Hospital," 16.
43. Cotton and Draper, "What Is Being Done for the Insane."
44. H. Cotton, J. W. Draper, and J. Lynch, "Intestinal Pathology in the Functional Psychoses: Preliminary Report of Surgical Findings, Procedures, and Results," *Medical Record* 97 (1920): 719–725.
45. *Annual Report of the New Jersey State Hospital*, 1921, 24.
46. Cotton, "Etiology and Treatment," 186.
47. Cotton, "The Relationship of Chronic Sepsis," 454, 457.
48. Cotton, "The Relationship of Chronic Sepsis," 457.
49. See the comments of Chalmers Watson and William Hunter in "Medico-Psychological Association of Great Britain and Ireland," 555–557.
50. Nicholas Kopeloff and Clarence Cheney, "Studies in Focal Infection: Its Presence and Elimination in the Functional Psychoses," *American Journal of Psychiatry* 79 (1922): 139–156.
51. Unsigned, "Discussion—Functional Psychoses," 205–207.
52. Unsigned, "Discussion—Functional Psychoses," 206.
53. "The Seventy-Eighth Annual Meeting of the American Psychiatric Association," Notes and Comment, *American Journal of Psychiatry* 79 (1922): 110–111.
54. Cotton had even gone so far as to name his second son after Meyer, with the great man's permission.
55. "Trenton State Hospital Survey—1924–1926," unpublished report by Dr. Phyllis Greenacre with the cooperation of Dr. Adolf Meyer. Only one copy of the original typescript has survived. It now resides in the Trenton State Hospital Archives, control number SZCTR002, box 6, New Jersey State Archives. Meyer made meticulous pencil notes of the meeting, which survive in his papers: Unit I / 767 (Henry A. Cotton), folder 23 (1925–1926), Adolf Meyer Correspondence, Chesney Archives. I also discussed the meeting with Phyllis Greenacre in an interview in New York on December 22, 1983.
56. William Hunter, "Chronic Sepsis as a Cause of Mental Disorder," *Journal of Mental Science* 73 (1927): 549–563, 551.
57. Henry Cotton, "Oral Infections," reprint of an address to the Fifty-Fifth Annual Meeting of the New Jersey Public Health and Sanity Association, Asbury Park, December 3, 1929, 10, Trenton State Hospital Archives.
58. Henry Cotton, "Gastro-Intestinal Stasis in the Psychoses," *Proceedings of the Fifth International Congress of Physiotherapy*, September 14–18, 1930, 8.
59. Cotton, "Gastro-Intestinal Stasis," 4–6.
60. Solomon Katzenelbogen, "Trenton State Hospital," unpublished typescript, n.d. [late 1930], Unit I / 2024 (S. Katzenelbogen), folder 20 (Miscellaneous ms. and notes), Adolf Meyer Correspondence, Chesney Archives, emphasis in the original. I am grateful to the late Gerald Grob for drawing my attention to this document and providing a photocopy of it.
61. Katzenelbogen, "Trenton State Hospital."
62. That, at least, was Phyllis Greenacre's conclusion when I interviewed her in December 1983, and it is of a piece with Meyer's general dislike of washing professional dirty laundry in public.
63. "Obituary: Henry A. Cotton," *Trenton Evening Times*, May 9, 1933.
64. Adolf Meyer, "In Memoriam: Henry A. Cotton," *American Journal of Psychiatry* 90 (1934): 921–923.
65. Emil Frankel, "Study of 'End Results' of 645 Major Operative Cases and 407 Non-operative Cases Treated at Trenton State Hospital, 1918–1932," unpublished report, 1932,

State of New Jersey, Department of Institutions and Agencies, copy in Trenton State Hospital Archives, control number SZCTR002, box 6, New Jersey State Archives.

CHAPTER 6 Body and Mind

1. Julius Wagner-Jauregg, "The Treatment of Dementia Paralytica by Malaria Inoculation," *Nobel Lectures: Physiology or Medicine, 1922–1941* (Amsterdam: Elsevier, 1965), 161. Robert Koch, who had discovered the bacillus that causes tuberculosis in 1882, developed tuberculin in 1890 from cultures of the bacillus. His initial claim that it would serve as an effective therapy against the disease was greeted with euphoria, but it was rapidly shown to be wrong. Koch's attempts to profit from his discovery had severely deleterious effects on his reputation, and he fled to Egypt. Remarkably, tuberculin did subsequently prove to have some use in medicine when employed as a diagnostic skin test for the presence of tuberculosis.

2. Five different *Plasmodium* parasites can cause malaria. So-called tertian malaria exists in two forms, benign and malignant. "Benign" tertian malaria is a serious disease but less immediately life-threatening than its counterpart, though the parasites that cause it can remain dormant in the liver, and symptoms can reappear months or years later. Malignant tertian malaria is much more dangerous, often life-threatening. Wagner-Jauregg was not always careful about which form of the disease he injected, a mistake that could and did have fatal consequences.

3. Quoted in Marc Roudebush, "A Battle of Nerves: Hysteria and Its Treatment in France during World War I," in *Traumatic Pasts,* ed. Mark Micale and Paul Lerner (Cambridge: Cambridge University Press, 2001), 253–279, 269; Clovis Vincent, *Le Traitement des phenomenes hysteriques par la reeducation intensive* (Tours: Arrault, 1916).

4. A. D. Adrian and L. R. Yealland, "Treatment of Some Common War Neuroses," *Lancet* 1 (1917): 867–871; L. R. Yealland, *Hysterical Disorders of Warfare* (London: Macmillan, 1918). For a detailed discussion of the Kaufmann cure, see Paul Lerner, *Hysterical Men: War, Psychiatry, and the Politics of Trauma* (Ithaca, NY: Cornell University Press, 2003), 102–113.

5. On Wagner-Jauregg's trial, see Hans-Georg Hofer, *Nervenschwäche und Krieg: Modernitätskritik und Krisenbewältigung in der österreichischen Psychiatrie (1880–1920)* (Vienna: Böhlau, 2004), 189–193.

6. George Kirby and H. A. Bunker, "Types of Therapeutic Response Observed in the Malaria Treatment of General Paralysis," *American Journal of Psychiatry* 83 (1926): 205–226.

7. J. Wagner-Jauregg, "The Treatment of General Paralysis by Inoculation of Malaria," *Journal of Nervous and Mental Disease* 55 (1922): 369–375, 375.

8. See White's correspondence on the matter, reprinted in Gerald Grob, *The Inner World of American Psychiatry, 1890–1940: Selected Correspondence* (New Brunswick, NJ: Rutgers University Press, 1985), 124–125.

9. Ernest Kusch, D. F. Milam, and W. K. Stratman-Thomas, "General Paresis Treated by Mosquito-Inoculated Vivax (Tertian) Malaria," *American Journal of Psychiatry* 93 (1936): 619–624, provided detailed instructions on how to set up and utilize such a mosquito colony, and the authors claimed that using live mosquitos was the therapeutically superior approach.

10. L. C. Cook, "The Place of Physical Treatments in Psychiatry," *Journal of Mental Science* 104 (1958): 933–942, 936. See also H. Goldsmith, "A Plea for Standardized and In-

tensive Treatment of the Neurosyphilitic and Paretic," *American Journal of Psychiatry* 72 (1925): 251–261, 256.

11. Joel Braslow, *Mental Ills and Bodily Cures* (Berkeley: University of California Press, 1997).

12. R. S. Carroll, "Aseptic Meningitis in Combatting the Dementia Praecox Problem," *New York Medical Journal* (October 3, 1923): 407–411; E. S. Barr and R. G. Barry, "The Effect of Producing Aseptic Meningitis upon Dementia Praecox," *New York State Journal of Medicine* 26 (1926): 89–92, 89.

13. For the first report of the use of such a device, see C. A. Neymann and S. L. Osbourne, "Artificial Fever Produced by High-Frequency Currents, Preliminary Reports," *Illinois Medical Journal* 56 (1929): 199–203. Seth Howes to Winfred Overholser, October 6, 1943, NARA RG 418 (St. Elizabeths Hospital), Entry 18 (Treatment Files), National Archives, Washington, DC.

14. Seth Howes to Winfred Overholser, October 6, 1943.

15. H. J. Macbride and W. L. Templeton, "The Treatment of General Paralysis of the Insane by Malaria," *Journal of Neurology and Psychopathology* 5 (1924): 13–27; L. E. Hinsie, "Malaria Treatment of Schizophrenia," *Psychiatric Quarterly* 1 (1927): 210–214.

16. One member of the jury that awarded the prize, the Swedish professor of psychiatry B. Gadelius, objected to the award to a "physician who injected malaria into a paralytic because he was in his eyes a criminal." Quoted in Edward Brown, "Why Wagner-Jauregg Won the Nobel Prize for Discovering Malaria Therapy for General Paresis of the Insane," *History of Psychiatry* 11 (2000): 371–382, 380.

17. J. R. Driver, J. A. Gammel, and L. J. Karnosh, "Malaria Treatment of Central Nervous System Syphilis," *Journal of the American Medical Association* 87 (1926): 1821–1827, 1822, 1827.

18. Gerald Grob, *Mental Illness and American Psychiatry, 1875–1940* (Princeton, NJ: Princeton University Press, 1983), 295. Such objections apply with equal force to the whole array of somatic treatments that appeared on the scene in the 1920s and 1930s.

19. These are Wagner-Jauregg's own reports of these patients' fates, as summarized in Magda Whitrow, "Wagner-Jauregg and Fever Therapy," *Medical History* 34 (1990): 294–310, 304, my emphasis.

20. Julius Wagner-Jauregg, *Fieber- und Infektionstherapie* (Vienna: Weidmann, 1936), 135.

21. Julius Wagner-Jauregg, *Lebenserinnerungen* (Vienna: Springer, 1950), 163–166.

22. Julius Wagner-Jauregg, "The History of the Malaria Treatment of General Paralysis," *American Journal of Psychiatry* 102 (1946): 577–582.

23. Francisco Lopez-Munoz, Ronaldo Ucha-Udabe, and Cecilio Alamo, "The History of Barbiturates a Century after Their Clinical Introduction," *Neuropsychiatric Disease and Treatment* 1 (2005): 329–343.

24. Jakob Kläsi, "Über Somnifen, eine medikamentose Therapie schizophrener Aufregungszustande," *Schweizer Archiv für Neurologie und Psychiatrie* 8 (1921): 131–134, 131, author's translation.

25. Jakob Kläsi, "Einiges über Schizophreniebehandlung," *Zeitschrift fur die gesamte Neurologie und Psychiatrie* 78 (1922): 606–620; Jakob Kläsi, "Über die therapeutische Anwendung der 'Dauernarkose' mittels Somnifens bei Schizophrenen," *Zeitschrift für die gesamte Neurologie und Psychiatrie* 74 (1922): 557–592.

26. See Kläsi, "Über die therapeutische Anwendug der 'Dauernarkose.'"

27. G. Windholz and L. H. Witherspoon, "Sleep as a Cure for Schizophrenia: A Historical Episode," *History of Psychiatry* 4 (1993): 83–93.

28. Windholz and Witherspoon, "Sleep as a Cure for Schizophrenia," 90. This paper provides a useful survey of the various attempts to make use of prolonged sleep therapy and the complications associated with all of them.
29. Edward Shorter and David Healy, *Shock Therapy: A History of Electroconvulsive Treatment in Mental Illness* (New Brunswick, NJ: Rutgers University Press, 2007), 16–17.
30. Elliot S. Valenstein, *Great and Desperate Cures* (New York: Basic Books, 1986), 56–57. Valenstein observed this scene at the Winter Veterans Administration Hospital in Topeka, Kansas. The memory remained vivid decades later: "With all these people—tossing, moaning, twitching, shouting, grasping—I felt as though I were in the midst of Hell as drawn by Gustav Doré for Dante's *Divine Comedy*" (57).
31. Isabel Wilson, *A Study of Hypoglycaemic Shock Treatment in Schizophrenia* (London: Her Majesty's Stationery Office, 1936), 10, 15, 31. Wilson gave the treatment her personal stamp of approval. Soon thereafter, English mental hospitals began to experiment with it.
32. L. Jessner and V. G. Ryan, *Shock Treatment in Psychiatry: A Manual* (New York: Grune and Stratton, 1941), 5–11.
33. Jay Shurley and Earl Bond, *Veterans Administration Technical Bulletin: Insulin Shock Therapy in Schizophrenia* (Washington, DC: Veterans Administration, 1946), 10.
34. Benjamin Wortis, translation of a lecture Sakel delivered in Paris at the University Clinic on July 11, 1937. The text was distributed when Sakel came to America later that year and lectured at the Harlem River State Hospital in New York on November 28, 1937. A copy is preserved in NARA RG 418 (St. Elizabeths Hospital), Entry 18 (Treatment Files), National Archives, Washington, DC.
35. These communications were subsequently issued as a small booklet of eleven pages, with a foreword endorsing his findings by Otto Pötzl, who had succeeded Wagner-Jauregg as professor of psychiatry at the University of Vienna.
36. Bernard Glueck, "The Hypoglycemic State in the Treatment of Schizophrenia," *Journal of the American Medical Association* 107 (1936): 1029–1031, quoted in Valenstein, *Great and Desperate Cures,* 48. This paragraph and portions of the next are indebted to Valenstein's account.
37. M. Sakel, "Origin and Nature of Hypoglycemic Therapy for the Psychoses," *Archives of Neurology and Psychiatry* 38 (1937): 188–203.
38. "Bedside Miracle," *Reader's Digest* 35 (November 1939): 73–75; "Insulin for Insanity," *Time,* January 25, 1937, 26, 28. Elliot Valenstein, in *Great and Desperate Cures,* 303n21, records many similar pronouncements in a wide variety of periodicals.
39. Shorter and Healy, *Shock Therapy,* 56. Having arrived in the United States essentially penniless, at his death Sakel was not only wealthy enough to provide for the establishment of a well-endowed foundation to promote insulin coma therapy, but he also had earned enough to leave the then-enormous sum of $2 million to his longtime companion, Marianne Englander. Much of this money came from the substantial sums he charged rich patients for insulin sub-coma "therapy," a homeopathic treatment he administered in his rooms at a hotel on Park Avenue. For those requiring actual comas, Sakel made arrangements to administer it, for larger sums still, at the private Slocum Clinic in Beacon, New York. See Edward Shorter, "Sakel versus Meduna," *Journal of ECT* 25 (2009): 12–14.
40. Benjamin Malzberg, "Outcome of Insulin Treatment of One Thousand Patients with Dementia Praecox," *Psychiatric Quarterly* 12 (1938): 528–553.

41. California Department of Institutions, *Report for Governor's Council on Activities during May 1939* (Sacramento: California State Printing Office, 1939).
42. Lawrence Kolb, *Shock Therapy Survey* (Washington, DC: US Public Health Service, 1941). The survey was prepared by Kolb, then assistant surgeon general. A summary appeared the next year: Lawrence Kolb and Victor Vogel, "The Use of Shock Therapy in 305 Mental Hospitals," *American Journal of Psychiatry* 99 (1942): 90–100.
43. Jessner and Ryan, *Shock Treatment in Psychiatry*, 53. That discrepancy emerged even though insulin treatment directly led to some deaths.
44. For examples of such studies, see J. S. Gottlieb and P. E. Huston, "Treatment of Schizophrenia: Follow-Up Results in Cases of Insulin Shock Therapy and in Control Cases," *Archives of Neurology and Psychiatry* 49 (1943): 266–271, based on experiences at Iowa Psychopathic Hospital; and W. Libertson, "A Critical Analysis of Insulin Therapy at Rochester State Hospital," *Psychiatric Quarterly* 15 (1941): 635–647, reporting on his results at Rochester State Hospital in upstate New York.
45. Manfred Sakel, "The Methodical Use of Hypoglycemia in the Treatment of Psychoses," *American Journal of Psychiatry* 94 (1937): 111–129, 121; D. Ruslander, "Observations in the Hypoglycemic Treatment of 55 Cases of Schizophrenia," *American Journal of Psychiatry* 94, (1938): 1337–1345, 1339; Lothar Kalinowsky and Paul Hoch, *Shock Treatments and Other Somatic Procedures in Psychiatry* (New York: Grune and Stratton, 1946), 93.
46. Charles Burlingame, "Insanity and Insulin Shock Treatment for Schizophrenia," *Forum* (1937): 98–102. Copy in Adolf Meyer Collection, Series I (Correspondence), Unit I / 557 (C. Charles Burlingame), folder 2 (correspondence with Meyer and related materials 1937–1938), Alan Mason Chesney Medical Archives, Johns Hopkins Medical Institutions.
47. A. B. Baker, "Cerebral Damage in Hypoglycemia: A Review," *American Journal of Psychiatry* 96 (1939): 109–127, 109.
48. Stanley Cobb, "Review of Neuropsychiatry," *Archives of Internal Medicine* 62 (1938): 883–899, 897.
49. Stanley Cobb, "Shock Therapy," *New England Journal of Medicine* 217 (1937): 195–196.
50. Stanley Cobb, "Review of Neuropsychiatry," *Archives of Internal Medicine* 60 (1937): 1098–1110; 62 (1938): 883–899; 64 (1939): 1328–1339; 66 (1940): 1341–1354; Cobb, "Shock Therapy."
51. See, for example, A. B. Baker, "Cerebral Lesions in Hypoglycemia," *Archives of Pathology* 26 (1938): 765–776; A. B. Baker and N. H. Lufkin, "Cerebral Lesions in Hypoglycemia," *Archives of Pathology* 23 (1937): 190–201; A. Ferraro and G. A. Jervis, "Pathologic Considerations on Insulin Treatment of Schizophrenia," *American Journal of Psychiatry* 96 (1939): 103–108; E. Gellhorn, "Hypoglycemia and Anoxia and Nervous System," *Archives of Neurology and Psychiatry* 40 (1938): 125–146; H. E. Himwich, K. M. Bowman, J. Wortis, and J. E. Farkas, "Brain Metabolism during the Hypoglycemic Treatment of Schizophrenia," *Science* 86 (1937): 271–272; B. L. Keyes, H. Freed, and H. E. Riggs, "Hypoglycemic Encephalopathy from Insulin Therapy," *Transactions of the American Neurological Association* (1937), 169; F. P. Moersch and J. W. Kernohan, "Hypoglycemia: Neurologic and Neuropathologic Studies," *Archives of Neurology and Psychiatry* 39 (1938): 242–257; J. Tannenberg, "Comparative Experimental Studies on Symptomatology and Anatomical Changes Produced by Anoxic and Insulin Shock," *Proceedings of the Society for Experimental Biology of New York* 40 (1939): 94–96; A. Weil, E. Liebert, and G. Heilbrunn,

"Histopathological Changes in the Brain in Experimental Hyperinsulinism," *Archives of Neurology and Psychiatry* 39 (1939): 467–481.

52. Kolb, *Shock Therapy Survey,* reporting on its use in 356 public and private mental hospitals. For the sources of psychiatrists' enthusiasm for the treatment, see D. B. Doroshow, "Performing a Cure for Schizophrenia: Insulin Coma Therapy on the Wards," *Journal of the History of Medicine and Allied Sciences* 62 (2007): 213–243.

53. Manfred Sakel, "The Nature and Origin of the Hypoglycemic Treatment of Schizophrenia," *American Journal of Psychiatry* 94 (Supplement) (1938): 24–40, 27; M. Sakel, "A New Treatment of Schizophrenia," *American Journal of Psychiatry* 93 (1937): 829–841, 832.

54. L. Bini, "Experimental Researches on Epileptic Attacks Induced by the Electric Current," *American Journal of Psychiatry* 94 (Supplement) (1938): 172–174, 174. For similar speculations, see F. Humbert and A. Friedemann, "Critique and Indications of Treatments in Schizophrenia," *American Journal of Psychiatry* 94 (Supplement) (1938): 174–183, 176; Jessner and Ryan, *Shock Treatment in Psychiatry,* 41.

55. Hans Hoch, "History of the Organic Treatment of Schizophrenia," in *Insulin Treatment in Psychiatry: Proceedings of the International Conference,* ed. Max Rinkel and Harold Himwich (New York: Philosophical Library, 1959), 11.

56. Ivan Bennett, "Hormonal and Other Blood Changes Occurring during Insulin Hypoglycemic Treatment," in *Insulin Treatment in Psychiatry,* ed. Max Rinkel and Harold Himwich (New York: Philosophical Library, 1959), 45. See also Manfred Sakel, "The Classical Sakel Shock Treatment: A Reappraisal," *Journal of Clinical and Experimental Psychopathology* 15 (1954): 255–316.

57. H. Himwich, F. Alexander, and B. Lipetz, "Effect of Acute Anoxia Produced by Breathing Nitrogen on the Course of Schizophrenia," *Proceedings of the Society for Experimental Biology and Medicine* 39 (1938): 367–369. Meduna later experimented along similar lines, though he used carbon dioxide instead. "I made initial mistakes: to my first few patients I gave 100 per cent carbon dioxide to inhale. The results were formidable: massive motor discharges, decerebrate fits on the objective level, and horrifying dream experiences on the subjective level. None of the patients ever permitted a second experiment. One of them, in fact, still in a half daze, jumped off the treatment table and began to run out of the laboratory at such speed that only at the door were we able to catch him and hold him by force until he had been calmed down enough to be permitted to leave the building." L. Meduna, "The Carbon Dioxide Treatment: A Review," *Journal of Clinical and Experimental Psychopathology* 15 (1954): 235–254, 236. Meduna reported here that he had also experimented with a long series of cyanide injections, abandoning this approach only after it, too, proved fruitless.

58. E. Gellhorn, "Effects of Hypoglycemia and Anoxia on the Central Nervous System," *Archives of Neurology and Psychiatry* 40 (1938): 125–146, 137.

59. Manfred Sakel, quoted in J. Wortis, "In Memoriam: Manfred Sakel, M.D.," *American Journal of Psychiatry* 115 (1958): 287–288.

60. William Lawrence, "Insulin Therapy," *New York Times,* August 8, 1943.

61. Harold Bourne, "The Insulin Myth," *Lancet* 262 (1953): 964–968, 965, 967.

62. See the letters published in the *Lancet,* November 21, 1953, and December 12, 1953.

63. H. P. David, "A Critique of Psychiatric and Psychological Research on Insulin Treatment in Schizophrenia," *American Journal of Psychiatry* 110 (1954): 774–776.

64. B. Ackner, A. Harris, and A. J. Oldham, "Insulin Treatment of Schizophrenia: A Controlled Study," *Lancet* 272 (1957): 607–611, 607, 611.

65. See Max Rinkel and Howard Himwich, eds., *Insulin Treatment in Psychiatry: Proceedings of the International Conference* (New York: Philosophical Library, 1959); William Sargant and Eliot Slater, *An Introduction to Physical Treatments in Psychiatry*, 5th ed. (London: Science House, 1972); D. Ewen Cameron, "Greetings," in Rinkel and Himwich, *Insulin Treatment in Psychiatry*, xxiii; Hans Hoch, "History of the Organic Treatment of Schizophrenia," 15, emphasis in the original.
66. O. H. Arnold, "Results and Efficacy of Insulin Shock Therapy," in Rinkel and Himwich, *Insulin Treatment in Psychiatry*, 215–216.
67. Dr. Hoch, "Discussion," in Rinkel and Himwich, *Insulin Treatment in Psychiatry*, 222.
68. D. N. Parfitt, "A Comment on Insulin Coma Therapy in Schizophrenia," *American Journal of Psychiatry* 113 (1956): 246–247, 247.
69. See Sylvia Nasar, *A Beautiful Mind* (New York: Simon and Schuster, 1998). Nash's partial recovery from his psychosis came about in 1990, many years after his confinement at Trenton State Hospital and is obviously unrelated to his insulin coma treatment. Other prominent figures who underwent insulin coma treatment include Albert Einstein's second son, Eduard, and Paul Robeson.

CHAPTER 7 Shocking the Brain

1. L. Meduna, "The Convulsive Treatment: A Reappraisal," *Journal of Clinical and Experimental Psychopathology* 15 (1954): 219–233, 219, 220.
2. L. Meduna, "Klinische und anatomische Beitrage zur Frage der genuinen Epilepsie," *Deutsche Zeitschrift für Nervenheilkunde* 129 (1932): 17–42. The claims about schizophrenic brains appeared only much later in one of Meduna's two unpublished autobiographical essays, as recorded in Edward Shorter and David Healy, *Shock Therapy: A History of Electroconvulsive Treatment in Mental Illness* (New Brunswick, NJ: Rutgers University Press, 2007), 304n54. Meduna was prone to historical revisionism about the origins of his work, giving varied accounts of his research that can hardly be treated as reliable.
3. Typically anecdotal is Robert Gaupp, "Die Frage der kombinierten Psychosen," *Archiv für Psychiatrie und Nervenkrankheiten* 76 (1926): 73–80.
4. Alfred Glaus, "Über Kombinationem von Schizophrenie und Epilepsie," *Zeitschrift für die gesamte Neurologie und Psychiatrie* 135 (1931): 450–500; Georg Müller, "Anfalle bei schizophrenen Erkrankungen," *Allgemeine Zeitschrift für Psychiatrie* 93 (1930): 235–240.
5. See, for example, S. Currie, K. Heathfield, R. Henson, and D. Scott, "Clinical Course and Prognosis of Temporal Lobe Epilepsy: A Survey of 666 Patients," *Brain* 94 (1971): 173–190; M. F. Mendez, J. L. Cummings, and D. F. Benson, "Schizophrenia in Epilepsy: Seizure and Psychosis Variables," *Neurology* 43 (1993): 1073–1077; Thomas M. Hyde and Daniel R. Weinberger, "Seizures and Schizophrenia," *Schizophrenia Bulletin* 23 (1997): 611–622 (an excellent overview of the issues); and Y. T. Chang, P. C. Chen, F. C. Sung, Z. N. Chin, H. T. Kuo, C. H. Tsai, and I. C. Chou, "Bi-directional Relation between Schizophrenia and Epilepsy," *Epilepsia* 52 (2011): 2036–2042. Skepticism about the claim that the two disorders could not coexist had surfaced less than a decade after Meduna began his work. See A. Yde, E. Lohse, and A. Faurbye, "On the Relations between Schizophrenia, Epilepsy, and Induced Convulsions," *Acta Psychiatrica et Neurologica Scandinavica* 16 (1941): 325–388.

6. Meduna may in part have been led to the idea of a mutual antagonism by the example of Wagner-Jauregg's use of malaria to drive out tertiary syphilis.
7. L. Meduna, "Über experimentelle Campherepilepsie," *Archiv für Psychiatrie* 102 (1934): 333–339.
8. "Autobiography of L. Meduna," unpublished manuscript, n.d., Meduna Papers, University of Illinois, quoted in Shorter and Healy, *Shock Therapy,* 26–27. Later, in 1954, Meduna wrote another unpublished autobiographical fragment that contained much more colorful detail, including patient conversations, apparently "remembered" twenty years after the fact, and designed to show the magical effects of the convulsions.
9. Lothar B. Kalinowsky, Hanns Hippius, and Helmfried E. Klein, *Biological Treatments in Psychiatry* (New York: Grune and Stratton, 1982), 217–218.
10. In much smaller doses, metrazol had previously been used to stimulate the circulatory and respiratory systems. Meduna recommended a dose of perhaps five times the usual tenth or two-tenths of a gram to transform this "ideal circulatory stimulant" into a drug that produced convulsions. See L. Meduna, "General Discussion of the Cardiazol Therapy," *American Journal of Psychiatry* 94 (1938): 40–50, 45. Cardiazol was the trade name of the drug metrazol in Europe.
11. Meduna, "General Discussion of Cardiazol Therapy," 50.
12. Solomon Katzenelbogen, "A Critical Appraisal of the 'Shock Therapies' in the Major Psychoses and Psychoneuroses; III. Convulsive Therapy," *Psychiatry* 3 (1940): 409–420, 410.
13. Meduna, "General Discussion of Cardiazol Therapy," 50.
14. L. Meduna, "Versuche über die biologische Beeinflussung des Ablaufes der Schizophrenie: Campher und Cardiazolkrämpfe," *Zeitschrift für die gesamte Neurologie und Psychiatrie* 152 (1935): 235–262; L. Meduna, *Die Konvulsionstherapie der Schizophrenie* (Halle: Marhold, 1937), 121. These claimed cure rates mirror those reported by Henry Cotton in his treatment of focal sepsis and Manfred Sakel in his use of insulin coma therapy. They should all be accorded the same degree of credibility.
15. Henry Rollin, *Festina Lente: A Psychiatric Odyssey* (London: British Medical Journal, 1990), 69; G. S. Nightingale, "Six Months' Experience with Cardiazol Therapy," *Journal of Mental Science* 84 (1938): 581–588.
16. S. McCrae, "'A Violent Thunderstorm': Cardiazol Treatment in British Mental Hospitals," *History of Psychiatry* 17 (2006): 67–90, 71–72, summarizing a 1937 account by Alexander Kennedy.
17. Katzenelbogen, "A Critical Appraisal of the Shock Therapies, III," 412.
18. Katzenelbogen, "A Critical Appraisal of the Shock Therapies, III," 412, 419; Kalinowsky, Hippius, and Klein, *Biological Treatments in Psychiatry,* 218–219. Patients' distress often led them to resist the treatment. Kalinowsky noted that "in cardiazol attacks [*sic*] . . . the patient often fights with the nurses, and has his muscles strongly contracted when the violent fit sets in"—something that supposedly made the physical complications of the treatment worse. L. Kalinowsky, "Electric-Convulsion Therapy in Schizophrenia," *Lancet* 234 (1939): 1232–1233, 1233.
19. P. Polatin, M. Friedman, M. Harris, and W. Horwitz, "Vertebral Fractures Produced by Metrazol-Induced Convulsions," *Journal of the American Medical Association* 112 (1939): 1684–1687.
20. Phyllis Greenacre to Adolf Meyer, November 21, 1937, Adolf Meyer Collection, Series I (Correspondence), Unit I/2128, folder 2 (Elmer Klein, corresp. with A. Meyer, encl.

Phyllis Greenacre), Alan Mason Chesney Medical Archives, Johns Hopkins Medical Institutions (hereafter Chesney Archives).

21. Meyer to Greenacre, November 22, 1937, Adolf Meyer Collection, Series XV (Confidential correspondence and medical records), Chesney Archives.

22. Ugo Cerletti, "Old and New Information about Electroshock," *American Journal of Psychiatry* 107 (1950): 87–94, 89.

23. Cerletti, "Old and New Information," 89–91.

24. Cerletti, "Old and New Information," 90–91.

25. There were several retrospective accounts of the discovery of ECT, many of which differ in detail as attempts were made by various people to claim some involvement in its first stages. The summary I have provided here is derived from that in Shorter and Healy, *Shock Therapy*, 37–46, which I judge the most reliable because it bases its account on Bini's contemporaneous notebooks. Similar problems plague the assessment of the earliest employment of ECT in the United States, with a variety of claimants to be the first to administer the treatment—a good indication, incidentally, of how positively the new therapy was viewed and how anxious psychiatrists were to secure credit for its introduction.

26. "Editorial: More Shocks," *Lancet* 234 (1939): 1373.

27. "Electrical Convulsion Treatment of Mental Disorders," editorial, *Journal of the American Medical Association* 115 (1940): 462–463, 462.

28. Lawrence Kolb and Victor Vogel, "The Use of Shock Therapy in 305 Mental Hospitals," *American Journal of Psychiatry* 99 (1942): 90–100, 92.

29. From the enormous Pilgrim State Hospital on Long Island, where Kalinowsky had been hired to give ECT, Worthing reported to the profession on the major advantages of ECT over its competitors. It was "cheap and easy to perform. . . . No continuous expense for drugs such as insulin or metrazol is necessary. After the purchase of the machine, the electric current is a negligible expense. . . . Thirty patients may be treated in half a day by one physician with the help of one nurse and two attendants." L. Kalinowsky and H. J. Worthing, "Results with Electric Convulsive Therapy in 200 Cases of Schizophrenia," *Psychiatric Quarterly* 17 (1943): 144–153.

30. Shorter and Healy provide a version of these events in *Shock Therapy*, 86–90.

CHAPTER 8 The Checkered Career of Electroconvulsive Therapy

1. ECT could also be administered on an outpatient basis. After the Second World War, as psychiatry moved from its base in the asylum out into the community, it provided an alternative to the expensive and drawn-out therapy of the psychoanalytic couch. Occasional published reports give a brief glimpse of this subterranean world. Abraham Myerson, "The Out-Patient Electric Shock Treatment of Manic-Depressive Psychoses," presentation to the Ninety-Seventh Meeting of the American Psychiatric Association, listed in "Proceedings of the Scientific Sessions," *American Journal of Psychiatry* 98 (1941): 280. See also P. Chodoff, O. Legault, and W. Freeman, "Ambulatory Shock Therapy: A 10-Year Survey," *Diseases of the Nervous System* 11 (1950): 195–201, 196. There is no way to reconstruct how widespread outpatient treatment became or how many patients volunteered for ECT treatment to head off admission to a mental hospital.

2. These developments are well summarized in Edward Shorter and David Healy, *Shock Therapy: A History of Electroconvulsive Treatment in Mental Illness* (New Brunswick, NJ: Rutgers University Press, 2007), 142–163.

3. By 1948, Gordon could list fifty different "theories" for why shock therapy worked. All were pure speculation, and none commanded broad support. See H. Gordon, "Fifty Shock Therapy Theories," *Military Surgeon* 103 (1948): 397–401.

4. Lucio Bini and T. Bazzi, "L'elettroshockterapia col metodo dell'annichilimento nelle forme gravi di psiconevrosi," *Rassegna di Neuropsichiatria e Scienze Affini* 1 (1947): 59–70; Joel T. Braslow, *Mental Ills and Bodily Cures* (Berkeley: University of California Press, 1997), 9.

5. Braslow, *Mental Ills and Bodily Cures*, 102–104.

6. The West Coast psychiatrist A. E. Bennett hypothesized that the resulting "organic confusional state makes the patient forget his worries and breaks up self-consciousness and obsessive thinking." A. E. Bennett, "Evaluation of Progress in Established Physiochemical Treatments in Neuropsychiatry: The Use of Electroshock in the Total Psychiatric Program," *Diseases of the Nervous System* 10 (1949): 195–206, 197.

7. Braslow, *Mental Ills and Bodily Cures*, 105–106.

8. J. H. Smith, J. Hughes, D. W. Hastings, and B. J. Alpers, "Electroshock Treatment in the Psychoses," *American Journal of Psychiatry* 98 (1942): 558–561, found no therapeutic advantage from using ECT in the treatment of schizophrenia, but the article praised the effects of ECT on morale among attendants.

9. H. J. Worthing, "A Report on Electric Shock Treatment at Pilgrim State Hospital," *Psychiatric Quarterly* 15 (Supplement) (1941): 306–309, 309. A. E. Bennett, writing from the Herrick Hospital in Berkeley, California, was equally frank: ECT had "no sustained value in true schizophrenia" but it was nonetheless useful "to control excited or aggressive paranoid patients"—especially when compared to the alternatives of continuous sedation, confinement in a "wet pack," or other forms of restraint. Bennett, "Evaluation of Progress in Established Physiochemical Treatments," 198.

10. J. H. Koenig and H. Feldman, "Non-standard Method of Electroshock Therapy," *Psychiatric Quarterly* 25 (1951): 65–72, 70–71.

11. F. Kino and F. Thorpe, "Electrical Convulsion Therapy in 500 Selected Psychotics," *Journal of Mental Science* 92 (1946): 138–145.

12. F. Thorpe, "Intensive Electrical Convulsion Therapy in Acute Mania," *Journal of Mental Science* 93 (1947): 89–92, 89, 91.

13. Walter Thompson, "Subconvulsive Shock Treatment of the Psychoses," *American Journal of Psychiatry* 99 (1942): 382–386, 382.

14. Thompson, "Subconvulsive Shock Treatment of the Psychoses," 382, 384, 385, 386. The machine delivered 30,000 volts at 5 milliamperes. The original report that prompted this experiment was Nathaniel Berkwitz, "Faradic Shock Treatment of the 'Functional' Psychoses: Preliminary Report," *Lancet* 59 (1939): 351–355.

15. In 1939 and 1940, at the Minneapolis General Hospital, at three private hospitals, and two state hospitals, Nathaniel Berkwitz had used a "Model T Ford spark coil to give fifteen volts of electricity to patients at half-second intervals, before injecting a barbiturate to produce unconsciousness, giving each patient 10–20 treatments." Seventy-three patients were subjected to this experiment over a three-month period. On Berkwitz's account, "Many of them objected to the treatment because of its unpleasant nature," but he dismissed their "excitement and fear" on the grounds that the pain he inflicted caused no lasting damage. Berkwitz, "Faradic Shock Treatment of the 'Functional' Psychoses"; N. Berkwitz, "Faradic Shock in Treatment of Functional Mental Disorders," *Archives of Neurology and Psychiatry* 44 (1940): 760–775, 774–775.

16. L. Kalinowsky and P. Hoch, *Shock Treatments* (New York: Grune and Stratton, 1946), 183. Reminiscing about his years as a junior psychiatrist during this period, albeit in an English mental hospital, Henry Rollin recalled that "the usual practice was to give each patient one ECT each week as a routine." Henry Rollin, "The Impact of ECT," in *Electroconvulsive Therapy: An Appraisal,* ed. Robert L. Palmer, 11–18 (Oxford: Oxford University Press, 1981), 16.
17. Kalinowsky and Hoch, *Shock Treatments,* 183–184.
18. Shorter and Healy, *Shock Therapy,* 93.
19. Peter Cranford, *But for the Grace of God: The Inside Story of the World's Largest Insane Asylum* (Augusta, GA: Great Pyramid Press, 1981), 88–89. Milledgeville was neither the largest nor the most overcrowded mental hospital at the time. That dubious honor went to Pilgrim State Hospital on Long Island, where Lothar Kalinowsky was in charge of administering ECT.
20. Shorter and Healy, *Shock Therapy,* 94.
21. P. H. Wilcox, "Electroshock Therapy: A Review of over 23,000 Treatments Using Unidirectional Currents," *American Journal of Psychiatry* 104 (1947): 100–112.
22. Ivan Belknap, *Human Problems of a State Mental Hospital* (New York: McGraw Hill, 1956).
23. Belknap, *Human Problems,* 191–194.
24. See L. Kalinowsky and H. J. Worthing, "Results with Electric Convulsive Therapy in 200 Cases of Schizophrenia," *Psychiatric Quarterly* 17 (1943): 144–153; Kalinowsky and Hoch, *Shock Treatments,* esp. 176, 179.
25. Koenig and Feldman, "Non-standard Method," 71.
26. W. Liddell Milligan, "Psychoneuroses Treated with Electrical Convulsions: The Intensive Method," *Lancet* 248 (1946): 516–520, 516. The paper contained a series of "illustrative cases" accompanied by claims that 51 percent of patients recovered and 46 percent were improved due to a series of interventions that "obliterated" the faulty electrical patterns in their brains, "allow[ing] the patient to be rehabilitated along correct lines" (520).
27. Clarence Neymann, "Commentary on 'Acute Excitement Induced by Electric Shock' by Vernon Evans," *Archives of Neurology and Psychiatry* 54 (1945): 68.
28. M. D. Kennedy and D. Anchell, "Regressive Electric-Shock in Schizophrenics Refractory to Other Shock Therapies," *Psychiatric Quarterly* 22 (1948): 317–320.
29. For early English examples, see L. Page and R. Russell, "Intensified Electrical Convulsion Therapy in the Treatment of Mental Disorders," *Lancet* 251 (1948): 597–598; P. L. Weil, "'Regressive' Electroplexy in Schizophrenics," *Journal of Mental Science* 56 (1950): 514–520.
30. M. Shoor and F. Adams, "The Intensive Electric Shock Therapy of Chronic Disturbed Psychotic Patients," *American Journal of Psychiatry* 107 (1950): 279–282.
31. Sylvia Cheng, H. Sinclair Tait, and Walter Freeman, "Transorbital Lobotomy versus Electroconvulsive Shock Therapy in the Treatment of Mentally Ill Tuberculous Patients," *American Journal of Psychiatry* 113 (1956): 32–35, 32.
32. D. Rothschild, D. J. Van Gordon, and A. Varjabedian, "Regressive Shock Therapy in Schizophrenia," *Diseases of the Nervous System* 12 (1951): 147–150, 147.
33. Koenig and Feldman, "Non-standard Method," 68, 71–72.
34. Koenig and Feldman, "Non-standard Method," 68. Tyler and Lowenbach claimed to have given multiple ECT treatments a day beginning as early as 1942 at the Duke University Hospital and North Carolina State Hospital at Dix Hill. They called

their method "polydiurnal electric shock," and they used the technique to create a period of prolonged confusion among their patients. Though they found "no significant residual impairment of function" in those they treated, they also acknowledged that the results were "not appreciably different from those found after other forms of shock therapy." E. A. Tyler and H. Lowenbach, "Polydiurnal Electric Shock Treatment in Mental Disorders," *North Carolina Medical Journal* 8 (1947): 577–582, 582.

35. Koenig and Feldman, "Non-standard Method," 66, 68.

36. E. S. Garrett and G. W. Mockbee, "Intensive Regressive Electroconvulsive Therapy in Treatment of Severely Regressed Schizophrenics," *Ohio State Medical Journal* 48 (1952): 505–509, 507, 509.

37. Rothschild, Van Gordon, and Varjabedian, "Regressive Shock Therapy in Schizophrenia," 147, 149, 150. The assurance about the absence of brain damage was, of course, worthless.

38. J. A. Brussel and J. Schneider, "The B.E.S.T. in the Treatment and Control of Chronically Disturbed Mental Patients—A Preliminary Report," *Psychiatric Quarterly* 25 (1951): 55–64, 59.

39. Brussel and Schneider, "The B.E.S.T. in the Treatment and Control of Chronically Disturbed Mental Patients," 59.

40. "The evidence as to whether ECT causes permanent cognitive impairment is inconclusive. The studies reported in the literature have not been controlled adequately for the assessment of such impairment." H. Goldman, F. E. Gomer, and D. L. Templer, "Long Term Effects on Electroconvulsive Therapy upon Memory and Perceptual-Motor Performance," *Journal of Clinical Psychology* 28 (1972): 32–34, 32. One early controlled study did find evidence of widespread, if "spotty," amnesia among ECT patients: "All of the ECT patients, as of approximately four weeks following the termination of treatment, exhibited clear-cut instances of retroactive amnesia." Irving Janis, "Psychologic Effects of Electric Convulsive Treatments: I. Post Treatment Amnesias," *Journal of Nervous and Mental Disease* 111 (1950): 359–382, 364. An authoritative review of these studies found that "Janis's work in many ways represents the most comprehensive assessment of personal memory ever undertaken in an ECT study." R. Weiner, "Does Electroconvulsive Therapy Cause Brain Damage?" *Behavioral and Brain Sciences* 7 (1984): 1–22, 12.

41. The issues are complex and difficult to disentangle, and they remained controversial and open to conflicting interpretations for decades. A long review of a wide range of such studies by Richard Weiner at Duke University concluded with a Scottish verdict of "not proven": "In summary, the results of numerous studies of memory performance after ECT still have not provided a definitive answer to the question of whether and if so, to what degree ECT is associated with persistent memory loss." Weiner, "Does Electroconvulsive Therapy Cause Brain Damage?," 15. But for papers that reach a guilty verdict, see also D. I. Templer, C. F. Ruff, and G. Armstrong, "Cognitive Functioning and Degree of Psychosis in Schizophrenics Given Many Electroconvulsive Treatments," *British Journal of Psychiatry* 123 (1973): 441–443; Larry Squire, "ECT and Memory Loss," *American Journal of Psychiatry* 134 (1977): 997–1001; and John Friedberg, "Shock Treatment, Brain Damage and Memory Loss," *American Journal of Psychiatry* 134 (1977): 1010–1014.

42. L. Kalinowsky, "Present State of Electric Shock Therapy," *Bulletin of the New York Academy of Medicine* 25 (1949): 541–553, 552; L. Kalinowsky, "Problems in Research on Electroconvulsive Therapy," *Behavioral and Brain Sciences* 7 (1984): 28–29, 28. Such chronic complainers, he concluded, ought never be given ECT.

43. L. Kalinowsky, "Organic Psychotic Syndromes Occurring during Electric Convulsive Therapy," *Archives of Neurology and Psychiatry* 53 (1945): 269–273, 269.
44. B. L. Pacella, S. Barrera, and L. Kalinowsky, "Variations in Electroencephalogram Associated with Electric Shock Therapy of Patients with Mental Disorders," *Archives of Neurology and Psychiatry* 47 (1942): 367–384.
45. Stanley Cobb, "Review of Neuropsychiatry for 1946," *Archives of Internal Medicine* 79 (1947): 113–126. Two slightly later animal studies were more carefully conducted and found some evidence of damage, though they equivocated about how serious it was. See A. Ferraro and L. Roizin, "Cerebral Morphologic Changes in Monkeys Subjected to a Large Number of Electrically Induced Convulsions," *American Journal of Psychiatry* 106 (1949): 270–284; and H. Hartelius, "Cerebral Changes Following Electrically Induced Convulsions," *Acta Psychiatrica Neurologica Scandinavica* 77 (Supplement) (1952): 1–128, which concluded, "The question of whether or not irreversible nerve cell changes may occur after ECT must be answered in the affirmative" (114).
46. Lothar Kalinowsky, "The Various Forms of Shock Therapy in Mental Disorders and Their Practical Importance," *New York State Journal of Medicine* 41 (1941): 2210–2215, 2211.
47. Kalinowsky and Hoch, *Shock Treatments*, 118.
48. Henry Rollin reminisced about his own experience: "The incidence of fractures was a staggering 35.4 per cent. . . . In order to minimise the occurrence of these orthopedic injuries a sizeable supporting cast of nurses was required. Nurses in pairs exerted pressure on the shoulders, hips, and legs of the patient and another nurse held the jaw firm. The patient had been instructed to bite hard on a rubber heel covered in gauze, previously inserted in the teeth." Rollin, "The Impact of ECT," 15.
49. H. Worthing and L. Kalinowsky, "The Question of Vertebral Fractures in Convulsive Therapy and in Epilepsy," *American Journal of Psychiatry* 98 (1942): 533–537, 534.
50. Shorter and Healy, *Shock Therapy*, 127.
51. I. Meschan, J. B. Scruggs, and J. D. Calhoun, "Convulsive Fractures of the Dorsal Spine Following Electroshock Therapy," *Radiology* 54 (1950): 180–193, 191. Compare Bennett: "The facts are that straight electroshock produces severe convulsive seizures, sufficient to fracture extremities and vertebrae and to cause muscular ruptures with dislocation." Bennett, "Evaluation of Progress in Established Physiochemical Treatments," 201. See also J. R. Lingley and L. Robbins "Fractures Following Electroshock Therapy," *Radiology* 48 (1947): 124–128.
52. See A. E. Bennett, "Preventing Traumatic Complication in Shock Therapy by Curare," *Journal of the American Medical Association* 114 (1940): 322–324; A. E. Bennett, "Curare: A Preventive of Traumatic Complications in Shock Therapy," *American Journal of Psychiatry* 97 (1941): 1040–1060.
53. The standard textbook on pharmacology by Gilman and Goodman, *The Pharmacological Basis of Therapeutics*, issued a stark warning: "In view of the extremely potent nature of curare and the ever-present danger of respiratory muscle paralysis . . . the safe employment of the drug is restricted, as a rule, to experienced anesthesiologists who are thoroughly acquainted with resuscitative procedure such as tubal intubation." Quoted in Shorter and Healy, *Shock Therapy*, 128.
54. Kalinowsky and Hoch, *Shock Treatments*, 151–152.
55. Bennett, "Evaluation of Progress in Established Physiochemical Treatments," 199–202.

56. See, for example, Byron Stewart, "Electro-Shock Therapy," *Bulletin of the Los Angeles Neurological Society* 7 (1942): 88–94.
57. G. Holmberg and S. Thesleff, "Succinyl-Choline-Iodide as a Muscular Relaxant in Electroshock Therapy," *American Journal of Psychiatry* 108 (1952): 842–846.
58. There are no systematic data on the adoption of modified ECT in the 1950s and 1960s. In his memoirs, the clinical psychologist Norman Endler provides a disturbing account of the continued use of unmodified ECT in a central Illinois state hospital in the late 1950s. Endler is unsparing in describing his revulsion at being forced to witness this spectacle during his training. Years later, Endler himself became seriously depressed. In desperation, he agreed to undergo modified ECT, and, according to him, this course of ECT restored him to health. In subsequent decades, many prominent figures who had undergone modified ECT for their own bouts with depression echoed his testimony. Norman Endler, *Holiday of Darkness: A Psychologist's Search for the Meanings of Madness* (Toronto: Wiley, 1982), 65–66.
59. B. C. Glueck, H. Reiss, and L. E. Bernard, "Regressive Electroshock Therapy: Preliminary Report on 100 Cases," *Psychiatric Quarterly* 31 (1957): 117–136. Experiments along these lines continued at Stony Lodge into the 1970s. See J. Exner and L. Murillo, "Effectiveness of Regressive ECT with Process Schizophrenia," *Diseases of the Nervous System* 34 (1973): 44–48.
60. For a discussion of Cameron and the CIA, see Jonathan Sadowsky, *Electroconvulsive Therapy in America: The Anatomy of a Medical Controversy* (London: Routledge, 2017), 53–61.
61. D. Ewen Cameron and S. K. Pande, "Treatment of the Paranoid Schizophrenic Patient," *Canadian Medical Association Journal* 78 (1958): 92–96.
62. D. Ewen Cameron, "Production of Differential Amnesia in the Treatment of Schizophrenia," *Comprehensive Psychiatry* 1 (1960): 26–34, 29, 28.
63. Once again, the patients were disproportionately female: twenty-one out of the thirty reported on.
64. Cameron, "Production of Differential Amnesia in the Treatment of Schizophrenia"; D. Ewen Cameron, J. G. Lohrenz, and K. A. Handcock, "The Depatterning Treatment of Schizophrenia," *Comprehensive Psychiatry* 3 (1962): 65–76.
65. Edward Shorter and David Healy contend, without supplying evidence, that Cameron "was driven from office for his use of regressive ECT." Shorter and Healy, *Shock Therapy*, 139.
66. Rebecca Lemov, "Brainwashing's Avatar: The Curious Career of Dr. Ewen Cameron," *Grey Room* 45 (Fall 2011): 60–87.
67. Baruch Silverman, "Dr. D. Ewen Cameron," *Canadian Medical Association Journal* 97 (1967): 985–986. A year before his death, Cameron returned to Montreal to receive an award for his many contributions to Canadians' mental health.
68. Anne Collins, *In the Sleep Room: The Story of the CIA Brainwashing Experiments in Canada* (Toronto: Lester and Orpen Dennys, 1988); Don Gillmor, *I Swear by Apollo: Dr. Ewen Cameron and the CIA-Brainwashing Experiments* (Montreal: Eden Press, 1987). Cameron may or may not have been aware that the CIA was the source of some of his funding.
69. On these developments, see Laura Bothwell and Scott Podolsky, "The Emergence of the Randomized, Controlled Trial," *New England Journal of Medicine* 375 (2016): 501–504; and Harry Marks, *The Progress of Experiment: Science and Therapeutic Reform in the United States, 1900–1990* (Cambridge: Cambridge University Press, 1997).

70. Hemingway made this remark to his biographer. See A. E. Hotchner, *Papa Hemingway: A Personal Memoir* (New York: Morrow, 1983), 280.
71. California's original legislation was passed in 1974 but was promptly suspended after objections from some psychiatrists. A revised statute went into effect in 1976 and still sharply limits the use of ECT in the state—a highly unusual legal intervention into medical practice.
72. Anthony Clare, "Ethical Issues Relating to ECT: A Medical View," in *Electroconvulsive Therapy: An Appraisal*, ed. Robert L. Palmer, 303–314 (New York: Oxford University Press, 1981), 312.
73. Shorter and Healy, *Shock Therapy*, 179–180.
74. Joyce Small and Ivor Small, "Electroconvulsive Therapy Update," *Psychopharmacology Bulletin* 17 (1981): 29–42, 29.

CHAPTER 9 Brain Surgery

1 Walter Freeman and James Watts, *Psychosurgery in the Treatment of Mental Disorders and Intractable Pain*, 2nd ed. (Springfield, IL: Charles Thomas, 1950), 125–127.
2. Freeman and Watts, *Psychosurgery*, 113, 115.
3. See Walter Freeman, "Autobiography," unpublished typescript, c. 1972, copy in possession of the author, courtesy of Freeman's son, Franklin Freeman, ch. 10, 6. On the poor quality of the medical students he taught, see ch. 10, 9–10.
4. Walter Freeman, *The Psychiatrist: Personality and Patterns* (New York: Grune and Stratton, 1968), 52.
5. A meningioma is a tumor that forms on the membranes covering the brain and spinal cord. While I find it somewhat bizarre that any brain tumor could be described as benign, about 90 percent of meningiomas are not malignant. In the early twentieth century, however, even benign tumors were difficult to remove and often proved fatal. For Brickner's earlier summary of his findings in this case, see R. M. Brickner, "An Interpretation of Frontal Lobe Function Based upon the Study of a Case of Partial Bilateral Frontal Lobectomy," *Research Publications of the Association for Research in Nervous and Mental Disease* 13 (1932): 259–351.
6. S. Ackerly, "Instinctive, Emotional and Mental Changes Following Prefrontal Lobe Extirpation," *American Journal of Psychiatry* 92 (1935): 717–729, 724–725.
7. On Cushing, see Michael Bliss, *Harvey Cushing: A Life in Surgery* (Oxford: Oxford University Press, 2007).
8. Percival Bailey, *Intercranial Tumors* (London: Tindall, 1933), 433.
9. Stanley Cobb subsequently provided a summary of Brickner's findings. "The patient showed: (1) limitation of the capacity to associate and synthesize, e.g., distractability with impairment of selection, retention and learning; (2) impairment of restraint of emotion with boasting, anger and hostility; (3) additional symptoms such as impairment of abstraction, judgment and initiative, with euphoria and increased slowness, stereotypy and compulsiveness." S. Cobb, "Review of Neuropsychiatry for 1940," *Archives of Internal Medicine* 66 (1940): 1341–1354, 1350.
10. Henri Claude, "Les functions des lobes frontaux," *Revue Neurologique* 65 (1936): 523.
11. Jacobsen explicitly noted that these defects paralleled those Brickner had observed in Patient A. Jacobsen sent Fulton a memorandum outlining the series of deficits he had observed following frontal lobe surgery on monkeys as early as November 27, 1933, and first reported the deficits in a publication the following year: C. F. Jacobsen, "The Effects

of Extirpation of the Frontal Association Areas in Monkeys upon Complex Adaptive Behavior," in "Proceedings of the American Physiological Society," *American Journal of Physiology* 109 (1934): 1–117, 59; C. Jacobsen, J. B. Wolfe, and T. A. Jackson, "An Experimental Analysis of the Functions of the Frontal Association Areas in Primates," *Journal of Nervous and Mental Disease* 82 (1935): 1–14.

12. M. P. Crawford, J. F. Fulton, C. F. Jacobsen, and J. B. Wolfe, "Frontal Lobe Ablation in Chimpanzees: A Resume of 'Becky' and 'Lucy,'" *Association for Research in Nervous and Mental Diseases* 27 (1948): 3–58, 57.

13. Jack D. Pressman, *Last Resort: Psychosurgery and the Limits of Medicine* (Cambridge: Cambridge University Press, 1998), 50–58.

14. "Leucotomy" was derived from the Greek *leukos,* meaning "white," since Moniz and Lima were targeting the white matter of the brain. Egas Moniz, "Prefrontal Leucotomy in the Treatment of Mental Disorders," *American Journal of Psychiatry* 93 (1937): 1379–1385, 1381, 1385. Egas Moniz, *Tentatives operatoires dans le traitement de certaines psychoses* (Paris: Masson, 1936). This assertion, that psychiatric progress would come from adopting an organic perspective on mental illness, was the central claim of the opening paragraph of his monograph.

15. Moniz, "Prefrontal Leucotomy," 1381.

16. See David Shutts, *Lobotomy: Resort to the Knife* (New York: Van Nostrand Reinhold, 1982), 54–55.

17. Egas Moniz, "Mein Weg zur Leukotomie," *Deutsche Medizinische Wochenschrift* 73 (1948): 581–583.

18. Cobb, "Review of Neuropsychiatry for 1940," 1353.

19. Watts to Fulton, March 11, 1935, Series I (General files), James Winston Watts, 1935–1955, box 178, folders 2643–2644, John Farquhar Fulton Papers, ms. 1236, Cushing/Whitney Medical Historical Library, Yale University Library (hereafter Fulton Papers, Yale University).

20. For Fulton's continuing interest in his protégé's career, see Watts to Fulton, January 4, 29, 1935; Fulton to Watts, January 7, 31, February 3, 7, 1935, Series I, James Winston Watts, Fulton Papers, Yale University.

21. Freeman and Watts, *Psychosurgery,* xviii.

22. Freeman and Watts, *Psychosurgery,* xix–xx.

23. Freeman to Fulton, October 24, 1936, Series I (General files), Walter Freeman, box 57, folder 829, Fulton Papers, Yale University. See also John Fulton's diary, November 16, 1936, Series IV (Personal and family papers), John F. Fulton, Diary, Fulton Papers, Yale University.

24. Freeman and Watts, *Psychosurgery,* 31.

25. Freeman and Watts, *Psychosurgery,* 31–32.

26. Freeman and Watts, *Psychosurgery,* 90–91.

27. Winfred Overholser, "Proceedings of the First Post-Graduate Course in Psychosurgery," *Digest of Neurology and Psychiatry* 17 (1949): 439.

28. Walter Freeman and James Watts, "Psychosurgery during 1936–1946," *Archives of Neurology and Psychiatry* 58 (1947): 417–425, 422–425. Schrader and Robinson pointed out later that "Freeman and Watts . . . operated for the most part on private patients who had not been ill long and who had not deteriorated." P. J. Schrader and M. F. Robinson, "Evaluation of Prefrontal Lobotomy through Ward Behavior," *Journal of Abnormal and Social Psychology* 40 (1945): 61–69, 63.

29. Jack Pressman's careful examination of the lobotomies performed at McLean Hospital in Boston showed that sixty-six of the eighty lobotomies (82.5 percent) were performed on women. Joel Braslow provides a long list of contemporary studies that document this gender disparity. See Joel T. Braslow, *Mental Ills and Bodily Cures* (Berkeley: University of California Press, 1997), 211n4. The pattern held internationally. In England, the Board of Control—the government entity charged with overseeing the nation's mental hospitals—reported that, of the 7,225 operations performed between 1948 and 1952, 4,468 (61.8 percent) were on women. Handwritten note in response to a Parliamentary question, found in the Public Record Office attached to a copy of Board of Control, "Prefrontal Leucotomy in 1000 Cases" (London: His Majesty's Stationery Office, 1947), and quoted in Ken Barrett, "Manhandling the Brain: Psychiatric Neurosurgery in the Mid-20th Century," *Neuropsychiatry News* (January 2016): 11. Among the 1,000 cases from before 1947, women outnumbered men in a ratio of 3:2.

30. Morton Kramer, "The 1951 Survey of the Use of Psychosurgery," in *Evaluation of Psychosurgery: Proceedings of the Third Research Conference on Psychosurgery, 1951,* ed. Fred Mettler and Winfred Overholser, US Public Health Service Publication 221 (Washington, DC: Public Health Service, 1954).

31. Charles Limburg, "A Survey on the Use of Psychosurgery with Mental Patients," in *Proceedings of the First Research Conference on Psychosurgery with Mental Patients,* ed. F. A. Mettler and N. Bigelow, US Public Health Service Publication 16 (Washington, DC: Government Printing Office, 1951), 65–173.

32. Andrew Scull and Diane Favreau, "'A Chance to Cut Is a Chance to Cure': Sexual Surgery for Psychosis in Three Nineteenth Century Societies," *Research in Law, Deviance, and Social Control* 8 (1986): 3–39; Carroll Smith-Rosenberg and Charles Rosenberg, "The Female Animal," *Journal of American History* 60 (1973): 332–356; Andrew Scull, *Madness in Civilization: A Cultural History of Insanity from the Bible to Freud, and from the Madhouse to Modern Medicine* (Princeton, NJ: Princeton University Press, 2015), 26–36.

33. Walter Freeman and James Watts, "Subcortical Prefrontal Lobotomy in the Treatment of Certain Psychoses," *Archives of Neurology and Psychiatry* 38 (1937): 225–229, 229.

34. H. S. Barahal, "1,000 Prefrontal Lobotomies: A Five to 10 Year Follow-up Study," *Psychiatric Quarterly* 32 (1958): 652–678, 654.

35. Braslow, *Mental Ills and Bodily Cures,* ch. 7, quotes on 155, 158, emphasis added.

36. Freeman and Watts, *Psychosurgery,* ix, 379.

37. The late Jack Pressman, who wrote the most extensive history of lobotomy to date, used this common trope as the title of his book.

38. T. P. Rees, "The Indications for a Pre-frontal Leucotomy," *Journal of Mental Science* 89 (1943): 161–164, 164.

39. Freeman, "Autobiography," ch. 9, 13.

40. William Alanson White to Smith Ely Jelliffe, August 7, 1936, box 22, Jelliffe Papers, Library of Congress, Washington, DC.

41. "Brain Surgery Feat Arouses Sharp Debate," *Baltimore Sun,* November 26, 1936, 1, 9.

42. Discussion of W. Freeman and J. Watts, "Prefrontal Lobotomy in the Treatment of Mental Disorders," *Southern Medical Journal* 30 (1937): 23–31, 31.

43. Freeman, "Autobiography," ch. 14, 5.

44. See Freeman to Meyer, November 17, 1937; Meyer to Freeman, November 18, 1937; Freeman to Meyer, March 1, 1938; Meyer to Freeman, March 2, 1938; Meyer to Freeman,

May 25, 1938; Freeman to Meyer, May 25, 1938; Freeman to Meyer, February 3, 1938; Freeman to Meyer, February 4, 1941; all in Adolf Meyer Collection, Series I (Correspondence), Unit I/1256 (Walter Freeman), folder 1 (1925–1938) and folder 2 (1939–1941), Alan Mason Chesney Medical Archives, Johns Hopkins Medical Institutions.

45. Discussion from Walter Freeman and James Watts, "Subcortical Prefrontal Lobotomy," 225–229; and the discussions of the audience reaction in Elliot Valenstein, *Great and Desperate Cures* (New York: Basic Books, 1986), 145; Pressman, *Last Resort,* 81–82; J. El Hai, *The Lobotomist* (Hoboken, NJ: Wiley, 2005), 137.

46. Walter Freeman to Egas Moniz, August 11, 1937, School of Medicine, MS 0803, W. Freeman and J. Watts Papers, Special Collections Research Center, George Washington University Libraries.

47. For the critical responses to Freeman's speech at the American Medical Association in Atlantic City in June 1937, see William L. Lawrence, "Surgery on the Soul-Sick: Relief of Obsessions Is Reported," *New York Times,* June 7, 1937, 1, 10.

48. Cobb, "Review of Neuropsychiatry for 1940," 1354. He had begun to doubt the value of lobotomy as early as 1937, complaining that "Freeman . . . reports treating in this way patients with neuroses and young people with schizophrenia. This mutilation of the brains of patients who might recover seems to me unjustifiable." S. Cobb, "Review of Neuropsychiatry for 1937," *Archives of Internal Medicine* 63 (1937): 1098–1110, 1098–1099.

49. "The Lobotomy Delusion," *Medical Record,* May 15, 1940, 335.

CHAPTER 10 Selling Psychosurgery

1. On Freeman and the press, see Gretchen Diefenbach, Donald Diefenbach, Alan Baumeister, and Mark West, "Portrayal of Lobotomy in the Popular Press," *Journal of the History of the Neurosciences* 8 (1999): 60–69. Freeman boasted about his ability to manipulate reporters. See Walter Freeman, "Autobiography," unpublished typescript, c. 1972, copy in possession of the author, courtesy of Freeman's son, Franklin Freeman, ch. 12, 7–8; and Walter Freeman, "Adventures in Lobotomy," unpublished manuscript, n.d., copy in possession of the author, ch. 4, "The Fourth Estate."

2. Thomas Henry, "Brain Operation by D.C. Doctors Aids Mental Ills," *Washington Evening Star,* November 20, 1936.

3. "Finds Surgery Aids Mental Cases," *New York Times,* November 26, 1936, 10.

4. William Laurence, "Surgery Used on the Soul-Sick," *New York Times,* June 6, 1937, 1, 10.

5. Stephen J. McDonough, "Brain Surgery Is Credited with Cure of 50 'Hopelessly' Insane Persons," *Houston Post,* June 6, 1941.

6. Waldemar Kaempffert, "Psychosurgery," *New York Times,* January 11, 1942; W. Kaempffert, "Turning the Mind Inside Out," *Saturday Evening Post,* May 24, 1941.

7. Elliot Valenstein has provided a sampling of these headlines: "Surgeon's Knife Restores Sanity to Nerve Victims"; "Wizardry of Surgery Restores Sanity to Fifty Raving Maniacs"; "Brain Surgery Is Credited with Cure of 50 Hopelessly Insane"; "Forgetting Operation Bleaches the Brain"; and "No Worse Than Removing a Tooth." Quoted in Elliot Valenstein, *Great and Desperate Cures* (New York: Basic Books, 1986), 157.

8. Fulton personally intervened in an executive session of the American Neurological Association to quash a censure motion. Watts wrote to thank Fulton for his help and to say he and Freeman finally had their book contract: "I feel that if it had not been for your letter to Mr. Thomas just before the Cushing Society meeting that he would prob-

ably have turned us down." Watts to Fulton, September 7, 1941, Series I (General files), James Winston Watts, 1935–1955, box 178, folders 2643–2644, John Farquhar Fulton Papers, ms. 1236, Cushing/Whitney Medical Historical Library, Yale University (hereafter Fulton Papers, Yale University).

9. During his time at Oxford, Fulton married a wealthy heiress, Lucia Wheatstone, whose money allowed him to indulge his passion for collecting old medical texts. This was an obsession he shared with Harvey Cushing, and their collections eventually formed the basis for the Yale Medical Historical Library. Fulton trained many famous students and won many accolades, yet he made no major scientific contributions of his own. However, his 1938 textbook, *Physiology of the Nervous System,* was an important synthesis of work by others.

10. John Fulton Diary, November 16, 1936, Series IV (Personal and family papers), John F. Fulton, Diary, Fulton Papers, Yale University.

11. John Fulton to Henry Viets, October 16, 1936, Series I (General files), Henry Viets, 1935–1943, box 173, folders 2586–2593, Fulton Papers, Yale University.

12. Here I concur with Jack Pressman, who extensively analyzes Fulton's role, which extended to weekly letters to his former student James Watts offering encouragement and support. See Jack D. Pressman, *Last Resort: Psychosurgery and the Limits of Medicine* (Cambridge: Cambridge University Press, 1998), esp. 85–96.

13. Elliot Valenstein, "A History of Psychosurgery," in *A History of Neurosurgery,* ed. Samuel Greenblatt, 499–516 (Park Ridge, IL: American Association of Neurosurgeons, 1977).

14. Freeman and Watts worried about this issue and tried to suggest it was not a problem. Attempting to measure the effect of lobotomy on IQ, they had the psychologist Thelma Hunt conduct psychological testing on some of their patients, reporting their results in *Psychosurgery.* See James Watts and Walter Freeman, "Intelligence Following Prefrontal Lobotomy in Obsession Tension States," *Journal of Neurosurgery* 1 (1944): 291–296. For contrary views, see Gösta Rylander, "Psychological Tests and Personality Analyses before and after Frontal Lobotomy," *Acta Psychiatrica Scandinavica* 22 (1947): 383–398; and Rosvold Enger and Mortimer Mishkin, "Evaluation of the Effects of Prefrontal Lobotomy on Intelligence," *Canadian Journal of Psychology* 4 (1950): 122–126.

15. Compare Freeman's claim that "prefrontal lobotomy can be a precise and very exact operation." Milton Greenblatt, R. Arnot, J. L. Poppen, and W. P. Chapman, "Report on Lobotomy Studies at the Boston Psychopathic Hospital," *American Journal of Psychiatry* 104 (1947): 361–368, 367.

16. As Fred Mettler later put it after examining brains of patients who had died in the Columbia Greystone experiments, "You are often very surprised to see what you do get because the anterior sweep very frequently produces much more extensive damage than the operator expects to produce." F. A. Mettler, in *Evaluation of Psychosurgery: Proceedings of the Third Research Conference on Psychosurgery, 1951,* ed. F. A. Mettler and W. Overholser, US Public Health Service Publication 221 (Washington, DC: Public Health Service, 1954); A. Meyer, E. Beck, and T. McLardy, "Prefrontal Leucotomy: A Neuroanatomical Report," *Brain* 70 (1947): 18–49, 18.

17. Pressman, *Last Resort,* 387–389, provides an excellent discussion of the literature on this issue.

18. Ward Halstead, H. T. Carmichael, and Paul Bucy, "Prefrontal Lobotomy: A Preliminary Appraisal of the Behavioral Results," *American Journal of Psychiatry* 103 (1946): 217–228.

19. State Hospital #4, *Biennial Report* 20 (Jefferson City, MO: Board of Managers of State Eleemosynary Institutions, 1939–1940), 6–9; Schrader to Walter Freeman, February 16, 1942, School of Medicine, MS 0803, W. Freeman and J. Watts Papers, Special Collections Research Center, George Washington University Libraries; L. H. Ziegler, "Bilateral Frontal Lobotomy: A Survey," *American Journal of Psychiatry* 100 (1943): 178–179.

20. The four other cases had likewise received an extraordinary range of physical treatments before being subjected to psychosurgery: endocrine therapy, narcosis treatment, and forty insulin comas; narcosis, fever therapy using typhoid vaccine, insulin comas, and sixteen metrazol convulsions; narcosis, forty-six insulin treatments, then a further series of insulin comas seven months later; and narcosis, then fifty-one insulin comas, which only led to further excitability and destructiveness, followed by lobotomy. E. A. Strecker, H. D. Palmer, and F. C. Grant, "A Study of Frontal Lobotomy," *American Journal of Psychiatry* 98 (1942): 524–532, 526, 528.

21. Strecker, Palmer, and Grant, "A Study of Frontal Lobotomy," 524.

22. Beginning in January 1941, Petersen and Buchstein undertook a series of forty-six lobotomies (eighteen males and twenty-eight females) at Willmar State Hospital in Minnesota, mostly under local anesthesia. M. G. Petersen and H. F. Buchstein, "Prefrontal Lobotomy in Chronic Psychosis," *American Journal of Psychiatry* 99 (1942): 426–430.

23. J. G. Lyerly, "Prefrontal Lobotomy in Involutional Melancholia," *Journal of the Florida Medical Association* 25 (1938): 225–229.

24. Wilder Penfield, "Bilateral Frontal Gyrectomy and Postoperative Intelligence," *Association for Research in Nervous and Mental Diseases* 27 (1948): 519–534, 534. For discussion of the vogue for alternative operations, see Valenstein, *Great and Desperate Cures*, 193–198, and the examples given in M. Greenblatt and H. S. Solomon, "Psychosurgery (Concluded)," part 2 of 2, *New England Journal of Medicine* 248, no. 2 (1953): 59–67, 61–63. Spurning knives, some experimented with freezing or burning brain tissue, or destroying it using radioactive material or ultrasound. One British surgeon, Cunningham Dax, who preferred curved cuts for some reason, proposed using a grapefruit knife. See A. Meyer and E. Beck, "Neuropathological Problems Arising from Prefrontal Leucotomy," *Journal of Mental Science* 91 (1945): 411–425, 422.

25. W. Freeman and J. Watts, *Psychosurgery in the Treatment of Mental Disorders and Intractable Pain*, 2nd ed. (Springfield, IL: C. C. Thomas, 1950), 62.

26. Freeman and Watts, *Psychosurgery*, 353.

27. Freeman and Watts, *Psychosurgery*, 331, 333.

28. L. Kalinowsky and P. Hoch, *Shock Treatments and Other Somatic Procedures in Psychiatry* (New York: Grune and Stratton, 1946), 224–225.

29. Kalinowsky and Hoch, *Shock Treatments*, 227.

30. Greenblatt, Arnot, Poppen, and Chapman, "Report on Lobotomy Studies," 367; J. L. Pool, R. G. Heath, and J. J. Weber, "Topectomy: Surgical Technique, Psychiatric Indications, and Postoperative Management," *Journal of Nervous and Mental Disease* 110 (1949): 464–477, 470, 474; Lothar Kalinowsky, "Principals Governing the Selection of Cases for Psychosurgery," *Surgery, Gynecology, and Obstetrics* 92 (1951), 602; Paul Hoch, "Evaluations of the Results of Topectomy Operations," *Surgery, Gynecology, and Obstetrics* 92 (1951): 609–611, 610.

31. Greenblatt and Solomon, "Psychosurgery (Concluded)," 59.

32. Mettler and Overholser, *Evaluation of Psychosurgery*, 143.
33. Greenblatt, Arnott, Poppen, and Chapman, "Report on Lobotomy Studies," 366.
34. "Discussion—Functional Psychoses," following Cotton, "Etiology and Treatment," and other papers, *American Journal of Psychiatry* 79 (1922): 199–205, 199, 204.
35. "Report of the Physician in Chief," *108th Annual Report of the Neuro-Psychiatric Institute of the Hartford Retreat* (1932): 8–9.
36. Certainly, Scoville's wife did not. She had suffered a mental breakdown when she discovered her husband was having an affair. As a guest at the Institute for Living, she was subsequently given hydrotherapy and pyrotherapy and then subjected to a series of ECT treatments. When her mental disturbance persisted, Scoville lobotomized her, had her released from the hospital, and then married a younger woman. See L. Dittrich, *Patient H.M.: A Story of Memory, Madness and Family Secrets* (New York: Random House, 2016), 375–378.
37. The first such facility devoted solely to psychosurgery, its opening was announced in *124th Annual Report of the Institute of Living* (Hartford, CT, 1948).
38. This commitment to somatic treatments, and to lobotomy in particular, cost Burlingame the presidency of the American Psychiatric Association after the Second World War, when the growing ranks of psychoanalysts and psychodynamically oriented psychiatrists united to reject his candidacy.
39. Charles Burlingame, "Psychiatric Sense and Nonsense," *Journal of the American Medical Association* 133 (1947): 971–972; *124th Annual Report of the Institute of Living*, 11.
40. Quoted in Pressman, *Last Resort*, 251.
41. Quoted in Pressman, *Last Resort*, 244–245. Pressman devotes an entire chapter to lobotomy at the McLean, and while I do not share his analytic framework, my discussion here draws on his pioneering research into the hospital records.
42. These details come from Alex Beam's study of the McLean, *Gracefully Insane* (New York: Public Affairs, 2001), 65–66.
43. *McLean Annual Report* (1945), 187.
44. Quoted in Pressman, *Last Resort*, 253.
45. Quoted in Pressman, *Last Resort*, 270.
46. A. M. Fiamberti, quoted in Valenstein, *Great and Desperate Cures*, 163.
47. W. Freeman, "Autobiography," ch. 8, 19–20.
48. Occasionally, the picture taking led to disastrous results. In one instance at least, while Freeman was admiring his handiwork and preparing his camera, one of the ice picks slid so deeply into the patient's brain that she never came round from the surgery and, within days, died.
49. Freeman and Watts, *Psychosurgery*, 56.
50. Freeman and Watts, *Psychosurgery*, x.
51. Freeman, "Adventures in Lobotomy," ch. 6, 59.
52. Freeman, "Adventures in Lobotomy," ch. 6, 59.
53. Freeman and Watts, *Psychosurgery*, 53, 55.
54. Walter Freeman to Egas Moniz, September 9, 1952, School of Medicine, MS 0803, W. Freeman and J. Watts Papers, Special Collections Research Center, George Washington University Libraries.
55. Lucille Cohen, "Surgery May Free 9 Mental Patients: Doctors Hope for 'Miracle' in New Operation," *Seattle Post-Intelligencer*, August 28, 1947.

56. Freeman's unpublished history of psychosurgery in the George Washington University archives, quoted in Valenstein, *Great and Desperate Cures,* 212.
57. John Fulton to Walter Freeman, October 2, 1947, Series I (General files), Walter Freeman, box 57, folder 829, Fulton Papers, Yale University.
58. Morison diary entry, December 13, 1947, Robert Swain Morison diary, 1947, RG 12 (Rockefeller Foundation Officers' Diaries), Rockefeller Foundation Records Collection, Rockefeller Archive Center, Sleepy Hollow, NY.
59. The conference proceedings were subsequently published in 1948 as a book, *The Frontal Lobes,* edited by John Fulton, Charles Airing, and Bernard Wortis (Baltimore, MD: Williams and Wilkins, 1948), a tome that ran to some 900 pages.
60. John Fulton, *Frontal Lobotomy and Affective Behavior: A Neurological Analysis* (New York: Norton, 1951), 95.
61. Walter Freeman to John Fulton, October 31, 1947, Series I (General files), Walter Freeman, box 57, folder 829, Fulton Papers, Yale University.
62. Gösta Rylander, "Personality Analysis before and after Frontal Lobotomy," *Frontal Lobes Association for Research in Nervous and Mental Disease* 27 (1948): 691–705, 692.
63. Rylander, "Personality Analysis before and after Frontal Lobotomy," 695, 692, 693, 696.
64. Discussion in Rylander, "Personality Analysis before and after Frontal Lobotomy," 703, 704.
65. Discussion in Rylander, "Personality Analysis before and after Frontal Lobotomy," 703.
66. For further details, see E. Fuller Torrey, *American Psychosis* (Oxford: Oxford University Press, 2013), ch. 1; and Lawrence Leamer, *The Kennedy Women* (New York: Ballantine, 1994).
67. For a report on the first 200 lobotomies in this program, see B. E. Moore, S. Friedman, B. Simon, and J. Farmer, "A Cooperative Clinical Study of Lobotomy," *Association for Research in Nervous and Mental Diseases* 27 (1947): 765–794.
68. Percival Bailey in his summary of developments in neurosurgery in 1948 clearly viewed Freeman's activities with disquiet: "The outstanding development of the year is transorbital lobotomy. This procedure is of a nature to distress the surgeon greatly. . . . Such a blind procedure is unjustifiable, and it is to be feared that its simplicity will cause it to be used too freely and bring disrepute on a potentially very useful measure." *1948 Year Book of Neurology, Psychiatry and Neurosurgery* (Chicago: Year Book Publishers, 1948), 493.
69. W. Freeman, in Mettler and Overholser, *Evaluation of Psychosurgery,* 61. At the same National Institute of Mental Health conference, Freeman continued to attack his neurosurgical rivals for refusing to endorse his operations: "My experience with the neurosurgeons has been a somewhat dog-in-a-manger proposition. They won't do it themselves and they don't let other people do it. . . . [T]he minor operation, transorbital lobotomy, is something that does not appeal to their imagination. It offends their surgical instincts. . . . It upsets them emotionally" (139).

CHAPTER 11 The End of the Affair

1. Walter Freeman, *The Psychiatrist: Personalities and Patterns* (New York: Grune and Stratton, 1968), 54.
2. Citation for 1949 award of the Nobel Prize in *Physiology or Medicine, in Nobel Lectures, Physiology or Medicine, 1942–1962* (New York: Elsevier, 1964), 243.

3. "Editorial," *Digest of Neurology and Psychiatry* 17 (1949): 668–669.
4. W. Freeman and J. Watts, *Psychosurgery in the Treatment of Mental Disorders and Intractable Pain*, 2nd ed. (Springfield, IL: C. C. Thomas, 1950), ix.
5. Fulton obtained the then-princely sum of $200,000 from the National Research Council's Committee on Veterans Medical Affairs to support his laboratory between 1948 and 1951.
6. J. L. Hoffman, "Clinical Observations Concerning Schizophrenic Patients Treated by Prefrontal Leukotomy," *New England Journal of Medicine* 241 (1949): 233–236.
7. "Lobotomy Disappointment," *Newsweek*, December 12, 1949, 523. This article was a report on a symposium on lobotomy held in Washington, DC, in November 1949.
8. Jay Hoffman, "A Clinical Appraisal of Frontal Lobotomy in the Treatment of the Psychoses," *Psychiatry* 13 (1950): 355–360, 357–358.
9. H. C. Solomon, "Psychiatric Evaluation of Results Following Unilateral Prefrontal Lobotomy under Direct Vision," *Surgery, Gynecology and Obstetrics* 91 (1951): 606–608.
10. M. Greenblatt and H. C. Solomon, "Psychosurgery (Concluded)," part 2 of 2, *New England Journal of Medicine* 248, no. 2 (1953): 59–67, 59. Note the narrow criteria for what constituted sufficient improvement to justify the operation.
11. Magnus Peterson and J. Grafton Love, "Graded Lobotomy," *American Journal of Psychiatry* 16 (1949): 65–68.
12. Freeman and Watts, *Psychosurgery*, 511.
13. Walter Freeman to George H. Stevenson, January 3, 1951, School of Medicine, MS 0803, W. Freeman and J. Watts Papers, Special Collections Research Center, George Washington University Libraries (hereafter, Freeman and Watts Papers, GWU).
14. Stevenson to Freeman, July 28, 1950; Freeman to Stevenson, December 12, 1950; January 3, 1951, Freeman and Watts Papers, GWU.
15. Stevenson to Freeman, April 4, 1941, and January 29, 1952, Freeman and Watts Papers, GWU.
16. Though forty ECT treatments was out of the ordinary, Freeman frequently resorted to repeated ECT treatments postoperatively. See Freeman and Watts, *Psychosurgery*, 157.
17. Freeman and Watts, *Psychosurgery*, 157.
18. Freeman and Watts, *Psychosurgery*, 437, 436.
19. Freeman and Watts, *Psychosurgery*, 437.
20. Freeman and Watts, *Psychosurgery*, 435–438. "Theoretically all patients can be relieved," Freeman claimed, "if the incisions are made far enough posteriorly, but often this cannot be done without sacrificing the skills that are required to maintain an independent existence or indeed any existence at all." W. Freeman, "Mass Action versus Mosaic Function of the Frontal Lobe," *Journal of Nervous and Mental Disease* 110 (1949): 413–418, 415.
21. Jonathan M. Williams, in "Proceedings of the First Post-graduate Course in Psychosurgery," *Digest of Neurology and Psychiatry* 17 (1949): 441.
22. Howard Dully and Charles Fleming, *My Lobotomy: A Memoir* (New York: Broadway Books, 2008).
23. Quoted in Jack El Hai, *The Lobotomist: A Maverick Medical Genius and His Tragic Quest to Rid the World of Mental Illness* (New York: Wiley, 2005), 245.
24. El Hai, *The Lobotomist*, 244–245.
25. Freeman to Moniz, September 9, 1952, Freeman and Watts Papers, GWU.
26. For Freeman's account of his West Virginia activities, see W. Freeman, H. W. Davis, I. C. East, H. S. Tait, S. O. Johnson, and W. B. Rogers, "West Virginia Lobotomy

Project," *Journal of the American Medical Association* 156 (1954): 939–943. Walter Freeman to Egas Moniz, September 9, 1953, Freeman and Watts Papers, GWU.

27. Paul Hoch, "Theoretical Aspects of Frontal Lobotomy and Similar Operations," *American Journal of Psychiatry* 106 (1949): 448–453, 451–452.

28. Freeman and Watts, *Psychosurgery*, 204.

29. M. Greenblatt, R. Arnot, J. L. Poppen, and W. P. Chapman, "Report on Lobotomy Studies at the Boston Psychopathic Hospital," *American Journal of Psychiatry* 104 (1947): 361–368, 367.

30. F. R. Ewald, W. Freeman, and J. W. Watts, "Psychosurgery: The Nursing Problem," *American Journal of Nursing* 47 (1947): 210–213, 210.

31. Mical Raz, *The Lobotomy Letters: The Making of American Psychosurgery* (Rochester, NY: University of Rochester Press, 2013), especially chs. 4, 5, and 6, makes use of the extensive correspondence between Freeman and his patients to document how these families coped, often with what were still extremely difficult and hard-to-manage patients.

32. Ewald, Freeman, and Watts, "Psychosurgery: The Nursing Problem," 212.

33. Ewald, Freeman, and Watts, "Psychosurgery: The Nursing Problem," 212.

34. Freeman and Watts, *Psychosurgery*, 198.

35. Freeman and Watts, *Psychosurgery*, 193.

36. Joel Braslow, *Mental Ills and Bodily Cures* (Berkeley: University of California Press, 1997), ch. 6., 135.

37. Greenblatt, Arnot, Poppen, and Chapman, "Report on Lobotomy Studies," 361.

38. P. J. Schrader and M. F. Robinson, "Evaluation of Prefrontal Lobotomy through Ward Behavior," *Journal of Abnormal and Social Psychology* 40 (1945): 61–69, 62, 63.

39. B. E. Moore, S. Friedman, B. Simon, and J. Farmer, "A Cooperative Clinical Study of Lobotomy," *Association for Research in Nervous and Mental Disease* 27 (1948): 769–794, 792, 793.

40. Winfred Overholser, "Proceedings of the First Post-graduate Course in Psychosurgery," *Digest of Neurology and Psychiatry* 17 (1949). 429. Of five patients who were listed as "not improved," he commented dryly that "in their case actually [this] means that their condition was worse." See also Drubin's similar findings with a sample of sixty-two male patients lobotomized in veterans hospitals. Lester Drubin, "Further Observations on Sixty-Two Lobotomized Male Veterans at the Veterans Hospital, Northport, New York," *Journal of Nervous and Mental Disease* 113 (1951): 247–256.

41. M. Rotov, "A History of Trenton State Hospital," unpublished paper, n.d., in possession of the author; Schrader and Robinson, "Evaluation of Prefrontal Lobotomy," 68. Internal documents at Trenton once again reveal that "difficult to manage patients and the criminally insane were the top priority for selection."

42. *Evaluation of Psychosurgery: Proceedings of the Third Research Conference on Psychosurgery*, ed. F. A. Mettler and W. Overholser (Washington, DC: Public Health Service, 1954), 135–137.

43. There had been some experimentation with lobotomies on institutionalized "mental defectives . . . low grade imbeciles and idiots." These operations were undertaken in the hopes of rendering these patients more easily managed and were abandoned when they produced the opposite results. See Greenblatt and Solomon, "Psychosurgery (Concluded)."

44. At one of England's two hospitals for the criminally insane, Rampton, lobotomies were the treatment of choice for patients whose "extreme behavior cannot be tolerated."

G. W. MacKay, "Leucotomy in the Treatment of Psychopathic Feeble-Minded Patients in a State Mental Deficiency Institution," *Journal of Mental Science* 94 (1948): 844–850. Fourteen of the twenty who were lobotomized were women.

45. L. C. Cook's presidential address to the Medico-Psychological Association in 1958 acknowledged that psychosurgery had "provoked bitter attacks" and proceeded to dismiss them as largely unfounded. "The Place of Physical Treatments in Psychiatry," *Journal of Mental Science* 104 (1958): 933–942, 938, 939.

46. As Jack Pressman has noted, "chemical lobotomy" was more than a catchy phrase. To tout the efficacy of the new drug, "hospital psychiatrists described its benefits through such signs as a noticeable reduction in disturbed behavior, less destruction to hospital property, better worker morale, a new therapeutic atmosphere, increased hopefulness in patients and families, and gains in public relations—the very same terms used previously to promote psychosurgery." Jack D. Pressman, *Last Resort: Psychosurgery and the Limits of Medicine* (Cambridge: Cambridge University Press, 1998), 422.

47. P. Bailey, "The Great Psychiatric Revolution," *American Journal of Psychiatry* 113 (1956): 387–388.

48. See "Leucotomy Today," editorial, *Lancet* 280 (1962): 1037–1038; J. Pippard, "Leucotomy in Britain Today," *Journal of Mental Science* 108 (1962): 249–255.

49. Alfred P. Noyes and Lawrence Kolb, *Modern Clinical Psychiatry,* 6th ed. (Philadelphia: Saunders, 1963), 550, 552. For evidence that this textbook was the most widely consulted basic text in psychiatry through the 1960s, see Joan Woods, Sam Pieper, and Shervert Frazier, "Basic Psychiatric Literature," *Bulletin of the Medical Library Association* 56 (1968): 295–309.

50. Trenton State Hospital in New Jersey performed nearly 800 lobotomies between 1948 and 1957. After the practice ended, Harold Magee, the superintendent, continued to insist that "the operation is a good procedure, giving unexpectedly good results when applied as a therapy to unresponding aggressive and assaultive patients." Dismayed by lobotomy's undeserved poor reputation, as he saw it, he predicted "this procedure will again be activated." Trenton State Hospital, *Annual Report* (1958), 8. Magee's retirement in 1959 ensured his prediction did not come to pass.

51. For a recent attempt by a leading American psychiatrist to assign these discredited therapies to a deserved grave and trumpet the scientific achievements of contemporary psychiatry, see Jeffrey Lieberman, *Shrinks: The Untold Story of Psychiatry* (Boston: Little, Brown, 2015). For a critique, see Andrew Scull, "Shrink Wrapped," *Times Literary Supplement,* June 8, 2016.

52. See Andrew Scull, *Madness in Civilization: A Cultural History of Insanity, from the Bible to Freud, and from Madhouses to Modern Medicine* (London: Thames and Hudson / Princeton, NJ: Princeton University Press, 2015).

53. The desperation of families, and their willingness to try anything that might help, comes across starkly in the correspondence reviewed in Raz, *The Lobotomy Letters,* ch. 4. See also Jack Pressman's account of the demands for lobotomy from the families of patients confined at the McLean Hospital. Pressman, *Last Resort,* 305–309.

54. See Harry Marks, "Notes from the Underground: The Social Organization of Therapeutic Research," in *Grand Rounds: One Hundred Years of Internal Medicine,* ed. R. Maulitz and D. Long, 297–336 (Philadelphia: University of Pennsylvania Press, 1988); and H. Marks, *The Progress of Medicine: Science and Therapeutic Reform in the United States, 1900–1990* (Cambridge: Cambridge University Press, 2000).

CHAPTER 12 Creating a New Psychiatry

1. Clifford Beers, *A Mind That Found Itself* (New York: Longmans, Green, 1908).
2. The concept of mental hygiene, which hinged on the idea that there were ways to head off mental illness, had been quite common in the nineteenth century. See Barbara Sicherman, *The Quest for Mental Health in America* (New York: Arno, 1980). Beers's contribution was to set up an organization devoted to the task of promoting mental hygiene. On his life, see Norman Dain, *Clifford W. Beers: Advocate for the Insane* (Pittsburgh, PA: University of Pittsburgh Press, 1980). On the mental hygiene movement, see J. C. Pols, "Managing the Mind: The Culture of American Mental Hygiene, 1910–1950" (PhD diss., University of Pennsylvania, 1997).
3. See Margot Horn, *Before It's Too Late: The Child Guidance Movement in the United States, 1922–1945* (Philadelphia: Temple University Press, 1989). The Commonwealth Fund established seven model child guidance clinics, with a staff consisting of a social worker, a psychologist, and a psychiatrist as director. Other clinics were founded in the late 1920s and early 1930s drawing on these examples, but many were forced to close their doors due to the lingering effects of the Great Depression.
4. See Gerald Grob, *Mental Illness and American Society 1875–1940* (Princeton, NJ: Princeton University Press, 1983), 234–265. Meyer to W. G. Morgan, July 2, 1930, quoted in Grob, *Mental Illness and American Society*, 165. Meyer had expressed similar doubts three years earlier. See Meyer to Abraham Flexner, April 20, 1927, Adolf Meyer Collection, Series I (Correspondence), Unit I/1186 (Abraham Flexner), folder 1 (1927), Alan Mason Chesney Medical Archives, Johns Hopkins Medical Institutions.
5. Gitelson to Earle Saxe, December 19, 1939, Maxwell Gitelson Papers, MS 22905, Manuscript Division, Library of Congress, Washington, DC.
6. Adolf Meyer to William Gerry Morgan, July 2, 1930, reproduced in Gerald Grob, *The Inner World of American Psychiatry: Selected Correspondence* (New Brunswick, NJ: Rutgers University Press, 1985), 181–183, 182.
7. Frederick Brown, "General Hospital Facilities for Mental Patients," *Mental Hygiene* 15 (1931): 378–384.
8. E. D. Bond, *Thomas Salmon, Psychiatrist* (New York: Norton, 1950).
9. Alan Gregg, diary entries for May 6, May 8, October 21, 1931; June 20, July 3, October 15, 1935, Alan Gregg diary, RG 12 (Rockefeller Foundation Officers' Diaries), Rockefeller Foundation Records Collection, Rockefeller Archive Center, Sleepy Hollow, NY (hereafter, Gregg diary). Under Gregg, NCMH's pleas for funding generally fell on deaf ears.
10. Howard S. Berliner, *A System of Scientific Medicine: Philanthropic Foundations in the Flexner Era* (New York: Tavistock, 1985); Richard Brown, *Rockefeller Medicine Men: Medicine and Capitalism in America* (Berkeley: University of California Press, 1979); Kenneth Ludmerer, *Learning to Heal: The Development of American Medical Education* (New York: Basic Books, 1985), ch. 10. The minutes of a staff conference held on January 13, 1930, reveal that the General Education Board had spent $77 million over the course of seventeen years on revamping medical education in the United States. Minutes, January 13, 1930, box 1, folder 4, RG 3 (Rockefeller Foundation Records, Administration, Program and Policy), series 906 (Medical Sciences), Rockefeller Archive Center, Sleepy Hollow, NY (hereafter, RG 3, series 906, Rockefeller Archive Center). Frederick Gates, who was the principal adviser to both John D. Rockefeller and his son, played a major role in pointing the foundation's philanthropy in this direction, persuading Rockefeller senior to endow

the Rockefeller Institute for Medical Research (now Rockefeller University) in 1901 and later building on the recommendations of a report by Abraham Flexner that had been funded by Andrew Carnegie. Abraham Flexner, *Medical Education in the United States and Canada* (Washington, DC: Science and Health Publications, 1910). Gates's position as a trustee at Johns Hopkins played an important role in convincing others of the importance of these reforms.

11. Thomas Duffy, "The Flexner Report—100 Years Later," *Yale Journal of Biology and Medicine* 84 (2011): 269–276.

12. David Edsall, "Memorandum Regarding Possible Psychiatric Developments," October 3, 1930, Program and Policy—Psychiatry—Reports—Pro Psy-1a, RG 3, series 906, Rockefeller Archive Center.

13. See Gregg diary, January 22, February 24, March 30, May 4, May 5, May 6, May 8, October 21, October 28, November 4, and December 18, 1931.

14. Alan Gregg, "The New Program for Intensive Development in the Fields of Psychiatry and Neurology," April 1933, box 1, folder 4, RG 3, series 906, Rockefeller Archive Center.

15. Writing about his colleague, Warren Weaver suggests that Mason's interest in psychiatry played some role in the Rockefeller Foundation's decision to focus on this specialty. See W. Weaver, *Max Mason 1877–1961, a Biographical Memoir* (Washington, DC: National Academy of Sciences, 1964); see also Wilder Penfield, *The Difficult Art of Giving: The Epic of Alan Gregg* (Boston: Little, Brown, 1967), 224; and William H. Schneider, "The Model Foundation Officer: Alan Gregg and the Rockefeller Foundation Medical Division," *Minerva* 41 (2003): 155–166. Schneider suggests that other foundation officers lent their support because of mental illness in their families.

16. Raymond Fosdick, *Chronicle of a Generation: An Autobiography* (New York: Harper and Brothers, 1958).

17. That not all the trustees were convinced that this decision was a good one is hinted at in the 1934 report of the appraisal committee, which bluntly stated that "in this field, as in all others in which the Foundation is engaged, the Trustees would expect officers to give them definite warning when, in their opinion, the threshold of diminishing returns is reached in relation to a particular approach. Definite sailing directions can always be changed if the course proves to be unwise." "Medical Sciences—Program and Policy—Psychiatry," box 2, folder 19, RG 3, series 906, Rockefeller Archive Center. Such doubts would resurface in more pointed form after the Second World War and toward the end of Gregg's tenure as director.

18. Worcester State Hospital, box 117, folder 1445, RG 1.1 (Rockefeller Foundation records: Projects, Grants), series 200 (United States), Rockefeller Archive Center; Gregg, "New Program for Intensive Development"; Alan Gregg, "The Emphasis on Psychiatry," Trustees Confidential Memorandum, October 1943, box 2, folder 18, RG 3, series 906, Rockefeller Archive Center

19. See Gregg, "New Program for Intensive Development."

20. Gregg, "The Emphasis on Psychiatry."

21. Staff conference minutes, October 7, 1930, box 1, folder 4, RG 3, series 906, Rockefeller Archive Center.

22. Susan Lamb and the late Jack Pressman have been particularly prominent exponents of the view that Adolf Meyer was the major influence here. See Jack D. Pressman, *Last Resort: Psychosurgery and the Limits of Medicine* (Cambridge: Cambridge University Press, 1998), 18–46; J. Pressman, "Psychiatry and Its Origins," *Bulletin of the History of*

Medicine 71 (1997): 129–139; J. Pressman, "Psychosomatic Medicine and the Mission of the Rockefeller Foundation," in *Greater Than the Parts: Holism in Biomedicine, 1920–1950*, ed. Christopher Lawrence and George Weisz, 189–208 (New York: Oxford University Press, 1998); Susan Lamb, *Pathologist of the Mind* (Baltimore, MD: Johns Hopkins University Press, 2014); and S. Lamb, "Social Skills: Adolf Meyer's Revision of Clinical Skill for the New Psychiatry of the Twentieth Century," *Medical History* 59 (2015): 443–464.

23. On Richter, see Jay Schulkin, *Curt Richter: A Life in the Laboratory* (Baltimore, MD: Johns Hopkins University Press, 2005); for Richter's relations with Adolf Meyer, see Andrew Scull and Jay Schulkin, "Psychobiology, Psychiatry and Psychoanalysis: The Intersecting Careers of Adolf Meyer, Phyllis Greenacre, and Curt Richter," *Medical History* 53 (2009): 5–36. Richter was nominated for a Nobel Prize and arguably should have won one. The Nobel Committee eventually recognized the importance of the circadian rhythm by awarding the 2017 Prize in Physiology or Medicine to Jeffrey Hall, Michael Rosbash, and Michael Young for uncovering the molecular mechanisms that control it.

24. Alan Gregg to Francis Blake, dean of Yale Medical School, December 3, 1942, box 120, folder 1484, RG 1.1, series 200, Rockefeller Archive Center.

25. Gregg to Meyer, June 2, 1941, box 95, folder 1145, RG 1.1, series 200, Rockefeller Archive Center.

26. Diane Paul, "The Rockefeller Foundation and the Origins of Behavioral Genetics," in *The Expansion of American Biology*, ed. Keith Benson, 262–283 (New Brunswick, NJ: Rutgers University Press, 1991), 270.

27. Gregg diary, May 28, 1935.

28. Staff conference minutes, October 7, 1930, Rockefeller Archive Center.

29. See Franz Alexander, "Psychoanalytic Training in the Past, the Present, and the Future," 4, Address to the Association of Candidates of the Chicago Institute for Psychoanalysis, Archives, Chicago Institute for Psychoanalysis. On the "hostility of men in the Medical Faculty in Chicago" and the sense of relief when Alexander's year at the university came to an end, see Gregg diary, May 8, 1931. Funding for his year in Boston at the Judge Baker Foundation was underwritten by the Rosenwald Fund, headed by his patient Alfred Stern.

30. Gregg diary, April 4, 1932. Max Mason, the president of the foundation, attended the meeting with Stern and Alexander. Gregg had made the same point about the need for a university affiliation for the institute three months early. See Gregg diary, February 14, 1933. At that meeting, Gregg also indicated it was not clear to him that psychoanalysts "know enough to have something definitely to contribute to the study of somatic diseases."

31. Gregg's draft authorization to fund the institute makes this abundantly clear, emphasizing the institute's promised focus on "the correlation of medical and physiological problems with the findings of psychoanalysis." Authorizing Resolution, 1934, p. 4, box 3, folder 28, RG 1.1 (Rockefeller Foundation records: Projects, Grants), series 216 (Illinois), subseries 216.A (Medical Sciences), Rockefeller Archive Center. Theodore Brown, "Alan Gregg and the Rockefeller Foundation's Support of Franz Alexander's Psychosomatic Research," *Bulletin of the History of Medicine* 61 (1987): 155–182, provides a detailed account of this episode, which I have supplemented with my own research in the Rockefeller Archive Center. In the Gregg diary, the entries for January 2 and February 4, 1934, document the meetings that led to his recommendation to provide support.

32. Theodore Brown notes "Alexander's not always subtle reorientation during his courtship of Gregg" as documented in a series of letters to Gregg. Brown, "Alan Gregg and the Rockefeller Foundation's Support," 172–173.

33. Stern helped reassure Gregg on this front, writing that the institute proposed to focus its research efforts on gastrointestinal disorders and on "psychological factors involved in ailments of the respiratory tract, hypertension, endocrine gland disturbances, and skin disorders." Stern to Gregg, June 26, 1934, box 3, folder 27, RG 1.1 (Rockefeller Foundation records: Projects, Grants), series 216 (Illinois), Rockefeller Archive Center.

34. Gregg diary, November 2, 1937. Gregg had earlier communicated his disquiet directly to Alexander. See Gregg to Alexander, March 24, 1937, box 3, folder 32, RG 1.1 (Rockefeller Foundation records: Projects, Grants), series 216 (Illinois), Rockefeller Archive Center.

35. R. A. Lambert to D. P. O'Brien, August 22, 1939; D. P. O'Brien to R. A. Lambert, August 30, 1939, box 3, folder 37, RG 1.1 (Rockefeller Foundation records: Projects, Grants), series 216 (Illinois), Rockefeller Archive Center.

36. Stern subsequently married Dodd and moved to New York, loosening his ties to Alexander. His new wife had a remarkable time in Berlin, sleeping with, among others, the first head of the Gestapo, the French ambassador, and the chief Russian spy in the city, with whom she was besotted. In the months leading up to her marriage, she carried on an affair with the Hollywood director Sidney Kauffman, and her love life did not diminish after marrying Stern. In Berlin, she became a Soviet spy, and she subsequently recruited the hapless Stern to the cause. Both were forced to flee in the McCarthy years, first to Mexico and then, on passports bought from Paraguay, to Prague, where they died in exile decades later.

37. Gregg diary, October 31, 1941.

38. Gregg diary, February 15, March 4, 1934.

CHAPTER 13 Talk Therapy

1. W. H. Auden, "For Sigmund Freud," *Kenyon Review* 2 (1940): 30–34, 33.

2. Edward Lazell, "The Psychology of War and Schizophrenia," *Psychoanalytic Review* 7 (1920): 224–245, 225. Lazell was on the staff of St. Elizabeths Hospital, which was run by William Alanson White, one of the few mainstream American psychiatrists sympathetic to psychoanalysis during this period. Lazell attempted to use psychoanalysis to treat some of his patients.

3. Lincoln Steffens, *The Autobiography of Lincoln Steffens* (New York: Harcourt, Brace, 1931), 655–656.

4. Hans Eulau, "Mover and Shaker: Walter Lippman as a Young Man," *Antioch Review* 11 (1951): 291–312; Robert Rosenstone, "The Salon of Mabel Dodge," in *Affairs of the Mind: The Salon in Europe and America from the 18th to the 20th Century*, ed. Peter Quennell, 131–151 (London: Weidenfeld and Nicolson, 1980).

5. Walter Lippman, "Freud and the Layman," *New Republic*, April 17, 1915, 9–10.

6. Max Eastman and Floyd Dell, "The Science of the Soul," *The Masses*, July 1915, 21–22; Eastman and Dell, "The Science of the Soul," *The Masses*, July 1916, 30.

7. Theodore Dreiser, *Neurotic America and the Sex Impulse: And Some Aspects of Our National Character* (Girard, KS: Haldeman-Julius, 1920). Proud of the piece, he promptly reprinted it in his collection of essays, T. Dreiser, *Hey Rub-a-Dub-Dub: A Book of the Mystery and Wonder and Terror of Life* (New York: Boni and Liveright, 1920).

8. Josef Breuer and Sigmund Freud, *The Standard Edition of the Complete Psychological Works of Sigmund Freud,* vol. 2: *Studies on Hysteria* (London: Hogarth Press, 1955), 160.

9. Max Eastman, "Exploring the Soul and Healing the Body," *Everybody's Magazine* 32, June 1915, 741.

10. This is the common English translation. Actually, Freud had spoken of a *Todestrieb,* or death drive. The idea was first proposed by Jung's patient and mistress Sabina Spielrein in her paper "Die Destruktion als Ursache des Werdens," *Jarhbuch für Psychianalitische und Psychopathologische Forschungen* 4 (1912): 465–503. Freud first adopted the idea in *Jenseits des Lustprinzips* (Beyond the pleasure principle) (Vienna: Internationaler Psychoanalytischer Verlag, 1920).

11. Quoted in W. David Sievers, *Freud on Broadway* (New York: Hermitage House, 1955), 65. The plodding plot summaries that Sievers provides of a host of forgotten plays document the period's embrace of psychoanalysis among the theater-going crowd.

12. Theodore Dreiser, *The Hand of the Potter* (New York: Boni and Liveright, 1918), 198–200.

13. *Long Day's Journey into Night* was, O'Neill wrote to his third wife Carlotta Monterey (to whom he dedicated the manuscript), "this play of old sorrow, written in[?] tears and blood." Three years after his death in 1953, she authorized its first public performance.

14. See, for example, David Lomas, *The Haunted Self: Surrealism, Psychoanalysis, Subjectivity* (New Haven, CT: Yale University Press, 2000).

15. Meyer to one of his patients, quoted in S. D. Lamb, *Pathologist of the Mind: Adolf Meyer and the Origins of American Psychiatry* (Baltimore, MD: Johns Hopkins University Press, 2014), 232. Chapter 6 of Lamb's book provides a useful analysis of Meyer's growing hostility to psychoanalysis. Meyer later claimed that, in relying on transference, the psychoanalyst was no better than a faith healer. Adolf Meyer, "Preparation for Psychiatry," *Archives of Neurology and Psychiatry* 30 (1933): 1111–1125, 1123.

16. Members of the Secret Committee, given ancient rings by Freud to signify their devotion to the cause, were Ernest Jones, Karl Abraham, Max Eitingon, Sándor Ferenczi, Otto Rank, and Hanns Sachs.

17. The Menninger Clinic was modeled in part on another successful family clinic in the Midwest, the Mayo Clinic, which enjoyed a stellar reputation in surgery.

18. Lawrence Friedman, *Menninger: The Family and the Clinic* (New York: Knopf, 1990), 59.

19. William Menninger also came back to Topeka more prepared to fight his brother, boasting that he had "a penis as big as Father and big brother" (Friedman, *Menninger,* 90), an endowment he put to vigorous use in a different setting in his nightly visits to the chief nurse, his longtime mistress.

20. Friedman, *Menninger,* 57–59, shows that Karl Menninger was well aware of the book's shortcomings.

21. Friedman, *Menninger,* 78.

22. I have drawn throughout this discussion on Friedman, *Menninger.* On the prevalence and extent of the sexual exploitation of the female staff, see 81–82. In its heyday in the early 1950s, the average cost of a stay at the Menninger Clinic was about $1,400 a month, roughly $185,000 a year in 2021 dollars. In the era of managed care, only a few of the private establishments survived. Menninger was not one of them.

23. At the Menninger Clinic, trainees were paid poorly and had to use a significant portion of their limited funds to pay for their training analysis. There was thus an incentive to extend the period of the training analysis, and so long as that analysis was in pro-

gress, the analysand was trapped. Concern that the information volunteered during the analytic sessions might be used against them was well founded: Karl Menninger regularly shared personal information about fellows in the program to their detriment. Friedman, *Menninger*, 186–191.

24. Douglas Kirsner, *Unfree Associations: Inside Psychoanalytic Institutes* (London: Process Press, 2000).

25. Suzanne Hollman, "White's Restraint and Progressive American Psychiatry at St. Elizabeths Hospital 1903–1937" (PhD diss., University College London, 2020).

26. Chestnut Lodge remained in the family for three generations but fell on hard times. In 1997 it was sold to an outside health company, and four years later it was bankrupt. It subsequently burned down.

27. Quoted in Friedman, *Menninger*, 118.

28. Friedman, *Menninger*, 116–119.

29. Friedman, *Menninger*, 134–135.

30. Fees at Austen Riggs now average over $300,000 a year for long-stay patients. Shorter stays cost in the region of $60,000 for six weeks. At the McLean, costs are higher still, averaging more than half a million dollars a year.

31. Nathan G. Hale, Jr., *The Rise and Crisis of Psychoanalysis in the United States: Freud and the Americans, 1917–1985* (New York: Knopf, 1995), 116–118.

32. Edward Gitre, "The Great Escape: World War II, Neo-Freudianism, and the Origins of U.S. Psychocultural Analysis," *Journal of the History of the Behavioral Sciences* 47 (2011): 18–43, 20.

33. Robert P. Knight, "The Present Status of Organized Psychoanalysis in the United States," *Journal of the American Psychoanalytical Association* 1 (1953): 197–221, 207.

34. Hale, *Rise and Crisis of Psychoanalysis*, 135. A decade earlier, the 60 members of the New York Psychoanalytic Society and Institute dispensed a grand total of 300 classical analyses (5 hours a week per patient) over the course of a year. Hale, *Rise and Crisis of Psychoanalysis*, 164.

35. Knight, "The Present Status of Organized Psychoanalysis," 207.

36. Operating in a deeply antisemitic university, Winternitz instructed his admissions committee to admit no more than five Jews in a given year (and only two Italian Catholics). See Michael A. Nevins, *Abraham Flexner: A Flawed American Icon* (New York: iUniverse, 2010), 70–71.

37. Quoted in Kirsner, *Unfree Associations*, 18. Lewin refused to collect the financial affidavits the immigration authorities required before they would sanction visas for European analysts.

38. George Makari, *Revolution in Mind: The Creation of Psychoanalysis* (New York: Norton, 2008), 462.

39. M. Mahler, *The Memoirs of Margaret Mahler*, ed. Paul E. Stepansky (New York: Free Press, 1988), 102.

40. Makari, *Revolution in Mind*, 474.

41. For details of the squabbles, which lasted from 1940 to 1945, see Makari, *Revolution in Mind*, 475–483. See also John Frosch, "The New York Psychoanalytic Civil War," *Journal of the American Psychoanalytic Association* 39 (1991): 1037–1064; and Kirsner, *Unfree Associations*, 20–26.

42. Before a new institute was admitted, it had to show it had at least ten trained analysts to serve as charter members, so the spread of the new societies indirectly maps the spread of psychoanalysis itself.

CHAPTER 14 War

1. Sullivan's salary of $7,500 was paid by the Rockefeller Foundation and disbursed through the William Alanson White Foundation. The Rockefeller Foundation expected Sullivan to spend his time instructing local draft boards, advisory physicians, and army medical officers about "the application of psychiatry to the work of recruitment." Memorandum, box 117, folder 1443, RG 1.1 (Rockefeller Foundation records: Projects, Grants), series 200 (United States), Rockefeller Archive Center, Sleepy Hollow, NY (hereafter, Rockefeller Archive Center).

2. Many came to regard the whole screening process as "little more than a farce." See discussion in Ellen Herman, *The Romance of American Psychology* (Berkeley: University of California Press, 1995), 88. See also Edgar Jones, Kenneth Hyams, and Simon Wessely, "Screening for Vulnerability to Psychological Disorders in the Military: An Historical Survey," *Journal of Medical Screening* 10 (2003): 40–46.

3. Allan Bérubé, *Coming Out under Fire: The History of Gay Men and Women in World War Two* (New York: Free Press, 1990), 9–11, 18–19.

4. For a discussion of the politics of this change of nomenclature, and its futility as a device for reducing the stigma surrounding soldiers' breakdowns, see Rebecca Jo Plant, "'Combat Exhaustion' vs. 'Psychoneurosis': American Psychiatrists and the Terminology of War Trauma during World War II," in *Gender and Trauma: 1945 to the Present,* ed. Paula Michaels and Christina Twomey (London: Bloomsbury, 2021), 79–95.

5. Malcolm J. Farrell, "Status and Development," in Medical Department, United States Army, Office of Medical History, *Neuropsychiatry in World War II,* vol. 1: *Zone of Interior* (Washington, DC: Office of the Surgeon General, 1966), 27–29.

6. Gerald Grob, "World War II and American Psychiatry," *Psychohistory Review* 19 (1990): 41–69, 54.

7. Herman, *The Romance of American Psychology,* 9.

8. Ben Shephard, *A War of Nerves: Soldiers and Psychiatrists in the Twentieth Century* (Cambridge, MA: Harvard University Press, 2001), 330.

9. Grob, "World War II and American Psychiatry," 59–60.

10. Hans Pols and Stephanie Oak, "War and Military Mental Health: The US Psychiatric Response in the 20th Century," *American Journal of Public Health* 97 (2007): 2132–2142, 2142.

11. For a similar assessment of the orientation of these military psychiatrists, see Plant, "'Combat Exhaustion.'"

12. Shephard, *A War of Nerves,* 219–220.

13. Ben Shephard quotes this assessment in *A War of Nerves,* 224. For the general picture, see Lloyd J. Thompson, Perry C. Talkington, and Alfred O. Ludwig, "Neuropsychology at Army and Divisional Levels," in Medical Department, United States Army, *Neuropsychiatry in World War II,* vol. 2: *Overseas Theaters,* 275–333 (Washington, DC: Office of the Surgeon General, 1973).

14. Franklin D. Jones, *Military Psychiatry since World War II* (Washington, DC: American Psychiatric Press, 2000), 9.

15. On the irrelevance of these "therapies" to the observed outcomes, see John C. Whitehorn, "Combat Exhaustion," 6, quoted in Gerald Grob, *From Asylum to Community: Mental Health Policy in Modern America* (Princeton, NJ: Princeton University Press, 1991), 16. Injections of sodium pentathol, which produced a dreamlike state, allowing for the recall of trauma, were widely employed without much success.

16. Quoted in Herman, *The Romance of American Psychology,* 118. As Herman dryly comments, "According to such definitions, virtually any human relationship could qualify as psychotherapeutic" (118).
17. Shephard, *A War of Nerves,* 218.
18. John Keegan, *The Face of Battle* (Harmondsworth, UK: Penguin, 1978), 335.
19. Shephard, *A War of Nerves,* 327.
20. Shephard, *A War of Nerves,* 332, 219.
21. Grob, *From Asylum to Community,* 15.
22. Herbert X. Spiegel, "Preventive Psychiatry with Combat Troops," *American Journal of Psychiatry* 101 (1944): 310–315.
23. Samuel Stouffer et al., *The American Soldier,* vol. 1: *Adjustment during Army Life* (Princeton, NJ: Princeton University Press, 1949).
24. Ellen Dwyer, "Psychiatry and Race during World War II," *Journal of the History of Medicine and Allied Sciences* 61 (2006): 117–143. During peacetime, Black mental patients were segregated from their white counterparts, either in separate wards or in separate mental hospitals where conditions were particularly noisome.
25. Albert J. Glass, "Lessons Learned," in Medical Department, United States Army, *Neuropsychiatry in World War II,* vol. 2, 999–1000.
26. Edgar Jones and Simon Wessely, *Shell Shock to PTSD: Military Psychiatry from 1900 to the Gulf War* (New York: Psychology Press, 2005), 25.
27. Quoted in Jones and Wessely, *Shell Shock to PTSD.*
28. See Roy Grinker and Herbert Speigel, *War Neuroses* (Philadelphia: Blakiston, 1945).
29. Jones and Wessely, *Shell Shock to PTSD,* 79. See also Shephard, *A War of Nerves;* and Pols and Oaks, "War and Military Mental Health."
30. Jones and Wessely, *Shell Shock to PTSD,* 80–81. These figures were from the Italian campaign. Separate figures for the period after the D-Day invasion largely conform to this pattern (84–85).
31. After the war, Douglas Bond, chair of psychiatry at Case Western Reserve University, shared with Robert Morison of the Rockefeller Foundation "his own experience in military psychiatry," denouncing "the uselessness of psychological selection procedures and psychiatric therapy as practised during the war" and promising a book on the subject. Robert Morison interview with Douglas Bond, August 26, 1948, memorandum, box 117, folder 1442, RG 1.1 (Rockefeller Foundation projects), series 200 (United States), Rockefeller Archive Center.
32. For the dismissive reaction of orthodox analysts, émigrés, and others, see Rebecca Plant, "William Menninger and American Psychoanalysis, 1946–48," *History of Psychiatry* 16 (2005): 181–202. David Rapaport, one of the émigré analysts, spoke scathingly of "a crew of pragmatic simplifiers," chief among them, of course, the Menningers, at whose clinic he then worked. Elizabeth Geleerd, another émigré analyst, was even blunter: the American popularizers had "watered down Freud's fundamental concepts considerably, even to the point where they are completely unrecognizable and almost hostile toward Freud's ideas." Letter from Elizabeth Geleerd to William Menninger, December 8, 1947, quotes in Plant, "William Menninger and American Psychanalysis," 191, 193.
33. William Menninger, *Psychiatry in a Troubled World* (New York: Viking, 1967), 451.
34. On the professional infighting, see Grob, *From Asylum to Community,* ch. 2.
35. See Joel Paris, *The Fall of an Icon: Psychoanalysis and Academic Psychiatry* (Toronto: University of Toronto Press, 2005); and Nathan Hale, *The Rise and Crisis of Psychoanalysis in the United States* (New York: Oxford University Press, 1998), esp. ch. 14.

36. Hale, *The Rise and Crisis of Psychoanalysis*, 246.
37. L. J. Friedman, "The 'Golden Years' of Psychoanalytic Psychiatry," *Psychohistory Review* 19 (1990), 5–40, 8.
38. Hale, *The Rise and Crisis of Psychoanalysis*, 222.
39. See Friedman, "The 'Golden Years' of Psychoanalytic Psychiatry." To put into perspective just how extraordinary it was for a single facility to attempt to train as many as 108 psychiatric residents, more than seventy years later, in 2012, the total number of first-year residents in psychiatry in the United States was 1,117. See A. Satiani, J. Niedermeier, B. Satiani, and D. P. Svendsen, "Projected Workforce of Psychiatrists in the United States: A Population Analysis," *Psychiatric Services* 19 (2018): 710–713, 711.
40. Hale, *The Rise and Crisis of Psychoanalysis*, 246, 248.
41. Hale, *The Rise and Crisis of Psychoanalysis*, 246–248, 253–256; John MacIver and Frederick C. Redlich, "Patterns of Psychiatric Practice," *American Journal of Psychiatry* 115 (1959): 692–697.

CHAPTER 15 Professional Transformations

1. Benjamin Spock, *The Commonsense Book of Baby and Child Care*, 2nd ed. (New York: Duell, Sloane and Pearce, 1957), 49–51, 245–260, 358–361.
2. Paul Green, *Jennifer Jones: The Life and Films* (Jefferson, NC: McFarland, 2011).
3. For a perspective on Freud in Hollywood by authors sympathetic to his ideas, see Krin Gabbard and Glen Gabbard, *Psychiatry and the Cinema* (Chicago: University of Chicago Press, 1987).
4. See Ellen Herman, *The Romance of American Psychology* (Berkeley: University of California Press, 1995), 247–250.
5. See L. Friedman, *Menninger: The Family and the Clinic* (New York: Knopf, 1990), 59, 84–85, 89–90, 163–166, 200–201.
6. William Menninger to Karl Menninger, November 17, 1939, box TIP, General Correspondence, 1930–1940, Menninger Archives, Kansas State Historical Society, Topeka, KS.
7. William Menninger, "Presentation to the American Psychoanalytic Association," folder PSA: "Correspondence re. Strengthening Organization," Menninger Archives, quoted in Rebecca Jo Plant, "William Menninger and American Psychoanalysis, 1946–48," *History of Psychiatry* 16 (2005): 181–202, 184–185.
8. Kenneth Eisold, *The Organizational Life of Psychoanalysis: Conflicts, Dilemmas, and the Future of the Profession* (London: Routledge, 2017), 59.
9. William Menninger, "Remarks on Accepting Nomination for Presidency of the American Psychoanalytical Association," *Psychoanalytic Quarterly* 15 (1946): 413–415, 413.
10. The journal had been edited by Clarence Farrar since 1931, and Farrar would continue in that post until 1965.
11. Quoted in Jack Pressman, *Last Resort: Psychosurgery and the Limits of Medicine* (New York: Cambridge University Press, 1998), 370.
12. Gerald Grob, *From Asylum to Community* (Princeton, NJ: Princeton University Press, 1991), 32–33.
13. Charles Burlingame, "Psychiatric Sense and Nonsense," *Journal of the American Medical Association* 133 (1947): 971–1044.
14. Jack Pressman has termed Bullard a "willing pawn" in a conspiracy hatched in Topeka by William Menninger and his allies. GAP wanted to hide what it was up to. Pressman, *Last Resort*, 368–369.

15. J. Berkeley Gordon, quoted in Robert Gordon, "In Memoriam: George S. Stevenson, 1892–1983," *American Journal of Psychiatry* 140 (1983): 1369–1371, 1371.
16. Burlingame complained bitterly to Robert Morison at the Rockefeller Foundation "about the publicity-seeking tactics of the Menningers, and the tendency of the Group for the Advancement of Psychiatry as a whole to oversell psychiatry to the American public." Morison responded dryly that "offhand it would seem that B. is throwing his stones from a rather fragile glass house." Morison diary entry for April 7, 1949, Robert Swain Morison diary, 1947, RG 12 (Rockefeller Foundation Officers' Diaries), Rockefeller Foundation Records Collection, Rockefeller Archive Center, Sleepy Hollow, NY (hereafter, Morison diary). Two years earlier, Morison had learned that Burlingame "has a public relations counsel not only for the Institute of Living but for himself personally." Morison's informant, Lydia Giberson (whom Morison described as "a lively intelligent, and perceptive person . . . [who] is not at all vindictive"), commented that "the professional staff at the Institute is ruthlessly exploited and G. confirms the opinion which RSM has heard elsewhere that the Institute comes very close to being a racket." Morison hastened to add: "The above adverse comments are, of course, confidential." Morison diary, September 25, 1947.
17. "Shock Therapy," *GAP Report No. 1,* September 15, 1947. This and other GAP reports are preserved in the William C. Menninger Archives, held at the Kansas Historical Society, Topeka KS, Unit ID 37226.
18. "Research on Frontal Lobotomy," *GAP Report No. 6,* June 1948, 2.
19. "Dr. Knight Assails 'Strong-Arm Cure': Psychiatrist Scores 'Beating Illness Out'—Urges 'Gentle' Approach to Patients," *New York Times,* June 6, 1948.
20. Quoted in Grob, *From Asylum to Community,* 28.
21. His manifestos are collected in William Menninger, *A Psychiatrist for a Troubled World: Selected Papers of William C. Menninger,* ed. B. H. Hall, 2 vols. (New York: Viking, 1967).
22. See Geddes Smith to Mr. Aldrich, memorandum, December 11, 1952, Commonwealth Fund records, Subgroup 1, Administration–Historical Files, series 1, Menninger Foundation, box 50, folder 863, Rockefeller Archive Center. For the Rockefeller Foundation's refusal of Menninger's requests, see William Menninger to Gregg, November 19, 1949; Gregg to Menninger, November 25, 1949; Menninger to Gregg, December 6, 1949; Gregg to Menninger, December 19, 1949; all in RG 1.1 (Rockefeller Foundation records: Projects, Grants), series 219 (Kansas), subseries 219.A (Medical Sciences), Menninger Foundation, Rockefeller Archive Center.
23. Commonwealth Fund records, Subgroup 1, Administration–Historical Files, series 1, Menninger Foundation, box 50, folder 863, Rockefeller Archive Center.
24. Commonwealth Fund records, Subgroup 1, Administration–Historical Files, series 1, Menninger Foundation, box 50, folder 863, Rockefeller Archive Center.
25. See note rejecting still another appeal from Menninger, December 10, 1953, in Rockefeller Brothers Fund records, RG 3 (Projects, Grants), Subgroup 1, Rockefeller Archive Center.
26. On Weaver, see Robert Kohler, "The Management of Science: The Experience of Warren Weaver and the Rockefeller Foundation Programme in Molecular Biology," *Minerva* 14 (1976): 279–306; and Pnina Abir-Am, "The Rockefeller Foundation and the Rise of Molecular Biology," *Nature Reviews Molecular Cell Biology* 3 (2002): 65–70. Abir-Am notes that Weaver's ties to Mason allowed Weaver to intrude on territory Gregg thought was rightfully his and to adopt a policy of granting many small and short-term grants—

something Gregg dismissed as distributing "chicken feed." It was chicken feed that had massive and lasting effects on the shape of the biological sciences.

27. Mina Rees, "Warren Weaver, July 17, 1894–November 24, 1978. A Biographical Memoir," *Biographical Memoirs*, vol. 57 (Washington, DC: National Academy of Sciences, 1987): 493–529, 504, http://www.nasonline.org/publications/biographical-memoirs/memoir-pdfs/weaver-warren.pdf.

28. Jack Pressman, "Psychosomatic Medicine and the Mission of the Rockefeller Foundation," in *Greater Than the Parts: Holism in Scientific Medicine, 1920–1950*, ed. Christopher Lawrence and George Weisz (New York: Oxford University Press, 1998), 202.

29. Robert S. Morison, interoffice memorandum, February 3, 1947, box 1, folder 5, RG 3 (Rockefeller Foundation Records, Administration, Program and Policy), series 906 (Medical Sciences), Rockefeller Archive Center, Sleepy Hollow, NY (hereafter, RG 3, series 906, Rockefeller Archive Center).

30. Chester Barnard to Alan Gregg and Robert S. Morison, August 9, 1948, box 2, folder 18, RG 3, series 906, Rockefeller Archive Center.

31. Barnard to Gregg and Morison, August 9, 1948.

32. Robert S. Morison to Chester Barnard, memorandum (marked also read by Alan Gregg), September 30, 1948, box 2, folder 18, RG 3, series 906, Rockefeller Archive Center.

33. Morison to Barnard, September 30, 1948.

34. Morison to Barnard, September 30, 1948.

35. Ronald Bayer, *Homosexuality and American Psychiatry: The Politics of Diagnosis* (Princeton, NJ: Princeton University Press, 1981).

36. Morison to Barnard, September 30, 1948.

37. Internal memorandum from Robert S. Morison, April 11, 1951, box 1, folder 5, RG 3, series 906, Rockefeller Archive Center.

38. Grob, *From Asylum to Community*, 42.

39. The first edition was published by Brunner in 1955 and the second by Basic Books in 1974. By the time the second edition appeared, Arieti had concluded that "only" a quarter of schizophrenics owed their disease directly to the malign influence of a refrigerator mother.

40. Frank L. Wright, ed., *Out of Sight, Out of Mind* (Philadelphia: National Mental Health Foundation, 1947). For recent commentary and critique, see Steven Taylor, *Acts of Conscience: World War II, Mental Institutions, and Religious Objectors* (Syracuse, NY: Syracuse University Press, 2009).

41. Alfred Q. Maisel, "Bedlam 1946," *Life Magazine*, May 6, 1946.

42. Harold Orlansky, "An American Death Camp," *Politics* 5 (1948): 162–168.

43. Albert Deutsch, *The Shame of the States* (New York: Harcourt, Brace, 1948), 41–42.

44. Deutsch, *The Shame of the States*, 28, 73, 75, 107.

45. Deutsch, *The Shame of the States*, 27.

46. Deutsch, *The Shame of the States*, 185.

47. Ivan Belknap, *Human Problems of a State Mental Hospital* (New York: McGraw Hill, 1956), xi, 212.

48. Erving Goffman, *Asylums: Essays on the Social Situation of Mental Patients and Other Inmates* (Garden City, NY: Doubleday, 1961). For other examples of studies reaching similar conclusions, see H. W. Dunham and S. K. Weinberg, *The Culture of the State Mental Hospital* (Detroit, MI: Wayne State University Press, 1960); and Robert Perrucci, *Circle of Madness: On Being Insane and Institutionalized in America* (Englewood Cliffs, NJ: Prentice-Hall, 1974).

CHAPTER 16 A Fragile Hegemony

1. "Psychosurgery," *Life*, March 3, 1947, quoted in Mical Raz, *The Lobotomy Letters: The Making of American Psychosurgery* (Rochester, NY: University of Rochester Press, 2013), 56–57. Raz's chapter "Between the Ego and the Ice Pick" reviews a series of Freudian accounts of why lobotomy worked. One particularly creative example explained that "the lobotomized schizophrenic is a more successful schizophrenic in that he has achieved what he has always wanted: that oral reunion with his mother previously feared as engulfing or devouring and threatening to his identity." Eugene Brody, "Superego, Introjected Mother, and Energy Discharge in Schizophrenia: Contribution to the Study of Anterior Lobotomy," *Journal of the American Psychoanalytic Association* 6 (1958): 481–501, 499. There were similar attempts to provide psychoanalytical rationales for electroconvulsive therapy and insulin coma therapy.

2. "The Social Responsibility of Psychiatry: A Statement of Orientation," *GAP Report no. 13*, 1950, 4–5. GAP's reports and records are preserved in the William C. Menninger Archives at the Kansas Historical Society, Topeka, KS.

3. M. Friedman and R. Rosenman, "Association of Specific Overt Behavior Pattern with Blood and Cardiovascular Findings," *Journal of the American Medical Association* 169 (1959): 1286–1296. This line of research was shown to be a tobacco industry–funded attempt to obscure the links between smoking and health. See M. P. Petticrew, K. Lee, and M. McKee, "Type A Behavior Pattern and Coronary Heart Disease: Philip Morris's 'Crown Jewel,'" *American Journal of Public Health* 102 (2012): 2018–2025; A. Landman, D. K. Cortese, and S. Glantz, "Tobacco Industry Sociological Programs to Influence Public Beliefs about Smoking," *Social Science and Medicine* 66 (2008): 970–981. There are reasons to doubt the type A phenomenon. See K. Smeigelskas, N. Zemaitiene, J. Julkunen, and J. Kauhanen, "Type A Behavior Pattern Is Not a Predictor of Premature Mortality," *International Journal of Behavioral Medicine* 22 (2015): 161–169. But belief in the association between personality type and susceptibility to heart attacks persists among laypeople.

4. See Franz Alexander, "The Influence of Psychologic Factors upon Gastrointestinal Disturbances," *Psychoanalytic Quarterly* 3 (1934): 501–588. Alexander expanded on these claims in the postwar period. See F. Alexander, *Psychosomatic Medicine* (New York: Norton, 1950). For an early twentieth-century attempt to link the emotions and gastric upset, see Walter B. Cannon, "The Influence of Emotional States on the Functions of the Alimentary Canal," *American Journal of Medical Science* 137 (1909): 480–486. In a longer historical perspective, the Hippocratic ideas that dominated Western medical thinking for two millennia had embraced such notions on a still-broader scale.

5. Barry Marshall and Robin Warren, "Unidentified Curved Bacilli in the Stomach of Patients with Gastritis and Peptic Ulceration," *Lancet* 323 (1984): 1311–1315.

6. Mark Jackson, *Asthma: The Biography* (Oxford: Oxford University Press, 2009), 142.

7. Helen Flanders Dunbar, *Mind and Body: Psychosomatic Medicine* (New York: Random House, 1947), 177, quoted in Jackson, *Asthma*, 143.

8. Jackson, *Asthma*. One of the leading advocates of parentectomy was Dr. Murray Pershkin of the National Home for Jewish Children in Denver, Colorado. See also Leonard Bernstein et al., "Pulmonary Function in Children: Preliminary Studies in Chronic Intractable Childhood Asthma," *Journal of Allergy* 30 (1959): 534–540.

9. See Edith Sheffer, *Asperger's Children: The Origins of Autism in Nazi Vienna* (New York: W. W. Norton, 2018).

10. Leo Kanner, "Autistic Disturbances of Human Contact," *Nervous Child* 2 (1943): 217–250.

11. The rapid rise in the number of children diagnosed with autism immediately followed the release of the fourth edition of the American Psychiatric Association's *Diagnostic and Statistical Manual* in 1994, something that the edition's editor, Allen Frances, has argued was the unintended consequence of an alteration in the criteria for the disorder. This is a highly controversial claim, fiercely rejected by parents of children with the diagnosis.

12. Kanner, "Autistic Disturbances of Affective Contact."

13. Leo Kanner, "Problems of Nosology and Psychodynamics of Early Infantile Autism," *American Journal of Orthopsychiatry* 19 (1949): 416–426.

14. "Medicine: The Child Is Father," *Time*, July 25, 1960.

15. Bruno Bettelheim, *The Empty Fortress: Infantile Autism and the Birth of the Self* (New York: Free Press, 1967), 125.

16. Memoirs by Bettelheim's former students document this abuse. See, for example, Stephen Eliot, *Not the Thing I Was: Thirteen Years at Bruno Bettelheim's Orthogenic School* (New York: St. Martin's Press, 2003); and Roberta Calar Redford, *Crazy: My Seven Years at Bruno Bettelheim's Orthogenic School* (New York: Trafford, 2010). Richard Pollak's biography of Bettelheim was inspired by his brother's mistreatment at the school: R. Pollak, *The Creation of Dr. B.: A Biography of Bruno Bettelheim* (New York: Simon and Schuster, 1997). See also Richard Bernstein, "Accusations of Abuse Haunt the Legacy of Dr. Bruno Bettelheim," *New York Times*, November 4, 1990.

17. For a sympathetic biography, see Gail Hornstein, *To Redeem One Person Is to Redeem the World: The Life of Frieda Fromm-Reichmann* (New York: Free Press, 2000).

18. Frieda Fromm-Reichmann, "Notes on the Development of Treatment of Schizophrenics by Psychoanalytic Psychotherapy," *Psychiatry* 11 (1948): 263–273, 265, 263–264.

19. Fromm-Reichmann, "Notes on the Development of Treatment," 271. Attempts to treat psychosis with psychoanalysis were associated with the psychoanalytic community in the Baltimore-Washington area, particularly the Sheppard and Enoch Pratt Hospital and Chestnut Lodge, where prominent analysts such as Harry Stack Sullivan and Silvano Arieti practiced. The Scottish psychiatrist R. D. Laing's early writings on schizophrenia, most notably *The Divided Self* (London: Tavistock, 1960) and (with Aaron Esterson) *Sanity, Madness and the Family* (London: Tavistock, 1964), also embraced a psychoanalytic account of the origins of schizophrenia.

20. The excommunication of Alfred Adler, Carl Jung, and Otto Rank established a pattern that would resurface in postwar America. See Douglas Kirsner, *Unfree Associations: Inside Psychoanalytic Institutes* (London: Process Press, 2000). For a contemporaneous comment on this phenomenon, see Morison's diary entry for August 1, 1947: "It is a never-failing source of interest to RSM to find how seldom experts on human relations apply their knowledge to situations in which they are personally involved." Robert Swain Morison diary, 1947, RG 12 (Rockefeller Foundation Officers' Diaries), Rockefeller Foundation Records Collection, Rockefeller Archive Center, Sleepy Hollow, NY (hereafter, Rockefeller Archive Center).

21. Writing to Gregg about the state of psychiatry, Morison commented, "It apals [*sic*] me how few departments have even one individual devoting the major part of his time to

developing new psychiatric concepts and validating current procedures." See also Morison to Gregg, April 5, 1948, box 117, folder 1442, RG 1.1 (Rockefeller Foundation records: Projects, Grants), series 200 (United States), Rockefeller Archive Center.

22. Internal memorandum from Robert S. Morison, April 11, 1951, box 1, folder 5, RG 3 (Rockefeller Foundation Records, Administration, Program and Policy), series 906 (Medical Sciences), Rockefeller Archive Center.

23. On this point, see Gerald Grob, *From Asylum to Community: Mental Health Policy in Modern America* (Princeton, NJ: Princeton University Press, 1991), esp. 61–65, on whose discussion I draw here.

24. For an early and comprehensive statement of these concerns, see Council of State Governments, *The Mental Health Programs of the Forty-Eight States: A Report to the Governors' Conference* (Chicago: Council of State Governments, 1950).

25. D. S. Levine and D. R. Levine, *The Cost of Mental Illness–1971*, ADM 76-265 (Rockville, MD: Department of Health, Education, and Welfare, 1975); Dorothy P. Rice, *The Economic Costs of Alcohol and Drug Abuse and Mental Illness*, ADM 90-1649 (Rockville, MD: Department of Health and Human Services, 1990).

26. Grob, *From Asylum to Community*, 68.

27. Jeanne L. Brand and Philip Sapir, eds., "An Historical Perspective on the National Institute of Mental Health," unpublished mimeograph, February 1964, 27.

28. S. Schneider, "National Institute of Mental Health," in *Encyclopedia of Psychology*, vol. 5 (New York: Oxford University Press, 2000), 394.

29. Brand and Sapir, "An Historical Perspective," 66–67.

30. Leo Madow, quoted in Rachel I. Rosner, "Psychotherapy Research and the National Institute of Mental Health," in *Psychology and the National Institute of Mental Health*, ed. W. Pickren and S. Schneider (Washington, DC: American Psychological Association, 2005), 134.

31. N. G. Hale, *The Rise and Crisis of Psychoanalysis in the United States* (New York: Oxford University Press, 1995), 252.

32. Lightner Witmer, "Clinical Psychology," *Psychological Clinic* 1 (1907): 1–9. An alternative term, widely employed for a time, was "orthogenics."

33. Leila Zenderland, *Measuring Minds: Henry Herbert Goddard and the Origins of American Intelligence Testing* (New York: Cambridge University Press, 1998); Donald Napoli, *Architects of Adjustment: The Practice and Professionalization of American Psychology, 1920–1945* (Port Washington, NY: Kennikat Press, 1981).

34. Charles Dana, "Psychiatry and Psychology," *Medical Record* 91 (1917): 265–267; Charles A. Dickinson et al., "The Relation between Psychiatry and Psychology: A Symposium," *Psychological Exchange* 2 (1933): 149–161; J. E. W. Wallin, *The Odyssey of a Psychologist* (Wilmington, DE: Author, 1955); and the discussion in Ingrid G. Farreras, "Before Boulder: Professionalizing Clinical Psychology, 1896–1949" (PhD diss., University of New Hampshire, 2001), chs. 1 and 2.

35. Farreras, "Before Boulder," 19; chapter 4 of Farreras's work documents how limited the employment prospects in the field were in the 1930s, with many unable to find positions.

36. S. L. Garfield, "Psychotherapy: A Forty Year Appraisal," *American Psychologist* 36 (1981): 174–183, 174.

37. On the war's impact on psychology, see James H. Capshew, "Psychology on the March: American Psychologists and World War II" (PhD diss., University of Pennsylvania, 1986).

38. Ellen Herman, *The Romance of American Psychology* (Berkeley: University of California Press, 1995), 84, 92, 94.
39. Andrew Abbott, *The System of Professions* (Chicago: University of Chicago Press, 1988).
40. Martin Seligman, *The Hope Circuit* (New York: Public Affairs, 2018), 179.
41. Ingrid Farreras, "The Historical Context for National Institute of Mental Health Support of American Psychological Association Training and Accreditation Efforts," in Pickren and Schneider, *Psychology and the National Institute of Mental Health*, 153–179, 169.
42. David B. Baker and Ludy T. Benjamin, Jr., "Creating a Profession: The National Institute of Mental Health and the Training of Psychologists, 1946–1954," in Pickren and Schneider, *Psychology and the National Institute of Mental Health*, 181–207.
43. Baker and Benjamin, "Creating a Profession"; see also Grob, *From Asylum to Community*, 65.
44. William C. Menninger, quoted in Grob, *From Asylum to Community*, 107–108.
45. Gitelson to Ives Hendrick, January 6, 1955, quoted in Hale, *The Rise and Crisis of Psychoanalysis*, 215.
46. Wolfle to D. C. Goldberg, May 18, 1949, Robert Sears Papers, Special Collections and University Archives Department, Green Library, Stanford University, Stanford, CA, quoted in Farreras, "The Historical Context," 172.
47. See, for example, A. T. Beck, *Cognitive Therapy and the Emotional Disorders* (New York: International Universities Press, 1976); and G. Butler, M. Fennell, and A. Hackmann, *Cognitive-Behavioral Therapy for Anxiety Disorders* (New York: Guilford Press, 2008).
48. C. M. Horgan, "The Demand for Ambulatory Mental Health Services," *Health Services Research* 21 (1986): 291–319.
49. C. Barry, H. Huskamp, and H. Goldman, "A Political History of Federal Mental Health and Addiction Insurance Parity," *Milbank Quarterly* 88 (2010): 404–433; Colleen Barry et al., "Design of Mental Health Insurance Coverage: Still Unequal after All These Years," *Health Affairs* 22 (2003): 127–137.
50. Christian Otte, "Cognitive Behavioral Therapy in Anxiety Disorders: Current State of the Evidence," *Dialogues in Clinical Neuroscience* 13 (2011): 413–421; S. G. Hoffmann and J. A. Smits, "Cognitive-Behavioral Therapy for Adult Anxiety Disorders: A Meta-Analysis of Randomized Placebo-Controlled Trials," *Journal of Clinical Psychiatry* 69 (2008): 621–632; R. E. Stewart and D. L. Chambless, "Cognitive-Behavioral Therapy for Adult Anxiety Disorders in Clinical Practice: A Meta-Analysis of Effectiveness Studies," *Journal of Consulting and Clinical Psychology* 77 (2009): 595–606; V. Hunot, R. Churchill, V. Teixeira, and M. Silva de Lima, "Psychological Therapies for Generalised Anxiety Disorder," *Cochrane Database of Systematic Reviews* (January 24, 2007), https://www.cochranelibrary.com/cdsr/doi/10.1002/14651858.CD001848.pub4/full.
51. Hunot et al., "Psychological Therapies," under "authors' conclusions."
52. A recent review of the use of CBT for treatment of children with anxiety disorders concluded with "Compared waitlist / no treatment, CBT probably increases post-treatment remission of primary anxiety diagnoses (CBT 49.4%; waitlist / no treatment 17.8%)" but noted that "confidence in these findings are limited due to concerns about the amount and quality of available evidence." A. C. James, T. Reardon, A. Soler, G. James, and C. Cresswell, "Cognitive Behavioural Therapy for Anxiety Disorders in Children

and Adolescents," *Cochrane Database of Systematic Reviews* (November 16, 2020), https://www.cochranelibrary.com/cdsr/doi/10.1002/14651858.CD013162.pub2/full. In a review of the usefulness of CBT for adults diagnosed with bipolar disorder, M. Oud et al., "Psychological Interventions for Adults with Bipolar Disorder: Systematic Review and Meta-Analysis," *British Journal of Psychiatry* 208 (2016): 213–222, it was likewise emphasized that much of the evidence reviewed is of low or very low quality. D. Lynch, K. R. Laws, and P. J. McKenna, "Cognitive Behavioural Therapy for Major Psychiatric Disorder: Does it Really Work? A Meta-Analytic Review of Well-Controlled Trials," *Psychological Medicine* 40 (2010): 9–24; C. Jones, D. Hacker, J. Xia, et al., "Cognitive Behavioural Therapy Plus Standard Care versus Standard Care for People with Schizophrenia," *Cochrane Database of Systematic Reviews* (December 20, 2018), https://www.cochranelibrary.com/cdsr/doi/10.1002/14651858.CD007964.pub2/full.

CHAPTER 17 The Birth of Psychopharmacology

1. Judith Swazey, *Chlorpromazine in Psychiatry* (Cambridge, MA: MIT Press, 1974), 160.
2. On the emergence of the German chemical industry, see J. P. Murman, *Knowledge and Comparative Advantage: The Coevolution of Firms, Technology, and National Institutions* (Cambridge: Cambridge University Press, 2003).
3. F. Lopez-Munoz et al., "History of the Discovery and Clinical Introduction of Chlorpromazine," *Annals of Psychiatry* 17 (2005): 113–135.
4. I rely here, and elsewhere in this chapter, on the informative account provided by Judith Swazey in *Chlorpromazine,* a discussion that is particularly useful given Swazey's access to internal Smith, Kline & French documents and to their executives at the time.
5. Quoted in Swazey, *Chlorpromazine,* 179.
6. Nicolas Rasmussen, *On Speed: The Many Lives of Amphetamine* (New York: New York University Press, 2008).
7. See Lester Grinspoon and Peter Hedbloom, *The Speed Culture: Amphetamine Use and Abuse in America* (Cambridge, MA: Harvard University Press, 1975).
8. Henri Laborit, "L'hibernation artificielle," *Acta Anaesthesiologica Belgica* (February 2, 1951): 24–29; H. Laborit, P. Huguenard, and R. Allaume, "Un nouveau stabilisateur vegetif (le 4560 RP)," *Press Medicale* 60 (1952): 206–208.
9. Pierre Deniker, "Experimental Neurological Syndromes and the New Drug Therapies in Psychiatry," *Comprehensive Psychiatry* 1 (1960): 92–102, 96.
10. Quoted in Swazey, *Chlorpromazine,* 136.
11. Swazey, *Chlorpromazine,* 137.
12. Heinz Lehmann, "The Introduction of Chlorpromazine to North America," *Psychiatric Journal of the University of Ottawa* 14 (1989): 263–265.
13. Stuart Kirk, Tomi Gomory, and David Cohen, *Mad Science: Psychiatric Coercion, Diagnosis, and Drugs* (New York: Routledge, 2013), 264.
14. Heinz Lehmann, interview with Judith Swazey, November 19, 1971, partially reprinted in Swazey, *Chlorpromazine,* 154–157. As Swazey fails to note, the persistence (and sometimes emergence) of major neurological side effects, not just parkinsonism, but also other serious movement disorders, is a common feature of exposure to Thorazine and other antipsychotics, one that is often a chronic iatrogenic consequence.

15. H. E. Lehmann and G. E. Hanrahan, "Chlorpromazine: New Inhibiting Agent for Psychomotor Excitement and Manic States," *Archives of Neurology and Psychiatry* 71 (1954): 227–237, 235.

16. Lehmann and Hanrahan, "Chlorpromazine," 230, 231. They reported that a handful of patients became depressed in the course of treatment but required "only four to six electroshocks to effect a lasting recovery" (233).

17. Swazey, *Chlorpromazine*, 170, 195–201. The costs of clinical tests on psychiatric patients at that date amounted to some $11,000 (189).

18. Swazey, *Chlorpromazine*, 171.

19. SK&F made $196 million in sales between 1954 and 1956, generating profits exceeding $75 million. Smith, Kline & French, *Annual Report*, 1956.

20. The advertisement appeared in *Diseases of the Nervous System* 16 (1955): 227.

21. Fortunately, Swazey reproduced what they had to say, providing us with a host of revealing details. The quotations that follow are drawn from this document. Swazey, *Chlorpromazine*, 201–207.

22. Quoted in Swazey, *Chlorpromazine*, 202–203.

23. Quoted in Swazey, *Chlorpromazine*, 202.

24. Quoted in Swazey, *Chlorpromazine*, 223.

25. Harold Eiber, "Chlorpromazine (Thorazine)-Rauwolfia Combination in Neuropsychiatry," *AMA Archives of Neurology and Psychiatry* 74 (1955): 36–39; L. H. Gahagan, "Ineffectiveness of Chlorpromazine and Rauwolfia Serpentina Preparations in the Treatment of Depression," *Diseases of the Nervous System* 18 (1957): 390–393.

26. Winfred Overholser, "Has Chlorpromazine Inaugurated a New Era in Mental Hospitals?" *Journal of Clinical and Experimental Psychopathology* 17 (1956): 197–201, 199. Overholser was superintendent of St. Elizabeths Hospital, the federal facility in Washington, DC. A host of similar reports were compiled by a specially convened California State Senate Interim Committee on the Treatment of Mental Illness, 1955–1956, following requests to superintendents across the country to comment on the impact of Thorazine in their institutions.

27. David Healy, *The Antidepressant Era* (Cambridge, MA: Harvard University Press, 1997), 46–47. National and international pharmacology societies were soon formed.

28. In a review of 962 papers dealing with the use of chlorpromazine in mental hospitals, "only ten papers mentioned controlled studies." I. F. Bennett, "Chemotherapy in Psychiatric Hospitals: Critical Review of the Literature and Research Trends," *Transactions of the First Research Conference on Chemotherapy in Psychiatry* (Washington, DC: Veterans Administration, 1956), 19.

29. The company supplied these data in response to a lawsuit, *Carter-Wallace, Inc. v. Wolins Pharmacology Corporation*, United States District Court for the Eastern District of New York, 326 F. Supp. 1299 (1971).

30. See Andrea Tone, *The Age of Anxiety: A History of America's Turbulent Affair with Tranquilizers* (New York: Basic Books, 2008).

31. For the details of Arthur Sackler's tactics, see Patrick Radden Keefe, *Empire of Pain: The Secret History of the Sackler Dynasty* (New York: Doubleday, 2021), 34–52. Sackler was posthumously inducted into the Medical Advertising Hall of Fame. A dubious honor, perhaps. In the words of the psychiatrist Allen Frances, "Most of the questionable practices that propelled the pharmaceutical industry into the scourge it is today can be attributed to Arthur Sackler." Quoted in Patrick Radden Keefe, "The Family That Built an Empire of Pain," *New Yorker*, October 23, 2017.

32. Gerald L. Klerman, "Assessing the Influence of the Hospital Milieu upon the Effectiveness of Psychiatric Drug Therapy," *Journal of Nervous and Mental Disease* 137 (1963): 143–154, 145.

33. See Joanna Moncrieff, "Magic Bullets for Mental Disorders: The Emergence of the Concept of an 'Antipsychotic' Drug," *Journal of the History of the Neurosciences* 22 (2013): 30–46; and J. Moncrieff, *The Bitterest Pills: The Troubling Story of Antipsychotic Drugs* (London: Palgrave Macmillan, 2013). As Moncrieff emphasizes, this changed rhetoric was political rather than scientific in nature.

34. National Institute of Mental Health Psychopharmacology Service Center Collaborative Study Group, "Phenothiazine Treatment in Acute Schizophrenia," *Archives of General Psychiatry* 10 (1964): 246–261, 257. The study authors attested to the safety of the three drugs on the basis of a six-week trial. Because antipsychotics tend to be prescribed for years, often for life, the decision to reach such a conclusion was remarkable. It would also prove remarkably misguided, as a subsequent study funded by the National Institute of Mental Health, published in 2005, would make clear.

35. John Cade, "The Story of Lithium," in *Discoveries in Biological Psychiatry*, ed. F. J. Ayd and B. Blackwell (Philadelphia: Lippincott, 1970), 218–229, 219; J. Cade, "Lithium Salts in the Treatment of Psychotic Excitement," *Medical Journal of Australia* 2 (September 3, 1949): 349–352.

36. Silas Weir Mitchell, "On the Use of Bromide of Lithium," *American Journal of the Medical Sciences* 60 (1870): 443–445; William Alexander Hamilton, *A Treatise on Diseases of the Nervous System* (New York: Appleton, 1871), 381.

37. This was an unsurprising choice: makers of patent medicines for gout had based their claims of therapeutic effect on lithium's ability to dissolve uric acid crystals. From this they extrapolated that the drug would dissolve the crystals that accumulated in the joints of gout sufferers. It did not.

38. David Healy reports, without citing a source, that in reality a number of Cade's patients died. David Healy, *The Creation of Psychopharmacology* (Cambridge, MA: Harvard University Press, 2002), 48.

39. Those interested in the details of the dispute can consult the account in Healy, *The Antidepressant Era*, 122–130. The key trial, in which members of Shepherd's group participated, was published in 1971. A. Coppen, R. Noguera, J. Bailey, et al., "Prophylactic Lithium in Affective Disorders: Controlled Trial," *Lancet* 298 (1971): 275–279.

40. A. Gignac, A. McGirr, R. W. Lam, and L. N. Yatham, "Recovery and Recurrence Following a First Episode of Mania: A Systematic Review and Meta-Analysis of Prospectively Characterized Cohorts," *Journal of Clinical Psychiatry* 76 (2015): 1241–1248.

41. T. G. Rhee et al., "20-Year Trends in the Pharmacologic Treatment of Bipolar Disorder by Psychiatrists in Outpatient Care Settings," *American Journal of Psychiatry* 177 (2020): 706–715. They note that, during the same period, the use of antipsychotics in bipolar disorder rose from 12.4 percent of those under treatment to 51.4 percent, and of antidepressants without a mood stabilizer from 17.9 percent to 40.9 percent. The rising use of antidepressants occurred "despite a lack of evidence for their efficacy in bipolar disorder" and considerable evidence of their use "increasing the risk of mania." Both second-generation antipsychotics (which received FDA approval for use in bipolar disorders in the 1990s) and antidepressants are, of course, far more profitable than lithium, and Rhee and colleagues suggest that aggressive marketing by the pharmaceutical industry lies behind at least some of these shifts.

42. Healy, *The Creation of Psychopharmacology*, 49.

CHAPTER 18 Community Care

1. Andrew Scull, *Psychiatry and Its Discontents* (Berkeley: University of California Press, 2019), ch. 4, "A Culture of Complaint."
2. Nellie Bly, *Ten Days in a Madhouse* (New York: Munro, 1887), reprinted from articles first published in Joseph Pulitzer's *New York World;* Albert Deutsch, *The Shame of the States* (New York: Harcourt, Brace, 1948).
3. Richard F. Mollica, "From Asylum to Community: The Threatened Disintegration of Public Psychiatry," *New England Journal of Medicine* 308 (1983): 367–373, 369, 371.
4. A. Wynter, *The Borderlands of Insanity* (New York: G. P. Putnam's Sons, 1875), 162–163.
5. Robert Felix, "A Model for Comprehensive Mental Health Centers," *American Journal of Public Health* 54 (1964): 1964–1969, 1965; see also R. Felix, "Community Mental Health: A Federal Perspective," *American Journal of Psychiatry* 121 (1964): 428–432.
6. Robert Felix, *Mental Illness: Progress and Prospects* (New York: Columbia University Press, 1967), 75.
7. Uri Aviram, Leonard Syme, and Judith Cohen, "The Effects of Policies and Programs on Reductions of Mental Hospitalization," *Social Science and Medicine* 10 (1976): 571–578; Joseph Morrissey, "Deinstitutionalizing the Mentally Ill: Processes, Outcomes, and New Directions," in *Deviance and Mental Illness,* ed. Walter Gove, 147–176 (Beverly Hills, CA: Sage, 1982); William Gronfein, "Psychotropic Drugs and the Origins of Deinstitutionalization," *Social Problems* 32 (1985): 437–454.
8. Henry Brill and Robert Patton, "Analysis of 1955–56 Population Fall in New York State Mental Hospitals during the First Year of Large-Scale Use of Tranquilizing Drugs," *American Journal of Psychiatry* 114 (1957): 509–517, 512, 513–514.
9. L. J. Epstein, "An Approach to the Effect of Ataraxic Drugs on Hospital Release Rates," *American Journal of Psychiatry* 119 (1962): 36–47, 42–44. These findings accord with analyses of data for Michigan by the state authorities, for Connecticut by Myers and Bean, and for Washington, DC, by Linn.
10. Paul Lerman, *Deinstitutionalization and the Welfare State* (New Brunswick, NJ: Rutgers University Press, 1982); Steven Segal and Uri Aviram, *The Mentally Ill in Community-Based Sheltered Care* (New York: Wiley, 1978); Steven Rose, "Deciphering Deinstitutionalization: Complexities in Policy and Program Analysis," *Milbank Memorial Fund Quarterly* 57 (1979): 429–460; Gronfein, "Psychotropic Drugs"; Gerald Grob, *From Asylum to Community: Mental Health Policy in Modern America* (Princeton, NJ: Princeton University Press, 1991). I first made this case in A. Scull, "The Decarceration of the Mentally Ill: A Critical View," *Politics and Society* 6 (1976): 173–212.
11. Senate Committee on Aging, *Trends in Long Term Care* (Washington, DC: Government Printing Office, 1981); Senate Committee on Aging, *The Role of Nursing Homes in Caring for Discharged Mental Patients* (Washington, DC: Government Printing Office, 1976); General Accounting Office, *The Mentally Ill in the Community: Government Needs to Do More* (Washington, DC: Government Printing Office, 1977); R. Reich and L. Siegal, "Psychiatry under Siege: The Mentally Ill Shuffle to Oblivion," *Psychiatric Annals* 3 (1973): 37–55.
12. National Institute of Mental Health Statistical Note #107, Table 2 (Bethesda, MD: Government Printing Office, 1974).
13. General Accounting Office, *The Mentally Ill in the Community,* 11.

14. The following paragraphs are drawn from a paper I wrote in the midst of these events: Scull, "A New Trade in Lunacy: The Re-commodification of the Mental Patient," *American Behavioral Scientist* 24 (1981): 741–754.
15. C. Aldrich and E. Mendikoff, "Relocation of the Aged and Disabled: A Mortality Study," *Journal of the American Geriatrics Society* 11 (1963): 105–194; K. Jasman, "Individualized versus Mass Transfers of Nonpsychotic Geriatric Patients from the Mental Hospital to the Nursing Home," *Journal of the American Geriatrics Society* 15 (1969): 280–284; E. Markus, M. Blenker, and T. Downs, "The Impact of Relocation upon Mortality Rates of Institutionalized Aged Persons," *Journal of Gerontology* 26 (1971): 537–541.
16. Markus, Blenker, and Downs, "The Impact of Relocation."
17. General Accounting Office, *The Mentally Ill in the Community*, 15–16.
18. Elaine Wolpert and Julian Wolpert, "From Asylum to Ghetto," *Antipode* 6 (1974): 63–76.
19. For the exposé, see *New York Times*, January 18, 21, 22, and February 4, 5, 6, 1975; and for the inquiry, see Charles Hynes, *Private Proprietary Homes for Adults: Their Administration, Control, Operation, Supervision, Funding, and Quality* (New York: Deputy Attorney General's Office, 1977).
20. "Why Were the Ill, Elderly, Poor, Sent into Unfit Boarding Homes," *Philadelphia Inquirer*, September 21, 1975.
21. S. Kirk and M. Thierren, "Community Mental Health Myths and the Fate of Formerly Hospitalized Mental Patients," *Psychiatry* 38 (1975): 209–217.
22. General Accounting Office, *The Mentally Ill in the Community*, 19.
23. Senate Committee on Aging, *The Role of Nursing Homes*, 724.
24. General Accounting Office, *The Mentally Ill in the Community*, 95.
25. Bruce J. Ennis, *Prisoners of Psychiatry* (New York: Harcourt, Brace, Jovanovich, 1972).
26. See Thomas Szasz, "Involuntary Psychiatry," *University of Cincinnati Law Review* 44 (1976): 347–365; and Stephen Morse, "A Preference for Liberty: The Case against the Involuntary Commitment of the Mentally Disordered," *California Law Review* 70 (1982): 54–106.
27. One recent suggestion along these lines is D. Sisti and E. Emanuel, "Improving Long-Term Psychiatric Care: Bringing Back the Asylum," *Journal of the American Medical Association* 313 (2015): 243–244. There has been no sign of any state moving to embrace its recommendations.
28. Thomas D. Reynolds, "Thinking about Schizophrenia," *Medical Annals of the District of Columbia* 41 (August 1972): 503, quoted in M. Summers, *Madness in the City of Magnificent Intentions* (Oxford: Oxford University Press, 2019), 297–298. The injectable form of Prolixin, an antipsychotic, has effects that last up to four weeks, and it was designed to get around the problem of patients failing to take their prescribed medications.
29. Another exception to this generalization is E. Fuller Torrey. See, for example, Torrey, *Nowhere to Go: The Tragic Odyssey of the Homeless Mentally Ill* (New York: HarperCollins, 1988). Torrey's critique has been marred by a sensationalizing of the links between violence and mental illness. See Torrey, *The Insanity Offense: How America's Failure to Treat the Seriously Mentally Ill Endangers Its Citizens* (New York: W. W. Norton, 2008). For a critique, see Richard Gosden and Sharon Beder, "Pharmaceutical Industry Agenda Setting in Mental Health Policies," *Ethical Human Sciences and Services* 3 (2001): 147–159.

30. For an analysis of these developments as they played out in California, see Joel Braslow et al., "Recovery in Context: Thirty Years of Mental Health Policy in California," *Perspectives in Biology and Medicine* 64 (2021): 82–102.

31. See William Fisher, Jeffrey Geler, and John Pandiani, "The Changing Role of the State Psychiatric Hospital," *Health Affairs* 28 (2009): 676–684. For the inpatient census numbers, see T. Lutterman, A. Berhane, B. Phelan, R. Shaw, and V. Rana, "Funding and Characteristics of State Mental Health Agencies," Health and Human Services Publication (SMA) 09-4424 (Rockville, MD: Center for Mental Health Services, 2007).

32. G. A. Greenberg and R. A. Rosenheck, "Jail Incarceration, Homelessness, and Mental Health: A National Study," *Psychiatric Services* 59 (2008): 170–177; P. A. Toro, "Toward an International Understanding of Homelessness," *Journal of Social Issues* 63 (2007): 461–481.

33. For a study of the role of jails and prisons, focused on developments in Pennsylvania, see Anne E. Parsons, *From Asylum to Prison: Deinstitutionalization and the Rise of Mass Incarceration after 1945* (Chapel Hill: University of North Carolina Press, 2018).

34. Braslow et al., "Recovery in Context."

35. Dignity and Power Now, "Impact of Disproportionate Incarceration of and Violence against Black People with Mental Conditions in the World's Largest Jail System," n.d., Supplementary Submission for the August 2014 CERD Committee Review of the United States, https://tbinternet.ohchr.org/Treaties/CERD/Shared%20Documents/USA/INT_CERD_NGO_USA_17740_E.pdf; Kim Hopper, John Jost, Susan Welber, and Gary Haugland, "Homelessness, Severe Mental Illness and the Institutional Circuit," *Psychiatric Services* 48 (1997): 659–664.

36. W. B. Hawthorne et al., "Incarceration among Adults Who Are in the Public Mental Health System," *Psychiatric Services* 63 (2012): 26–32. On the impact of race and class on imprisonment, see Becky Pettit and Bruce Western, "Mass Imprisonment and the Life Course: Race and Class Inequality in Incarceration," *American Sociological Review* 69 (2004): 151–169.

37. Summers, *Madness in the City*, 266.

38. In 2019, the median white family owned about $184,000 in wealth. By comparison, the median Black family owned $23,000. The median Hispanic family owned $38,000. Ana Hernández Kent and Lowell R. Ricketts, "Wealth Gaps between White, Black, and Hispanic Families in 2019," *St. Louis Fed on the Economy* blog, Federal Reserve Bank of St. Louis, January 5, 2021, https://www.stlouisfed.org/on-the-economy/2021/january/wealth-gaps-white-black-hispanic-families-2019.

39. S. Assari, "Unequal Gain of Equal Resources across Racial Groups," *International Journal of Health Policy Management* 7 (2018): 3–9.

40. D. S. Massey, "American Apartheid: Segregation and the Making of the Underclass," *American Journal of Sociology* 96 (1990): 329–357; D. T. Lichter, D. Parisi, and M. C. Taquino, "The Geography of Exclusion: Race, Segregation, and Concentrated Poverty," *Social Problems* 59 (2012): 364–388.

41. Figures are from the National Institute of Justice.

42. Leila Morsy and Richard Rothstein, "Toxic Stress and Children's Outcomes," Economic Policy Institute, May 1, 2019; D. Umberson et al., "Death of Family Members as an Overlooked Source of Racial Disadvantage in the United States," *Proceedings of the National Academy of Sciences* 114 (2017): 915–920; M. E. Smith et al., "The Impact of Exposure to Gun Violence Fatality on Mental Health Outcomes in Four Urban US

Settings," *Social Science and Medicine* 246 (2020): 112587; F. Edwards, H. Lee, and M. Esposito, "Risk of Being Killed by Police Use of Force in the United States by Age, Race-Ethnicity, and Sex," *Proceedings of the National Academy of Sciences* 116 (2019): 16793–16798.

43. Deidre M. Anglin et al., "From Womb to Neighborhood: A Racial Analysis of Social Determinants of Psychosis in the United States," *American Journal of Psychiatry* 178 (2021): 599–610. I am grateful to Robin Murray for drawing my attention to this article.

44. For a splendid ethnographic account of the realities of deinstitutionalization in present-day Los Angeles, see Neil Gong, "Mind and Matter: Madness and Inequality in Los Angeles" (PhD diss., University of California, Los Angeles, 2019).

45. J. F. Hayes, L. Matson, K. Waters, M. King, and D. Osborn, "Mortality Gap for People with Bipolar Disorder and Schizophrenia: UK-Based Cohort Study 2000–2014," *British Journal of Psychiatry* 211 (2017): 175–181. Hayes and his colleagues report that the gap between the seriously mentally ill and the general population has worsened over time. See also T. M. Laursen, "Life Expectancy among Persons with Schizophrenia or Bipolar Affective Disorder," *Schizophrenia Research* 131 (2011): 101–104; S. Saha, D. Dent, and J. McGrath, "A Systematic Review of Mortality in Schizophrenia: Is the Mortality Gap Worsening over Time?" *Archives of General Psychiatry* 64 (2007): 1123–1131; and S. Brown, "Excess Mortality of Schizophrenia: A Meta-Analysis," *British Journal of Psychiatry* 171 (1997): 502–508.

CHAPTER 19 Diagnosing Mental Illness

1. R. D. Laing, *The Politics of Experience and the Bird of Paradise* (London: Penguin, 1967). For a sympathetic reappraisal of Laing by another Scottish psychiatrist, see Allan Beveridge, "R. D. Laing Revisited," *Psychiatric Bulletin* 22 (1998): 452–456.

2. Thomas Scheff, "The Societal Reaction to Deviance: Ascriptive Elements in the Psychiatric Screening of Mental Patients in a Midwestern State," *Social Problems* 11 (1964): 401–413.

3. These notions were articulated most prominently by Erving Goffman, in his best-selling *Asylums: Essays on the Social Situation of Mental Patients and Other Inmates* (New York: Doubleday, 1961), which dismissed psychiatrists as practitioners of a tinkering trade, and the facilities they presided over as total institutions akin to prisons and concentration camps; and by Thomas Scheff in *Being Mentally Ill: A Sociological Theory* (Chicago: Aldine, 1966).

4. Bruce J. Ennis and Thomas R. Litwack, "Psychiatry and the Presumption of Expertise: Flipping Coins in the Courtroom," *California Law Review* 62 (1974): 693–752.

5. The film swept all five major Academy Awards and appears on most lists of the outstanding American films of the twentieth century.

6. Benjamin Pasamanick, Simon Dinitz, and Mark Lefton, "Psychiatric Orientation and Its Relation to Diagnosis and Treatment in a Mental Hospital," *American Journal of Psychiatry* 116 (1959): 127–132, 127, 131.

7. Aaron T. Beck, "Reliability of Psychiatric Diagnoses: 1. A Critique of Systematic Studies," *American Journal of Psychiatry* 119 (1962): 210–216, 213. See also A. T. Beck, C. H. Ward, M. Mendelson, J. E. Mock, and J. K. Erbaugh, "Reliability of Psychiatric Diagnoses: 2. A Study of Consistency of Clinical Judgments and Ratings," *American Journal of Psychiatry* 119 (1962): 351–357; and Robert L. Spitzer, J. Cohen, J. Fleiss, and J. Endicott,

"Quantification of Agreement in Psychiatric Diagnosis," *Archives of General Psychiatry* 17 (1967): 83–87.

8. Karl Menninger, *The Vital Balance: The Life Process in Mental Health and Illness* (New York: Penguin, 1963).

9. Quoted in Jeffrey Lieberman, *Shrinks: The Untold Story of Psychiatry* (Boston: Little, Brown, 2015), 142.

10. Gerald Grob, "The Origins of DSM-I: A Study in Appearance and Reality," *American Journal of Psychiatry* 148 (1991): 421–431.

11. J. E. Cooper, R. E. Kendell, and B. J. Gurland, *Psychiatric Diagnosis in New York and London: A Comparative Study of Mental Hospital Admissions* (Oxford: Oxford University Press, 1972).

12. David Rosenhan, "On Being Sane in Insane Places," *Science* 179 (January 19, 1973): 250–258. A recent survey of twelve leading abnormal psychology textbooks, for example, found that half of them still gave extensive attention to Rosenhan's paper, summarizing its design and endorsing its basic findings. Only two of them acknowledged any criticisms of the study, though methodological critiques have emerged over the years since the article's initial publication. See Jared Bartels and Daniel Peters, "Coverage of Rosenhan's 'On Being Sane in Insane Places' in Abnormal Psychology Textbooks," *Teaching of Psychology* 44 (2017): 169–173. As another example of the study's extraordinary half-life, the *Washington Post* ran a wholly uncritical article summarizing Rosenhan's findings: Nathaniel Morris, "This Secret Experiment Tricked Psychiatrists into Diagnosing Sane People as Having Schizophrenia," *Washington Post*, January 1, 2018.

13. Rosenhan held an appointment at Haverford College from 1960 to 1962 and thereafter was employed as a part-time lecturer at the University of Pennsylvania and then Princeton University.

14. Rosenhan, "On Being Sane in Insane Places," 252.

15. No fewer than fifteen letters from psychiatrists were published in *Science* 180 (April 27, 1973): 356–369, along with a rejoinder from Rosenhan.

16. See Susannah Cahalan, *The Great Pretender* (New York: Grand Central Publishing, 2019).

17. Haverford State Hospital intake notes for "David Lurie," aka David Rosenhan, February 6, 1969. I am grateful to Susannah Cahalan for sharing these materials with me. They are also quoted in Cahalan, *The Great Pretender*, 183, 184.

18. Hannah S. Decker, *The Making of DSM III: A Diagnostic Manual's Conquest of American Psychiatry* (Oxford: Oxford University Press, 2013), 141–142.

19. See Ronald Bayer, *Homosexuality and American Psychiatry: The Politics of Diagnosis* (Princeton, NJ: Princeton University Press, 1987).

20. Spitzer's efforts are recorded in almost mind-numbing detail in Decker, *The Making of DSM III*.

21. See the paper written by one of their leading lights, Samuel Guze, "The Need for Toughmindedness in Psychiatric Thinking," *Southern Medical Journal* 63 (1970): 662–671, which is dismissive of the "tender-minded" psychoanalysts. The obscure publication in which this paper appeared testifies to the outcast status the Washington University department occupied in the psychiatry of the day. The NIMH regularly refused to fund its faculty's clinical research programs through the end of the 1960s.

22. John Feighner, "The Advent of the Feighner Criteria," *Current Contents* 43 (October 23, 1989): 14, http://garfield.library.upenn.edu/classics1989/A1989AU44300001.pdf.

23. J. Feighner, E. Robins, S. Guze, R. Woodruff, Jr., G. Winokur, and R. Munoz, "Diagnostic Criteria for Use in Psychiatric Research," *Archives of General Psychiatry* 26 (1972): 57–63.

24. Decker, *The Making of DSM-III*, 56.

25. Ronald Bayer and Robert Spitzer, "Neurosis, Psychodynamics, and DSM III: A History of the Controversy," *Archives of General Psychiatry* 42 (1985): 187–196, 188.

26. Quoted in David Healy, *The Psychopharmacologists: III: Interviews* (New York: Oxford University Press, 2000), 407.

27. Quoted in Edward Shorter, *Before Prozac: The Troubled History of Mood Disorders in Psychiatry* (New York: Oxford University Press, 2009), 157.

28. Peter Bourne, "Military Psychiatry and the Viet Nam Experience," *American Journal of Psychiatry* 127 (1970): 481–488, 487. Bourne was head of the army's psychiatric research team.

29. See E. Jones and S. Wessely, *Shell Shock to PTSD* (New York: Psychology Press, 2005), 131; Allan Horwitz, *PTSD: A Short History* (Baltimore, MD: Johns Hopkins University Press, 2018), 88–98.

30. Chaim Shatan, "Post Vietnam Syndrome," *New York Times*, May 6, 1972.

31. Horwitz, *PTSD*, 94–98.

32. Wilbur Scott, "PTSD in DSM III: A Case in the Politics of Diagnosis and Disease," *Social Problems* 37 (1990): 294–310, 308.

33. Paul Eberle and Shirley Eberle, *The Abuse of Innocence: The McMartin Preschool Trial* (New York: Prometheus, 1993).

34. Richard Ofshe and Ethan Watters, *Making Monsters: False Memories, Psychotherapy, and Sexual Hysteria* (Berkeley: University of California Press, 1996).

35. Horwitz, *PTSD*, 132–134. For a useful autopsy of the phenomenon, see Richard J. McNally, *Remembering Trauma* (Cambridge, MA: Belknap Press of Harvard University Press, 2003).

36. *Diagnostic and Statistical Manual of Mental Disorders*, 3rd ed. (Washington, DC: American Psychiatric Association, 1980), 238.

37. Lori Zoellner et al., "The Evolving Construct of Posttraumatic Stress Disorder (PTSD): DSM-5 Criteria Changes and Legal Implications," *Psychological Injury and Law* 6 (2013): 277–289; Horwitz, *PTSD*, 135–164.

38. Horwitz, *PTSD*, 138. *DSM* 5 both tightened and expanded these boundaries. It removed witnessing a traumatic event in the mass media as a trigger for PTSD, but it lowered the diagnostic threshold for children and for first responders who were exposed to repeated reports of traumatic events.

39. Quoted in James Davies, "How Voting and Consensus Created the Diagnostic and Statistical Manual of Mental Disorders (DSM-III)," *Anthropology and Medicine* 24 (2017): 32–46, 38.

40. Howard Berk, quoted in Bayer and Spitzer, "Neurosis," 189.

41. L. Rockland, "Some Thoughts on the Subject: Should Psychodynamics Be Included in the DSM-III?," unpublished paper, quoted in Bayer and Spitzer, "Neurosis," 191.

42. Quoted in Bayer and Spitzer, "Neurosis," 189–191.

43. For detailed accounts of all this political maneuvering, see Decker, *The Making of DSM-III*, and, from the perspective of the winning side, Bayer and Spitzer, "Neurosis."

44. Thomas Hackett, "The Psychiatrist: In the Mainstream or on the Banks of Medicine," *American Journal of Psychiatry* 134 (1977): 432–435.

45. Nathan G. Hale, Jr., *The Rise and Crisis of Psychoanalysis in the United States: Freud and the Americans 1917–1985* (Oxford: Oxford University Press, 1995), 330–331.
46. Hale, *The Rise and Crisis of Psychoanalysis*, 332.
47. For examples of psychoanalysts themselves expressing concerns about these trends, see Van Buren O. Hammett, "A Consideration of Psychoanalysis in Relation to Psychiatry Generally, circa 1965," *American Journal of Psychiatry* 122 (1965): 42–54; and George Engle, "Some Obstacles to the Development of Research in Psychoanalysis," *Journal of Abnormal Psychology* 16 (1968): 195–204.
48. Hale, *The Rise and Crisis of Psychoanalysis*, 304.
49. See the discussion in Douglas Kirsner, *Unfree Associations: Inside Psychoanalytic Institutes* (London: Process Press, 2000), 59–62, 127–134.
50. The lawsuit was settled in November 1988, following an agreement by the institutes not to discriminate against psychologists and other nonmedically qualified applicants. See Robert Pear, "M.D.'s Make Room for Others in the Ranks of Psychoanalysts," *New York Times*, August 19, 1992.
51. Hale, *The Rise and Crisis of Psychoanalysis*, 302.
52. Debra A. Katz and Marcia Kaplan, "Can Psychiatry Residents Be Attracted to Analytic Training? A Survey of Five Residency Programs," *Journal of the American Psychoanalytic Association* 58 (2010): 927–952, 927–928.
53. Rick Mayes and Allan Horwitz succinctly summarize the problem: "Insurance companies viewed psychotherapies as a financial 'bottomless pit' requiring potentially uncontrollable resources; patients could spend years in psychoanalytic therapy." R. Mayes and A. Horwitz, "DSM III and the Revolution in the Classification of Mental Illness," *Journal of the History of the Behavioral Sciences* 41 (2005): 249–267, 253. In time, the insurance industry would become sympathetic to the therapies offered by clinical psychologists, not only because they charged lower rates, but also because their interventions promised measurable results and were time-limited. For discussion, see Andrew Scull, *Psychiatry and Its Discontents* (Berkeley: University of California Press, 2019), ch. 12.
54. Clinical psychology and social work are heavily gendered fields, dominated by women. This reinforced the tendency to pay less for their services, as women are notoriously discriminated against in the labor market.
55. Nancy C. Andreasen, "DSM and the Death of Phenomenology in America: An Example of Unintended Consequences," *Schizophrenia Bulletin* 33 (2007): 108–112, 111.
56. The Reagan administration repeatedly threatened budget and personnel cuts and quickly brought the institute to heel. See J. L. Fox, "NIMH Faces Renewed Uncertainties," *Science* 225 (1984): 148–149. About half the positions at the NIMH headquarters had been cut by 1982, and the administration sought to end community support programs and support for clinical training programs, though those plans faced congressional opposition.
57. I owe this last point to an anonymous reviewer for *Psychological Medicine*.
58. R. Blashfield, J. Keeley, E. Flanagan, and S. Miles, "The Cycle of Classification: DSM-I through DSM-5," *Annual Review of Clinical Psychology* 10 (2014): 25–51, 32.
59. "Presidential Proclamation: The Decade of the Brain, 1990–1999," *Federal Register* 55 (July 17, 1990): 29553.
60. Steven Sharfstein, "Big Pharma and Psychiatry: The Good, the Bad, and the Ugly," *Psychiatric News*, August 19, 2005.
61. Leon Eisenberg, "Mindlessness and Brainlessness in Psychiatry," *British Journal of Psychiatry* 148 (1986): 497–508, 500.

CHAPTER 20 The Complexities of Psychopharmacology

1. For their discovery, see David Healy, *The Antidepressant Era* (Cambridge, MA: Harvard University Press, 1997), 43–77.
2. T. Kalman and M. Goldstein, "Satisfaction of Manhattan Psychiatrists with Private Practice: Assessing the Impact of Managed Care," *Journal of Psychotherapy Practice and Research* 7 (1998): 250–258; M. Henneberger, "Managed Care Changing Practice of Psychotherapy," *New York Times,* October 9, 1994; E. J. Pollock, "Managed Care's Focus on Psychiatric Drugs Alarms Many Doctors," *Wall Street Journal,* December 1, 1995; Robert Schreter, "Earning a Living: A Blueprint for Psychiatrists," *Psychiatric Services* 46 (1995): 1233–1235.
3. See, for example, John Fulton, *The Physiology of the Nervous System,* 3rd ed. (New York: Oxford University Press, 1949); and Clifford Morgan and Eliot Stellar, *Physiological Psychology,* 2nd ed. (New York: McGraw Hill, 1950).
4. See Arvid Carlsson, "Early Psychopharmacology and the Rise of Modern Brain Research," *Journal of Psychopharmacology* 4 (1990): 120–126.
5. For the early history of the brain sciences, see Edwin Clarke and L. S. Jacyna, *Nineteenth Century Origins of Neuroscientific Concepts* (Berkeley: University of California Press, 1987).
6. For a useful summary of this history, on which I have drawn here, see Alison Abbott, "Levodopa: The Story So Far," *Nature* 466 (August 26, 2010): S6–S7.
7. For the persistence of this idea in some quarters, see Pierre Deniker, "Discovery of the Clinical Use of Neuroleptics," in *Discoveries in Psychopharmacology,* ed. M. Parnham and J. Bruinvels, vol. 1, 163–180 (Amsterdam: Elsevier, 1983).
8. The crucial early papers were P. Seeman, T. Lee, M. Chau-Wong, and K. Wong, "Anti-psychotic Drug Doses and Neuroleptic/Dopamine Receptors," *Nature* 261 (1976): 177–179; and I. Creese, D. Burt, and S. Snyder, "Dopamine Receptor Binding Predicts Clinical and Pharmacological Potencies of Antischizophrenic Drugs," *Science* 192 (1976): 481–483.
9. For an early suggestion, see J. Van Rossum, "The Significance of Dopamine Receptor Blockade for the Action of Neuroleptic Drugs," in *Proceedings of the Fifth Collegium Internationale Neuorpharmacologicum,* ed. H. Brill, 321–329 (Amsterdam: Excerpta Medica Foundation, 1967). For the first widely cited exposition of the theory, see S. Matthysse, "Antipsychotic Drug Actions: A Clue to the Neuropathology of Schizophrenia?," *Federation Proceedings* 32 (1973): 200–205. See also S. Matthysse and J. Lipinski, "Biochemical Aspects of Schizophrenia," *Annual Review of Medicine* 26 (1975): 551–565.
10. These are reviewed by Elliot Valenstein in *Blaming the Brain: The Truth about Drugs and Mental Health* (New York: Free Press, 1998); and more recently by Kenneth Kendler and Kenneth Schaffner, "The Dopamine Hypothesis of Schizophrenia: An Historical and Philosophical Analysis," *Philosophy, Psychiatry, and Psychology* 18 (2011): 41–63.
11. Joanna Moncrieff, *The Bitterest Pills: The Troubling Story of Antipsychotic Drugs* (Basingstoke, UK: Palgrave Macmillan, 2013), 65.
12. Joanna Moncrieff suggests that a powerful reason that psychiatrists have embraced the dopamine hypothesis, in the face of much disconfirming evidence, is that it "suggests, as many psychiatrists have wanted to believe for a long time, that psychiatric conditions are real diseases with tangible and specific biological origins, and that antipsychotic drugs constitute a genuine and innocuous medical treatment, which counteract the underlying defect in a highly-targeted manner." Moncrieff, *The Bitterest Pills,* 61.

13. Kendler and Schaffner, "The Dopamine Hypothesis," 54.
14. J. R. Lacasse and J. Leo, "Serotonin and Depression: A Disconnect between the Advertisements and the Scientific Literature," *PLoS Medicine* 2 (2005): e392; Nathan Greenslit and Ted Kaptchuk, "Antidepressants and Advertising: Pharmaceuticals in Crisis," *Yale Journal of Biology and Medicine* 85 (2012): 153–158.
15. Quoted in Greenslit and Kaptchuk, "Antidepressants and Advertising," 157.
16. J. M. Donahue, M. Cevasco, and M. B. Rosenthal, "A Decade of Direct-to-Consumer Advertising of Prescription Drugs," *New England Journal of Medicine* 357 (2007): 673–681.
17. Peter Kramer, *Listening to Prozac* (New York: Viking, 1993); Elizabeth Wurtzel, *Prozac Nation: Young and Depressed in America* (Boston: Houghton Mifflin, 1994).
18. A Pfizer advertisement for Zoloft, and a GlaxoSmithKline advertisement for Paxil, both quoted in Lacasse and Leo, "Serotonin and Depression." The authors quote many other pharmaceutical advertisements that promote the chemical-imbalance theory to consumers.
19. Lacasse and Leo, "Serotonin and Depression"; M. Wilkes, B. Doblin, and M. Schapiro, "Pharmaceutical Advertisements in Leading Medical Journals: Experts' Assessment," *Annals of Internal Medicine* 116 (1992): 912–919.
20. See reviews of these findings in Valenstein, *Blaming the Brain*, ch. 4; and in David Healy, *Let Them Eat Prozac* (New York: New York University Press, 2004).
21. M. Purgato et al., "Paroxetine versus Other Anti-depressive Agents for Depression," *Cochrane Database of Systematic Reviews* (April 3, 2014), https://www.cochranelibrary.com/cdsr/doi/10.1002/14651858.CD006531.pub2/full; R. Kavoussi, R. Segraves, A. Hughes, J. Ascher, and J. Johnston, "Double-Blind Comparison of Bupropion Sustained Release and Sertraline in Depressed Outpatients," *Journal of Clinical Psychiatry* 58 (1997): 532–537.
22. "Beyond Prozac," *Newsweek*, February 7, 1994.
23. "'You Have to Get Help': Frightening Experience Now a Tool to Help Others," *USA Today*, May 7, 1999.
24. F. Freyhan, "Clinical and Investigative Impacts," in *Pharmacology Frontiers*, ed. Nathan Kline (London: Churchill, 1959), 10; F. Freyhan, "Psychomotility and Parkinsonism in Treatment with Neuroleptic Drugs," *Archives of Neurology and Psychiatry* 78 (1957): 465–472.
25. P. Deniker, "Experimental Neurological Syndromes and the New Drug Therapies in Psychiatry," *Comprehensive Psychiatry* 1 (1960): 92–102, 100.
26. National Institute of Mental Health Pharmacology Service Center Collaborative Study Group, "Phenothiazine Treatment in Acute Schizophrenia," *Archives of General Psychiatry* 10 (1964): 246–261, 255.
27. Information on percentages is from D. Jeste, M. Caligiuri, and J. Paulsen, "Risk of Tardive Dyskinesia in Older Patients: A Prospective Longitudinal Study of 266 Outpatients," *Archives of General Psychiatry* 52 (1995): 756–765. For a careful recent review of the issues, see Stanley Caroff, "Overcoming Barriers to Effective Management of Tardive Dyskinesia," *Neuropsychiatric Disease and Treatment* 15 (2019): 785–794. In 2018, the FDA approved two drugs, valbenazine and deutetrabenazine, which act to reduce tardive dyskinesia symptoms to some degree. However, "their impact on real-world long-term outcomes remains to be investigated," Caroff notes (790).
28. Nathan Kline, "On the Rarity of 'Irreversible' Oral Dyskinesias Following Phenothiazines," *American Journal of Psychiatry* 124 (1968): 48–54, 51.
29. Daniel X. Freedman, "Neurological Syndromes Associated with Antipsychotic Drug Use," *Archives of General Psychiatry* 28 (1973): 463–467.

30. George Crane, "Clinical Psychopharmacology in Its Twentieth Year: Late, Unanticipated Effects of Neuroleptics May Limit Their Use in Psychiatry," *Science* 181 (1973): 124–128. For the earlier papers, see G. E. Crane, "Tardive Dyskinesia in Patients Treated with Neuroleptic Drugs: A Review of the Literature," *American Journal of Psychiatry* 124 (supp.) (1968): 40–48; and G. E. Crane, "Dyskinesia and Neuroleptics," *Archives of General Psychiatry* 19 (1968): 700–703.

31. G. Gardos and J. O. Cole, "Overview: Public Health Issues in Tardive Dyskinesia," *American Journal of Psychiatry* 137 (1980): 776–781, 776. Gardos and Cole warned that "safe, effective, and specific treatment methods are lacking. Non-specific measures such as dose reduction may be thought of as belated efforts at prevention with no certainty of success" (779).

32. *Tardive Dyskinesia: Report of the American Psychiatric Association Task Force on Late Neurological Effects of Antipsychotic Drugs* (Washington, DC: American Psychiatric Association, 1980). These data now appear to underestimate the scale of the problem. In addition to Jeste et al., "Risk of Tardive Dyskinesia," see Pierre-Michel Llorca et al., "Tardive Dyskinesia and Antipsychotics: A Review," *European Psychiatry* 17 (2002): 129–138; and M. Carbon, C. Hsieh, J. Kane, and C. Cornell, "Tardive Dyskinesia Prevalence in the Period of Second-Generation Antipsychotic Use: A Meta-Analysis," *Journal of Clinical Psychiatry* 78 (2017): 264–278.

33. For commentary on the increasing reliance on high-dose, high-potency antipsychotics in this period, see Ross Baldessarini, Bruce Cohen, and Martin Teicher, "Significance of Neuroleptic Dose and Plasma Level in the Pharmacological Treatment of Psychoses," *Archives of General Psychiatry* 45 (1988): 79–91.

34. Peter Weiden, John Mann, Gretchen Haas, Marlin Mattson, and Allen Frances, "Clinical Nonrecognition of Neuroleptic-Induced Movement Disorders: A Cautionary Study," *American Journal of Psychiatry* 144 (1987): 1148–1153.

35. See J. Kane et al., "Clozapine for the Treatment-Resistant Schizophrenic: A Double-Blind Comparison with Chlorpromazine," *Archives of General Psychiatry* 45 (1988): 789–796.

36. John Crilly, "The History of Clozapine and Its Emergence in the US Market: A Review and Analysis," *History of Psychiatry* 18 (2007): 39–60.

37. Obviously, Clozapine was emphatically not a second-generation drug, but its novelty on the North American market allowed it to be sold as part of this new class of pills.

38. Allan V. Horwitz, *DSM: A History of Psychiatry's Bible* (Baltimore, MD: Johns Hopkins University Press, 2021), 72.

39. Quoted in Edward Shorter, *Before Prozac: The Troubled History of Mood Disorders in Psychiatry* (New York: Oxford University Press, 2009), 164.

40. Shorter, *Before Prozac*, 164.

41. David Flockhart, "Dietary Restrictions and Drug Interactions with Monoamine Oxidase Inhibitors: An Update," *Journal of Clinical Psychiatry* 73 (supp.) (2012): 17–24.

42. Jordan Moraczewski and Kapil Aedma, "Tricyclic Antidepressants," StatPearls, December 7, 2020, https://www.ncbi.nlm.nih.gov/books/NBK557791/.

43. For an analysis of how Eli Lilly created the market for Prozac, see Margaret L. Eaton and Mark Xu, *Developing and Marketing a Blockbuster Drug: Lessons from Eli Lilly's Experience with Prozac* (Cambridge, MA: Harvard Business Press, 2005).

44. M. Domino and M. Swartz, "Who Are the New Users of Antipsychotic Medications," *Psychiatric Services* 59 (2008): 507–514.

45. L. Pratt, D. Broady, and Q. Gu, "Antidepressant Use among Persons 12 and Over: United States, 2011–2014," CDC, National Center for Health Statistics, Data Brief 283, August 2017, https://www.cdc.gov/nchs/data/databriefs/db283.pdf.

46. For a case study of this process, see Glen Spielmans, "The Promotion of Olanzapine in Primary Care: An Examination of Internal Industry Documents," *Social Science and Medicine* 69 (2009): 14–20. It is clear from the acknowledgments and wording in the text at some points that fears of lawsuits brought by Eli Lilly prompted some very cautious language: "I acknowledge the careful review of British attorneys in helping to shape the final version of this article" (19). Ironically, the paper's highly revealing data exposing the machinations of Lilly were the result of the discovery process in a lawsuit against the company, which forced the production of internal company documents. Those documents are available at http://blog.dida.library.ucsf.edu/2014/08/zyprexa-documents-and-tamiflu-csrs-added.html.

47. Spielmans, "The Promotion of Olanzapine," 15, 16, quoting company documents. Zyprexa achieved sales of $4 billion in 2002 and $5 billion by 2007.

48. See G. C. Alexander et al., "Increasing Off-Label Use of Antipsychotic Medications in the United States, 1995–2008," *Pharmacoepidemiology and Drug Safety* 20 (2011): 177–184; and H. Verdoux, M. Tournier, and B. Bergaud, "Antipsychotic Prescribing Trends," *Acta Psychiatrica Scandinavica* 121 (2010): 4–10.

49. On the increasing unapproved use of antipsychotic medications in children, see J. N. Harrison, F. Cluxton-Keller, and D. Gross, "Antipsychotic Medication Prescribing Trends in Children and Adolescents," *Pediatric Health Care* 26 (2012): 139–145. For the picture with antidepressants, see E. Lee et al., "Off-Label Prescribing Patterns of Antidepressants in Children and Adolescents," *Pharmacoepidemiology and Drug Safety* 21 (2012): 137–144.

50. Dinci Pennap et al., "Patterns of Early Mental Health Diagnosis and Medication Treatment in a Medicaid-Insured Birth Cohort," *JAMA Pediatrics* 172 (2018): 576–584.

51. Brian Piper et al., "Trends in Use of Prescription Stimulants in the United States and Territories, 2006 to 2016," *PLoS One* 13 (2018), e0206100.

52. S. Almashat, C. Preston, T. Waterman, and S. Wolfe, "Rapidly Increasing Criminal and Civil Monetary Penalties against the Pharmaceutical Industry, 1991 to 2010," Public Citizens' Health Research Group, December 16, 2010, https://www.citizen.org/wp-content/uploads/migration/rapidlyincreasingcriminalandcivilpenalties.pdf. This article documents many of the off-label promotion activities that have come to light and led to successful legal action. See also A. S. Kesselheim et al., "False Claims Act Prosecution Did Not Deter Off-Label Drug Use in the Case of Neurontin," *Health Affairs* 30 (2011): 2318–2327; and A. S. Kesselheim, "Off-Label Drug Use and Promotion: Balancing Public Health Goals and Commercial Speech," *American Journal of Law and Medicine* 37 (2011): 225–257.

53. D. Healy and J. Le Noury, "Pediatric Bipolar Disorder: An Object Study in the Creation of an Illness," *International Journal of Risk and Safety in Medicine* 19 (2007): 209–221.

54. J. C. Blader and G. A. Carlson, "Increased Rates of Bipolar Disorder Diagnoses among US Child, Adolescent, and Adult Inpatients, 1996–2004," *Biological Psychiatry* 62 (2006): 107–114; I. Harpaz-Rotem and R. Rosenheck, "Changes in Outpatient Psychiatric Diagnosis in Privately Insured Children and Adolescents from 1995 to 2000," *Child Psychiatry and Human Development* 34 (2004): 329–340, 339–340; C. Moreno et al., "Na-

tional Trends in Outpatient Diagnosis and Treatment of Bipolar Disorder in Youth," *Archives of General Psychiatry* 64 (2007): 1032–1039.

55. Joseph Biederman et al., "Attention-Deficit Hyperactivity Disorder and Juvenile Mania: An Overlooked Co-morbidity," *Journal of the American Academy of Child and Adolescent Psychiatry* 35 (1996): 997–1008; J. Bostic et al., "Juvenile Mood Disorders and Office Psychopharmacology," *Pediatric Clinics of North America* 44 (1997): 1487–1503; S. Faraone, J. Biederman, et al., "Bipolar and Antisocial Disorders among Relatives of ADHD Children," *American Journal of Medical Genetics* 81 (1998): 108–116; J. Biederman et al., "Open-Label, Eight Week Trial of Olanzapine and Risperidone for the Treatment of Bipolar Disorder in Preschool Age Children," *Biological Psychiatry* 58 (2005): 589–594. For a list of the antidepressant and antipsychotic drugs that had not received approval for use in pediatric patients by 2009, see Ian Larkin et al., "Restrictions on Pharmaceutical Detailing Reduced Off-Label Prescribing of Antidepressants and Antipsychotics in Children," *Health Affairs* 33 (2104): 1014–1023.

56. Data provided to Senator Charles Grassley showed that Biederman and his colleague Timothy Wilens had received consulting fees totaling at least $1.6 million each between 2000 and 2007. Only a fraction of this income had been reported to Harvard. "Psychiatric Group Faces Scrutiny over Drug Industry Ties," *New York Times,* July 12, 2008.

57. Alan Schwarz, "The Selling of Attention Deficit Disorder," *New York Times,* December 14, 2013. Schwarz notes that there were no advertisements for ADHD drugs in the *Journal of the American Academy of Child and Adolescent Psychiatry* between 1990 and 1993. About a decade later, there were a hundred pages a year.

58. Gardiner Harris, "Research Center Tied to Drug Company," *New York Times,* November 24, 2008.

59. Gardiner Harris, "Drug Maker Told Studies Would Aid It, Papers Say," *New York Times,* March 19, 2009. Biederman testified that he personally prepared the slides containing these statements. At a deposition taken on February 26, 2008, an interesting exchange occurred. Asked what rank he held at Harvard, Biederman replied, "Full professor." "What's after that?" the lawyer asked. "God," Biederman responded. "Did you say God?" the lawyer asked. "Yeah," Biederman said.

60. Joseph Biederman et al., "Open-Label, 8-Week Trial of Olanzapine and Risperidone for the Treatment of Bipolar Disorder in Preschool Age Children," *Biological Psychiatry* 58 (2005): 589–594. This trial was conducted on children between four and six years of age. The authors note that the usefulness of risperidone in the treatment of depressive symptoms was equally evident in a trial with older children and adolescents. See Joseph Biederman et al., "An Open-Label Trial of Risperidone in Children and Adolescents with Bipolar Disorder," *Journal of Child and Adolescent Psychopharmacology* 15 (2005): 311–317. Both studies, it should be said, are scientifically worthless. Both did show, however, disturbing patterns of weight gain among the children, despite the brief duration of the studies.

61. Quoted in Gary Greenberg, *The Book of Woe: The DSM and the Unmaking of Psychiatry* (New York: Blue Rider Press, 2013), 147.

62. "Expert or Shill," editorial, *New York Times,* November 29, 2009.

63. "Credibility Crisis in Pediatric Psychiatry," editorial, *Nature Neuroscience* 11 (2008): 93.

64. Joseph Biederman, "Credibility Crisis in Pediatric Psychiatry," letter to editor, *Nature Neuroscience* 11 (2008): 1233. The psychiatrist E. Fuller Torrey was unimpressed,

saying Biederman "has taken the pharmaceutical company corruption of psychiatry to new depths." E. Fuller Torrey, "Psychoprostitution," *Public Citizen*, May 28, 2009.

65. Thomas Sullivan, "Harvard Psychiatrists Sanctioned for Failing to File Proper Disclosure," *Policy and Medicine*, July 2011.

66. I. Katz, B. Rovner, and L. Schneider, "Use of Psychoactive Drugs in Nursing Homes," *New England Journal of Medicine* 327 (1992): 1392–1393. That percentage has undoubtedly risen since then. In 2016 and 2017, Human Rights Watch visited 109 nursing homes in six states to examine local practices. They estimated that in an average week, more than 179,000 nursing home residents were being given unapproved antipsychotic medications. "'They Want Docile': How Nursing Homes in the United States Overmedicate People with Dementia," report, Human Rights Watch, New York, February 5, 2018. For a recent estimate of the resultant mortality, see Donovan Maust, "Antipsychotics, Other Psychotropics, and the Risk of Death in Patients with Dementia," *JAMA Psychiatry* 72 (2015): 438–445.

67. Spielmans, "The Promotion of Olanzapine," 17–18.

68. L. Schneider, K. Dagerman, and P. Insel, "Risk of Death with Atypical Antipsychotic Drug Treatment for Dementia: A Meta-Analysis of Randomized Placebo-Controlled Trials," *Journal of the American Medical Association* 294 (2005): 1934–1943.

CHAPTER 21 Genetics, Neuroscience, and Mental Illness

1. The major exception to the postwar neglect of genetics was an English psychiatrist, Eliot Slater, who studied in Munich under Ernst Rüdin and endorsed the Nazi policy of forced sterilization of mental patients. Though he later denounced Nazi *Rassenhygiene* and married a Jew, he continued to refer approvingly to Rüdin's work and began twin studies of his own. With Willi Mayer-Gross, he wrote the standard textbook *Clinical Psychiatry*, used by generations of British psychiatry trainees. He became an evangelical proponent of lobotomies and insulin comas and persisted in recommending both into the 1970s, long after they had been discredited. Slater advocated for a monogenic theory of schizophrenia, insisting that the disease was caused by a single, yet-to-be-discovered gene.

2. Eric Kandel, "A New Intellectual Framework for Psychiatry," *American Journal of Psychiatry* 155 (1998): 457–469, 459, emphasis in the original.

3. See D. Rosenthal, "Some Factors Associated with Concordance and Discordance with Respect to Schizophrenia in Monozygotic Twins," *Journal of Nervous and Mental Disease* 129 (1959): 1–10; D. Rosenthal, "Problems of Sampling and Diagnosis in the Major Twin Studies of Schizophrenia," *Journal of Psychiatric Research* 1 (1962): 116–134.

4. G. Allen, "Twin Studies: Problems and Prospects," *Progress in Medical Genetics* 4 (1965): 242–269.

5. K. S. Kendler and S. R. Diehl, "The Genetics of Schizophrenia: A Current Genetic-Epidemiological Perspective," *Schizophrenia Bulletin* 19 (1993): 261–285; K. Kendler, "Twin Studies of Psychiatric Illness," *Archives of General Psychiatry* 50 (1993): 905–915; B. Riley and R. Williamson, "Sane Genetics for Schizophrenia," *Nature Medicine* 6 (2000): 253–255.

6. As the prominent psychiatric geneticist Kenneth Kendler has acknowledged, "the claim that [family, twin, and adoption studies] 'prove' that a disorder is 'biological' is weak." K. S. Kendler, "What Psychiatric Genetics Has Taught Us about the Nature of Psychiatric Illness and What Is Left to Learn," *Molecular Psychiatry* 18 (2013): 1058–1066,

1058. For discussion of this literature, see M. Henriksen, J. Nordgaard, and L. Jansson, "Genetics of Schizophrenia: Overview of Methods, Findings and Limitations," *Frontiers of Human Neuroscience* 11 (2017), article 322.

7. Huntington's chorea is a horrific inherited disorder that begins with subtle shifts in mood, then alterations of gait, followed by uncontrollable spasmodic and jerky movements (chorea) before it proceeds inexorably to a host of neurological and psychiatric catastrophes, ending in dementia and premature death.

8. Quoted in Gary Greenberg, *The Book of Woe: The DSM and the Unmaking of Psychiatry* (New York: Blue Rider Press, 2013), 60. Prominent contributors to this effort subsequently published an extensive manifesto, laying out how the revisions ought to proceed. See D. J. Kupfer, M. B. First, and D. A. Regier, *A Research Agenda for DSM-V* (Washington, DC: American Psychiatric Association, 2002). Kupfer and Regier were named chair and vice chair of the *DSM 5* Task Force by Steven Sharfstein, the then president of the APA, in 2006. The complete Task Force was not assembled until July 2007.

9. D. A. Regier, W. E. Narrow, E. A. Kuhl, and D. J. Kupfer, "The Conceptual Development of DSM-V," *American Journal of Psychiatry* 166 (2009): 645–650, 646; D. J. Kupfer, E. A. Kuhl, W. E. Narrow, and D. A. Regier, "On the Road to DSM," *Cerebrum*, Dana Foundation, https://www.dana.org/article/updating-the-diagnostic-and-statistical-manual-of-mental-disorders/.

10. Regier, Narrow, Kuhl, and Kupfer, "The Conceptual Development of DSM-V," 646. See also a later articulation of this position: D. A. Regier, W. E. Narrow, D. S. Rae, and M. Rubio-Stiper, "Advancing from Reliability to Validity: The Challenge for DSM / ICD 11 Revisions," in *Psychopathology in the Genome and Neuroscience Era*, ed. C. Zorumski and C. Rubin, 85–96 (Washington, DC: American Psychiatric Association, 2005).

11. Regier, Narrow, Kuhl, and Kupfer, "The Conceptual Development of DSM-V," 649.

12. For a recent acknowledgment that psychiatric disorders are not reducible to alleles at specific genetic loci, see Steven Hyman, "Psychiatric Disorders: Grounded in Human Biology, but Not Natural Kinds," *Perspectives on Biology and Medicine* 64 (2021): 6–28.

13. For an early, widely publicized example, see R. Sherrington et al., "Localization of a Susceptibility Locus for Schizophrenia on Chromosome 5," *Nature* 336 (1988): 164–167. Their conclusions were partially retracted by an overlapping research team the following year: S. D. Detera-Wadleigh et al., "Exclusion of Linkage to 5q11–13 in Families with Schizophrenia and Other Psychiatric Disorders," *Nature* 340 (1989): 391–393. The original results were not replicated in R. Crowe et al., "Lack of Linkage to Chromosome 5q11–q13 Markers in Six Schizophrenic Pedigrees," *Archives of General Psychiatry* 48 (1991): 357–361; and D. St Clair et al., "No Linkage of Chromosome 5q11–q13 Markers to Schizophrenia in Scottish Families," *Nature* 339 (1989): 305–309. Leon Eisenberg noted the "undiscriminating enthusiasm that greeted alleged 'breakthroughs' in psychiatric genetics" and the subsequent recognition that what had been trumpeted was, in fact, a "false positive." Leon Eisenberg, "The Social Construction of the Human Brain," *American Journal of Psychiatry* 152 (1995): 1563–1575, 1565.

14. Patrick F. Sullivan, "Spurious Genetic Associations," *Biological Psychiatry* 61 (2007): 1121–1126, 1121.

15. M. S. Farrell et al., "Evaluating Historical Candidate Genes for Schizophrenia," *Molecular Psychiatry* 20 (2015): 555–562; E. C. Johnson et al., "No Evidence that Schizophrenia Candidate Genes Are More Associated with Schizophrenia than Non-candidate

Genes," *Biological Psychiatry* 82 (2017): 702–708; Richard Border et al., "No Support for Historical Candidate Gene or Candidate Gene-by-Interaction Hypotheses for Major Depression," *American Journal of Psychiatry* 176 (2019): 376–387.

16. K. S. Kendler and M. C. O'Donovan, "A Breakthrough in Schizophrenia Genetics," *JAMA Psychiatry* 71 (2014): 1319–1320.

17. Kendler, "What Psychiatric Genetics Has Taught Us," 1065.

18. Kendler, "What Psychiatric Genetics Has Taught Us," 1059.

19. Schizophrenia Working Group, "Biological Insights from 108 Schizophrenia-Associated Genetic Loci," *Nature* 511 (2014): 421–427. Another group examined GWAS data on links to five major psychiatric disorders: schizophrenia, bipolar disorder, major depressive disorder, attention-deficit hyperactivity disorder, and autism spectrum disorder. This group, too, conceded that "the effect sizes of the genome-wide significant loci are individually quite small and the variance they [collectively] account for is insufficient for predictive or diagnostic usefulness." Cross-Disorder Group of the Psychiatric Genomics Consortium, "Identification of Risk Loci with Shared Effects on Five Major Psychiatric Disorders: A Genome-Wide Analysis," *Lancet* 381 (2013): 1371–1379, 1378.

20. Sophie E. Legge et al., "Genetic Architecture of Schizophrenia: A Review of Major Advancements," *Psychological Medicine* 51 (2021): 1–10.

21. Hyman, "Psychiatric Disorders," 21.

22. Major Depressive Disorder Working Group of the Psychiatric GWAS Consortium, "A Mega-Analysis of Genome-Wide Association Studies for Major Depressive Disorder," *Molecular Psychiatry* 18 (2013): 497–511, 497.

23. Hyman, "Psychiatric Disorders," 14.

24. Johnson et al., "No Evidence," 706.

25. Border et al., "No Support," 376.

26. R. Uher and M. Rutter, "Basing Psychiatric Classification on Scientific Foundations: Problems and Prospects," *International Review of Psychiatry* 24 (2012): 591–605, 594. Compare this conclusion with Stephen Lawrie's claim that the Psychiatric Genomics Consortium's findings were a "remarkable success . . . a game changer" because they showed "at least some biological validity of the clinical concept that is schizophrenia." S. Lawrie, "Clinical Risk Prediction in Schizophrenia," *Lancet Psychiatry* 1 (2014): 406–408, 406. If this is a success, one wonders what failure would look like. A more sober conclusion is offered by Henriksen, Nordgaard, and Jansson, "Genetics of Schizophrenia," who note ruefully that "if common genetic variation implicates an intractable amount of genes of only very small individual effect alleles, we may find ourselves in a situation, where, as Goldstein put it, 'in pointing at everything, genetics would point at nothing.'" Here they are referencing D. B. Goldstein, "Common Genetic Variation and Human Traits," *New England Journal of Medicine* 360 (2009): 1696–1698.

27. S. E. Hyman, "An Interview with Steven Hyman," *Trends in Cognitive Sciences* 16 (2012): 3–5, 4.

28. S. M. Purcell et al., "Common Polygenic Variation Contributes to Risk of Schizophrenia and Bipolar Disorder," *Nature* 460 (2009): 748–752; S. H. Lee et al., "Genetic Relationship between Five Psychiatric Disorders Estimated from Genome-Wide SNPs," *Nature Genetics* 45 (2013): 984–994; M. J. Gandal et al., "Shared Molecular Neuropathology across Major Psychiatric Disorders Parallels Polygenetic Overlap," *Science* 359 (2018): 693–697; Cross-Disorder Group of the Psychiatric Genetics Consortium, "Genomic Relationships, Novel Loci, and Pleiotropic Mechanisms across Eight Psychiatric Disorders," *Cell* 179 (2019): 1469–1482; Alastair Cardno and Michael Owen, "Genetic Relationships

between Schizophrenia, Bipolar Disorder, and Schizoaffective Disorder," *Schizophrenia Bulletin* 40 (2014): 504–515.

29. Brainstorm Consortium, "Analysis of Shared Heritability in Common Disorders of the Brain," *Science* 360 (June 22, 2018), eaap8757, quote from "Research Article Summary," 1313. As the authors note in the research paper, this situation contrasts markedly with what one finds for neurological disorders like Alzheimer's and Parkinson's diseases. The consortium is a cross-national collaboration among researchers at Harvard, Stanford, Oxford, and Cambridge, among others.

30. Henry Maudsley, *The Physiology and Pathology of Mind* (London: Macmillan, 1868), 204–205.

31. For a similar assessment of the destabilizing effects of findings from genetics, see Allan V. Horwitz, *DSM: A History of Psychiatry's Bible* (Baltimore, MD: Johns Hopkins University Press, 2021), ch. 6.

32. I already described the unlikely combination of disorders lumped together under the category of major depression. It is worth remembering that the Swiss psychiatrist Eugen Bleuler, who coined the term "schizophrenia," preferred to speak of "the schizophrenias," implicitly acknowledging the heterogeneity of the disorders he sought to make sense of under this label.

33. I owe this point to an anonymous reviewer for *Psychological Medicine*.

34. Graham K. Murray et al., "Could Polygenic Risk Scores Be Useful in Psychiatry?" *JAMA Psychiatry* 78 (2020): 210–219.

35. To be fair, the authors acknowledge this problem, citing a study where some patients were informed that they had a high genetic risk for depression, and others were told that their risk was low. The first group then reported a heightened number of depressive symptoms, even though both groups had been deceived: nothing was known about either group's genetic risk for the disorder. See M. S. Lebowitz and W. K. Ahn, "Testing Positive for a Genetic Predisposition to Depression Magnifies Retrospective Memory for Depressive Symptoms," *Journal of Consulting and Clinical Psychology* 85 (2017): 1052–1063.

36. Brainstorm Consortium, "Analysis of Shared Heritability."

37. Uher and Rutter, "Basing Psychiatric Classification," 595.

38. For a prescient warning that antipsychotic drugs might affect brain structure, see David Marsden, "Cerebral Atrophy and Cognitive Impairment in Chronic Schizophrenia," *Lancet* 308 (1976): 1132. For some of the studies demonstrating the drug-brain connection, see Jeffrey Lieberman et al., "Antipsychotic Drug Effects on Brain Morphology in First-Episode Psychosis," *Archives of General Psychiatry* 62 (2005): 361–370; K. A. Dorph-Petersen et al., "The Influence of Chronic Exposure to Antipsychotic Medications on Brain Size before and after Tissue Fixation: A Comparison of Haloperidol and Olanzapine in Macaque Monkeys," *Neuropsychopharmacology* 30 (2005): 1649–1661; B. C. Ho et al., "Progressive Structural Brain Abnormalities and Their Relationship to Clinical Outcome," *Archives of General Psychiatry* 60 (2003): 585–594; Joanna Moncrieff and Jonathan Leo, "A Systematic Review of the Effects of Antipsychotic Drugs on Brain Volume," *Psychological Medicine* 40 (2010): 1–14; R. B. Zipursky, T. J. Reilly, and R. M. Murray, "The Myth of Schizophrenia as a Progressive Brain Disease," *Schizophrenia Bulletin* 39 (2013): 1363–1372; and B.-C. Ho, N. Andreasen, S. Ziebell, R. P. Pierson, and V. Magnotta, "Long-Term Antipsychotic Treatment and Brain Volumes: A Longitudinal Study of First-Episode Schizophrenia," *Archives of General Psychiatry* 68 (2011): 128–137.

39. Robin Murray, "Mistakes I Have Made in My Research Career," *Schizophrenia Bulletin* 43 (2017): 253–256, 253–254.

40. There are, it turns out, huge methodological errors that vitiate the findings of much fMRI research. As one devastating survey of these problems (originally titled "Voodoo Correlations in Social Neuroscience") concluded, "a disturbingly large, and quite prominent, segment of fMRI research on emotion, personality, and social cognition is using seriously defective research methods and producing a profusion of numbers that should not be believed . . . the questionable analysis methods discussed here are also widespread in fields that use fMRI to study individual differences, such as cognitive neuroscience, clinical neuroscience, and neurogenetics." Edward Vul, Christine Harris, Piotr Winkielman, and Harold Pashler, "Puzzlingly High Correlations in fMRI Study of Emotion, Personality, and Social Cognition," *Perspectives on Psychological Science* 4 (2009): 274–290, 285.

41. J. B. Savitz, S. L. Rauch, and W. C. Drevets, "Clinical Application of Brain Imaging for the Diagnosis of Mood Disorders: The Current State of Play," *Molecular Psychiatry* 18 (2013): 528–539, 528. Nor does there seem to be much prospect of future progress. Savitz and colleagues comment on the manifold difficulties and confounding factors confronting researchers.

42. S. E. Hyman, "Can Neuroscience Be Integrated into the DSM-V?" *Nature Reviews Neuroscience* 8 (2007): 725–732. Hyman thought that perhaps the answer was yes in a small subset of mental disorders. That proved to be an overstatement.

43. Pliny Earle to Clark Bell, April 16, 1886, box 1, folder 3, Pliny Earle Papers, American Antiquarian Society, Worcester, MA.

44. Benedict Carey, "New Definition of Autism Will Exclude Many, Study Suggests," *New York Times*, January 19, 2012.

45. *Diagnostic and Statistical Manual of Mental Disorders* 5 (Washington DC: American Psychiatric Association, 2013), 51.

46. See, for example, Joel Lexchin, Lisa Bero, Benjamin Djulbegovich, and Otavio Clark, "Pharmaceutical Industry Sponsorship and Research Outcome and Quality: A Systematic Review," *British Medical Journal* 326 (2003): 1167–1170; Gisela Schott, Henry Pachl, Ulrich Limbach, Ursula Gundert-Remy, Wolf-Dieter Ludwig, and Klaus Lieb, "The Financing of Drug Trials by Pharmaceutical Companies and Its Consequences, Part 1," *Deutsches Arzteblatt International* 107 (2010): 279–285; and Andreas Lundh, Joel Lexchin, Barbara Mintzes, Jeppe Schroll, and Lisa Bero, "Industry Sponsorship and Research Outcome," *Cochrane Database of Systematic Reviews*, February 16, 2017, https://www.cochranelibrary.com/cdsr/doi/10.1002/14651858.MR000033.pub3/full.

47. H. Ledford, "Psychiatry Manual Revisions Spark Row," *Nature* 460 (2009): 445.

48. Hannah S. Decker, *The Making of DSM-III: A Diagnostic Manual's Conquest of American Psychiatry* (Oxford: Oxford University Press, 2013).

49. I am grateful to Will Carpenter of the University of Maryland Medical Center, who urged me to clarify this point.

50. For example, see Allan Horwitz and Jerome Wakefield, *The Loss of Sadness: How Psychiatry Transformed Normal Sorrow into Depressive Disorder* (Oxford: Oxford University Press, 2007). Michael Taylor blamed much of the increase in diagnoses on the drug industry: "The pharmaceutical industry adores the explosion of conditions, because as 'medical diagnoses' the *DSM* categories provide the rationale for prescribing drugs." Michael Alan Taylor, *Hippocrates Cried: The Decline of American Psychiatry* (Oxford: Oxford University Press, 2013), 64.

51. Allen Frances, "DSM-5 Is a Guide, Not a Bible: Simply Ignore Its Ten Worst Changes," *Psychiatric Times*, December 5, 2012.

52. See, for example, Robert Spitzer, "DSM-V Transparency: Fact or Rhetoric?," *Psychiatric Times,* March 7, 2009; and then Allen Frances, "Whither DSM-V?," *British Journal of Psychiatry* 195 (2009): 391–392; A. Frances, "Opening Pandora's Box: The 19 Worst Suggestions for DSM-5," *Psychiatric Times,* January 1, 2010; A. Frances, "DSM-5 Badly Flunks the Writing Test," *Psychiatric Times,* June 11, 2013; and A. Frances, "Psychosis Risk Syndrome Is Back," *Psychiatric Times,* November 26, 2013.

53. A. Schatzberg, J. H. Scully, Jr., D. J. Kupfer, and D. A. Regier, "Setting the Record Straight: A Response to Frances [*sic*] Comment on DSM-V," *Psychiatric Times,* July 1, 2009.

54. Michael First, a leading participant in the construction of *DSM IV,* quoted in Greenberg, *The Book of Woe,* 92.

55. Quoted in Horwitz, *DSM,* 122.

56. Schatzberg had co-founded a start-up, Corcept Therapeutics, and had stock in the company variously valued at between $4.5 million and $6 million. (His employer, Stanford University, had also invested in the enterprise.) His grant was to investigate the use of mifepristone, a product produced by his company, for possible use as an antidepressant. Contacted by the staff of Senator Chuck Grassley, who was investigating corrupt practices in psychiatry, Stanford at first protested that Schatzberg hadn't violated their conflict-of-interest policies, then asked him to step down temporarily as principal investigator. He later resumed his position. In 2010, Schatzberg relinquished his chairmanship of Stanford's department of psychiatry. See "Stanford Psychiatrist Removed from Drug Study," *Nature* 454 (2008): 679. For commentary by an investigator for Senator Grassley, see Paul Thacker, "The Ugly Underbelly of Medical Research," Project on Government Oversight (POGO) blog, January 13, 2011, https://pogoblog.typepad.com/pogo/2011/01/the-ugly-underbelly-of-medical-research.html.

57. Quoted in Horwitz, *DSM,* 125.

58. Quoted in Horwitz, *DSM,* 126. Horwitz's book is likely to be regarded, with good reason, as the definitive account of classification and American psychiatry.

59. Horwitz, *DSM,* 149, emphasis in the original.

60. See Assembly of the American Psychiatric Association, "Motion on Cross-Cutting Dimensions and Severity Scales," American Psychiatric Association Annual Meeting, New York, 2012; Owen Whooley and Allan Horwitz, "The Paradox of Professional Success: Grand Ambition, Furious Resistance, and the Derailment of the DSM-5 Review Process," in *Making the DSM-5: Concepts and Controversies,* ed. J. Paris and J. Phillips, 75–92 (New York: Springer, 2013); and P. Zachar and M. First, "Transitioning to a Dimensional Model of Personality Disorder in DSM 5.1 and Beyond," *Current Opinion in Psychiatry* 28 (2015): 66–72.

61. On kappa, see A. J. Viera and J. M. Garrett, "Understanding Interobserver Agreement: The Kappa Statistic," *Family Medicine* 37 (2005): 360–363.

62. Allen Frances, "Two Fallacies Invalidate the DSM-5 Field Trials," *Psychiatric Times,* January 10, 2012.

63. Helena Kraemer et al., "DSM-5: How Reliable Is Reliable Enough?," *American Journal of Psychiatry* 169 (2012): 13–15, 13.

64. I am grateful once more to Will Carpenter for this point.

65. Kraemer et al., "DSM-5." The authors indicated that the standard for evaluating the trials should be the reliability of diagnoses in other areas of medicine, and then they set the bar way below that. For criticism, see K. D. Jones, "A Critique of the DSM-5 Field Trials," *Journal of Nervous and Mental Disease* 200 (2012): 517–519; and Samuel Lieblich,

"High Heterogeneity and Low Reliability in the Diagnosis of Major Depression Will Impair the Development of New Drugs," *British Journal of Psychiatry Open* 1 (2015), e5–e7.

66. D. A. Regier et al., "DSM-5 Field Trials in the United States and Canada, Part II: Test-Retest Reliability of Selected Categorical Diagnoses," *American Journal of Psychiatry* 170 (2013): 59–70. As Regier and colleagues recognize, "If the diagnostic criteria defining the disorder in a given group of patients cannot be assessed reliably by two or more clinicians, then patients with those diagnoses cannot be expected to have common treatment responses or similar etiological and laboratory findings" (60).

67. Quoted in Pam Belluck and Benedict Carey, "Psychiatry's Guide Is Out of Touch with Science, Experts Say," *New York Times,* May 6, 2013.

68. Thomas Insel, "Transforming Diagnosis," NIMH Director's blog, April 29, 2013, no longer available online. See Belluck and Carey, "Psychiatry's Guide," for the Bible quotation.

69. Quoted in Greenberg, *The Book of Woe,* 340. Steven Hyman had earlier expressed similar incredulity that "grant reviewers, journal referees, editors, and regulatory agencies [all] acted as if DSM criteria . . . pick out real human diseases." Hyman, "Interview," 4.

70. Insel, "Transforming Diagnosis."

71. Quoted in Greenberg, *The Book of Woe,* 340.

72. See, for example, "The NIMH Withdraws Support for DSM 5," *Psychology Today,* May 4, 2013; Gary Greenberg, "The Rats of NIMH," *New Yorker,* May 16, 2013; "Psychiatry in Crisis! Mental Health Director Rejects Psychiatric 'Bible' and Replaces with Nothing," *Scientific American,* May 4, 2013; "NIMH and APA Clash over Upcoming DSM-5," *Medscape,* May 7, 2013; "Psychiatry Divided as Mental Health 'Bible' Denounced," *New Scientist,* May 3, 2013; M. McCarthy, "Director of Top Research Organization for Mental Health Criticizes DSM for Lack of Validity," *British Medical Journal* 346 (2013), f2954; and Belluck and Carey, "Psychiatry's Guide."

73. Christopher Ross and Russell Margolis, in "Research Domain Criteria: Strengths, Weaknesses, and Potential Alternatives for Future Psychiatric Research," *Molecular Neuropsychiatry* 5 (2019): 218–235, argue that RDoC is "fundamentally flawed" (220). See also D. R. Weinberger et al., "Whither Research Domain Criteria (RDoC)? The Good, the Bad, and the Ugly," *JAMA Psychiatry* 72 (2015): 1161–1162.

74. Thomas Insel and Jeffrey Lieberman, "DSM-5 and RDoC: Shared Interests," news release, American Psychiatric Association, Washington, DC, May 14, 2013.

CHAPTER 22 The Crisis of Contemporary Psychiatry

1. Robin Murray, "Mistakes I Have Made in My Research Career," *Schizophrenia Bulletin* 43 (2017): 253–256, 256.

2. S. Guloksuz and J. van Os, "The Slow Death of the Concept of Schizophrenia and the Painful Birth of the Psychosis Syndrome," *Psychological Medicine* 48 (2018): 229–244.

3. Kenneth Kendler, "What Psychiatric Genetics Has Taught Us about the Nature of Psychiatric Illness and What Is Left to Learn," *Molecular Psychiatry* 18 (2013): 1058–1066, 1059. I think "give no guarantee" should read as "are highly unlikely to correspond to a single definable pathophysiology."

4. H. Christian Fibiger, "Psychiatry, the Pharmaceutical Industry and the Road to Better Therapeutics," *Schizophrenia Bulletin* 38 (2012): 649–650. Fibiger had served as vice president of neuroscience at both Eli Lilly and Amgen before moving to academia.

5. Antonio Regalado, "Q. and A. with Tom Insel on His Decision to Join Alphabet," *MIT Technology Review*, September 21, 2015.

6. See Alex Berenson, Gardiner Harris, Barry Meier, and Andrew Pollack, "Despite Warnings, Drug Giant Took Long Path to Vioxx Recall," *New York Times*, November 14, 2004; and Jim Giles, "Drug Giant Merck Accused of Deaths Cover-Up," *New Scientist*, April 15, 2008.

7. See Patrick Radden Keefe, *Empire of Pain: The Secret History of the Sackler Dynasty* (New York: Doubleday, 2021).

8. Katie Thomas and Michael Schmidt, "Glaxo Agrees to Pay $3 Billion in Fraud Settlement," *New York Times*, July 2, 2012.

9. Katie Thomas, "J.&J. to Pay $2.2 Billion in Risperdal Settlement," *New York Times*, November 4, 2013.

10. Tracy Staton, "Pfizer Adds Another $325M to Neurontin Settlement. Total $945M," *Fierce Pharma*, June 2, 2014. For the range of company misconduct in this case, see C. S. Langefeld and M. A. Steinman, "The Neurontin Legacy: Marketing through Misinformation and Manipulation," *New England Journal of Medicine* 360 (2009): 103–106.

11. On Biederman, see Gardiner Harris and Benedict Carey, "Researchers Fail to Reveal Full Drug Pay," *New York Times*, June 8, 2008; Gardiner Harris, "Research Center Tied to Drug Company," *New York Times*, November 28, 2008; and Jeanne Lenzer, "Review Launched after Harvard Psychiatrist Failed to Disclose Industry Funding," *British Medical Journal* 336 (2008): 1327. On Schatzberg, see Jacob Goldstein, "Grassley Questions Stanford Psychiatrist's Industry Ties," *Wall Street Journal*, June 25, 2008.

12. Gardiner Harris, "Top Psychiatrist Didn't Report Drug Makers' Pay," *New York Times*, October 3, 2008.

13. Susan Chimonas, Frederica Stahl, and David Rothman, "Exposing Conflict of Interest in Psychiatry: Does Transparency Matter?" *International Journal of Law and Psychiatry* 35 (2012): 490–495, 493; Charles E. Grassley, Ranking Member, Senate Finance Committee, to James W. Wagner, President, Emory University, September 16, 2008, https://www.finance.senate.gov/imo/media/doc/Grassley%20Letter%20to%20Emory%20University%20(Sep%2016,%202008).pdf; Paul Thacker, "How an Ethically Challenged Researcher Found a Home at the University of Miami," *Forbes*, September 13, 2011. Thacker was Grassley's chief investigator.

14. David Armstrong, "Medical Journal Editor Steps Down over Undisclosed Ties," *Wall Street Journal*, August 28, 2006.

15. Harris, "Top Psychiatrist Didn't Report Drug Makers' Pay."

16. Jocelyn Kaiser, "Senate Probe of Research Psychiatrists," *Science* 325 (2009): 30. Gary Greenberg uncovered a revealing email from a Johnson & Johnson marketing executive, warning his superiors about the danger of curtailing even a tiny fraction of Biederman's financial support. The company had rejected Biederman's request for an honorarium, a mere $3,000. "I have never seen someone so angry. Since that time, our business has become non-existent in his area of control." The decision not to pay his speaking fee must be reversed, or "I am truly afraid of the consequences." Quoted in Gary Greenberg, *The Book of Woe: The DSM and the Unmaking of Psychiatry* (New York: Blue Rider Press, 2013), 87. Following the Harvard review that found them guilty of violating conflict-of-interest policies, Biederman and his two colleagues responded: "Our mistakes were honest ones. We always believed that we were complying in good faith with institutional policies. We now recognize that we should have devoted more time and attention to the

detailed requirements of these policies and to their underlying objectives." Richard Knox, "Harvard Punishes 3 Psychiatrists over Undisclosed Industry Pay," National Public Radio, July 2, 2011, https://www.npr.org/sections/health-shots/2011/07/02/137572941/harvard-punishes-3-psychiatrists-over-undisclosed-industry-pay.

17. For Emory's findings and proposed remedies, see "Emory Announces Actions Following Investigation," news release, Emory University, December 22, 2008, http://shared.web.emory.edu/emory/news/releases/2008/12/conflict_of_interest_action.html. For later developments, see "More Than One Bad Apple," *Nature* 455 (2008): 835. The University of Miami administration had sought and received assurances from NIMH director Thomas Insel that there would be no barrier to Nemeroff receiving grant money. That was strictly not necessary, because NIMH sanctions attach to the institution, not the individual. Nemeroff has acknowledged, "I made mistakes in the area of conflict of interest for which I am sorry and remorseful. However," he added, "the mistakes I made were honest mistakes." Thacker, "How an Ethically Challenged Researcher Found a Home."

18. On this development, see Harry Marks, *The Progress of Experiment: Science and Therapeutic Reform in the United States, 1900–1990* (Cambridge: Cambridge University Press, 2000). As David Healy has pointed out in *Pharmageddon* (Berkeley: University of California Press, 2012), controlled trials provide a valuable means of exposing treatments that do not work but can and have been misused to introduce treatments of marginal efficacy.

19. Released first in Europe as a treatment for anxiety, insomnia, and morning sickness, thalidomide had not yet been marketed in the United States when it became apparent that it produced major birth defects.

20. Patrizia Cavazzoni, "FDA's Decision to Approve a New Treatment for Alzheimer's Disease," US Food and Drug Administration, June 7, 2021, https://www.fda.gov/drugs/news-events-human-drugs/fdas-decision-approve-new-treatment-alzheimers-disease.

21. See Pam Belluck and Rebecca Robbins, "FDA Approves Alzheimer's Drug Despite Fierce Debate over Whether It Works," *New York Times*, June 7, 2021; and Laurie McGinley, "Furor Rages over FDA Approval of Controversial Alzheimer's Drug," *Washington Post*, June 17, 2021. Weeks after its initial blanket approval of the drug, the FDA issued new guidance, restricting its approval to the use of early, milder cases of Alzheimer's. See Matthew Perrone, "FDA Trims Use of Contentious Alzheimer's Drug amid Backlash," Associated Press News, July 8, 2021, https://apnews.com/article/science-business-health-8c3f0633453fcb8e87b83baa452753b3.

22. Rebecca Robbins and Pam Belluck, "Alzheimer's Drug Is Bonanza for Biogen, Most Likely at Taxpayer Expense," *New York Times*, June 8, 2021.

23. H. Melander, J. Ahlqvist-Rastad, G. Meijer, and B. Beermann, "Evidence B[i]ased Medicine—Selective Reporting from Studies Sponsored by Pharmaceutical Industry," *British Medical Journal* 326 (2003): 1171–1173; J. Jureidini, L. McHenry, and P. Mansfield, "Clinical Trials and Drug Promotion: Selective Reporting of Study 329," *International Journal of Risk and Safety in Medicine* 20 (2008): 73–81; C. S. Landefeld and M. A. Steinman, "The Neurontin Legacy: Marketing through Misinformation and Manipulation," *New England Journal of Medicine* 360 (2009): 103–106.

24. See, for example, Epic Systems Corporation v Lewis (138 S. Ct. 1612), a 5–4 decision handed down in 2018. The further rightward drift of the Supreme Court will probably lead to additional restrictions on class-action lawsuits, one of the few means of penetrating the veil of corporate secrecy.

25. Glen Spielmans and Peter Parry, "From Evidence-Based Medicine to Marketing-Based Medicine: Evidence from Internal Industry Documents," *Bioethical Inquiry* 7 (2010): 13–29, 14.

26. Healy, *Pharmageddon*, 104–113; S. Sismondo, "Ghost Management: How Much of the Medical Literature Is Shaped behind the Scenes by the Pharmaceutical Industry?," *PLOS Medicine* 4 (2007): 1429–1433; J. Wislar, A. Flanagin, P. Fontanarosa, and C. DeAngelis, "Honorary and Ghost Authorship in High Impact Biomedical Journals," *British Medical Journal* 343 (2011), d6128. For an exposure of the massive use of ghostwriting in the Vioxx scandal, see Daniel Cressey, "Merck Accused of Disguising Its Role in Research: Drug Company Used Ghost Writers for Papers Published on Vioxx Trials," *Nature* 452 (2008): 791; and Catherine DeAngelis and Paul Bebbington, "Impugning the Integrity of Medical Science: The Adverse Effects of Industry Influence," *Journal of the American Medical Association* 299 (2008): 1833–1835.

27. See John Geddes, Nick Freemantle, Paul Harrison, and Paul Bebbington, "Atypical Antipsychotics in the Treatment of Schizophrenia: Systematic Overview and Meta-Regression Analysis," *British Medical Journal* 321 (2000): 1371–1376, who conclude that once the excessive doses of the first-generation drug (generally haloperidol) are taken into account, "differences in efficacy and overall tolerability disappear, suggesting that many of the perceived benefits of atypical antipsychotics are really due to excessive doses of the comparator drug used in the trials." The same point is made forcefully in Peter Tyrer and Tim Kendall, "The Spurious Advance of Antipsychotic Drug Therapy," *Lancet* 373 (2009): 4–5.

28. Joanna Moncrieff, Simon Wessely, and Rebecca Hardy, "Meta-Analysis of Trials Comparing Antidepressants with Active Placebos," *British Journal of Psychiatry* 172 (1998): 227–231, use a review of early trials using an active comparator and conclude that these show that "the specific effects of antidepressants may therefore be smaller than is generally believed, with the placebo accounting for more of the clinical improvement than is already known to be the case" (230).

29. Geddes et al., "Atypical Antipsychotics." Significantly, this meta-analysis was not funded by the pharmaceutical industry.

30. Jeffrey Lieberman et al., "Effectiveness of Antipsychotic Drugs in Patients with Chronic Schizophrenia," *New England Journal of Medicine* 353 (2005): 1209–1223, 1210.

31. J. Kim, E. Macmaster, and T. L. Schwartz, "Tardive Dyskinesia in Patients Treated with Atypical Antipsychotics: Case Series and Brief Review of Etiologic and Treatment Considerations," *Drugs Context* 9 (2014), 212259; B. J. Kinon et al., "Incidence of Tardive Dyskinesia in Older Adult Patients Treated with Olanzapine or Conventional Antipsychotics," *Journal of Geriatric Psychiatry and Neurology* 28 (2015): 67–79; E. M. Cornett et al., "Medication-Induced Tardive Dyskinesia: A Review and Update," *Ochsner Journal* 17 (2017): 162–174.

32. The most prominent advantage of atypical drugs was claimed to be the reduction in akathisia or movement disorders. Thus, the finding that "the proportion of patients with extrapyramidal symptoms did not differ substantially among those who received first-generation and second-generation drugs in our study" (Lieberman et al., "Effectiveness of Antipsychotic Drugs," 1218) was particularly unwelcome. I suspect this outcome reflected the fact that previous studies had manufactured an advantage by prescribing an excessively high dose of haloperidol. Two subsequent large-scale meta-reviews confirmed that, while there was variability in the side-effect profiles, it was fallacious to claim that

second-generation antipsychotics had a lesser propensity to cause extrapyramidal side effects. See Stefan Leucht et al., "Second-Generation versus First-Generation Antipsychotic Drugs for Schizophrenia: A Meta-Analysis," *Lancet* 373 (2009): 31–41, 41; and Stefan Leucht et al., "Comparative Efficacy and Tolerability of 15 Antipsychotic Drugs in Schizophrenia: A Multiple-Treatments Meta-Analysis," *Lancet* 382 (2013): 951–962.

33. Su Ling Young, Mark Taylor, and Stephen Lawrie, "'First Do No Harm': A Systematic Review of the Prevalence and Management of Antipsychotic Adverse Effects," *Journal of Psychopharmacology* 29 (2015): 353–362, 360.

34. Lieberman et al., "Effectiveness of Antipsychotic Drugs," 1215, 1218, 1215.

35. Lieberman et al., "Effectiveness of Antipsychotic Drugs," 1209, 1215, 1218.

36. See K. Wahlbeck, A. Tuunainen, A. Ahokas, and S. Leucht, "Dropout Rates in Randomized Antipsychotic Drug Trials," *Psychopharmacology* 155 (2001): 230–233.

37. Peter B. Jones et al., "Randomized Controlled Trial of the Effect on Quality of Life of Second- vs. First-Generation Antipsychotics in Schizophrenia: Cost Utility of the Latest Antipsychotic Drugs in Schizophrenia (CUtLASS 1)," *Archives of General Psychiatry* 63 (2006): 1079–1087, 1085.

38. S. E. Hyman, "An Interview with Steven Hyman," *Trends in Cognitive Sciences* 16 (2012): 3–5, 4.

39. See S. Erhart, S. Marder, and W. Carpenter, "Treatment of Schizophrenia Negative Symptoms: Future Prospects," *Schizophrenia Bulletin* 32 (2006): 234–237.

40. Stefan Leucht et al., "Antipsychotic Drugs versus Placebo for Relapse Prevention in Schizophrenia: A Systematic Review and Meta-Analysis," *Lancet* 379 (2012): 2063–2071, 2063.

41. Stefan Leucht et al., "Sixty Years of Placebo-Controlled Antipsychotic Drug Trials in Acute Schizophrenia: Systematic Review, Bayesian Meta-Analysis, and Meta-Regression of Efficacy Predictors," *American Journal of Psychiatry* 174 (2017): 927–942, 927. See also Peter Haddad and Christopher Correll, "The Acute Efficacy of Antipsychotics in Schizophrenia: A Review of Recent Meta-Analyses," *Therapeutic Advances in Psychopharmacology* 8 (2018): 303–318.

42. Johan Ormel et al., "The Antidepressant Standoff: Why It Continues and How to Resolve It," *Psychological Medicine* 50 (2020): 177–182, 177.

43. B. T. Walsh, S. N. Seidman, R. Sysk, and M. Gould, "Placebo Response in Cases of Major Depression: Variable, Substantial and Growing," *Journal of the American Medical Association* 287 (2002): 1840–1847; F. Haour, "Mechanisms of the Placebo Effect and Conditioning," *Neuroimmunomodulation* 12 (2005): 195–200.

44. Bret Rutherford, Tor Wager, and Steven Roose, "Expectancy and the Treatment of Depression: A Review of Experimental Methodology and Effects on Patient Outcome," *Current Psychiatry Review* 6 (2010): 1–10. The NIMH trial is reported in A. J. Rush et al., "Acute and Longer-Term Outcomes in Depressed Outpatients Requiring One or Several Treatment Steps: A Star*D Report," *American Journal of Psychiatry* 163 (2006): 1905–1917.

45. Erick Turner, Annette Matthews, Eftihia Linardatos, Robert Tell, and Robert Rosenthal, "Selective Publication of Antidepressant Trials and Its Influence on Apparent Efficacy," *New England Journal of Medicine* 358 (2008): 252–260, 252, 252, 259.

46. D. Goldberg et al., "The Effects of Detection and Treatment on the Outcome of Major Depression in Primary Care: A Naturalistic Study in 15 Cities," *British Journal of General Practice* 48 (1998): 1840–1844; J. Sareen et al., "Common Mental Disorder Diagnosis and Need for Treatment Are Not the Same," *Psychological Medicine* 43 (2013): 1941–

1951; H. A. Whiteford, "Estimating Remission from Untreated Major Depression: A Systematic Review and Meta-Analysis," *Psychological Medicine* 43 (2013): 1569–1585.

47. Irving Kirsch, "Placebo Effect in the Treatment of Depression and Anxiety," *Frontiers in Psychiatry* 10 (2019), art. 407; M. Stone, S. Kalaria, K. Richardson, and B. Miller, "Components and Trends in Treatment Effects in Randomized, Placebo-Controlled Trials in Major Depressive Disorder from 1979 to 2016," paper presented at the American Society of Clinical Psychopharmacology, Miami, FL, 2018. See also Janus Jakobsen et al., "Selective Serotonin Reuptake Inhibitors versus Placebo in Patients with Major Depressive Disorder: A Systematic Review with Meta-Analysis and Trial Sequential Analysis," *BMC Psychiatry* 17 (2017): 58; and Arif Khan and Walter Brown, "Antidepressants versus Placebo in Major Depression: An Overview," *World Psychiatry* 57 (2015): 294–300.

48. Arif Khan et al., "Why Has the Antidepressant-Placebo Difference in Antidepressant Clinical Trials Diminished over the Past Three Decades?," *CNS Neuroscience and Therapeutics* 16 (2012): 217–226, 224, 225.

49. Khan and Brown, "Antidepressants versus Placebo," 296, 299. The authors make another very important point: the high rate of placebo response is not confined to depression but is also found in "other common disorders, such as hypertension, asthma and diabetes" (299).

50. James Ferguson, "SSRI Antidepressant Medications: Adverse Effects and Tolerability," *Primary Care Companion of the Journal of Clinical Psychiatry* 3 (2001): 22–27; R. Rosen, R. Lane, and M. Menza, "Effects of SSRIs on Sexual Function: A Critical Review," *Journal of Clinical Psychopharmacology* 19 (1999): 67–85; A. S Bahrick, "Persistence of Sexual Dysfunction Side Effects after Discontinuation of Antidepressant Medications: Emerging Evidence," *Open Psychology Journal* 1 (2008): 42–50; S. H. Kennedy and S. Rizvi, "Sexual Dysfunction, Depression, and the Impact of Antidepressants," *Journal of Clinical Psychopharmacology* 29 (2009): 157–164.

51. See Healy, *Pharmageddon*, 109–113; Jureidini, McHenry, and Mansfield, "Clinical Trials and Drug Promotion"; and L. McHenry and J. Jureidini, "Industry-Sponsored Ghost Writing in Clinical Trial Reporting: A Case Study," *Accountability in Research* 15 (2008): 152–167.

52. J. Davies and J. Read, "A Systematic Review into the Incidence, Severity and Duration of Anti-depressant Withdrawal Effects," *Addictive Behaviors* 97 (2018): 111–121; G. A. Fava et al., "Withdrawal Symptoms after Selective Serotonin Reuptake Inhibitor Discontinuation: A Systematic Review," *Psychotherapy and Psychosomatics* 84 (2015): 72–81.

53. J. Small and I. Small, "Electroconvulsive Therapy Update," *Psychopharmacology Bulletin* 17 (1981): 29–42, 33.

54. UK ECT Review Group, "Efficacy and Safety of Electroconvulsive Therapy in Depressive Disorders: A Systematic Review and Meta-Analysis," *Lancet* 361 (2003): 799–808; C. H. Kellner et al., "Bilateral, Bitemporal and Right Unilateral Electrode Placement in ECT: Randomised Trial," *British Journal of Psychiatry* 196 (2010): 226–234.

55. See, for example, Leonard Roy Frank, *The History of Shock Treatment* (San Francisco: L. R. Frank, 1978); Burton Rouche, "As Empty as Eve," *New Yorker*, September 9, 1974, 84–100 (the pseudonymous account of Marilyn Rice); Marion Milner, *The Hands of the Living God* (London: Virago Press, 1988); and Linda Andre, *Doctors of Deception: What They Don't Want You to Know about Shock Therapy* (New York: Random House, 2005). Among psychiatrists, Peter Breggin's *Brain-Disabling Treatment in Psychiatry* (New York: Springer, 2008) stands out. The late Thomas Szasz claimed that "what the rack and the stake were to the inquisition, what the concentration camp and the gas

chamber were to National Socialism, the mental hospital and electroshock are to institutional psychiatry." Quoted in Edward Shorter and David Healy, *Shock Therapy: A History of Electroconvulsive Treatment in Mental Illness* (New Brunswick, NJ: Rutgers University Press, 2007), 187. In his later years, Szasz closely allied himself with Scientology, an organization that regularly denounced psychiatry as "an industry of death" and made ECT one of its foremost targets. For recent critique of the research on ECT, see J. Read, I. Kirsch, and L. McGrath, "Electroconvulsive Therapy for Depression: A Review of the Quality of ECT versus Sham ECT Trials and Meta-Analyses," *Ethical Human Psychology and Psychiatry* 21 (2019): 64–103.

56. For example, see Andy Behrman, *Electroboy: A Memoir of Mania* (New York: Random House, 2002); Carrie Fisher, *Shockaholic* (New York: Simon and Schuster, 2011); and Simon Winchester, *The Man with the Electrified Brain* (San Francisco: Byliner, 2013).

57. See the careful review of some of this literature in Jonathan Sadowsky, *Electroconvulsive Therapy in America: The Anatomy of a Medical Controversy* (New York: Routledge, 2017), 139–143.

58. John Geddes et al., "Efficacy and Safety of Electroconvulsive Therapy in Depressive Disorders: A Systematic Review and Meta-Analysis," *Lancet* 361 (2003): 799–808. For a sharp dissent from this conclusion regarding the effectiveness of ECT, criticizing the quality of the underlying studies, see Read, Kirsch, and McGrath, "Electroconvulsive Therapy for Depression." There is some evidence, incidentally, that Blacks are given ECT less often than whites. Mark Williams et al., "Outcome of ECT by Race in the CORE Multi-site Study," *Journal of ECT* 24 (2008): 117–121; W. R. Breakey and G. J. Dunn, "Racial Disparity in the Use of ECT for Affective Disorders," *American Journal of Psychiatry* 161 (2004): 1635–1641.

59. Ana Jelovac, Erik Kolshus, and Declan McLoughlin, "Relapse Following Successful Electroconvulsive Therapy for Major Depression: A Meta-Analysis," *Neuropsychopharmacology* 38 (2013): 2467–2474, 2472.

60. Max Fink, *Electroshock: Restoring the Mind* (New York: Oxford University Press, 1999), 81. An anonymous reviewer of this book speculated that, in acknowledging damage caused in the past by ECT, Fink may have been constructing a "progress narrative" intended to increase confidence in shock therapy as applied today. Fink's claims on behalf of ECT's benignness are sharply at odds with the conclusion of the most recent American Psychiatric Association report on the treatment, published in 2001, which informs clinicians that "in many patients the recovery from retrograde amnesia will be incomplete, and there is evidence that ECT can result in persistent or permanent memory loss." American Psychiatric Association, *The Practice of ECT,* 2nd ed. (Washington, DC: American Psychiatric Press, 2001), 71.

61. H. Sackeim, "Memory and ECT: From Polarization to Reconciliation," editorial, *Convulsive Therapy: The Journal of ECT* 16 (2000): 87–96, 87, 88. Sackeim noted that "prospective patients, family members, and the public often want to know the frequency with which patients report substantial memory impairment following ECT. While we believe that such reports are infrequent, there is little objective evidence to support this judgment or to even broadly estimate base rates" (95)—a statement some might view as rather astonishing.

62. Harold Sackeim et al., "Cognitive Effects of Electroconvulsive Therapy in Community Settings," *Neuropsychopharmacology* 32 (2007): 244–254, 252. Sackeim and colleagues found increased amounts of amnesia and cognitive decline in those treated with bilateral as compared with unilateral ECT, and comparison of those treated with bilat-

eral ECT with a group of healthy patients suggested that "the deficits were substantial" and persistent.

63. J. Blaine and S. Clark, "Report of the NIMH-NIH Consensus Development Conference on Electroconvulsive Therapy," *Psychopharmacology Bulletin* 22 (1986): 442–502; American Psychiatric Association, *The Practice of Electroconvulsive Therapy: A Task Force Report of the American Psychiatric Association* (Washington, DC: American Psychiatric Association, 1990).

64. Shorter and Healy, *Shock Therapy*, 235.

65. Blaine and Clark, "Report of the NIMH-NIH Consensus Development Conference," 452. These attempts to develop a professional consensus have demonstrably not resolved the issue. Some psychiatrists continue to call for the complete disuse of the treatment. See H. A. Youssef and F. A. Youssef, "Time to Abandon Electroconvulsion as a Treatment in Modern Psychiatry," *Advances in Therapy* 16 (1999): 29–38. As for actual practice, a survey conducted by the American Psychiatric Association reported that only 7.4 percent of American psychiatrists prescribed ECT. L. M. Koran, "Electroconvulsive Therapy," *Psychiatric Services* 47 (1996): 121–129, 123. It is not obvious that the proportion has risen much since that survey was done. Indeed, Eric Slade and colleagues report that "ECT is either not available or not used as an inpatient procedure in nearly 9 out of 10 US hospitals" and "its availability is limited and declining." Eric Slade et al., "Association of Electroconvulsive Therapy with Psychiatric Readmissions in US Hospitals," *JAMA Psychiatry* 74 (2017): 798–804, 799.

66. T. G. Rhee et al., "20-Year Trends in the Pharmacologic Treatment of Bipolar Disorder by Psychiatrists in Outpatient Care Settings," *American Journal of Psychiatry* 177 (2020): 706–715. I discussed possible sources of this seemingly puzzling trend in Chapter 14.

67. G. Soomro, D. G. Altman, S. Rajagopal, and M. Browne, "Selective Serotonin Reuptake Inhibitors (SSRIs) versus Placebo for Obsessive Compulsive Disorder (OCD)," *Cochrane Database of Systematic Reviews* (January 23, 2008), https://www.cochranelibrary.com/cdsr/doi/10.1002/14651858.CD001765.pub3/full; Edna Foa, "Cognitive Behavioral Therapy of Obsessive-Compulsive Disorder," *Dialogues in Clinical Neuroscience* 12 (2010): 199–207.

68. S. Leucht et al., "Second-Generation versus First-Generation Antipsychotic Drugs," 37.

69. Sanjay Kukreja, Gurvinger Kalra, Nilesh Shah, and Amresh Shrivastave, "Polypharmacy in Psychiatry: A Review," *Mens Sana Monographs* 11 (2013): 82–99; H. Ito, A. Koyama, and T. Higuchi, "Polypharmacy and Excessive Dosing: Psychiatrists' Perceptions of Antipsychotic Drug Prescription," *British Journal of Psychiatry* 187 (2005): 243–247; S. M. Stahl, "Antipsychotic Polypharmacy: Evidence Based or Eminence Based?" *Acta Scandinavica Psychiatrica* 106 (2002): 321–322.

70. Jeffrey Lieberman, *Shrinks: The Untold Story of Psychiatry* (Boston: Little, Brown, 2015), 12.

71. Interpersonal therapy focuses on relieving symptoms by improving the patient's relations with others and, as with CBT, seems to work for some patients with mild or moderate symptoms of depression and anxiety. See Pim Cuijpers et al., "Interpersonal Psychotherapy for Mental Health Problems: A Comprehensive Meta-Analysis," *American Journal of Psychiatry* 173 (2016): 680–687. A more hesitant endorsement of these interventions is Sarah Hetrick et al., "Cognitive Behavioural Therapy (CBT), Third-Wave CBT and Interpersonal Therapy (IPT) Based Interventions for Preventing Depression in

Children and Adolescents," *Cochrane Database of Systematic Reviews* (August 9, 2016), https://www.cochranelibrary.com/cdsr/doi/10.1002/14651858.CD003380.pub4/full.

72. S. Jauhar et al., "Cognitive-Behavioural Therapy for the Symptoms of Schizophrenia: Systematic Review and Meta-Analysis with Examination of Potential Bias," *British Journal of Psychiatry* 204 (2014): 20–29. For CBT's failure to reduce the chances of relapse, see P. A. Garety et al., "Cognitive-Behavioural Therapy and Family Intervention for Relapse Prevention and Symptom Reduction in Psychosis: Randomised Controlled Trial," *British Journal of Psychiatry* 192 (2008): 412–423.

73. Keith R. Laws et al., "Cognitive Behavioral Therapy for Schizophrenia—Outcomes for Functioning, Distress and Quality of Life: A Meta-Analysis," *BMC Psychology* 6 (2018): 32. Laws and colleagues rightly criticize NICE for failing to update its clinical guidance since 2008 and for neglecting to account for evidence accumulated since then that runs counter to its recommendations. Other studies have been similarly pessimistic about CBT's effects on the emotional blunting, social withdrawal, and apathy associated with schizophrenia. E. Velthorst et al., "Adaptive Cognitive-Behavioural Therapy Required for Targeting Negative Symptoms in Schizophrenia: Meta-Analysis and Meta-Regression," *Psychological Medicine* 45 (2015): 453–465.

74. Christopher Jones et al., "Cognitive Behavioural Therapy versus Other Psychosocial Treatments for Schizophrenia," *Cochrane Database of Systematic Reviews* (April 18, 2012), https://www.cochranelibrary.com/cdsr/doi/10.1002/14651858.CD008712.pub2/full. See also Jauhar et al., "Cognitive Behavioural Therapy for the Symptoms of Schizophrenia," and Laws et al., "Cognitive Behavioral Therapy for Schizophrenia."

75. Bi-Yu Ye et al., "Effectiveness of Cognitive Behavioral Therapy in Bipolar Disorder: An Updated Meta-Analysis with Randomized Controlled Trials," *Psychiatry and Clinical Neurosciences* 70 (2016): 351–361. M. Oud et al., "Psychological Interventions for Adults with Bipolar Disorder: Systematic Review and Meta-Analysis," *British Journal of Psychiatry* 208 (2016): 213–222, is slightly more optimistic but emphasizes that much of the evidence reviewed is of low or very low quality.

76. D. Lynch, K. R. Laws, and P. J. McKenna, "Cognitive Behavioural Therapy for Major Psychiatry Disorder: Does It Really Work? A Meta-Analytical Review of Well-Controlled Trials," *Psychological Medicine* 40 (2010): 9–24. This study also emphasizes CBT's limitations in the treatment of schizophrenia and bipolar disorder. See also I. Elkin et al., "Treatment of Depression Collaborative Research Program: General Effectiveness of Treatments," *Archives of General Psychiatry* 46 (1989): 971–982.

Epilogue

1. Nancy Andreasen, "DSM and the Death of Phenomenology in America: An Example of Unintended Consequences," *Schizophrenia Bulletin* 33 (2007): 108–112, 111.

2. Matthew Noor, Yunzhe Liu, and Raymond Dolan, "Functional Neuroimaging in Psychiatry and the Case for Failing Better," *Neuron* 110 (2022): 2524.

3. H. Christian Fibiger, "Psychiatry, the Pharmaceutical Industry, and the Road to Better Therapeutics," *Schizophrenia Bulletin* 38 (2012): 649–650, 649.

4. Sten Stovall, "R&D Cuts Curb Brain-Drug Pipeline," *Wall Street Journal*, March 27, 2011.

5. Steven Hyman suggests that this abandonment of the field "reflects a widely shared view that the underlying science remains immature and that therapeutic development in psychiatry is simply too difficult and too risky." Steven E. Hyman, "Psychiatric Drug De-

velopment: Diagnosing a Crisis," *Cerebrum*, Dana Foundation, April 2013. See also David Nutt and Guy Goodwin, "ECPN Summit on the Future of CNS Drug Research in Europe," *European Neuropsychopharmacology* 21 (2011): 495–499; and Colin Hendrie and Alasdair Pickles, "The Failure of the Antidepressant Drug Discovery Process Is Systemic," *Journal of Psychopharmacology* 27 (2013): 407–416.

6. Andrew Abbott, *The System of Professions* (Chicago: University of Chicago Press, 1988).

7. The contrast between the mind-twist and brain-spot enthusiasts was coined by the Harvard psychiatrist Elmer Ernest Southard as long ago as 1914. E. E. Southard, "The Mind Twist and Brain Spot Hypotheses in Psychopathology," *Psychological Bulletin* 11 (1914): 117–130.

8. For a useful summary of some relevant research, see Bruce Wexler, *Brain and Culture: Neurobiology, Ideology, and Social Change* (Cambridge, MA: MIT Press, 2006).

9. Leon Eisenberg, "The Social Construction of the Human Brain," *American Journal of Psychiatry* 152 (1995): 1563–1575, 1571.

10. Andreasen, "DSM and the Death of Phenomenology," 111.

11. Michael MacDonald, *Mystical Bedlam: Madness, Anxiety, and Healing in Seventeenth Century England* (Cambridge: Cambridge University Press, 1981), 1.

12. See, for example: Nirmita Panchal, Rabah Kamal, Cynthia Cox, and Rachel Garfield, "The Implications of COVID-19 for Mental Health and Substance Use," issue brief, Kaiser Family Foundation, February 10, 2021, https://www.kff.org/coronavirus-covid-19/issue-brief/the-implications-of-covid-19-for-mental-health-and-substance-use/; Tianchen Wu et al., "Prevalence of Mental Health Problems during the COVID-19 Pandemic: A Systematic Review and Meta-Analysis," *Journal of Affective Disorders* 281 (2021): 91–98, https://doi.org/10.1016/j.jad.2020.11.117; Osea Giuntella, Kelly Hyde, Silvia Saccardo, and Sally Sadoff, "Lifestyle and Mental Health Disruptions during COVID-19," *Proceedings of the National Academy of Sciences* 118, no. 9 (2021), e2016632118, https://doi.org/10.1073/pnas.2016632118; and for a historical perspective, Kishen Neelam et al., "Pandemics and Pre-Existing Mental Illness: A Systematic Review and Meta-Analysis," *Brain Behavior Immunity–Health* 10 (2021), 100177, https://doi.org/10.1016/j.bbih.2020.100177.

13. "Tackling the Mental Health Impact of the COVID-19 Crisis," OECD Policy Responses to Coronavirus (COVID-19), May 12, 2021, https://www.oecd.org/coronavirus/policy-responses/tackling-the-mental-health-impact-of-the-covid-19-crisis-an-integrated-whole-of-society-response-0ccafa0b/.

14. Betty Pfefferbaum and Carol S. North, "Mental Health and the Covid-19 Pandemic," *New England Journal of Medicine* 383, no. 6 (2020): 510–512, 510, https://doi.org/10.1056/NEJMp2008017.

ACKNOWLEDGMENTS

My research on this subject spans four decades. Back in the early 1980s, I thought about writing a book on the somatic therapies employed by American psychiatrists. I was spending the year at what was then the Wellcome Institute for the History of Medicine on a Guggenheim fellowship. Like all historians of medicine, I deeply lament the closure of the institute (or "centre," as it was renamed after its ill-starred move to University College London). The Wellcome Library and its archival collections remain a key resource for all of us working in the field. I am extremely grateful to the then director of the institute, William Bynum, for making my 1981 visit, and many subsequent ones, so enjoyable and productive. I am also grateful to those in charge of its Contemporary Medical Archives, who made it possible to consult the papers of England's leading lobotomist and enthusiast for physical treatments for mental illness, William Sargant.

During the intervening decades, I have had reason to be grateful to many archives and archivists for the extraordinary assistance they provided. Particularly important were the National Archives and the Library of Congress in Washington, DC; the Rockefeller Archives; the Yale Medical Library's historical collections; the archives of the New Jersey State Hospital in Trenton; and the Alan Mason Chesney Archives at the Johns Hopkins University. The libraries at the University of California campuses at San Diego and San Francisco have been vital resources over the many years it has taken to complete this book, and I owe them my thanks as well.

Over the course of a long career studying the history of mental illness, and the lives and practices of those who have taken charge of managing it,

I have written a good many books on the subject, some more specialized than this one. But throughout my career, two major projects have challenged and provoked me. I've thought hard about the issues they raised for me, and I tried to transform vague initial hunches and preliminary reflections into what I hope are disciplined, serious, and sustained examinations of each of them.

The first of these books, *Madness in Civilization,* which I published seven years ago now, was an exercise in chutzpah, an attempt to provide a cultural history of madness in civilization, ranging from ancient Greece and China to the modern world, and encompassing religion and medicine, the plastic and visual arts, politics and folk beliefs, music and the movies, and much else besides. By comparison, this book, which confines its attention to the period between the early 1800s and the present, and focuses mostly on what American psychiatrists have thought and done about mental illness, would seem to have a simpler tale to tell. But that is not true. Both books depended on a lifetime of research and reflection; both would have been impossible to write without it.

Like its predecessor, *Desperate Remedies* is driven by both my sympathy for the victims of what constitute some of the most profound forms of human suffering and my concern with the intractable puzzles of how to account for mental illness and how to treat it. Over the many years thinking about the development and practices of American psychiatry, my initial concern with events in the 1910s, 1920s, and 1930s has broadened into an attempt to make sense of psychiatry from its first beginnings until today and to analyze the existential crisis that I believe now confronts the profession—and, by extension, those who seek its help.

Desperate Remedies, as its title suggests, is in many respects a highly critical examination of the psychiatric enterprise. I am therefore particularly grateful to the psychiatrists who have assisted me along the way and who have, in a number of cases, read portions of the manuscript and offered commentary on what I have had to say about their profession. In North America, these include Joel Braslow, George Makari, Will Carpenter, and David Healy; and in Britain, Sir Robin Murray and Sir Simon Wessely. They have not been shy about making clear where they disagree with me, but I have listened and learned much from what they have had to say, and I deeply appreciate their willingness to engage with my critique. I'm grateful, too, to the historian and psychoanalyst Daniel Pick, with whom I've had some fruitful conversations and whose work on degeneration I admire. I'm also indebted to the foremost historian of American medicine, Charles Rosenberg. The two of us were colleagues many years ago, and I have admired his wide-ranging scholarship over the years since my time as a junior faculty member at the University of Pennsylvania. The late Gerald Grob and I had a highly contentious relationship in print, which belied our cordial friendship in person. With typical generosity, Gerry provided valuable help with my research on several occasions. Before his untimely death, Jack Pressman and I had spirited conversations about how to interpret lobotomy. We worked in many of the same archives, and while I differ from some of his conclusions, I have great respect for his scholarship.

Franklin Freeman, Walter Freeman's son, was kind enough to share some family memorabilia with me, including a copy of the unpublished autobiography Walter Freeman wrote for his children. Conversations with Phyllis Greenacre and her son Peter Richter gave me valuable insights into Henry Cotton's work at Trenton State Hospital, as did a lengthy interview with the institution's longtime dentist, Ferdearle Fischer, who pulled hundreds of thousands of teeth in a vain attempt to cure mental illness. Among sociologists, Allan Horwitz was kind enough to share his work on psychiatry's diagnostic manuals before it appeared in print. He also provided a close, careful, and valuable reading of several of the later chapters in this book.

Over the years, the University of California at San Diego, my academic home for over forty years, has provided me with funds to travel to archives. A number of foundations have also given invaluable assistance: the Guggenheim Foundation, the American Council of Learned Societies, the American Philosophical Society, the Rockefeller Foundation, and the Commonwealth Fund. I hope this book in some small way repays their generosity and confidence in my scholarship.

Portions of Chapter 12, "Creating a New Psychiatry," first appeared in *History of Psychiatry* and was then reprinted in a collection of my essays, *Psychiatry and Its Discontents,* published by the University of California Press. I am grateful to Sage and the press for permission to reuse those materials here. A different version of Chapter 18, "Community Care," was published in *Perspectives in Biology and Medicine* in 2021, and I thank Johns Hopkins University Press for permission to reprint the passages that appeared in that article here. Chapter 18 also contains a few paragraphs from a paper of mine written in the early stages of the deinstitutionalization movement, "A New Trade in Lunacy: The Recommodification of the Mental

Patient," which first appeared in *American Behavioral Scientist*. Once again, I am grateful to Sage for permission to reuse those materials. Finally, Chapters 20, 21, and 22 build on ideas first presented in "American Psychiatry in the New Millennium: A Critical Appraisal," published in *Psychological Medicine* in 2021, and I would like to thank Cambridge University Press for permitting reuse of those materials.

My agent, Caroline Dawnay, has been an enthusiastic supporter of my work and has worked tirelessly to make sure that this book reaches a broad audience. I worked with her first many years ago and am delighted that she has chosen to represent me again. No author could ask for a better advocate. Thanks to Caroline, I have been privileged to work with two terrific publishers. Casiana Ionita of Allen Lane and Penguin Books provided some most helpful editorial comments on an earlier version of this manuscript. Subsequently, Joy de Menil of Harvard University Press devoted much time and effort to an extraordinarily detailed review of the manuscript. I am immensely grateful to her not only for going through the book line by line, but also for suggesting a change in the way I structured the argument that follows. Quite how Joy found the time to invest so much in a detailed dialogue with me I cannot fathom, but the book has benefited greatly from her engagement with it.

As always, some of my deepest debts are to two friends of many years' standing, both blessed with superb editorial skills and the kindness to share them with no more recompense than such inadequate thanks as these. Amy Forrest and Stephen Cox read drafts of every chapter of this book, and it is immeasurably the better for their unsparing and insightful comments and criticism. My wife, Nancy, has added her own critical voice and has put up with my obsession with the irrational for fifty years and more. In 2016, we shared the tragic loss of our youngest son, a loss that only our love and the love of our other children and grandchildren have allowed us to endure. It is to our dearest Alex that this book is dedicated.

INDEX